Second Edition

Walk a Mile

A Journey Towards Justice and Equity in Canadian Society

THERESA ANZOVINO
Niagara College

JAMIE ORESAR
Niagara College

DEBORAH BOUTILIER
Niagara College

with contributions by Yale Belanger, *University of Lethbridge*, and Samah Marei, *Niagara College*

NELSON

NELSON

Walk A Mile: A Journey Towards Justice and Equity in Canadian Society, Second Edition
by Theresa Anzovino, Jamie Oresar, and Deborah Boutillier

VP, Product Solutions, K–20:
Claudine O'Donnell

Publisher, Digital and Print Content:
Leanna MacLean

Marketing Manager:
Claire Varley

Content Manager:
Toni Chahley

Photo and Permissions Researcher:
Carrie McGregor

Production Project Manager:
Jaime Smith

Production Service:
MPS Limited

Copy Editor:
Linda Szostak

Proofreader:
MPS Limited

Indexer:
MPS Limited

Design Director:
Ken Phipps

Higher Education Design PM:
Pamela Johnston

Interior Design:
Dave Murphy

Cover Design:
Colleen Nicholson

Compositor:
MPS Limited

**Library and Archives Canada
Cataloguing in Publication Data**

Anzovino, Theresa, author
    Walk a mile : a journey towards
justice and equity in Canadian
society / Theresa Anzovino, Jamie
Oresar, and Deborah Boutilier.
—Second edition.

Includes index. Subtitle for first
edition was: Experiencing and
understanding diversity in Canada.
ISBN 978-0-17-673027-7 (softcover)

    1. Multiculturalism—
Canada—Textbooks. 2. Ethnic
groups—Canada—Textbooks.
3. Canada—Civilization—Textbooks.
4. Textbooks. I. Oresar, Jamie, author
II. Boutilier, Deborah, author III. Title.

HM1271.A59 2018   305.800971
C2017-904222-X

ISBN-13: 978-0-17-673027-7
ISBN-10: 0-17-673027-3

To my family and friends, for your love and support—Theresa

For my daughter, Joey – may you experience a world united by difference—Jamie

For my wild and wonderfully diverse family, with love—Deborah

Foreword by Craig and Marc Kielburger

For us, "walking a mile in someone else's shoes" is only a partial metaphor.

Our international development work through WE Charity often leads us to remote communities where the only way to travel is by foot. In the Chimborazo region of Ecuador, for example, the road stops halfway up the Andean mountains, and we walk the rest of the way. Beside us are our friends from the mountaintop villages, and the mules carrying cement, lumber, and other school-building materials. We would offer to trade our comfortable hiking shoes for our friends' bare feet, but we know if they accepted we'd never keep up.

An even more challenging walk is with groups of Maasai women and children in Kenya to collect their families' water from the nearest source stream. On the return trip—often two kilometres or more—we balance a 40-litre jerry can of water on our heads like the others, some of whom are as young as six. It's a powerful experience that reaffirms our commitment to clean water projects in communities like theirs, to ensure their walk is much shorter and the children can go to school.

We've learned innumerable lessons by travelling great and difficult distances, literally and figuratively, with our diverse overseas partners. Understanding other ways of living, thinking, and doing has enriched our own life experience and made our work projects more effective and impactful.

It has also inspired us to share these experiences with as many people as possible because we are convinced that a whole generation of diversity-competent global citizens can change the world.

We've now travelled with thousands of young people who want not simply to appreciate diversity, but to live it. On our overseas volunteer trips, one-third of participants' time is dedicated to cultural immersion in the communities we visit. We eat traditional local meals, participate in centuries-old celebrations, and contribute to daily chores like the water walk.

Most importantly, when they return home, our volunteers pursue a better world with renewed vigour. Some have even organized a fundraising "Mamas' Water Walk," where students from various local schools collected pledges and completed a two-kilometre walk with water jugs on their backs.

Leaving their comfort zones to truly connect with their fellow human beings, our volunteer travellers gain a deeper appreciation of our world's diversity; an understanding of how that diversity plays out in power, privilege, and hardship; and a lifelong sense of empathy, compassion, and dedication to justice and equity.

Now, living diversity doesn't necessarily require international travel. We Canadians have the great fortune of living in one of the world's most diverse nations, brimming with opportunities to meet people with different backgrounds and experiences from our own. It's not far outside our comfort boxes that we can find culturally different foods, music, or community festivals. Even take a look around your own neighbourhood, workplace, or social group, and see people of different gender, age, ability, sexual orientation, income level, and countless other identities than yours. For the truly adventurous, visit a local temple, mosque, or church and get a glimpse into a new way of seeing the world that will expand your mind.

There are also more tangible benefits to becoming diversity competent, in the enhanced prospects for your career and social life. Imagine opening your job search to the world—we regularly encounter fellow Canadians working in banks in China, overseeing construction projects in Africa, or running development programs in South America. A 2008 University of California–Berkeley study even showed that people with culturally diverse social groups have lower levels of the stress hormone cortisol, which lowers their risk of cancer, heart disease, and Type 2 diabetes.

For all these reasons, we are ecstatic to see this book in classrooms across Canada, and we congratulate Professor Boutilier and Professor Anzovino for translating their decades of experiential learning and teaching into a practical guide to diversity competency. The knowledge and active reflection provoked in these pages will help a generation of young Canadians to understand their own identity, the place of identity in our relationships with our fellow global citizens, and our responsibility to take action for a more equitable and just world.

We therefore urge you to see this book as just the beginning. Let it be your guide to a world of opportunity, of understanding, of experiences, of action. You don't have to walk a mile with a jug of water on your head, but don't let the title of this book remain a simple metaphor. Try not to just see the world through someone else's eyes, but to live it. Live diversity, and it won't be just your life that is better for it.

Craig Kielburger and Marc Kielburger, Co-Founders of WE

Source: Photo courtesy of Marc and Craig Kielburger

About the Authors

THERESA ANZOVINO

Theresa Anzovino has completed a Masters of Arts degree in Sociology (York University, 1994), Bachelor of Arts Degree in Sociology (University of Waterloo, 1985) and Teaching Adults Certificate (Niagara, 1989). Her academic areas of interest include feminist jurisprudence, migration, diversity, human rights, and universal design for learning. She is a proud mother to son Daniel. Anzovino is a professor in the School of Liberal Arts and Sciences at Niagara College and the 2013 recipient of the Teaching Excellence Award at Niagara College. Prior to this, she worked as a CEO for a large organization within the non-profit sector dedicated to refugee protection and resettlement. This work earned her numerous humanitarian and leadership awards. When she invites you to walk a mile in her shoes, it will be barefoot on a beach connecting with the earth.

JAMIE ORESAR

Jamie Oresar has completed a Masters of Teaching degree (Griffith University, 2004) and a Bachelor of Arts degree in Sociology (Western University, 2001). Her academic areas of interest include diversity and social inclusion, criminology, globalization and sustainable development, and gender studies. Oresar is a professor of Sociology in the School of Liberal Arts and Sciences at Niagara College. Prior to this, she has taught at the secondary level in Ontario, as well as internationally in China, Australia, and Mexico. When she is not teaching, Oresar loves to travel and spend time outside with her partner, Joe, daughter, Joey, and dog, Bentley. She invites you to walk a mile in the shoes that have led her across three continents.

DEBORAH BOUTILIER

Deborah Boutilier holds a Doctorate of Education degree in Sociology and Equity Studies (University of Toronto, 2008); Master's degrees in Sociology (State University of New York at Buffalo, 1987) and Education (Brock University, 1998), and a Baking Certificate (Niagara College, 2013). Her academic areas of interest include diversity and social inclusion, social constructions of gender and homicide, and the learning processes of information technology in the practice of cross-cultural computer-mediated exchanges. When she is not teaching in the School of Liberal Arts and Sciences at Niagara College, she loves reading, writing, baking, and walking her dog, Harley. She invites you to take a walk in her favourite shoes, but warns you that they leak!

About Contributing Authors

YALE BELANGER

Dr. Yale D. Belanger holds a doctorate in Indigenous Studies (Trent University), and is Professor of Political Science at the University of Lethbridge (Alberta). His areas of study include First Nations casino and economic development, Aboriginal gambling, housing and homelessness, and local community development. He has written and edited several books and dozens of articles and book chapters, and remains a regular contributor to international, national, and regional media. When not stuck writing at a computer, he enjoys hiking in the coulees with Loki and Walt the dogs.

SAMAH MAREI

Samah Marei earned her degree in History from UCLA and is currently completing post-graduate work at Oxford. She has taught locally at the elementary level as well as at Niagara College, and is a teacher of Arabic at the Qasid Institute. In addition to teaching history and sociology, Samah has spent the last 15 years presenting workshops on Diversity and Inclusion in both the workforce and the community. Samah sits on the Board of Gillian's Place in St. Catharines and spends her time volunteering at various shelters and with her family on their Niagara hobby farm.

Brief Table of Contents

Detailed Table of Contents

A Unique Learning System

Walk A Mile employs a unique approach that integrates academic and experiential learning tools to help motivate students to actively engage with the text and issues of social justice and equity in the real world. This learning will help prepare students to challenge oppression and injustice when they experience it in their personal and professional lives. Students come to understand that the enterprise of diversity is to construct a society where *all* people can experience the world as just; where *all* people have equivalent access to opportunities; where *all* people, including those historically underserved and underrepresented, feel valued, respected, and able to live lives free of oppression; and where *all* people can fully participate in the social institutions that affect their lives. Most reviewers agree that *Walk A Mile* is unlike any other resource currently available for diversity courses.

THREE TYPES OF BOXES

In Their Shoes uses students' stories to give an authentic voice to the lived experience of "real" people that other students can identify with.

Picture This ... uses carefully chosen photographs to speak to the undiscovered themes in each chapter as students consider the historical and future implications of each photograph.

Agent of Change uses examples of Canadians, some famous and others not, to highlight the ways in which positive social change can impact families, neighbourhoods, communities, Canadian society, and beyond.

KEY TERMS

Every term is carefully defined and conveniently located in the text margins beside the section where the term is first introduced. A complete glossary of all key terms is included at the end of the text.

END-OF-CHAPTER SKILL-BUILDING MATERIAL

Each chapter ends with a reading, followed by several discussion questions that instructors can use as the basis for a written assignment, an oral discussion, or a class debate. The KWIP feature closes each chapter by leading students through a process of reflective questioning using the following steps:

K: Know it and own it—What do I bring to this?

W: Walk the talk—How can I learn from this?

I: It is what it is—Is this inside or outside my comfort zone?

P: Put it in play—How can I use this?

COURSEMATE

Nelson Education's *Walk a Mile* **CourseMate** brings course concepts to life with interactive learning and exam preparation tools that integrate with the printed textbook. Students activate their knowledge through quizzes, games, and flashcards, among many other tools.

Interactive Teaching and Learning Tools include interactive teaching and learning tools:

- Flashcards

- Multiple Choice Quizzes

- *Picture This* with reflection questions

- *Diversity in the Media* with critical thinking questions

- The KWIP framework with reflection questions ... and more!

Preface

A JOURNEY TOWARD JUSTICE AND EQUITY IN CANADIAN SOCIETY

The second edition of *Walk A Mile* uses an anti-oppression framework with the goal of promoting justice and equity for members of diverse populations while challenging patterns of oppression and discrimination. The journey toward justice and equity acknowledges the historical oppression and social exclusion of affected communities in Canada—placing the experience of affected communities at the centre of analysis and embracing the hope of creating a society where the dignity and intrinsic worth of every human being is respected and valued.

Ten years ago, we struggled to find a resource for teaching and learning that was written at a level that would engage our students and that provided a *balance* of theory *and* content that would work in experiential and active learning environments. So we created a course manual called *Walk a Mile*. Thanks to the vision and tenacity of Jillian Kerr, Senior Learning Solutions Consultant at Nelson Education, the course manual evolved into the first edition, *Walk a Mile: Experiencing and Understanding Diversity in Canada*.

As the new subtitle suggests, *Walk a Mile: A Journey Toward Justice and Equity in Canadian Society* moves the content of this second edition in the direction of justice and equity.

Our Reviewers Write…
   "I love the new title and the focus on equity and justice. It takes it to a deeper level."

When we were approached to write a second edition by Nelson Education Ltd. publisher Leanna McLean, we were grateful for the opportunity to update the pedagogical elements and research. We were equally grateful for the opportunity to refocus and expand upon concepts like oppression, power, privilege, intersectionality, social justice, and equity. This second edition has also been enriched by the thoughtful, thorough suggestions provided by reviewers from across the country. Their feedback has been instrumental in recreating this resource, which aims to balance student engagement activities, theoretical material, and critical thinking and self-reflection exercises, written in an inviting style so that every learner becomes part of the learning experience.

Our Reviewers Write …
   I have been reviewing books for a new … Diversity course and this manual is the first textbook that attempts to go "beneath the tip of the iceberg." It is not just a text full of theory and concepts, but attempts to capture the real life application and significance of the people whose lives they are supposed to address.

GOALS OF THIS BOOK: WHEN STUDENTS WALK A MILE, THEY WILL BE ABLE TO …

Our goals as authors of this textbook were to create a text that would help students to

- define diversity as a framework that acknowledges difference, power, and privilege using principles of social equity, social justice, and anti-oppression

- actively engage in examining issues of diversity, including social inequality, race, ethnicity, immigration, religion, gender, sexuality, ability, age, and family in ways that are relevant to the lives of students today

- extend learning beyond the classroom to the real world and diverse experiences of affected persons and communities

- enjoy the interactive experience of learning about diversity in a non-traditional manner

- understand diversity through greater self-awareness, knowledge, and empathy for those who experience prejudice and discrimination

- model ways of being in the world that help to promote awareness, respect, and inclusiveness in building positive relationships with diverse communities

- critically analyze roots of oppression and inequality for historically disadvantaged and underrepresented communities and make connections with systemic discrimination experienced by these communities in contemporary society

- devise sustainable and inclusive strategies to eliminate barriers to full participation of diverse communities

WALK A MILE IN THE CLASSROOM

As educators, we make many decisions—including if and how we want to use a textbook. One of the features that we often look at is the textbook's fit with our course plan. *Walk a Mile* was designed to provide you, as an instructor, with a balanced presentation of information in steps that mirror the process of becoming diversity competent. The intention of its design was to give you ideas to help with the development of course plans, including course goals, student learning objectives, assessment plans, units of instruction, and course schedule. For those instructors with an established syllabus, *Walk a Mile* can be customized to include specific chapters to meet your needs and provide you with the content required to cover your course's topics.

> *Our Reviewers Write …*
> *The materials, concepts, and pedagogical techniques in this text are everything I would want to include in a diversity course.*

Walk a Mile welcomes your students into an invitational learning environment where content and pedagogy interact in non-traditional ways. More than a compendium of intellectual content, this text will engage your students through an active learning approach that makes diversity relevant to their personal and professional lives. Students "must talk about what they are learning, write about it, relate it to past experiences, and apply it to their daily lives. They must make what they learn part of themselves" (Chickering & Gamson, March 1987). *Walk a Mile* uses pedagogical elements to tap into the voices, experiences, creativity, and passion of postsecondary students, infusing it with content relevant to students' own lives. *Walk a Mile* is shaped by a belief that experiential activities are among some of the most powerful teaching and learning tools available. For example, the pedagogical element entitled In Their Shoes shares powerful experience narratives of students as tools for empathy, while the pedagogical element entitled KWIP facilitates reflection on experience and action. Together, these elements awaken in students an understanding that learning scaffolds on experience and reflection. Most of our reviewers identify *Walk a*

Mile's uniqueness and strength as its ability to engage students through its active learning approach.

> *Our Reviewers Write …*
> *I especially like this text because it takes the hassle out of teaching. These authors have done a lot of the work for faculty: the book is full of good ideas for developing and supporting a syllabus, it has plenty of in-class activity ideas, and opportunities for students to engage with the material outside of the classroom and through the Internet.*

PEDAGOGICAL FRAMEWORK

Walk a Mile is a text that supports the implementation of active learning based upon research on best practices in learning environments (Michael 2006). The hallmarks of active learning include: a holistic approach that involves cognitive and affective domains of learning; embracing students' unique ways of knowing, learning, and experiencing; making connections between academic knowledge and practice; emphasizing reflective practice; and active engagement in critical thinking processes that involve analysis of concepts, forming opinions, synthesizing ideas, questioning, problem solving, and evaluating information (McKeachie & Svinicki 2006).

> *Our Reviewers Write …*
> *The pedagogy integrating academic and experiential learning tools is inspiring. It has the ability to motivate students and teachers to actively engage (think, reflect, and reflex) with the text and issues of diversity in the real world.*

Walk a Mile: A Journey Toward Justice and Equity in Canadian Society requires learning in the cognitive and affective domains. It is not a book that focuses exclusively on the intellectual journey and content because learning about diverse communities also requires empathy and understanding that are culled from experience.

Opening Lyrics

Each chapter begins with a line from a song that relates to the specific theme in each chapter. Why do we use song lyrics and not a quotation from a book or article? Music unites people and builds a commonality at the outset of the chapter. Music also makes us feel—its lyrics help us relate to the experience as we take in the emotional aspects of a song. And feeling, as we know, is an important part of the process of becoming diversity competent.

In Their Shoes

In Their Shoes is a pedagogical tool that uses students' stories to give an authentic voice to the lived experience of "real" people that other students can identify with. In Their Shoes consists of a piece of original work written by a college or university student on the specific theme of each chapter. The student authors come from a variety of different programs and institutions. With courage and integrity, they shared their personal narratives so that readers might grow in empathy, understanding, and knowledge as they walk a mile "in their shoes." Many of the student authors saw the writing of their stories as an opportunity to engage in the political act of storytelling. Their hope has been that a dialogue might begin about the validity of lived experience as a form of knowledge, and about the importance of teaching and learning empathy. Student readers actively engage with these stories because they are relevant to their lives here and now.

Picture *This*...

Does a picture say a thousand words? The pictures in *Walk a Mile* have been selected to evoke meaningful critical thought in the minds of students and in their discussions with their peers. *Picture This...* uses carefully chosen photographs to speak to the undiscovered themes in each chapter as students consider the historical and future implications of each photograph. This feature provides a perfect opportunity for students to use their sociological imaginations.

Agent of Change

This is a new pedagogical tool added to the second edition. Each chapter features a person who has been a catalyst for positive social change. Some of those featured are national figures whose visionary leadership, innovation, philanthropy, social entrepreneurship and activism are well known throughout Canada and beyond. Other agents of change we chose to feature are so-called "everyday Canadians" who are doing extraordinary things to challenge oppression and fight against discrimination. Common among all agents of change is a spark that is fuelled by a leap of faith, a passion, and an energy to create a more socially just world. The point? The belief that this spark is within all of us.

KWIP

KWIP is a pedagogical tool located at the end of each chapter that engages students in active learning. The four-step KWIP process can facilitate reflective and reflexive practice, asking students to consider the following questions: What do I bring to this? How can I learn from this? Is this inside or outside my comfort zone? and How can I use this in my own life?

K: Know it and own it—What do I bring to this? The "K" in the KWIP process is about knowing and owning who you are as a person as the first step in understanding and respecting others and where they are coming from. It involves as much learning about oneself as it does learning about others. As part of this process, you are asked to examine and reflect upon aspects of your identity and social location, and accept yourself for who you are and what you can bring to this discussion.

W: Walk the talk—How can I learn from this? The "W" in the KWIP process establishes the intention to learn about the complexities of oppression. As you set a purposeful and curious intention to learn about oppression as a variable and multidimensional concept, you are challenged to "walk the talk" through problem-based learning.

I: It is what it is—Is this inside or outside my comfort zone? The "I" in the KWIP process requires you to honestly confront the fact that we get uncomfortable when we encounter perspectives that are different from our own. More than that, our discomfort is often rooted in the various privileges each of us hold. Anti-oppressive practice requires a consciousness of these various privileges and the ability to see our selves systemically. This activity invites you to practise critical self-reflection and institutional reflection as you explore the complexities of power and privilege. Increasing awareness of both our own oppression(s) and our roles as oppressors of others opens the possibility for changing oppressive practices in our personal and professional lives.

P: Put it in play—How can I use this? The "P" in the KWIP process involves examining how others are practising diversity and how you might use this. It is a call to move from social analysis to social action—by knowing and owning who you are; by establishing the intention to learn about new things; and by choosing to become involved in anti-oppressive practice. How can this broader perspective enrich our lives?

Readings and Discussion Questions

Our experience has been that readings can form the basis of lively classroom discussions, so every chapter includes a reading carefully chosen to encourage students to make connections between the chapter's themes and its key concepts and ideas. Each reading is followed by several discussion questions that instructors can use as the basis for a written assignment, an oral discussion, or a class debate. We have found that students who work on these active learning exercises are better able to process what they've read and to focus on important information.

CHAPTER HIGHLIGHTS

The key issues and topics, inspired by students, covered in each chapter are outlined below. These highlights begin with action verbs reflective of the active process of learning required in becoming diversity competent.

NEW TO THIS EDITION

New features and key changes to the second edition include the following:

- **Walk A Mile's new subtitle**, *A Journey Toward Justice and Equity in Canadian Society*, reflects the movement of content in the second edition in the direction of justice and equity. The second edition provides more comprehensive coverage of concepts such as oppression, anti-oppression, power, privilege, intersectionality, social justice, and equity.

- **NEW chapters on Gender and Sexuality.** Based on reviewer feedback, the second edition divides the topics of Gender and Sexuality into two separate chapters to allow for more comprehensive coverage. By separating these chapters, the authors take the discussion to a deeper level and include new material on transgender issues, homophobia, biphobia, transphobia, Gay Pride, sexual identities, compulsive heterosexuality and heteronormativity, the social construction of gender, sexual assault on campus, hegemonic masculinity and emphasized femininity, and the gender neutrality movement.

- **NEW contributed chapter on Indigenous Peoples** authored by Dr. Yale Belanger who holds a doctorate in Indigenous Studies (Trent University), and is Professor of Political Science at the University of Lethbridge (Alberta). This chapter includes new content on residential schools, Murdered and Missing Indigenous Women, Idle No More, and social media engagement.

- **Agent of Change** is a NEW pedagogical tool contained in each chapter that highlights a Canadian who has been a catalyst for positive social change. These change agents provide students with concrete examples of Canadians who are promoting justice and equity for members of diverse populations while challenging patterns of oppression and discrimination.

KWIP has been expanded. Based on reviewer feedback, KWIP has been redesigned to improve student engagement through critical thinking and self-reflection exercises, and problem-based learning activities. The new KWIP activities are designed to help students arrive at a better understanding of how their own ideas and realities have been shaped by their social location and experiences to date.

CHAPTER-BY-CHAPTER CHANGES

Chapter 1 – Diversity, Oppression, and Privilege

- NEW Agent of Change – Carol and Amanda Todd

- Takes an approach that focuses on oppression/anti-oppression, privilege, equity, and social justice

- NEW Reading – *"But You Are Different: In Conversation with a Friend"* by Sabra Desai

- NEW Figure 1.1 – PCS Model of Three Levels of Oppression

- NEW Figure 1.2 – Intersectionality: A Fun Guide

- NEW Figure 1.3 – Matrix of Oppression

Chapter 2 – Forms of Oppression

- NEW Agent of Change – Charlene Heckman

- NEW Reading *"How to Become an Ally"*

- NEW Figure 2.1 – Stereotypes, prejudice, and discrimination are distinct but interrelated concepts that have different impacts on human thought, feeling, and actions.

- NEW Figure 2.3 – Are you an ally? Try reading each of these characteristics and see how many apply to you.

Chapter 3 – Social Inequality

- NEW Agent of Change – Marc and Craig Kielberger

- NEW section on Intersectionality

- Enhanced coverage of various types of poverty to include transitional, marginal, and residual poverty

- Enhanced coverage of homelessness, including mental health
- NEW Table 3.1 – Low Income Cut-Offs, After Tax, 2014

Chapter 4 – Gender

- NEW Agent of Change – Dr. Feridun Hamdullahpur
- NEW chapter dealing specifically with Gender
- NEW In Their Shoes
- NEW section on Intersectionality
- NEW material on the social construction of gender, agents of gender socialization, and gender roles
- NEW section on hegemonic masculinity and emphasized femininity
- Updated data on gender inequality in education, work, and politics
- NEW section on violence against women
- Figure 4.1 – The Genderbread Person v.3.3

Chapter 5 – Sexuality

- NEW Agent of Change – Matt Boles
- NEW chapter dealing specifically with Sexuality
- NEW section on Intersectionality
- NEW material on homophobia, biphobia, and transphobia
- NEW material on Gay Pride
- NEW material on social construction of sexuality
- NEW material on sexual identities
- NEW material on compulsive heterosexuality and heteronormativity
- NEW material on violence against women on postsecondary campuses
- NEW Table 5.1 – Ontario Human Development and Sexual Health Curriculum
- NEW Reading Rights Fight Prompts New Trans-Inclusive Rules for Ontario Hockey

Chapter 6 – Race and Racialization

- NEW Agent of Change – Silas Balabyekkubo
- NEW Figure 6.1 – The Racialization of Poverty in Canada
- NEW Reading – Broken Circle: The Dark Legacy of Indian Residential Schools. "The Menage" by Theodore Fontaine
- NEW material on forms of racism
- NEW material on the practice of carding

Chapter 7 – Indigenous Peoples

- NEW In Their Shoes
- NEW Agent of Change – Cindy Blackstock
- NEW contributed chapter by Yale Belanger
- NEW section on Idle No More and social media engagement
- NEW section on Murdered and Missing Indigenous Women

Chapter 8 – Immigration

- NEW Agent of Change – Dr. Jean Placide Rubabaza
- NEW In Their Shoes
- NEW section on Syrian refugees
- NEW Reading – "The Thinnest Line – When Does a Refugee Stop Becoming a Refugee?"
- NEW Table 8.1 – Canada Point System for Federal Skilled Workers
- NEW Table 8.3 – Top Host Countries for Refugees Worldwide in 2015
- Updated section on ways to immigrate to Canada reflects changes in legislation, policy, and programs

Chapter 9 – Multiculturalism

- NEW Agent of Change – The Right Honourable Prime Minister Justin Trudeau
- NEW Picture This…

Chapter 10 – Religion

- NEW Agent of Change – Tendesai Cromwell
- NEW Figure 10.1 – Major World Religions
- NEW Figure 10.2 – Major World Religions, by Country
- NEW Table 10.1 – Religious Affiliation in Canada, 2011
- NEW Table 10.2 – Police-Reported Hate Crimes in Canada, 2013
- NEW Table 10.3 – Police-Reported Religiously Motivated Hate Crimes in Canada, 2012–2013
- NEW Figure 10.3 – Trends in Canadian Religious Disaffiliation, by Generation
- NEW Reading – Native Spirituality and Christian Faith – Beyond Two Solitudes
- NEW section on intersectionality

Chapter 11 – Ability

- NEW In Their Shoes
- NEW Agent of Change – Bill MacPhee
- NEW material on visible and invisible disabilities, history of disability issues in Canada, deinstitutionalization and anti-psychiatry movement
- Enhanced material on mental health and stigma, and universal design
- NEW Figure 11.1 – Disability Rights in Canada
- NEW Figure 11.3 – Grounds of Discrimination Complaints Received by Canadian Human Rights Commission 2015
- NEW Figure 11.4 – Prevalence of Disabilities by Type, Canadians Aged 15 Years or Older
- NEW Figure 11.3 – Universal Design for Learning Guidelines
- NEW Reading – Living with Post-Traumatic Stress Disorder By Vesna Plazacic

Chapter 12 – Age

- NEW Agent of Change – Moses Znaimer
- Revised chapter approach to focus on age

- Topics include: Age Stratification, Ageism and Employment, Ageism and Youth, Ageism and the Aging Population, Combatting Ageism
- NEW Figure 12.1 – Generations at a Glance
- NEW Figure 12.2 – Evolution of Music Technology
- NEW Figure 12.3 – 10 Signs You Are Experiencing Adultism
- NEW Reading – Keeping an Eye Out: How Adults Perceive Students
- NEW section on Intersectionality

Chapter 13 – Families

- NEW Agent of Change – Lucas Medina
- Chapter revised and restructured
- NEW Reading – Growing Up with Same-Sex Parents
- NEW section on Intersectionality
- NEW discussion of homogamy, endogamy, exogamy, monogamy, polygamy, cluttered/closed nest, empty nest, sandwich generation, skip-generation, and multi-generational households
- NEW sections on singlehood, family violence, foster care, assisted human reproduction, co-parenting, and childfree families

ANCILLARIES

 The **Nelson Education Teaching Advantage** (NETA) program delivers research-based instructor resources that promote student engagement and higher-order thinking to enable the success of Canadian students and educators. Visit Nelson Education's **Inspired Instruction** website at www.nelson.com/inspired/ to find out more about NETA.

The following instructor resources have been created for *Walk A Mile: A Journey Toward Justice and Equity in Canadian Society*, Second Edition. Access these ultimate tools for customizing lectures and presentations at www.nelson.com/instructor.

NETA Test Bank

This resource was written by Adele Caruso of Niagara College. It includes over 350 multiple-choice questions written according to NETA guidelines for effective construction and development of higher-order questions.

NETA PowerPoint

Microsoft® PowerPoint ® lecture slides for every chapter have been created by Anastasia Bake of the University of Windsor. There is an average of 35 slides per chapter, many featuring key figures, tables, and photographs from *Walk a Mile*. NETA principles of clear design and engaging content have been incorporated throughout, making it simple for instructors to customize the deck for their courses.

Image Library

This resource consists of digital copies of figures, short tables, and photographs used in the book. Instructors may use these jpegs to customize the NETA Power-Point or create their own PowerPoint presentations. An Image Library Key describes the images and lists the codes under which the jpegs are saved. Codes normally reflect the Chapter number (e.g., C01 for Chapter 1), the Figure or Photo number (e.g., F15 for Figure 15), and the page in the textbook. C01-F15-pg26 corresponds to Figure 1-15 on page 26.

NETA Instructor Guide

This resource was written by Adele Caruso of Niagara College. It is organized according to the textbook chapters and addresses key educational concerns, such as typical stumbling blocks that students face and how to address them. Other features include suggested answers and points to consider for the KWIP activities at the end of the chapter.

CourseMate Nelson Education's CourseMate for Walk A Mile brings course concepts to life with interactive learning and exam preparation tools that integrate with the printed textbook. Students activate their knowledge through quizzes, games, and flashcards, among many other tools.

CourseMate provides immediate feedback that enables students to connect results to the work they have just produced, increasing their learning efficiency. It encourages contact between students and faculty: you can select to monitor your students' level of engagement with CourseMate, correlating their efforts to their outcomes. You can even use CourseMate's quizzes to practise "Just in Time" teaching by tracking results in the Engagement Tracker and customizing your lesson plans to address their learning needs.

Watch student comprehension and engagement soar as your class engages with CourseMate. Ask your Nelson sales representative for a demo today.

ACKNOWLEDGMENTS

Special thanks are owing to Jillian Kerr, Senior Learning Solutions Consultant at Nelson Education, who believed in the possibilities for *Walk a Mile* when it was nothing more than a course manual. Your belief in this book, your efforts in facilitating its development, and your support for us as authors made this happen, and we are truly grateful.

In Their Shoes places the experience of affected community members at the centre of analysis. Heartfelt thanks to our students who have influenced and contributed to the creation of this textbook; you were the spark that inspired us to create something different. Your feedback, both inside and outside of the classroom, helped us to refine our content in ways relevant to your lives, here and now. With courage and integrity, you shared the stories of your lived experience so that we might grow in empathy, understanding, and knowledge as we walked in your shoes. This book was not only written *for* students, but also was written *with* students. Our thanks to Martina, Daniel, Cameron, John, Jessica, Alexander, Mia, Anton, Susan, Patti, Cindy, Shannon, and Sarah.

We're very grateful to contributing authors Dr. Yale Belanger from the University of Lethbridge (Alberta) and Samah Marei from Niagara College, whose knowledge and unique insight have so greatly enriched this text.

A very special thank you to the Agents of Change who have made our world a better place where everyone can belong—Carol and Amanda Todd, Charlene Heckman, Marc and Craig Kielberger, Dr. Feridun Hamdullahpur, Matt Boles, Cindy Blackstock, Dr. Jean-Placide Rubabaza, Silas Balabyekkubo, The Right Honourable Justin Trudeau, Tendesai Cromwell, Bill McPhee, Moses Znaimer, and Lucas Medina.

We would like to express our gratitude to the following reviewers, who provided constructive and candid feedback that helped to shape the focus, content, and pedagogical elements of the second edition of *Walk a Mile*:

Howard Bloom, Georgian College

Tara Gauld, Confederation College

Marie-Sophia Grabowiecka, Vanier College

Wendi Hadd, John Abbot College

Ivanka Knezevic, University of Toronto

Kalyani Thurairajah, MacEwan University

Mike Winacott, Georgian College

We would also like to thank the instructors who helped shape our first edition:

Michele Lemon, Sheridan College

Patricia Kaye, Fanshawe College

Blake Lambert, Humber College

Stephen Decator, St. Clair College

Sean Ashley, Simon Fraser University

James R. Vanderwoerd, Redeemer University College

Francis Adu-Febiri, Camosun College and University of Victoria

David Aliaga Rossel, Vancouver Island University

Cindy Haig, Fleming College

Tara Gauld, Confederation College

Anastasia Blake, St. Clair College of Applied Arts and Technology

In addition, we acknowledge Niagara College of Applied Arts and Technology and our colleagues who have supported us on this journey.

We would like to acknowledge the extraordinary talent and dedication of everyone we worked with at Nelson Education Ltd. Special thanks go to Jillian Kerr, Senior Learning Solutions Consultant at Nelson Education. We are grateful to publisher Leanna MacLean who championed the work of this second edition and whose support was invaluable along the way. We are so appreciative of the talent, skill, passion, and professionalism of Content Development Manager Toni Chahley, who was absolutely inspiring to work with and whose encouragement saw us through to the end. We are very grateful to the rest of the team at Nelson: Jaime Smith, Production Project Manager; Claire Varley, Marketing Manager; Carrie McGregor, Freelance Permissions Researcher; Lynn McLeod, Permissions Manager; Linda Szostak, Copy Editor; and Megha Bhardwaj, Project Manager.

We are especially honoured to have Craig and Marc Kielburger author the foreword to *Walk a Mile*. How inspiring would it be to *walk a mile* in the shoes of the Kielburger brothers, Marc and Craig? They are the embodiment of what it means to be socially conscious global citizens committed to changing the world through social action premised on equity, justice, diversity, and inclusion. As educators, we have had the opportunity to witness the transformative change they inspire in young people's lives.

We wish to acknowledge the support given by our family members, friends, and colleagues during this process. Writing a book requires more than knowledge, experience, passion, and a good computer.

It often means locking yourself away for hours at a time while family, friends, and colleagues pick up the pieces. So we would like to acknowledge those who have supported us on this journey.

I (Jamie) would like to thank my partner, Joe, for your understanding and patience, and for always being in my corner. An enormous thank you to my parents and in-laws for your support and endless days of babysitting; my daughter, Joey, who inspires me daily; my strong circle of women, for your guidance and support and for always building me up; and my students and colleagues who challenge and motivate me to learn and grow as an educator every day.

I (Theresa) wish to acknowledge my "circle of support" who have helped me throughout the process of writing the first and second editions of *Walk a Mile*. First, to my son Daniel, who inspires me each and every day with his courage to face enormous challenges, and his resilience and strength to live the life he has imagined. I am so proud of you. I am grateful for the blessing of my extraordinary brothers, Paddy and Mike, whose love and support is ever present. Truly, these books would never have come to fruition without the emotional support and caregiving provided by my aunt, Dr. Mary O'Reilly, who is the embodiment of benevolence in our lives and one of the most inspiring women I know. Of equal blessing is my uncle, John Pearson, whose indelible support makes him the person we have always counted on (the dishwasher saved my life). I owe special thanks and gratitude to my nephew, Jacob, who came to live with us when Daniel got sick. To my niece, Katie, who helped us with the chores of daily life and whose love enriches my life daily; to my sister-in-laws, Heather and Cathy, my nieces, Katie and Julia, my nephew and godson, Aidan, who fill my life with love and joy and have supported us throughout this journey; to my aunt, Sr. Veronica O'Reilly, and my uncles, Pat and Mike O'Reilly, who have had an enormous and wonderful influence on my life and who took on the role of caregivers. Thank you to my friend and co-author, Dr. Deborah Boutilier, who did a great deal of the heavy lifting in the first edition, cooked meals for us, and provided constant emotional support during Daniel's surgery and treatment. Thank you to my friend and co-author, Jaime Oresar, for stepping in for the second edition and for making this process so much fun! To my friends, Kristen, John, Charlene, and Debbie— I thank you for all you have done to support me in this endeavour. I would also like to express my heartfelt gratitude to President Dan Patterson and my NC family for your incredible support. Last, but certainly not least, I am so thankful for the blessing of being born and raised by a mother and father who worked every day to create a socially just world. Thanks, Mom—this is for you.

PREFACE REFERENCES

Chickering, A. W., & Gamson, Z. F. (March 1987). Seven principles for good practice. *AAHE Bulletin* 39 (7), 3–7.

Collins, P. H. (2000). *Black feminist thought: Knowledge, consciousness and the politics of empowerment* (2nd ed.). New York: Routledge.

Keller, H. (1903). *Optimism: An essay.* New York: Crowell and Company.

McKeachie, W., & Svinicki, M. (2006). *Teaching tips: Strategies, research, and theory for college and university teachers.* Belmont, CA: Wadsworth.

Michael, J. (2006). Where's the evidence that active learning works? *Advances in Physiology Education, 30*(4), 159–167.

Diversity, Oppression, and Privilege

> *"Cause they don't even know you, all they see is scars, they don't see the angel living in your heart."*
>
> *(Sixx:AM, Skin, 2011)*

LEARNING OUTCOMES

By mastering this unit, students will gain the skills and ability to:

- demonstrate an understanding of the historical and contextual dimensions of social exclusion, privilege, and oppression

- identify and discuss the nature and dynamics of oppression and reflect upon the importance of creating an anti-oppressive environment

- assess the ways in which oppression is linked to issues of diversity, the social construction of "other" in identity formation, and the privileging of some identities over others

- analyze the complexity, multiplicity, intersectionality, fluidity, and contextuality of personal and social identity

- evaluate the role of power in the social construction of knowledge, privilege, and bias

Rawpixel.com/Shutterstock

When Canada's Prime Minister Justin Trudeau spoke at the World Economic Forum in Davos, Switzerland, he spoke about diversity as a source of strength as we contemplate a world at the brink of enormous change—now described as a fourth Industrial Revolution. "We need societies that recognize diversity as a source of strength, not a source of weakness … Diversity is the engine of invention. It generates creativity that enriches the world. We know this in Canada" (Trudeau, 2016). To those who might question diversity and suggest it creates discord and insecurity, Trudeau (2016) suggests we "embrace diversity and the new ideas that spring from it, while simultaneously fostering a shared identity and shared values in safe, stable communities that work." While the concept of diversity has evolved over time, it is an enduring Canadian value and a framework that has idealized the facilitation of social inclusion and equity, and the creation of a sense of belonging and shared identity.

THE CONCEPT OF DIVERSITY: KEY CHALLENGES

The fact is that as human beings, we are an incredibly diverse species. To prove this, Timothy Allen, a photographer for the BBC series *Human Planet* (2011), captured in stunning photography the diversity of human experience as he travelled to over 40 countries—from the BaAka villagers in Congo Valley in the Central African Republic to the Kazakh hunter in the remote mountain region of western Mongolia, from the Laotian fishers on the raging Mekong River to the Bajan Laut people of Malaysia who don't set foot on land. The reality of our human diversity portrayed through his images can leave a person speechless.

Today, we use the concept of diversity to refer to the social construction of differences or a special event, a human resource strategy or a value in a mission statement, a corporate strategic objective or a policy, the name of a course or a vocational learning outcome for your college program. But if we utilize diversity as a framework for social change, we need more than a celebration of human diversity and more than the acceptance of differences that exist among us. There is a need for a framework that can promote justice and equity, and challenge patterns of oppression and discrimination. No issue of justice or equity hangs on the appreciation of difference or the celebration of diversity.

Difference: In a social context, a term used to refer to difference in social characteristics.

The Celebratory Approach

One of the key challenges of contextualizing diversity in Canada has been its use as a celebratory paradigm that fails to address unequal power relations and histories of social exclusion, discrimination, privilege, and oppression. Upon the death of one of this century's greatest moral leaders, Nelson Mandela, we recall the inspiration he found in Canada's respect for diversity. In Mandela's first address to Canadian parliament (McQuigge, 2013), he remarked, "Your respect for diversity within your own society and your tolerant and civilized manner of dealing with the challenges of difference and diversity had always been our inspiration." There have been many examples where Canada as a nation has faced the challenges of diversity. In July 2005, the House of Commons passed the Civil Marriage Act, making Canada the fourth country in the world to legalize same-sex marriage. According to Irwin Cotler (2015), former Minister of Justice and Attorney General of Canada and emeritus professor of law at McGill University, not only was this an achievement of marriage equity; it was also an example of Canadian leadership in matters of equity, freedom, and justice, and a remembrance of the "virtues of debating serious issues in a manner becoming of a vibrant and open democracy." In 1982, the Canadian Charter of Rights and Freedoms, considered the highest law of Canada, guaranteed rights and freedoms such as freedom of expression, rights of Indigenous Peoples, the right to equality, language rights, and the protection of Canada's multicultural heritage. We were, in fact, the first country in the world to adopt multiculturalism as an official policy (Government of Canada, 2017). But criticism of celebratory diversity emerged when racialized and ethnic communities were treated as archetypes, not individuals—viewing superficial differences as exotic. Diversity initiatives designed to "celebrate difference" fell prey to the same trivialization that "celebrating multiculturalism" in Canada in the 1970s did. We have come to realize that diversity needs to mean a great deal more than celebrating difference on special days and at special events, just as multiculturalism means a great deal more than saris, samosas, and steel bands.

The Difference Approach

Another key challenge of contextualizing diversity in Canada is rooted in the social construction of **differences** in a manner that serves to keep the standards of the dominant group intact. As a concept, diversity is often used to refer to salient aspects of our identity to which we attach social meaning, such as class, gender, sexual orientation, race, ethnicity, class, age, ability and so on. The danger here is that certain characteristics become valued and create privilege, while other characteristics are devalued and result in marginalization, thereby creating oppression. It is in this manner that approaches to diversity, rooted in a sameness/ difference dichotomy, reinforce and devalue difference as "other." According to law professor Catherine MacKinnon (1989), neither a sameness nor a difference approach works:

> You can't change the relationship between those who are equal and those who are unequal by giving them the same things … the relation between the two stays the same, and it is the relation that defines the inequality. The dominant measure is set by advantaged people … sameness and difference is not the issue of inequality. It never has been. To make this the issue conceals, among other things, the way that the dominant group becomes the measure of everything, including the measure of the disadvantaged group's entitlement to equal treatment. (p. 5)

Diversity initiatives that utilize the sameness and difference approach can be problematic when the goal is to help people learn how to "appreciate and accept difference," as it implies a need to learn how to put up with things that are different from the dominant group. This approach fails to address unequal power relations and is often amnesic of histories of social exclusion and oppression in Canada. It can be hard to acknowledge that social power and privilege provide advantages (often unearned) to some people and disadvantages to others. But once this analysis begins, a person can no longer say, "I'm not really comfortable talking about the role oppression has played in people's lives, so we will just pretend we are all the same."

An Anti-Oppression Approach

What if instead we reconceptualized **diversity** as a framework whose strategies promote equity, justice, and inclusion while challenging patterns of oppression and discrimination? Such a framework would acknowledge that power and privilege are central in the social construction of our identities. In other words, the concept of diversity would recognize that people will experience certain advantages and disadvantages based on the socially constructed meanings assigned to aspects of identity such as race, class, gender, ability, age, sexual orientation, family status, religion, spirituality, language, accent, ethnicity, citizenship, and so on.

Practitioners employing this framework would acknowledge the historical oppression and social exclusion of affected communities. They would have a new consciousness of how power works, that is grounded in the affected community's definition and understanding of their own reality. This new consciousness could shake the foundation of what the practitioner had accepted as truth and reality. But with the experience of affected communities at the centre of analysis, practitioners could begin to challenge dominant viewpoints and perspectives that contribute to the oppression and marginalization of others. This framework needs to ensure anti-oppressive practices are consistent with the needs of diverse populations, addressing issues of power, privilege, self-definition, leadership, and participation. This might be accomplished through strategies whereby marginalized communities share power and leadership in creating social change for affected communities. This might also be accomplished through the strategies of "becoming an ally," outlined by author Anne Bishop (2002) in Chapter 2. Key to any strategy used is the need for affected communities to define their own issues, develop their own solutions, and determine their own leadership structures. It is not enough simply to invite representatives of diverse populations to belong. Nor can those in power create a solution *for* diverse populations.

ANTI-OPPRESSION AND CRITICAL SOCIAL THEORY

There is no one theory of oppression, nor is there one singular approach to anti-oppressive practice. As Mullaly (2010) notes, "currently, there is much discussion on the nature, dynamics, forms, functions, and causes of oppression, but there is no dominant theory of oppression or dominant approach to anti-oppression." Arguably, there is no imperative for one theory nor one approach to explain all aspects of oppression (Mullaly, 2010). The discussion of oppression and anti-oppression used in this text falls within a perspective known as critical social theory and is used because of its practical and political dimension that advocates a transformation of society for the purpose of liberating those who are oppressed (Agger, 1989; Kellner 1989; Leonard, 1990). **Critical social theory** is a cluster

> **Diversity:** An anti-oppression framework built on principles that value social equity, social justice, and social inclusion.
>
> **Critical social theory (also known as critical theory):** A macro theory interested in those who are oppressed, which critiques social structures that exploit and marginalize members of a society and whose goal is liberation from oppression.

of theoretical perspectives, and as Mullaly (2010) notes, includes some forms of feminist theory (patriarchy as a form of oppression), critical race theory (racism as a cause of oppression), queer theory (heterosexism as a cause of homophobia), liberation theology, cultural studies theorists (oppression result of dominant culture), structural theorists (oppression caused by social structures that privilege dominant groups over subordinate groups), and anti-oppression theorists (all subordinate groups are consciously and/or unconsciously oppressed on personal, cultural and institutional levels by visible and invisible structures). Common among these approaches is an interest in those who are oppressed, a critique of social structures that exploit and marginalize members of a society, and the goal of changing the world to be free of domination and oppression (Leonard, 1990; Mullaly, 2010). Ben Agger (1989) once wrote that critical social theory "conceives human liberation as the highest purpose of intellectual activity." A majority of helping professionals will acknowledge oppression exists; assist targeted individuals and groups to deal with the effects of oppression on their lives; and make changes to existing structures, institutions, policies, and practices to ameliorate the effects of domination and marginalization (Mullaly, 2010). Fewer helping professionals adopt an approach whose goal is transformative social change for the liberation of all from oppression (Mullaly, 2010).

THE NATURE AND DYNAMICS OF OPPRESSION

Oppression is a complex and multidimensional social phenomenon that involves the intentional and unintentional **domination** of **non-dominant groups** in society by powerful **dominant groups** and occurs on individual, cultural, and structural levels (Mullaly, 2010). At some point in our lives, all of us will experience restrictions on our personal freedom, but this does not necessarily constitute oppression. Oppression occurs when a person's real or perceived membership in a group results in their exclusion from full

Oppression: The intentional and unintentional domination of NON-dominant groups by powerful dominant groups that occurs on individual, cultural, and structural levels in society.

Domination: The systematic and continuous exertion of power by dominant groups over non-dominant groups.

Non-dominant groups: Groups of people in a society without (or with less) power and privilege.

Dominant groups: Groups of people in a society who have power and privilege.

participation and active citizenship in society, and oppression also occurs through the denial of opportunities and rights that the dominant group takes for granted (Mullaly, 2010).

Some of the ideologies that support the oppression of non-dominant groups in a society include a belief that people are responsible for their own oppression (known as victim blaming); a belief that all members of a group are the same (known as stereotyping); that a ranking of those who are superior and inferior is part of a natural hierarchy and natural outcome of human nature (known as essentialism); the belief that majority always rules, even if at the expense of non-dominant group members (known as might is right); the belief that if everyone works hard and takes advantage of opportunities available, they can succeed in life (myth of meritocracy); and the belief in the survival of the fittest—that human beings are naturally competitive for scarce resources.

There are several dynamics of oppression that are important to anti-oppressive practice generally and the interpretation of Table 1.1 specifically. First, oppression is not a static concept; our experiences with oppression are constantly changing over the course of our lives. Second, by trying to determine a singular category of oppression or by ranking forms of oppression in a hierarchical structure of importance, we create an either/or dichotomy, with one side being privileged and the other being subjugated as "other" (Collins, 1993). These kinds of reductionist monocausal explanations fail to acknowledge the complex and multidimensional experience of oppression. Third, we are all oppressors and oppressed. Depending on social and historical contexts, we can be members of an oppressed group or we can be oppressors, or we can be both, sometimes even simultaneously. As Anne Bishop (1994, p. 61) states, "we are oppressors in some parts of our identity and oppressed in others." Within the matrix of domination, there are "few pure victims or oppressors. Each individual derives varying amounts of penalty and privilege from the multiple systems of oppression which frame everyone's lives" (Collins, 1993, p. 619). Fourth, it is also important to consider all three levels where oppression occurs within society as they are all potential sites of resistance and liberation.

Multiple Levels of Oppression

People experience and resist oppression that occurs on three levels: the personal biography level, the group or community level of the cultural context,

TABLE 1.1

Dominant and Non-Dominant Groups in Canada Based on Aspects of Social Identity

Social Identity	Dominant Group	Non-Dominant Group
Age	Ages 30–60	Youth and seniors
Ability	Able-bodied people	People who are mentally and/or physically disabled
Language of Origin	English Language	Non-English languages
Faith/Religion	Christianity	Non-Christians
Gender	Male and cisgender*	Female and transgender
Socio-economic status	Middle- and upper-class people	Working class, low-income people; people living in poverty
Race	White people	People of colour and Indigenous people
Sexual orientation	Heterosexual people	Lesbian, gay, and bisexual people

* The term cisgender means having a gender perception that matches one's physical gender.

Source: Ontario Council of Agencies Serving Immigrants (2012). Positive Spaces Initiative's Online Training Course. Retrieved from Ontario Learn: http://learnetwork.ca

and the structural level (Collins, 1993; Mullaly, 2010). Thompson (1997; 2002) refers to this as the PCS model of analysis, suggesting that any time we look at issues of discrimination and oppression, we need to consider all three of these levels and how they interact and influence one another. Within this model, all three levels of oppression continually interact with each other (see Figure 1.1).

Oppression at the personal level usually involves negative stereotypes or prejudice toward members of a non-dominant group. These negative thoughts, attitudes, and behaviours can be blatant and intentional acts of aggression and/or hatred, or they can be hidden and unconscious acts of aversion or avoidance (Mullaly, 2010). Intentional acts of aggression or hatred can include bullying, harassment, graffiti, name-calling, violence, or threats of violence that are all used to intimidate and denigrate non-dominant group members as second-class citizens (Mullaly, 2010). Many acts of oppression at the personal level have gone underground and are expressed as covert forms of domination. This might include intentionally avoiding interaction, avoiding eye contact, maintaining physical distance, or using a hostile tone of voice with persons of particular non-dominant groups.

FIGURE 1.1

PCS Model of Three Levels of Oppression

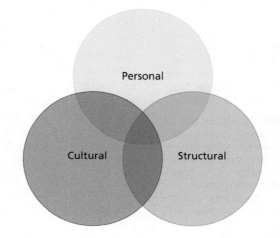

As you reflect upon an experience with oppression in your own life, how might you resist this oppression at the personal, the cultural, and the structural level?

Source: Adapted from Mullaly, B. (2010). *Challenging Oppression and Confronting Privilege* (2nd ed.). Don Mills: Oxford University Press, p. 62.

Oppression at the group or community level of the cultural context involves dominant groups creating and maintaining hierarchical divisions of race, class, gender, ability, and so on, to universalize and impose its own culture as a superior culture (Mullaly, 2010). Culture does not refer to ethnicity and is distinct from oppression at the structural level in this analysis. **Culture** refers to everything in our social environment that we learn through socialization, including cultural norms, values, beliefs, ideals, attitudes, customs, practices, and symbols. For example, language and discourse as a culture's symbolic system plays a role in keeping subordinate groups subservient to the dominant group. Using **inclusive language** "avoids exclusion and stereotyping and is free from descriptors that portray individuals or groups of people as dependent, powerless, or less valued than others" (Queen's University, 2017). Members of a dominant group will often rail against the use of inclusive language as political correctness gone awry, but this viewpoint fails to consider the fact that language and discourse are reflective of social power structures within society.

Oppression at the structural level comprises the ways in which oppression is institutionalized and legitimated in society (Mullaly, 2010). At this level, oppression consists of the ways in which a society's laws, social policies and practices, social institutions, and economic and political systems work together to favour the dominant group at the expense of the non-dominant group (Mullaly, 2010). The majority of structural oppression is covert and hidden, making resistance and social change challenging. It is important that any analysis of oppression consider the ways in which oppression is more than just a personal issue, and examine the cultural and structural levels of oppression.

Culture: The total of everything in our social environment that we learn through socialization, that is passed down from one generation to the next, and that continues to change throughout our lives.

Inclusive language: The deliberate selection and use of vocabulary that avoids the exclusion of particular groups and that avoids the use of false generic terms (Canadian Race Relations Foundation, 2017).

Internalized oppression: Occurs when targeted people internalize (or begin to believe) the negative stereotypes and misinformation that the larger society communicates to them, either as individuals or as part of a larger group.

Intersectionality: A concept used to describe the ways in which various aspects of identity interconnect on multiple and often simultaneous levels and can form interlocking systems of oppression.

Internalized Oppression

Internalized oppression occurs when targeted people internalize (or begin to believe) the negative stereotypes and misinformation that the larger society communicates to them, either as individuals or part of a larger group. Internalized oppression that operates at the *personal* level results in individuals believing and internalizing the negative messages others communicate and then acting in self-defeating ways that reinforce these stereotypes. Some examples of the ways in which internalized oppression at a personal level might manifest itself include the following: a person who will not contribute in a meeting because they have internalized negative messages that lead them to believe their contribution is not important or valid; or a student who has internalized negative messages about their abilities and gives up learning and pursuing their dreams. Internalized oppression that operates at a *community* level results in people from the same groups believing the negative messaging that society communicates about other members of their group. This can result in people from the same groups engaging in behaviour that is destructive to members of their group. An example of this is when a woman joins men in a conversation about women being unable to perform certain jobs. It is important to remember that not all targeted groups or individuals who experience oppression will internalize it, but for those who do, internalized oppression can have serious effects. At a personal level, internalized oppression can cause people to see themselves as unworthy, damage self-image, and be a source of self-destructive behaviour and affect personal health and well-being. At a community level, internalized oppression can result in distrust and blaming, making it difficult to build alliances for resistance and creating change.

Intersectionality and the Matrix of Oppression

> One day our descendants will think it incredible that we paid so much attention to things like the amount of melanin in our skin or the shape of our eyes or our gender instead of the unique identities of each of us as complex human beings. (Franklin Thomas)

Our identities are complex, and the ways in which aspects of our identity intersect can influence the ways we experience oppression. The concept of **intersectionality** is used to describe how forms of oppression associated with aspects of social identity, such as racism, classism, sexism, heterosexism, ableism, ethnocentrism, ageism and so on, are viewed as interlocking systems rooted in cultural and historic contexts (Collins, 1993). An intersectional paradigm illuminates the ways in which different forms of oppression are all interconnected and therefore *cannot* be viewed separately from one another (see Figure 1.2).

Intersectionality theory was first coined by Kimberle Crenshaw when she discussed the intersectionality experience of black women as more powerful than the sum of their race and gender (Crenshaw, 1991). The concept of intersectionality was also integral to sociologist Patricia Hill Collins' discussion of black feminism. "Black feminist thought … fosters an enhanced theoretical understanding of how race, gender, and class oppression are part of a single, historically created system" (Collins, 1993, p. 618). Legal remedies for discrimination have often forced people to select one protected category to characterize this experience. For example, someone files a human rights complaint alleging they were discriminated against because they are a woman (protected category being gender). But this singular approach to oppression fails to consider that we have complex and multiple identities that intersect with one another. For example, sexism experienced by an Asian woman will be different than sexism experienced by an Indigenous woman, a poor white woman, or a woman with schizophrenia. In response to criticism of this singular approach, some have adopted additive models of oppression that use the sum of different forms of oppression but maintain their separateness. An example of this is a person making a human rights complaint who adds another protected category, alleging they were discriminated against based on gender and disability. But these additive models of oppression still treat each form of oppression separately, as sexism + ableism, rather than as a new combined experience. Additive models are also characterized by either/or dichotomous thinking where one side of the dichotomy is privileged and the "Other" side is subjugated (Collins, 1993). For example, persons who identify as gender fluid are often asked, "What are you?" in an attempt to categorize them according to a male/female gender dichotomy.

It is important to interpret Table 1.1 using an intersectional paradigm that helps us to understand that there is not one form of oppression but rather interlocking systems of oppression that work together to produce injustice (Collins, 1993; 2000). In embracing *a both/and conceptual stance*, Collins (1993) suggest we can move beyond an understanding of oppression as additive and separate systems to what she sees as "the more fundamental issue

FIGURE 1.2

Intersectionality: A Fun Guide

INTERSECTIONALITY a fun guide

this is Bob.

Bob is a stripey blue triangle! AND SHOULD BE PROUD.

Hi!

yay! me

SADLY SOME PEOPLE DO NOT LIKE BOB. BOB FACES OPPRESSION FOR BEING A TRIANGLE, & FOR HAVING STRIPES. DOWN WITH STRIPES GOD HATES TRIANGLES

LUCKILY, THERE ARE LIBERATION GROUPS! BUT THEY AREN'T INTERSECTIONAL. SO THEY LOOK LIKE THIS

WELCOME As! WELCOME Es!

THEY DON'T TALK TO EACH OTHER. IN FACT, THEY COMPETE.

BOB CAN'T WORK OUT WHERE TO GO. AM I MORE STRIPE OR TRIANGLE?

I'M MORE OPPRESSED! NO, I AM! I DESERVE MORE!

BOB WISHES THAT THE TRIANGLES AND STRIPES COULD WORK TOGETHER

OPPRESSION OF ONE AFFECTS US ALL! No LIBERATION WITHOUT EQUAL REPRESENTATION!

INTERSECTIONALITY IS THE BELIEF THAT OPPRESSIONS ARE INTERLINKED AND CANNOT BE SOLVED ALONE.

OPPRESSIONS ARE NOT ISOLATED. INTERSECTIONALITY NOW!

Source: By Dr. Miriam Dobson, https://miriamdobson.com/2013/04/24/intersectionality-a-fun-guide. Reprinted with permission.

of social relations of domination. Race, class, and gender constitute axes of oppression that characterize Black women's experiences with a more generalized matrix of domination." The Matrix of Oppression

FIGURE 1.3

Matrix of Oppression

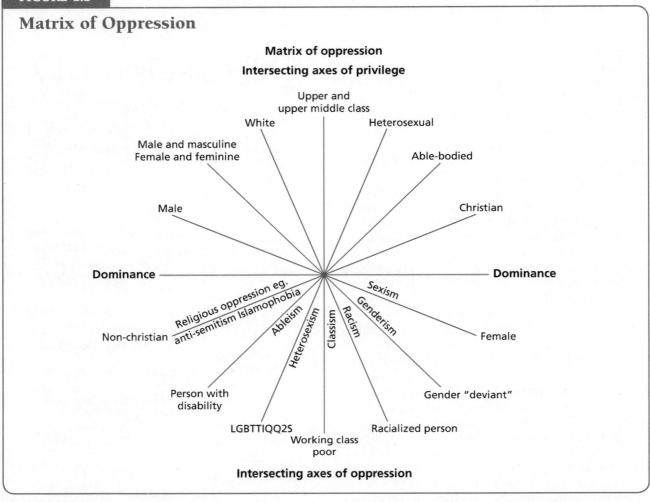

Matrix of oppression

Intersecting axes of privilege

Upper and upper middle class

White · Heterosexual

Male and masculine / Female and feminine · Able-bodied

Male · Christian

Dominance —————— Dominance

Religious oppression eg. anti-semitism Islamophobia · Sexism · Genderism · Racism · Classism · Ableism · Heterosexism

Non-christian · Female

Person with disability · Gender "deviant"

LGBTTIQQ2S · Racialized person

Working class poor

Intersecting axes of oppression

Source: Adapted from Morgan, K. (1996). "Describing the Emperor's New Clothes: Three Myths of Educational (In-)Equity." In A. Diller, B. Houston, M. Ayim, & K. Morgan (Eds.), The Gender Question in Education: Theory, Pedagogy, and Politics (pp. 105–122). Boulder, CO: Westview."

(Figure 1.3), based on Patricia Hill Collins' **matrix of domination**, is built on a foundation of ideological beliefs about domination and how it is socially constructed. There are multiple dimensions on the matrix, such as race, gender, age, ability, religion, sexual orientation, and class, but the overarching relationship among all is domination (Collins, 1993). There is no hierarchy along the matrix in terms of one axis being more important than another. What is important is that they are intersecting axes that represent interlocking systems of oppression. We know injustices "intersect" through people's multidimensional experiences. The flip side of axes of oppression are axes of privilege.

> **Matrix of domination:** Term associated with the work of Patricia Hill Collins that refers to forms of oppression and resistance based on socially constructed differences shaped by cultural and historic contexts where an individual or group can experience both oppression and privilege as a result of their combined identities.

Heterogeneity Within Oppressed Groups

It is important to consider that not all members of a particular non-dominant group will share the same experiences with oppression. For example, women do not all share the same experience of sexism. People do not have one single identity; they have multiple identities that change over time, place, and social context.

Heterogeneity is part of every form of oppression, and every group experiences it (Mullaly, 2010). To use a singular approach to oppression, such as anti-racism, presumes that all racialized groups and individuals experience racism in the same way. The singular approach to oppression fails to consider the fact that we all have multiple identities from different social categories that intersect with one another (Mullaly, 2010). There are many forms and experiences of racism, sexism, classism, ableism, heterosexism, and so on.

Bullying as Oppression

Bullying is a widespread, complex, and evolving social problem in Canada. **Bullying** is defined as a form of direct or indirect aggression that involves a real or perceived power imbalance (Mishna & Van Wert, 2015). Traditional face-to-face bullying includes verbal, social, and physical bullying. Verbal bullying involves behaviours such as teasing, name-calling, and threats. Social bullying, also known as relational bullying, includes behaviours that intentionally exclude or socially isolate, such as spreading malicious rumours or encouraging people not to be friends with someone. Physical bullying includes behaviours such as spitting, pushing, punching, kicking, tripping, and so on. As our interface with information and communication technologies increases, **cyberbullying** has emerged as a new form of online aggression that has unique implications for prevention and intervention (Mishna & Van Wert, 2015). Often times, this form of bullying takes place on popular social media platforms such as Facebook, Twitter, and Yik Yak. Following the death of Amanda Todd, the work done by her mother Carol (featured as the Agent of Change in this chapter) has contributed to greater awareness and prevention strategies in Canada.

Biased-based bullying is a form of bullying motivated by intolerance or hatred toward others due to real or perceived aspects of their identity, such as race, ethnicity, religion, gender, sexuality, ability, age, and so on (Mishna & Van Wert, 2015). For example, bias against persons who identify as lesbian, gay, bisexual, transgender, queer, and questioning is often expressed through homophobic and transphobic bullying, which is a serious issue in Canadian schools (Mishna & Van Wert, 2015). A Canadian climate survey conducted in 2009 found the following: 59 perent of LGBTQ high school students were verbally harrassed, compared to 7 percent of non-LGBTQ students; 25 percent of LGBTQ high school students were physically harrassed, compared

to 8 percent of non- LGBTQ students; 31 percent of LGBTQ high school students were harrassed online, compared to 8 percent of non- LGBTQ students; 73 percent of LGBTQ high school students did not feel safe at school, compared to 20 percent of non-LGBTQ students; and 51 percent of LGBTQ high school students did not feel accepted, compared to 19 percent of non-LGBTQ students (Taylor, et al., 2009). Biased-based bullying of students with disabilities is also a serious issue in Canadian high schools, colleges, and universities. In a recent report by the Canadian Human Rights Commission (2017), it was found that 27 percent of Canadian high school, college, and university students with a disability are being bullied because of their disability.

UNPACKING OUR PRIVILEGE

So what is privilege? Often referred to as the flip side of oppression, privilege is the unearned advantages that confer dominance and is based on membership in a particular social group (McIntosh, 2000; Heldke and O'Connor, 2004). Privilege does not include advantages that are earned, and the failure to distinguish between earned and unearned advantages allows privileged groups to say that all of their privilege is earned (Mullaly, 2010). Privilege also confers dominance by giving one group power over another group. Anti-oppressive practice examines the role of power and privilege in people's lives. Oppression and privilege are viewed as interconnected; we all experience both privilege and oppression to some degree (Mullaly, 2010). Each of us has multiple and intersecting identities whereby we experience a combination of unearned disadvantage and unearned advantage because of visible and invisible aspects of these identities. For oppression to exist, there needs to be an oppressed group and a privileged group that benefits from oppression. Privilege opens doors of opportunity and oppression slams those doors shut (Johnson, 2006).

Heterogeneity: Means having dissimilar characteristics. The opposite is homogeneity, which means having the same characteristics.

Bullying: A form of direct or indirect aggression that involves a real or perceived power imbalance.

Cyberbullying: Involves the use of information and communication technologies such as the Internet, social networking sites, websites, email, text messaging, and instant messaging to intimidate or harass others (Royal Canadian Mounted Police, 2016).

Biased-based bullying: Bullying resulting from bias against someone because of a real or perceived aspect of their identity.

AGENT OF CHANGE

Carol and Amanda Todd

Photo courtesy of Carol Todd

Carol and Amanda Todd

In each chapter throughout the textbook, we have featured Canadians as AGENTS OF CHANGE—doing extraordinary things to create social change in their neighbourhoods, their communities, and their country, and across the world. They are advocates against micro and macro levels of oppression. They are motivated by principles of social justice and equity.

We begin with Carol Todd, a Canadian icon for resiliency, strength, and hope. When you meet Carol, some of your first impressions are of a woman of quiet strength and depth, a mother with a heart filled with love, a tech-savvy educator with a passion for teaching, and a stalwart anti-bullying advocate.

Her reach is far and wide; I am never sure when we talk if she is at home in British Columbia or another part of the world speaking in elementary schools, high schools, colleges, universities, and other public venues about anti-bullying, mental health, and cyber safety initiatives. What is constant is the impact Carol has on those she speaks to. It never ceases to amaze me the number of students I teach who talk about having been bullied and reaching out to Carol and finding support and help when they did. It speaks to me of Carol's authenticity as an agent of change and of the impact of her daughter Amanda's legacy.

On September 7, 2012, Amanda Todd shared with the world an eight-minute YouTube video entitled *My Story: Struggling, bullying, suicide, self-harm* that was a silent flash-card narrative of her own experiences of being bullied. The message posted by Amanda below the video reads:

I'm struggling to stay in this world, because everything just touches me so deeply. I'm not doing this for attention. I'm doing this to be an inspiration and to show that I can be strong. I did things to myself to make pain go away, because I'd rather hurt myself than someone else. Haters are haters but please don't hate, although I'm sure I'll get them. I hope I can show you guys that everyone has a story, and everyone's future will be bright one day, you just gotta pull through. I'm still here aren't I?*

On October 10, 2012, Amanda took her own life. Following her death, Amanda's YouTube video went viral and her story has received international media coverage. In the years since this tragedy, Carol has courageously continued the conversation Amanda started by speaking to students, parents, educators, and helping professionals around the world. Carol has appeared on the *Dr. Phil Show* (2014) and the *Fifth Estate*'s documentaries, "The Sextortion of Amanda Todd" (2013) and "The Stalking of Amanda Todd: The Man in the Shadows" (2014). The list of keynote addresses and public presentations found on the Amanda Todd Legacy Society website is staggering.

Carol is the recipient of numerous awards for her anti-bullying and mental health activism, including the Me to We Award for Social Action, TELUS WISE Outstanding Canadian Citizen Award, BC Community Achievement Award, and more. While very deserving of this recognition, what matters most to Carol is the ability to continue the conversation that Amanda started so that we can create communities and neighbourhoods that practise kindness and give a sense of belonging and inclusion for all. As Carol prepared to travel to the Netherlands in 2017 for the trial of Aydin Coban, her grief was palpable as she wrote of Amanda, "what I would give up to be able to see her, feel her, touch her, cuddle her and kiss her once again." And still Carol channels this grief to save others from the kind of bullying and emotional pain Amanda suffered.

Students describe Carol as inspirational … authentic … strong … a voice for change.

Through Carol, Amanda's legacy continues to be one of transformational social change for all of us as Canadians.

To learn more, visit the Amanda Todd Legacy Society at http://www.amandatoddlegacy.org/

*Source: http://www.amandatoddlegacy.org. Reprinted with permission.

Morris, T. (2015, May 22). On A Plate: A Short Story About Privilege. The Pencilsword #10.

"Privilege increases the odds of having things your own way, of being able to set the agenda in a social situation and determine the rules and standards and how they're applied. Privilege grants the cultural authority to make judgements about others and to have those judgements stick. It allows people to define reality and to have prevailing definitions of reality fit their experience. Privilege means being able to decide who gets taken seriously, who receives attention, who is accountable to whom and for what. And it grants a presumption of superiority and social permission to act on that presumption without having to worry about being challenged." (Johnson, 2006, p. 33)

The "luxury of obliviousness" is a term Johnson uses to describe not having to think about privilege. People are often resistant to talking about power and privilege because of the fear that it makes dominant groups uncomfortable or pits groups against each other (Johnson, 2006). The process of unpacking our knapsack of invisible privilege is often met by hostility and resistance. So we tend to ignore the issue of privilege, as it is easier to look at the oppression of groups rather than implicating ourselves through our role as an oppressor (Mullaly, 2010). But denying the existence of privilege makes social change difficult (Johnson, 2006). The fact is, we all have varying amounts of privilege based on aspects of our identity, and we all experience oppression based on other aspects of identity. So when someone tells us that we have privilege, it does not mean that we intentionally oppress members of a non-dominant group or that we wilfully engage in discriminatory behaviour. We are also likely a member of a marginalized group based on some aspect of our identity that has caused us suffering or pain. "As with oppression, one of the features of privilege is that various privileges intersect and interact with other privileges and with different forms of oppression at the same time" (Mullaly, 2010, p. 293). When we are able to recognize that our own privilege can be invisible to us, as well as the marginalization of non-dominant groups, we can begin to break the cycle of oppression. When we are able to acknowledge our privilege, we are able to see the unearned advantages that come from our membership in a dominant group. Activities, like those contained in KWIP at the end of each chapter, that

Social justice: Concept that challenges the social structures, processes, and practices associated with inequalities that lead to oppression.

facilitate recognition of and reflection upon privilege, are not about making you personally apologetic for certain privileges you have. These activities can help us to recognize the interconnectedness of our own personal histories, stories, and experiences to the larger systems within society. Then, the recognition and acknowledgement of privilege can be the starting point for a larger conversation about how we "unlearn" oppression (Bishop, 2002).

JOURNEY TOWARD JUSTICE AND EQUITY

So how do we construct a society where all people can experience the world as just? Where all people have equivalent access to opportunities? Where all people, including those historically underserved and under-represented, feel valued, respected, and able to live lives free of oppression? And where all people can fully participate in the social institutions that affect their lives?

How is it just or equitable that being born in one part of Canada means you don't have access to safe, clean drinking water? How is it just or equitable that being born in one part of the world might mean you could be forced to work at four years of age? How is it just or equitable that being born in one family over another makes the difference in being able to afford tuition for your post-secondary education? To begin to find the answers to these questions, the concepts of social justice and equity and their relationship to oppression and anti-oppression must be understood.

Social Justice

When analyzing oppression as a **social justice** issue, it is important to consider the social structures, processes, and practices associated with class, gender, sexuality, race, age, ability, and other patterns of inequality that contribute to oppression in the first place. Obviously, when someone needs immediate help to alleviate suffering and severe deprivation, it is important that there are programs and services that can provide for basic needs such as money, food, clothing, medical care, or shelter; this is a priority for any group or program seeking social justice (Mullaly, 2010). For example, a mother comes to a service agency to get food to feed her children. As part of its social justice mandate, the agency provides groceries or meals to help meet the basic food needs of this family. Unfortunately, many conventional approaches end here. But when your goal is social justice, you try to also change the structural policies or practices that created the inequities in the first place (Marullo & Edwards, 2000). This means asking why poverty exists, and what structural issues and policies within society are making it necessary for agencies to meet food security needs through food banks or soup kitchens.

Distributive and redistributive social justice is concerned with the fair allocation of resources among diverse groups in a community (Maiese, 2013). This concept of social justice is one that is commonly used today and fits well with a legal rights approach. Distributive and redistributive social justice is concerned with one group having more material resources than another (like wealth and income), or having more non-material social goods, like rights and opportunities, than another group (Mullaly, 2010). But Iris Marion Young (1990) points to two major limitations with the distributive/redistributive notions of social justice. First, this approach ignores the social structures, processes, and practices that allow resources to go to one group in the first place. Second, Young (1990) argues that rights and opportunities cannot be distributed and redistributed in the same manner as material resources. For example, when dominant groups extend rights to non-dominant groups, there is no guarantee that they can exercise these rights. A person living in poverty may have a right to legal representation but be financially unable to hire proper legal counsel (Mullaly, 2010).

The approach used in this text is one that broadly defines social justice as "the elimination of institutionalized domination and oppression" (Young, 1990, p. 15). It is an approach to social justice that considers the following to be key definitional elements: oppression as a central concern; consideration of the social structures, processes and practices that allocate resources to one group over another; full and active participation of all members, which includes the voices and perspectives of persons historically **underserved** and marginalized. But as long as groups with power are reaching out to marginalized communities to get affected members to embrace their agendas, then oppression can continue. As a leading figure in the racial justice movement, Rinku Sen uses a critique of **tokenism** that gets "white organizations reaching out to racialized communities to get communities of colour to embrace white agenda and leadership" (2010). She notes that the problem is about shared power and equity. What is required is for marginalized communities to share both the power and the leadership to shape the agenda that moves forward a strategy that has results for the affected community (Sen, 2010). Justice and equity can only be achieved when marginalized communities are fully participatory in creating change.

Equity versus Equality

The conservative approach to anti-oppression is fairness and justice through equality, meaning that everyone is treated the same way. The liberal approach to anti-oppression sees diversity as difference, meaning that everyone is included despite differences in social characteristics. Neither of these approaches is about equity (Sen, 2010). So, can we use the concept of diversity as more than an approach that recognizes differences—instead, as a framework for achieving social equity? To accomplish this requires an understanding of what social equity means and how it is differentiated from the concept of equality. Both equality and equity try to achieve a fair outcome for all people. **Equality** implies that everyone gets the same thing; but because everyone is different, treating everyone the same in all situations is not necessarily fair and can lead to unequal results. Social **equity** is trying to achieve equitable outcomes for all people by ensuring everyone gets what is *right for them*. To provide an applied definition as an example, equity in health can be defined as "the absence of systematic disparities in health (or in the major social determinants of health) between social groups who have different levels of underlying social advantage/disadvantage" (Braveman & Gruskin, 2003). You can see from this definition that equity is a "more flexible measure allowing for equivalency" that accounts for the effects of social disadvantage and does not use the socially advantaged position as the measure of sameness (University of Melbourne, 2013). The process for determining and achieving this equivalency requires self-definition and the ability for marginalized communities to share power and leadership in creating change.

> **Distributive and redistributive justice:** A social justice model that is concerned with the fair distribution or redistribution of material and non-material resources between different groups within a society.
>
> **Underserved:** Disadvantaged because of structural barriers and disparities.
>
> **Tokenism:** The practice of including one or a small number of members of a minority group to create the appearance of representation, inclusion, and non-discrimination, without ever giving these members access to power.
>
> **Equality:** Fairness and justice achieved through same treatment.
>
> **Equity:** Principle based on fairness, justice, access, opportunity, and advancement for everyone, while recognizing historically underserved and unrepresented populations, identifying conditions needed to provide effective opportunities for all groups, and eliminating barriers to their full participation.

Interaction Institute for Social Change | Artist: Angus Maguire

EQUALITY

EQUITY

Using the image above, describe how equality differs from equity.

Walk a Mile in My Shoes:
One Size Never Fits All

A good illustration of the difference between the concepts of equality and equity came from a professor at the University of Waterloo. In class one day, he invited everyone to sit in a circle. He then asked each student to remove his or her right shoe and throw it into a pile in the centre of the circle. The professor then randomly redistributed the shoes, one shoe to each student, and asked the students to put on their new shoes. Of course, everyone was somewhat perplexed, and very few students were actually happy with the shoe they had been given. A student then asked the professor to explain the point of this exercise.

"I want to treat you all equally and I have, because everyone has one shoe for their right foot and one

shoe for their left foot," the professor said. "What's the problem with this? Seems fair."

The student responded, "I think the point you are making is that the solution is never one size fits all. We need the shoes that each fit us best individually."

The inference was not lost on this student, who had summed up the difference between equality and equity. *Equality* means that in order to be fair and just, you give all people the same thing or treat them in the same way, regardless of their individual needs. This approach presumes that everyone is on a level playing field. In the classroom exercise, everyone ended up with the same thing (two shoes), regardless of their individual needs (like shoe size, width, comfort, etc.), so they were treated equally. *Equity* also aims to promote fairness and justice, but it does so by trying to understand and give people what they need individually. In order for the classroom exercise to demonstrate equity, it is no longer about everyone getting "a" shoe; it is about getting the shoe that fits the individual student.

At the end of this classroom experiment, the professor asked students to return the right shoes into the centre of the circle. "Now I want *you* to choose the shoe that fits you best individually," he said.

Chaos erupted as students ran to grab their own shoes. In the end, everyone was happy to have chosen his or her own shoe. Then something quite extraordinary happened: Two students with the same size shoes began to barter with one another, and a pair of stilettos was traded for a pair of runners. Seeing that both parties were happy, a few more trades ensued. Everyone left the class happy with the shoes they were wearing. Not only were students' choices in this exercise self-determined, but they also had the power to collaborate on change if they wished. Ultimately, the second exercise was fair because students all got the shoes that fit their individual needs and because they were given the power to define what that was. While the process in the second part of the exercise was a more chaotic one, everyone was a lot happier with the result.

Starting the Journey: The Power of Narratives

One of the bases for establishing just and equitable relationships is the understanding that there is not just one *who* nor just one *how to*. Who you are, how you interpret the world around you, how you live your life, how you know what you know, and how you've come to value and believe in certain things over others are all relative to particular forms of social power and privilege.

Personal narratives (stories) are a powerful way of learning and knowing. As Feridun Hamdullahpur, President of the University of Waterloo, says, the power of human stories are at the heart of our experiences:

> We live in a quantitative age. So much is measurable, from our economies to our technologies. Big data, in fact, holds so much promise for everything from direct marketing to public policy-making that it seems life is just a series of algorithms and correlations. Except it isn't ... At the heart of every technical, scientific and entrepreneurial venture are human experiences fuelled by the intangibles in life: hope, ambition, concern, competition. We find meaning and energy through these journeys. It's important to remember: we're not instruments

of science or technology or economy. We have stories. We have narratives and futures and aspirations. (Hamdullahpur, 2016)

Placing the narratives of non-dominant group members at the centre of our analysis offers fresh insight and new ways of knowing about anti-oppression that allows non-dominant groups to define their own realities (Collins, 1993). Reading this textbook, you will find in each chapter the personal stories that come from other Canadian post-secondary students who ask you to walk a mile *in their shoes*. Engaging in the political act of storytelling, these students shared their narratives in the hopes that a dialogue will begin about the validity of lived experience and the importance of empathy. Their narratives are a poignant reminder that the enterprise of diversity is to construct a society where *all* people can experience the world as just: where *all* people have equivalent access to opportunities; where *all* people, including those historically underserved and under-represented, feel valued, respected, and able to live lives free of oppression; and where *all* people can be fully participatory in the social institutions that affect their lives.

KNOWING YOUR OWN STORY

When beginning the intellectual journey of analyzing concepts like oppression, power, privilege, justice, and equity, a good place to start is by examining your own **identity**. This includes knowing your own story and where it fits in your personal map of the world.

The term **personal identity** (or individual identity) is used to refer to a person's self-concept based on personal attributes (Hogg & Tindale, 2005). For example, if a person says, "I am well-educated," they are referring to personal identity. The term **social identity** (sometimes referred to as group identity or community identity) is a person's self-concept based on the attributes of a group he or she aligns with (Hogg & Tindale, 2005). Saying "we are a happy family" refers to social identity.

Personal and social identities are complex, multidimensional, fluid, and intersectional; there is no one single social category that can define us. The lived experience of the affluent, agnostic

Identity: Social construction of a person's sense of self that is based upon social categories that influence self-perception and the perception of others.

Personal identity: The part of a person's identity determined by their individual attributes and characteristics.

Social identity: The part of a person's identity that is determined by attributes and characteristics of groups the person aligns themselves with.

lesbian living in rural Manitoba will undoubtedly be different from a middle-class, Christian, heterosexual woman living in Newfoundland. "Reducing people to a single dimension of who they are separates and excludes them, marks them as "other," as different from "normal" people … and therefore as inferior" (Johnson, 2006). Identity is also not a static concept; it is fluid. It can change depending on the social context you are in and who you are with. Each time our social context changes, the negotiation of new identities begins again. This change also occurs over time, creating moments when certain aspects of one's personal or social identity may be more **salient** than others.

It is important to recognize that all social forces and social positions are relational, cross-cutting one another (Collins, 2000). This concept is referred to as *intersectionality* (Collins, 2000). Our identities are multi-faceted and multidimensional, arising out of our lived experience in different social contexts, which are defined by different forms of power. Therefore, there is no homogeneous identity that can define a collective standpoint (Collins, 2000). Someone can't say that because she is a woman, she speaks on behalf of all women everywhere. The concept of a universal sisterhood that constructs gender oppression as sole determinant is deficient. Our social identities are too complex for this kind of analysis.

It is one dimensional to suggest some groups are more oppressed than others. We have seen this argument emerge in the arena of competing human rights. The Supreme Court of Canada has been clear that there is no hierarchy of Charter rights, stating that no right is inherently superior to another right (*Dagenais v. Canadian Broadcasting Corp.*, 1994). It is similarly one dimensional to suggest that a white woman living in poverty is a privileged person. Since each person is not just one thing, our identities interact and impact each other. Some aspects of our identity can also be invisible, so we may enjoy social power, provided that we do not disclose these parts of our identity in social contexts that might discriminate if they were visible.

While certain aspects of our identity may be more salient than others, there is never one single identifier that can define who we are. We have multiple social identities that change over time and at different stages of our life. As you walk a mile in the shoes of the student who shares his story in the In Their Shoes feature, consider just how **fluid** our identities can be and how quickly they can change. While he faces his own mortality, he remains driven by a call to something larger than himself, to realize a dream of being a healer for others. His struggle to survive is interwoven with a struggle to live life defined by more than a master status ascribed to him by his brain tumour. He wants to be recognized as more than "you know, the guy with the tumour." Not only does he face changes in his social context, but he also faces the possibility of personality changes from treatment, all of which leaves him wondering who he will be when it is all over. Whoever that is, he will fight to have others recognize that we are always more than our physical and cognitive abilities.

What is common to all forms of identity is that they are socially constructed. Society constructs the social meaning that is given to the characteristics and attributes that are used to define us. Even aspects of our identity that we believe we are born with, like race, are in fact social constructs. The concept of race is a biological myth, but real in terms of its social consequences. Certain aspects of our identity may afford us the privilege of belonging to the dominant group within a society. Other aspects of our identity may have the effect of "othering us." The social construction of difference as "other" has the effect of marginalizing individuals and groups of people within a society.

ENDING THOUGHTS

We need to move forward in Canada with an anti-oppression framework that can promote justice and equity for members of diverse populations and that can also challenge patterns of oppression and discrimination. Justice and equity are not achieved through the celebration of diversity nor the appreciation of differences. Our identities are complex, and the ways in which aspects of our identity intersect can influence the ways we experience oppression. Some aspects of our identity can result in us being oppressed at the same time as other aspects can result in us experiencing privilege. Oppression is a complex and multidimensional phenomenon that occurs on a personal, cultural, and institutional level within society—all levels being potential sites of resistance and liberation. And if we hope to create a society where the dignity and intrinsic worth of every human being is respected and valued, we will need to be prepared to actively challenge oppression in its many forms.

Salient: Characteristic of identity that describes an aspect that is most noticeable or most important.

Fluid: Characteristic of identity that describes it as something that can change and be shaped.

IN THEIR SHOES

If a picture can say a thousand words, imagine the stories your shoes could tell! Try this student story on for size – have you walked in this student's shoes?

I invite you to walk a mile in my shoes, or in my head, as my story shows how your personal and social identity can change within a day. On February 17, 2013, I am a 21-year-old post-secondary student studying kinesiology and dreaming of practising rehabilitative medicine one day. My friends and family nicknamed me "Mr. Healthy" because of my lifestyle. Some people call me "driven," but I live my life according to the performance goals I have set for myself. A year and a half ago, I began experiencing headaches and was referred to a neurologist who misdiagnosed me with migraines. On February 17, 2013, I experience what I think is a migraine, and then I lose all feeling on my left side of my body. Suspecting something neurological, the doctor at Urgent Care orders a CT scan for the morning—in retrospect, this is part of what saved my life. I spend Family Day with my mom sitting in the emergency room of my local hospital waiting for the CT scan. Once it is finished, I wait for the doctor who comes in and tells me very directly that I have a mass. I say, "I have what?" He repeats that I have a "mass." Translation: I have a brain tumour. He tells me it is being analyzed by a neurosurgeon as we speak, and that I will have to go to Hamilton General Hospital tomorrow for a consultation with a neurosurgeon. In two minutes, my life changed. Was this real or a dream?

I meet with a great neurosurgeon the next day at Hamilton General. His words are, "We don't have a choice. We have to remove this tumour. If left to continue … bottom line is you won't survive." Everything else seemed irrelevant but the words "you have a tumour and you won't survive this *unless*." You face your own mortality like never before and everything changes. My family and friends are very emotional—not so much for me—I think I was pretty clear on what had to be done and there was no choice but to face it head on. It sounds cliché, but it is all the support from my wonderful family and great friends that sees me through.

The following day is game time. As I am rolled into the OR, my MRI results are up on the screen and the tumour is massive. The neurosurgeon is explaining to his team the strategy—people are moving quickly. I do remember the OR prep, but after that there is nothing. No transition to sleep or death. That portion of memory does not exist. Wake up in recovery and everything is a haze. Wave at my mom and my uncle. Wake up again later in the step-down unit. Surgeon says he is happy with the outcome and I feel more at ease.

Death is not something I was afraid of. I knew it was a possibility but it wasn't something I could control. If I survive, great. If I die, so be it. Things will play out as they will. You just find peace. It makes you strong because the anxiety of life and death does not wear on you. People associate death with this negative aura and it does not feel like that to me. The recovery from neurosurgery is pretty intense. My concentration was limited; my short-term memory had declined. In order to reduce brain edema, I have to take a corticosteroid and I get almost every side effect possible from this drug. Most discouraging is the muscle atrophy—years spent in the gym feel like they are gone in a matter of weeks. I was known in the gym as the power and strength guy—my max bench press was 385 lbs, leg press 1300 lbs, deadlift 500 lbs, squat 500. Now I can only dream of getting under a load like that. I was also a fairly gifted sprinter. Running is joy for me and I conditioned for 200 and 400 sprints. Now I am winded just walking up the stairs.

People look at me differently and they are often uncomfortable around me. I feel like this tumour is starting to define me. It's almost dehumanizing in a sense. People don't look at me like me anymore—I'm the tumour boy. Yes, I know my disability is very visible at the moment, with this incision on the side of my head. I've literally watched people walk into doors as they stare at me.

I feel like this journey has dramatically changed who I am and the context by which others now define me. But I am still me—just because I have this tumour doesn't mean I am any different than who I was. As soon as you tell someone though, or they see your scar, it's like you are no longer normal. On a physiological level, this is partly true, but I am still me. It's like accidental discrimination.

I've come to learn that brain tumours don't discriminate—they can affect anyone, at any age. It has become hard to plan for a future that you may not have. What I do know is that I have more neurosurgery in a few weeks, followed by six weeks of radiation therapy. I will lose my short-term memory. There is a good chance my personality will change. Brain seizures mean I can't drive right now, so I have lost some independence. But I am alive and determined. I am inspired by my neurosurgeon, Dr. T. Gunnarsson, and know that one day I will find a way to use my knowledge and skills to help heal others. I know it is what I was destined to do in life. I have learned that "who we are" can change in a matter of seconds. The "ability" to learn at whatever college or university you are studying at is truly a privilege. I dream of the day when I can join you there again.

EXCERPTED FROM *BUT YOU ARE DIFFERENT: IN CONVERSATION WITH A FRIEND*

By Sabra Desai

Sabra: So, you think I'm not like the rest of them.

Alex: Yes, you are different. Well, you know what I mean.

Sabra: No, I don't. Tell me exactly what you do mean.

Alex: Well, when I see you, I don't see your colour. I don't see you as a South Asian. You're not like the rest of them. I'd like to think that I judge you through my own personal experiences with you. I don't judge you on the basis of your culture, colour, or class for that matter. I refuse to see you as being different. You're just another human being.

Sabra: First you tell me that I'm different, and then you say that you refused to see me as being different. So, which is it? Let's try to unravel this.

Alex: Well, I meant that you're more like me, you know, like one of us.

Sabra: Oh, so, I'm more like you and less like, should I say it, "a real South Asian." You see, although you're not saying it, your statement reveals that you have some preconceived ideas of South Asians, the people that I'm supposed to be so unlike. This means that whatever your preconceived ideas are of South Asians, they make South Asians less acceptable, less attractive, and less appealing to you than I. Well, this is not just stereotyping, this is racist stereotyping.

Alex: Well, just wait a minute, Sabra, you're accusing me of being a racist. You're taking my statement entirely out of context, and you know that isn't what I meant. I really don't appreciate the implication that I am being racist.

Sabra: Before we get to unravelling the implications of the statement "but you are different," let me ask: are you upset that I'm challenging your thinking? Would you feel better if I were grateful for being "accepted" by you and less analytical or critical of your reasons for doing so? I think that part of what might be upsetting you is that for all your willingness to "accept" me, you're not willing to accept me as a South Asian.

Alex: By saying that you are different, I assumed that the senseless stereotypes of South Asians do not apply to you. I thought you'd appreciate my comment, but you surprised me. I thought that coming from a society where you were very much defined by race, you'd want people to ignore differences and treat you the same as everyone else.

Sabra: You obviously don't recall asking me this very question once before. You told me once that you thought that having come from a country as racist and segregated as South Africa, I would endorse the concept of "melting pot." To that I said: in spite of the fact that I left such a racist society where race was all that mattered, in my mind to say that one's ethnicity or race does not matter is still racist.

So goes one of the many conversations I have had with Alex and a number of other friends. One tires of these exchanges after a while, for it seems that I can never be accepted as the person that I am, but only as what my friends have made me out to be. I at least expect my "differentness" to be acknowledged, although I really want it to be appreciated or, if they genuinely care, to be explored. I wish that Alex and company would recognize that I am a member of a marginalized minority group and that as "mainstreamers," they have a tendency to negate, romanticize, or stereotype our experiences and differentness. I am a South Asian human being, but there appears to be some difficulty or reluctance on their part to accept me as such, to try to understand what being South Asian is to me and then to try and come to grips with the reality rather than attempting to make themselves comfortable with some sanitized image.

I think that one may be "different" by degree within the context of one's ethnocultural group, but it is not possible to be different from something that, by definition, one is a part of. So, perhaps because I do not fit Alex's preconceived notion of what a South Asian should be, she doesn't think of me as one; she sees me as the preferred anomaly. My ethnicity, culture, and colour do not matter to her. She says that she sees me only as a human being. For me, the implication of this is that being a South Asian somehow devalues me and lessens my humanness. So, in order to relate to me, she remakes me into her image; she whitens me so that I can be like her. In other words, she is suggesting that if I change the colour of my skin, I could be just like her, that I grew up in South Africa, where every aspect of my life was mediated by the colour of my skin and my ethnicity. As a person classified Indian, and of course non-white, whether it was the privileges that were denied or the ones granted me by both statutory and informal laws, they were all arbitrarily applied to me because of my colour and ethnicity; nothing else mattered. For example, where I could go to school, where I could live, where I could receive medical services, or for that matter be hospitalized, and what jobs I could aspire to were all dictated by the laws of the country. Where and with whom I could play were all largely predetermined for me because of my skin. What my white counterparts—yes, I dare say counterparts—could take for granted, I could not. Where I could shop, whether I could try on the garment before buying it, and what restaurants I could eat at, were all considerations to be taken seriously. In apartheid, pre-Mandela South Africa, one small infraction could mean that I was breaking the law.

All of this is part of me. It is central to who I am, even though I have now spent more than half my life here in Canada. I am a product of that social context; it is part of my cultural identity. Saying that I am different does not change the way in which my ethnicity and race are used to set me apart from whites. I say that because, in my experience, the labels of ethnicity are reserved to describe the "others," most often "minorities" who remain across the divide—the ones referred to as "them," "they," "those people." We are different when some people want to exclude us on the grounds that we do not fit, we do not know the rules, we do not speak the language, or we are not desired as neighbours. Yet, paradoxically, some people perceive us to be the "same," particularly when we are to receive certain long-overdue rights and privileges such as equal access to jobs, housing, culturally sensitive social services, education, and recognition of our heritage languages.

Recognizing my differences is in a way acknowledging the irrefutable and undeniable power of racial construction and culture on all of our lives. Therefore, words like "Black" and "white," referring to the race of a people, cannot be simply dismissed to suggest that one is colour-blind. The very act of not referring to my race or ethnicity, when it is central to the discussion or context, renders me invisible. The omission delivers a powerful message concerning what must be ignored—that is, colour, culture, and ethnicity.

In saying this, I can hear you ask the question, as several have so often asked: "But don't you think that this constant reference to skin colour or one's racial identity by minorities reinforces the racism in our society? It might, in fact, reinforce the existing stereotypes." To this I say, not recognizing that things such as gender, skin colour, race, and ethnicity mediate one's life chances in a society largely differentiated on the basis of these variables is like burying one's head in the sand. Any denial of the significance of race or skin colour implies that skin colour poses an insurmountable obstacle for white people when they interact or deal with people of colour.

Wishing that we live in a colour-blind society is not necessarily a virtue. Colour-blindness, as I come across it, means denial of the differentness in culture, identity, and experiences of those who are less valued in society. This denial reflects the tendency of the oppressor to minimize and deny the impact of the oppression. The oppressors can afford the enormous luxury of ignoring the formative and fundamental influences of social context and history on "other" people's lives while remaining keenly aware and protective of the benefits and privileges of these same influences on their lives and the lives of their descendants. An African-Canadian friend of mine once said, "When you see me and do not see my colour, it is just as problematic as when you see me and all you see is my colour." Colour-blindness is an illusion in the minds of those Canadians who like to think as a nation we are not racially hierarchical. Until very recently, despite the hundreds of years of domination, exploitation, and oppression of Aboriginal peoples, the prevailing idea concerning race and ethnic relations was that racism did not exist in Canada. It has been documented and ought to be obvious at this stage that social divisions based on colour and ethnicity are a fundamental characteristic of the political and socioeconomic configuration of our country.

Source: Sabra Desai, "But You Are Different: In Conversation with A Friend," from *Talking About Identity*, Carl E. James and Adrienne Shadd, eds. (Toronto: Between the Lines, 2001), 241-249. Reprinted by permission of Between the Lines.

DISCUSSION QUESTIONS

1. How does Alex's privilege affect her conversation with Sabra?
2. How would you respond to the question, "But don't you think that this constant reference to skin colour or one's racial identity by minorities reinforces the racism in our society?"
3. How does Sabra interpret the concept of differentness from her conversation with Alex?

KWIP

KNOW IT AND OWN IT: WHAT DO I BRING TO THIS?

The "K" in the KWIP process is about knowing and owning who you are as a person as the first step in understanding and respecting others and where they are coming from. It involves as much learning about oneself as it does learning about others. As part of this process, you are asked to examine and reflect upon aspects of your identity and social location, and accept yourself for who you are and what you can bring to this discussion.

Choose *one* of the following journal activities based upon your personal experiences with verbal, social, physical, or cyberbullying:

1. **Letter to My Bully** – Write a letter to a person or group who bullied you at some point during your life. Tell them how they made you feel and how their bullying has affected you.

2. **Letter to the Person I Bullied** – Write a letter to a person or group that you bullied at some point during your life. Tell them why you bullied them and how this behaviour affected your life.

3. **Letter to the Person I Saw Bullied** – Write a letter to a person or group that you witnessed being bullied. Tell them why you did not intervene, or tell them why you did. How did this make you feel?

   Note: This journal activity is intended to help you to reflect upon times in your life when you have experienced oppression or when you have been the oppressor. *This activity does not require that you actually send the letter.*

WALKING THE TALK: HOW CAN I LEARN FROM THIS?

The "W" in the KWIP process establishes the intention to learn about the complexities of oppression. As you set a purposeful and curious intention to learn about oppression as a variable and multidimensional concept, you are challenged to "walk the talk" through problem-based learning.

Allies or Enemies

Carolyn and Heather, although from different geographical locations, have similar backgrounds. Both were raised in poverty and were exposed to abuse and alcoholism in their families. They and their families received constant visits from child and family services workers, police, and representatives of other regulatory agencies. They were victims of harsh and discriminatory treatment at their respective schools. Both also worked hard, struggled to get an education, received a few breaks, and eventually graduated from social work programs. Here the similarities end. Heather would bend over backward for the service users with whom she worked, especially if they came from conditions of poverty. She had tremendous empathy for them and a keen understanding of their situations. Carolyn, on the other hand, became one of the most punitive and moralistic social workers in the agency, especially toward those who were poor. She also treated the clerical staff and others in subservient positions in the agency in an overbearing and heavy-handed manner.

1. What would explain the difference in approach used by Heather and Carolyn?

2. Using Carolyn as an exemplar, in what ways can professionalism be oppressive?

3. How might an understanding of poverty as a form of oppression at a cultural and structural level in society help to change Carolyn's behaviour?

Mullaly, B. (2010). *Challenging Oppression and Confronting Privilege* (2 ed.). Don Mills: Oxford University Press, p. 58.

IT IS WHAT IT IS: IS THIS INSIDE OR OUTSIDE MY COMFORT ZONE?

The "I" in the KWIP process requires you to honestly confront the fact that we get uncomfortable when we encounter perspectives that are different from our own. More than that, our discomfort is often rooted in the various privileges each of us hold. Anti-oppressive practice requires a consciousness of these various privileges and the ability to see ourselves systemically.

This activity invites you to practise critical self-reflection and institutional reflection as you explore the complexities of power and privilege. Increasing awareness of both our own oppression(s) and our roles as oppressors of others opens the possibility for changing oppressive practices in our personal and professional lives.

ASKING: Do I have privilege?

1. I can move about at school or work without fear of being bullied because of some aspect of my identity.

2. I can read a newspaper and understand its contents.

3. I can be accepted into any postsecondary institution of my choosing and be able to afford the tuition.

PUT IT IN PLAY: HOW CAN I USE THIS?

The "P" in the KWIP process involves examining how others are practising diversity and how you might use this. It is a call to move from social analysis to social action—by knowing

4. I can be sure that people are not going to make negative comments about my appearance, weight, or size (body-shaming).

REFLECTING: Honouring Our Privilege

Describe two traits that you possess that are disempowering to you or that disadvantage you. Then describe two traits that privilege you over someone else. How might you use this social privilege to address the oppression of others?

and owning who you are, by establishing the intention to learn about new things, and by choosing to become involved in anti-oppressive practice. How can this broader perspective enrich our lives?

ACTIVITY: CALL TO ACTION

REUTERS/Chris Wattie

Gord Downie, lead singer of the Tragically Hip, was honoured at the Assembly of First Nations for his work highlighting the impact of residential schools.

Lead singer of the Tragically Hip, Gord Downie, is being honoured in this picture by the Assembly of First Nations for his project, *Secret Path* (Tasker, 2016). It tells the story of 12-year-old Chanie Wenjack, who died in 1966 in flight from the Cecilia Jeffrey Indian Residential School near Kenora, Ontario, walking home to the family he was taken from over 400 miles away (Tasker, 2016). As he stood in front of 600 Indigenous Chiefs at a special assembly in Quebec, he was given an eagle feather as a gift from the Creator, wrapped in a star blanket, and given a Lakota spirit name meaning "man who walks among the stars" (Tasker, 2016). Downie is leading the way down a path of reconciliation with Indigenous people.

Located at www.nelson.com/student

Study Tools
CHAPTER 1

- Review Key terms with interactive **flash cards**
- Check your Comprehension by completing **chapter review quizzes**
- Gauge your understanding with *Picture This* and accompanying short answer questions
- Develop your critical thinking/reading skills through compelling **Readings** and accompanying short answer questions
- Apply your understanding to your own experience with **Connect A Concept** activities
- Evaluate Diversity in the Media with engaging *Video Activities*
- Reflect on your Understanding with *KWIP* activities

REFERENCES

Agger, B. (1989). *Socio(ontology): A disciplinary reading.* Urbana, IL: University of Illinois Press.

Agger, B. (2006). *Critical social theories: An introduction* (2nd ed.). Boulder, CO: Westview Press.

Alberta Health Services: Healthy diverse populations. (2008). *Best practices in diversity competency.* Alberta, Canada. Retrieved from http://www.calgaryhealthregion.ca/programs/diversity/diversity_resources/research_publications/diversity_comp_rprt_2009.pdf

Allen, T. Audio Slideshow: Human Planet. *BBC Human Planet.* Retrieved from BBC News World.

Amanda Todd Legacy Society. (2016). *Home.* Retrieved from Amanda Todd Legacy Society: http://www .amandatoddlegacy.org/

Anderson, S., & Middleton, V. (2005). *Explorations in privilege, oppression and diversity.* Belmont, CA: Thomson, Brooks/Cole.

Bishop, A. (2002). Step 5: Becoming an ally. In *Becoming an ally: Breaking the cycle of oppression in people* (2nd ed.). Halifax: Fernwood Publishing, 114–120.

Bishop, A. (2015). *Becoming an ally: Breaking the cycle of oppression in people* (3rd ed.). Halifax: Fernwood Publishing.

Bishop, A. (n.d.). *Tools for achieving equity in people and institutions.* Retrieved 2017, from Becoming An Ally: http:// www.becominganally.ca/Becoming_an_Ally/Home.html

Braveman, P., & Gruskin, S. (2003). Defining equity in health. *Journal of Epidemiology and Community Health,* 57, 254–258.

Bregman, P. (2012). Diversity training doesn't work. *Harvard Business Review.* Retrieved from http://blogs.hbr.org/ bregman/2012/03/diversity-training-doesnt-work.html

Canadian Human Rights Commission. (2017). *Left out: Challenges faced by persons with disabilities in Canada's schools.* Ottawa, ON: Canadian Human Rights Commission.

Canadian Race Relations Foundation. (2017). *CRRF glossary of terms.* Retrieved from Canadian Race Relations Foundation: http://www.crrf-fcrr.ca/en/resources/ glossary-a-terms-en-gb-1

Collins, P. H. (1990). *Black feminist thought: Knowledge, consciousness, and the politics of empowerment.* London: HarperCollins.

Collins, P. H. (1993). Black feminist thought in the matrix of domination. In *Social theory: The multicultural and classic readings,* Westview Press, 615–625.

Collins, P. H. (2000). *Black feminist thought: Knowledge, consciousness, and the politics of empowerment* (2nd ed.). New York: Routledge.

Cotler, I. (2015, July 2). How Canada led the way on same-sex marriage. Retrieved from thestar.com: https:// www.thestar.com/opinion/commentary/2015/07/02/ how-canada-led-the-way-on-same-sex-marriage.html

Crenshaw, K. (1991, July). Mapping the margins: Intersectionality, identity politics, and violence against women of color. *Stanford Law Review, 43*(6), 1241–1299.

Cross T., Bazron, B., Dennis, K., & Isaacs, M. (1989). *Towards a culturally competent system of care.* Volume I. Washington, D.C.: Georgetown University Child Development Center, CASSP Technical Assistance Center.

Dagenais v. Canadian Broadcasting Corp. [1994]. 3 S.C.R. 835.

Desai, S. (2001). But you are different: In conversation with a friend. In C. James & A. Shadd (Eds.), *Talking about identity: Encounters in race, ethnicity and language.* Toronto: Between the Lines, 241–249.

Diller, J. V. (2007). *Cultural diversity: A primer for human services* (3rd ed.). Belmont, CA: Thomson Brooks/Cole.

Fearon, J. D. (1999). *What is identity (as we now use the word)?* Stanford, CA: Stanford University.

Government of Canada. (2017, March 15). *What is the Canadian Charter of Rights and Freedoms?* Retrieved from Your Guide to the Canadian Charter of Rights and Freedoms: http://canada.pch.gc.ca/eng/1468851006026#a1

Hamdullahpur, F. (2016, Fall). Human stories are the heart of Waterloo: Nourishing the individual spirit is integral to the university experience. Retrieved from *University of Waterloo Magazine:* https:// uwaterloo.ca/magazine/fall-2016/presidents-message/ human-stories-are-heart-waterloo

Heldke, L., & O'Connor, P. (2004). *Oppression, privilege and resistance: Theoretical perspectives on racism, sexism and heterosexism.* New York: McGraw-Hill.

Hogg, M., & Tindale, R. (2005). Social identity, influence and communication in small groups. In J. Hartwood & H. Giles, *Intergroup communication: Multiple perspectives.* New York: Peter Lang, 141–164.

Home, A. (2002). Challenging hidden oppression: Mothers caring for children with disabilities. *Critical Social Work: An Interdisciplinary Journal Dedicated to Social Justice,* 3(1).

Johnson, A. G. (2005, February). *Who me?* Retrieved from http://www.agjohnson.us/essays/whome/

Johnson, A. G. (2006). *Privilege, power, and difference* (2nd ed.). New York: McGraw-Hill.

Kellner, D. (1989). *Critical theory, Marxism, and modernity.* Baltimore, MD: John Hopkins University Press.

Kimmel, M., & Ferber, A. (2003). *Privilege: A reader.* Boulder, CO: Westview Press.

Leonard, S. T. (1990). *Critical theory in political practice.* Princeton, NJ: Princeton University Press.

Loden, M. (1995). *Implementing diversity: Best practices for making diversity work in your organization.* McGraw-Hill Education.

MacKinnon, C. (1989). *Equality rights: An overview of equality theories.* Ottawa, ON: National Meeting of Equality Seeking Groups.

Maiese, M. (2013, June). *Distributive justice.* Retrieved from Beyond Intractability: http://www.beyondintractability .org/essay/distributive-justice

Marullo, S., & Edwards, B. (2000). From charity to justice: The potential of university-community collaboration for social change. *American Behavioral Scientist,* 895–912.

McIntosh, P. (2000). White privilege and male privilege: A personal account of coming to see correspondences through work in women's studies. In A. Minsa, *Gender basics: Feminist perspectives on women and men* (2nd ed.). Belmont, CA: Wadsworth.

McQuigge, M. (2013, December 5). Mandela said he found inspiration in Canadian respect for diversity. Retrieved from *The Gazette*: http://www.montrealgazette.com/news/Mandela+said+found+inspiration+Canadian+respect/9252452/story.html

Mishna, F., & Van Wert, M. (2015). *Bullying in Canada*. Don Mills, ON: Oxford University Press.

Morgan, K. (1996). Describing the emperor's new clothes: Three myths of educational (in-)equity. In A. Diller, B. Houston, M. Ayim, & K. Morgan (Eds.), *The gender question in education: Theory, pedagogy, and politics*. Boulder, CO: Westview, 105–122.

Morris, T. (2015, May 22). On a plate: A short story about privilege. *The Pencilsword #10*. The Wireless. Retrieved from http://thewireless.co.nz/articles/the-pencilsword-on-a-plate

Mullaly, B. (2010). *Challenging oppression and confronting privilege* (2nd ed.). Don Mills, ON: Oxford University Press.

National Vet E-Learning. (2013, May 22). Barriers to effective communication: Stereotypes. Retrieved from *Flexible Learning Toolboxes*: http://toolboxes.flexiblelearning.net.au/demosites/series9/903/content/resources/03_effective_communication/08_barriers/page_002.htm

Nelson Mandela Centre of Memory. (2013). Life and times of Nelson Mandela. Retrieved from http://www.nelsonmandela.org/content/page/biography

Ontario Council of Agencies Serving Immigrants. (2012). Positive Spaces Initiative's online training course. Retrieved from http://learnatwork.ca

Queen's University. (2017). *Inclusive language guidelines*. Retrieved from Style Guide: http://queensu.ca/styleguide/inclusivelanguage

Royal Canadian Mounted Police. (2016, June 14). *Bullying and cyberbullying*. Retrieved from Royal Canadian Mounted Police Centre for Youth Crime Prevention: http://www.rcmp-grc.gc.ca/cycp-cpcj/bull-inti/index-eng.htm

Sen, R. (2010). Popularizing racial justice: Building clarity, unity and strategy to move us forward [Facing Race 2010's Closing Plenary]. Chicago.

Statistics Canada (2005). Study: Canada's visible minority population in 2017. Retrieved from http://www.statcan.gc.ca/daily-quotidien/050322/dq050322b-eng.htm

Tasker, J. P. (2016, Dec 6). *Man who walks among the stars: AFN honours tearful Gord Downie*. Retrieved from CBC News Politics: http://www.cbc.ca/news/politics/gord-downie-tragically-hip-afn-honoured-1.3883618

Taylor, C., Peter, T., Schachter, K., Paquin, S., Beldom, S., Gross, Z., & McMinn, T. (2009). *Youth speak up about homophobia and transphobia: The first national climate survey on homophobia in Canadian schools. Phase one report*. Toronto, Ontario: Egale Canada Human Rights Trust.

Thompson, N. (1997). *Anti-discriminatory practice* (2nd ed.). London: Macmillan.

Thompson, N. (2002). Developing anti-discriminatory practice. In D. R. Tomlinson & W. Trew (Eds.), *Equalizing opportunities, minimizing oppression: A critical review of anti-discriminatory policies in health and social welfare*. London: Routledge, 41–55.

Todd, A. (2012, September 07). *My story: Struggling, bullying, suicide, self-harm* [Video file]. Retrieved from https://www.youtube.com/watch?v=vOHXGNx-E7E

Todd, C. (2017, January 11). *Support Carol in her journey to the Netherlands - Comments*. Retrieved from You Caring Compassionate Crowdfunding: https://www.youcaring.com/caroltodd-732452

Trudeau, J., The Right Honourable. (2016, January 20). *The Canadian opportunity*. Address presented at World Economic Forum in Switzerland, Davos.

Trudeau, J., The Right Honourable. (2016, January 21). *Global shapers on pluralism*. Address presented at World Economic Forum in Switzerland, Davos.

United Way of the Columbia-Willamette. (2013, July 10). *Measuring up: Assessing ourselves on equity*. Retrieved from United Way of the Columbia-Willamette: http://uwpdx.blogspot.ca/search/label/allhandsraised

University of Melbourne. (2013, April 19). *What is social equity?* (B. McSherry, Editor) Retrieved from Melbourne Social Equity Institute: http://www.socialequity.unimelb.edu.au/what-is-social-equity/

Young, I. (1990). *Justice and the politics of difference*. Princeton, NJ: Princeton University Press.

Forms of Oppression

"I got my army; nothing can harm me...fall back."

ESMA (2014)

LEARNING OUTCOMES

By mastering this unit, students will gain the skills and ability to:

- identify the formal and informal mechanisms within society (both past and present) that cause or reinforce different forms of oppression

- examine the social, psychological, and political effects of stereotypes, prejudice, discrimination, and oppression on diverse groups

- understand the intersectionality of systems of privilege through the "ism" prism, and consider the implications for social change

- develop strategies for the ongoing examination of oppressive assumptions, biases, and prejudice; learning how to change oppressive systems; and creating more socially just and equitable societies

The women's suffrage movement in Canada arose as a response to oppression. The goal of this movement was to address equity, justice, and human rights issues for the purpose of improving lives in Canada. In addition to political justice, including women's right to vote and run for office, this social movement was also concerned about women's access to education, improvements in healthcare, and the end of violence against women and children (Status of Women Canada, 2009).

Even though women nearly equalled men in numbers, they did not have the right to vote in federal elections in Canada until 1918. The right for women to vote in provincial elections spanned many years—with women first winning this right in Manitoba in 1916, and with Quebec being the last province to grant this right to women in 1940 (Status of Women Canada, 2009). Even then, there were questions as to how far women's rights extended: under Canadian law, were they "persons" entitled to sit in the Senate? So, in 1927, a group of women from Alberta known as the "Famous Five"— Nellie McClung, Emily Murphy, Irene Parlby, Louise McKinney, and Henrietta Muir Edwards—asked the Supreme Court of Canada to answer the question, "Does the word 'person' in Section 24 of the B.N.A. [British North America] Act include female persons?" (Status of Women Canada, 2009). Known now in Canadian history as the "Persons Case," they sought to have women legally considered a "person" under the law so that a woman could be eligible for appointment to the Canadian Senate. After weeks of debate, the Supreme Court of Canada decided that for the purpose of eligibility to be appointed to the Senate, the word "person" did *not* include women. Though shocked by the decision of the court, the Famous Five didn't give up their battle. Instead, they took their case to the Privy Council in Britain, at that time the highest court of appeal for Canadians. The Privy Council overturned the Supreme Court decision on October 18, 1929, and in 1930, Cairine Wilson became the first woman appointed to the Senate in Canada (Status of Women Canada, 2009).

Fast forward 87 years to January 27, 2017, as an estimated five million women and men worldwide and over one million in Washington, D.C. joined together with coalitions of marchers across the world in an international movement committed to equality, diversity, inclusion, and the recognition of women's rights as human rights (Women's March Global, 2017). From coast to coast across Canada, approximately 120 000 people participated in over 35 sister marches (Women's March Canada, 2017). So why was everyone marching? The mission of Women's March Global was to bring together persons wanting to defend women's rights in the face of what they describe as "the rising rhetoric of far-right populism around the world" (Women's March Global, 2017). Following the march, the ensuing hatred for feminism and feminists expressed through social media and the use of newspaper slogans such as "She the People" caused many in Canada to pause and wonder how far we had come in terms of ending gender oppression since the Privy Council's decision defined women as "persons" under the law. One of the goals of the women's suffrage movement in Canada was to end violence against women. Yet, some 80 years later, violence against women remains a significant social problem in Canada (see Chapter 4 for further discussion). Cyber violence, which includes online stalking, harassment, and threats, has emerged as a new form of violence against women.

How, then, do we end gender oppression and all forms of oppression in Canadian society? We know that oppression is dynamic and multidimensional; it can vary according to historical and social contexts. As such, we need to be cautious in offering single and universal causal explanations and solutions for oppression, as it is a complex social phenomenon (Mullaly, 2010). The practical and political dimensions of critical social theory can be useful as they advocate a transformation of society for the purpose of liberating those who are oppressed (Leonard, 1990). According to Agger (1989), critical social theory needs to include the following: awareness about current forms of oppression to envision a future emancipated from domination and exploitation; recognition of oppression as a structural phenomenon that affects people's everyday lives and is reproduced through the conscious and unconscious internalization of dominant-subordinate relationships; the power of individual and collective agency to create social change in people's everyday lives, making them responsible for their own liberation; and recognition that liberation is not achieved through the oppression of others, including oppressors (Mullaly, 2010).

FIVE FACES OF OPPRESSION

Iris Young (1990) argued that most forms of oppression today are embedded in the everyday practice of well-intentioned liberal societies; both the oppressed and the oppressors are often unaware of the dynamics of domination. Some tend to think that oppression happens because of the deliberate and malevolent intentions of dominant groups within a society. But oppression can often remain unidentified because "it mostly occurs through the systemic constraints on subordinate groups, which take the form of unquestioned norms, behaviours, and symbols, and in the underlying assumptions of institutional rules" (Mullaly, 2010).

Young (2004) suggests that oppression has five forms, all involving the abuse of power. All oppressed groups will experience at least one or more of these forms of oppression:

1. **Exploitation** involves the unfair use of people's time or labour and failure to compensate them fairly.
2. **Marginalization** refers to an act of exclusion that forces minority groups to the fringes of society.
3. **Powerlessness** occurs when the dominant group leaves the subordinate group with virtually no access to the rights and privileges enjoyed by the powerful. When the subordinate group comes to believe that they deserve this unfair treatment, its members are described as living in a "culture of silence."
4. **Cultural imperialism** describes the condition when the dominant group has made their beliefs and values the norms of a society.
5. **Violence** can involve physical attacks, constant fear that violence will occur, or stigmatization of members of a subordinate group through harassment, ridicule, and intimidation (Young, 2004).

STEREOTYPES, PREJUDICE, AND DISCRIMINATION

Exploitation: The unfair use of people's time or labour without compensating them fairly.

Marginalization: The process of pushing groups or individuals with less social power to the margins of society.

Powerlessness: Occurs when the dominant group has left the subordinate group with virtually no access to the rights and privileges that the dominant group enjoys.

Cultural imperialism: A form of oppression where the dominant group has made their beliefs and values the norms of a society.

Violence: The intentional use of physical force or power to cause injury, harm or death.

Stereotype: A label that may have some basis in fact, but that has been grossly over-generalized and applied to a particular segment of the population or situation.

The consequences of power are manifest in every interaction in society, at all levels. Regardless of whether power is expressed between individuals (a bully abusing a weaker individual), or by a dominant group imposing its beliefs and values on a minority group (society celebrating Christian holidays to the exclusion of other religions), the end result is the same: unfair and in many cases illegal and often inhumane treatment. Stereotypes, prejudice, and discrimination grow from the unequal distribution of power, creating dominant and subordinate groups in society and allowing the subjugation of one in favour of the other. The relationships that exist between groups may be real in and of themselves, but all of the players within the groups create or "socially construct" their own realities of those relationships. Those who hold power in society are able to stereotype and discriminate against weaker individuals or groups, whether that weakness is real or perceived.

Stereotypes

We routinely use labels to categorize people, places, and things. However, **stereotypes** generalize about the behaviours and characteristics associated with members of those categories. We often use stereotypes because doing so is easy: we do not have to think as hard when we use them. For example, if you were asked for words to describe rock concerts, senior citizens, tattoos, and school, you would likely use stereotypes ("rowdy," "frail," "edgy," "cliquey") as part of your descriptions.

So what's wrong with using stereotypes? Well, stereotypes have been described as "'mental cookie cutters'—they force simple patterns on a complex mass and assign a limited number of characteristics to all members of a group" (Nachbar & Lause, 1992). While stereotypes may have some basis in fact, they have been grossly overgeneralized and applied to an entire segment of the population or an entire situation. When we stereotype, we ignore all individual characteristics about the individuals in that group or about the specific situation. For example, if we say that women are bad drivers or sociology lectures are boring, we fail to take into account the number of women who are excellent drivers, and the sociology lectures that are fun and exciting.

People learn about the world through varying forms of the media, such as television, magazines, movies, and the Internet, and these can play a role in creating stereotypes. The collective media shape young minds today through overgeneralized, value-laden images that perpetuate stereotypical myths about different cultural groups (Chung, 2007). For example, did you grow up watching Disney movies? They have been a part of our lives for what seems like forever. How old were you when you began to realize that the characters in those movies had certain predictable qualities? One group of researchers analyzed 26 popular Disney movies and found consistent use of stereotypes in the presentation of male and female characters. Disney's males express their emotions through physical actions; they cannot restrain their sexual responses; they are strong and heroic by nature; they do not perform domestic work; and if overweight, they are slow and unintelligent. In contrast, Disney's females are valued more for their beauty than their intellect; they are helpless and need male protection;

FIGURE 2.1

Stereotypes, prejudice, and discrimination are distinct but interrelated concepts that have different impacts on human thought, feeling, and actions. Try changing the picture in the middle and the corresponding dialogue bubbles.

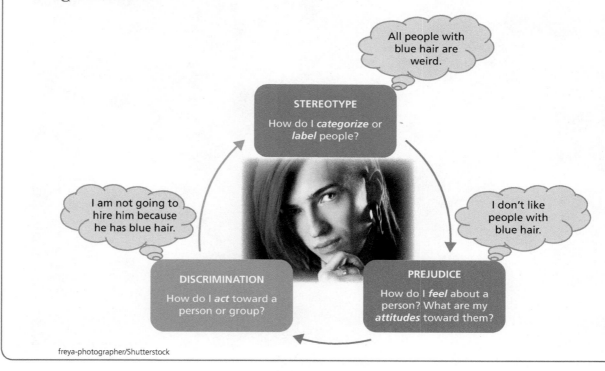

freya-photographer/Shutterstock

their work focuses on household chores, and their goals focus on marriage; and if overweight, they are mean, ugly, and unattached (Towbin, Haddock, Zimmerman, Lund, & Tanner, 2003). Though stereotypes seem harmless, when children grow up watching these images repeatedly, they can begin to believe that every woman should be like Cinderella, Belle, or Snow White, and every man should act like Aladdin or John Smith, because these images help to establish what pass for norms in our society. The mass media, in all their various forms, teach society what the acceptable and normal roles are for both men and women (Hammer, 2009).

Prejudice

In his seminal book *The Nature of Prejudice*, American psychologist Gordon Allport defines prejudice as "an antipathy [negative feeling] based on faulty and inflexible generalization. It may be felt or expressed. It may be directed toward a group or an individual of that group" (Allport, 1954, p. 9). **Prejudice** is very similar to stereotyping but it involves a prejudging component, and prejudice is often a precursor to discriminatory behaviour. When individuals are prejudiced, they have preconceived notions—usually negative—about a group of people, based on their physical, cultural, or social characteristics. So, if you have a problem with your roommate because he or she is messy ("Chris is a slob"), you are likely not guilty of prejudice. However, if you stereotype your roommate because of his or her religion, gender, or ethnicity, that can be a form of prejudice.

Sometimes prejudice can result from ethnocentric attitudes, or **ethnocentrism**. This refers to the practice of assuming that the standards of your

Prejudice: A negative attitude based on learned notions about members of selected groups, based on their physical, social, or cultural characteristics.

Ethnocentrism: Refers to a tendency to regard one's own culture and group as the standard, and thus superior, whereas all other groups are seen as inferior.

Lauren Elisabeth/Shutterstock

Many feel that Disney princess movies perpetuate stereotypical gender roles. Do you agree? Why, or why not? Are the main characters in newer Disney princess movies, such as *The Princess and the Frog, Tangled,* and *Brave,* less stereotypical?

own culture are universally normal and superior to other cultures. Do you know Dr. Seuss's story of the Sneetches? The Sneetches were yellow bird-like creatures who lived on the beach. Some of them had green stars on their bellies and others were plain-bellied. The star-bellied Sneetches thought they were much better than the plain-bellied Sneetches and treated them rather unfairly. They called them names and forbade them from attending their marshmallow roasts. One day Mr. McBean came to the beach with a machine that could put stars on the plain-belled Sneetches—for a fee, of course. The original star-bellied Sneetches then decided it was much better to be plain-bellied, and Mr. McBean obliged in removing their stars—for a larger fee, of course. The foolishness of adding and removing stars went on until the Sneetches had no money left at all. At the end of the story, all of the Sneetches realized the folly of their poor treatment of their fellow Sneetches and agreed it was much better to accept each other as they were, regardless of the number or placement of their stars (Seuss, 1961).

Ethnocentrism is just one of the many "isms" used to refer to negative feelings and attitudes. "These types of attitudes can be expressed as "isms" (ageism, sexism, racism, etc.) and refer to a way of thinking about other persons based on negative stereotypes about race, age, sex, etc." (Ontario Human Rights Commission, 2008). While ageism, sexism, racism, ableism, and other forms of "isms" don't always result in discriminatory action, they are often the cause of discrimination and harassment (Ontario Human Rights Commission, 2008). Therefore, it is important from a human rights perspective to address acts of discrimination and also the "isms" that contribute to unfair treatment.

In writing *The Sneetches,* Theodore Geisel (Dr. Seuss) created a children's book that became a timeless classic. More importantly, he told a story of prejudice and discrimination that reflected the experience of Jews in Nazi Germany and German-occupied nations through the 1930s and World War II: they were forced to wear yellow stars to differentiate them from their neighbours.

FIGURE 2.2

The "Ism" Prism

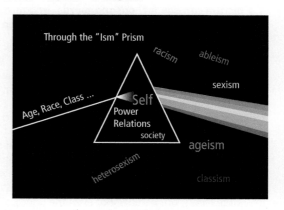

Injustices are like rays of light through the lens of a prism; they refract in a set of "isms" that can contribute to the oppression of non-dominant groups.

The "Ism" Prism

The injustices done by the powerful to the powerless can be seen as rays of light through the lens of a prism: we see them refract into a set of "isms" that can contribute to the oppression of non-dominant groups (see Figure 2.2). For example, age, when refracted through the prism, becomes ageism. It transforms from a concept to an "ism," altered by the influences and biases of social interactions, self-perceptions, power relations, and a host of other effects that determine how people relate to each other based on age. Characteristics such as class, race, ability, gender, and size go through the same process, passing through the prism to become "isms" in their own way, determined perhaps by family relations, ethnicity, religion, and (always) power. The "isms" that exist in society multiply in number as dominant groups increasingly marginalize groups that are different from the norm, groups that lack the power and resources to resist the stereotyping and prejudice that lead to discrimination. The examples listed below are merely a few of the common "isms" that exist today.

Ableism

Ableism refers to the set of ideas and attitudes that define "normal" abilities of people and that allocates inferior status and value to individuals who have developmental, emotional, physical, or psychiatric disabilities. This can result in discrimination against those perceived to be "disabled" (see Chapter 11). The authors of one study looked at over 300 000 records of employment and found that it is not just the disability itself that causes discrimination, but also the stigma that surrounds it. These authors found that it did not matter whether the disability was apparent or existed in a record of employment: it still had an effect on workplace discrimination. Workers with current or previous disabilities, including mental illness, epilepsy, diabetes, and PTSD, experienced discrimination that affected hiring, promotion, reinstatement, and reasonable accommodation (Draper, Hawley, McMahon, & Reid, 2012).

Ageism

The term **ageism** refers to the stereotyping of and prejudice against individuals or groups because of their age. Examples include the tendency to view seniors as unable to work, confused, and fragile, and young people as undisciplined and erratic. Ageism can lead to discriminating against people based on their age, which is illegal, whether

Ableism: Refers to the set of ideas and attitudes that define "normal" abilities of people, and that allocates inferior status and value to individuals who have developmental, emotional, physical, or psychiatric disabilities.

Ageism: Stereotyping, prejudice, and discrimination against individuals or groups because of their age; for example, the tendency to view seniors as unable to work, confused, and fragile.

IN THEIR SHOES

If a picture can say a thousand words, imagine the stories your shoes could tell! Try this student story on for size – have you walked in this student's shoes?

My name is Cameron. I am told by my grandmother that I was a very happy child until about the age of 5 or so. Then I became very depressed and very withdrawn. This was because of a major depressive episode that ended up lasting 11 years. This was brought on by two mentally and physically abusive family members and severe bullying at school. I would get kicked down two flights of concrete stairs by other students several times a day. I would be physically assaulted sometimes in the middle of class and often I would be chased home with knives and baseball bats, and no one ever tried to stop it—not teachers, not parents, not even the principal.

At age 10, a neighbour began to abuse me physically, mentally, and sexually—this went on most days for four years. I never said anything because he and his friends said they would frame me for the same thing if I ever spoke about it. In Grade 8, I went to a private school to get away from my public school problems. The principal hated little boys and decided to take this out on me every morning all school year long. She would haul me into her office and scream at me for about an hour because she hated me. After which point I would be locked in a windowless broom closet for the rest of the day without bathroom breaks until one of the secretaries would unlock the door and say, "I've distracted her, run boy, the last bus for home is leaving."

I always had a rough home life. My mom would make it her mission to get us all to fight every day and then patch things up by the end of the evening, only to get into an unwarranted screaming match the next day. She would often call up my father at work and tell him I had done something bad, which I had not. He would come racing home from work and scream at me and sometimes threaten to beat me up. My brother who is much older was just as manipulative as my mom. He moved out young and I would visit him most weekends. He spent the entire time I was with him telling me how he was planning on taking as much cash as he could from my parents and then just disappearing into the night, and he expected me to do the same eventually. He left four years ago. I speak to him maybe four times a year for a few minutes each time.

I had a series of abusive relationships with other men. All but one was abusive in many different ways. One had attempted to rape me and get me to contract AIDS. I would have been 14 at the time. Fortunately, I was able to escape him. When I was 16, I developed wildly fluctuating bipolar disorder. Also that year, I had what is known medically as a first episode psychosis or, as it's more commonly known, a complete mental breakdown. I spent six weeks in the psychiatric ward of my local hospital where I attempted suicide several times. I had my second overdose at this point (my first was at age 10). I began self-mutilating in addition to the anorexia I had had since Grade 6. This gave me a way to punish myself. I felt if this many people had tried to hurt me, I must have done something to deserve it. I was diagnosed with bipolar type 2 with psychotic features, a rather severe case of post-traumatic stress disorder, generalized anxiety disorder. It took six months for everything to stabilize, and then an additional 3.5 years spent in a "walking coma" before I was fully healed.

This is extremely hard to write all of this. It is hard to remember everyone I've told my life story to—absolutely hundreds of mental health professionals, teachers, professors, and everyday people, and it never gets any easier. I am shaking from head to toe and I feel sick, but this has to be mentioned. I refuse to live in silence. It is unfortunate how we live in a society where when someone gets a physical illness that almost ends their lives, people automatically rush in with love and support, but if I was to tell people something small, like that I suffer from bipolar disorder or I take psychiatric medications, they will always look at me with fear and hate, like I am a beast who must live his days in exile. All this over what other people willingly and consciously chose to do to me starting at a very young age. Society had me marked for death.

So please, I ask just one thing of you: I sincerely hope and pray that after reading this, you will be kind when someone seems a little strange, or if they mention they suffer from mental illness, be a kind ear and a warm heart, then maybe my words, and my life, will not have been in vain.

it involves a landlord who refuses to rent an apartment to a young couple, or a business that refuses to hire a qualified older person for a job. Age is protected by the Canadian Charter of Rights and Freedoms.

Age discrimination is difficult to prove in a court of law, but that doesn't stop those being treated unfairly from trying. In London, Ontario, seven women aged 60 or more are taking an age-discrimination suit against their former employer, InStore Focus, to the Ontario Human Rights Tribunal. Employed as "brand ambassadors" who offered product samples to customers, the women had between 3 and 15 years of service with the company. However, they were let go after being told that they didn't fit the image the store wanted.

According to one woman, "My manager called me and said, 'I have some bad news, (the store) is profiling and they no longer want to use you. They want soccer moms, they feel soccer moms shop for children'" (Hensen, 2012). The women lost needed income, suffered serious blows to their self-esteem, and now face the daunting task of finding new employment.

However, age discrimination is not limited to older adults. The President and CEO of Volunteer Canada reports that young volunteers today feel the sting of discrimination "while seeking, and receiving volunteer responsibilities. They feel discounted by other volunteers and say they are given the simpler types of tasks that nobody else would want to do" (Volunteer Canada, 2012).

Classism

Classism refers to the systematic oppression of subordinate classes by dominant class groups in order to gain advantage and strengthen their own positions. Examples of classist attitudes include the belief that one occupation is better than another, that poor people are impoverished because they don't budget their money properly, and that working-class people are less intelligent than upper-class people (OPIRG, 2012). Over 30 years ago, economist Bradley R. Schiller found empirical evidence of classism when he measured the relative socio-economic achievements of black and white adolescents. Schiller's research showed that class discrimination was potentially as evident and harmful as racial discrimination to black youths who came from impoverished backgrounds (Schiller, 1971).

Classist attitudes are the basis for class discrimination, and they affect everyone. For example, have you ever had to complete a peer evaluation at work or school? In one study, almost 250 students answered a range of questions aimed at measuring their class biases, including the statement, "During an in-class presentation, I would rate a poor person's performance lower than a wealthier person's performance." Analysis of the results of the study found that some students demonstrated negative biases toward classmates they perceived to be poor. In addition, students evaluating their peers allocated lower grades to students who were financially poor but higher performing—grades that differed greatly than assigned by the teachers, while female students were more likely to grade harshly than males (Sadler & Good, 2006; Moorman & Wicks-Smith, 2012).

Heterosexism

Heterosexism refers to the belief in the natural superiority of heterosexuality as a way of life and its logical right to social dominance. Comprised of a system of ideas and institutionalized beliefs, it leads to the oppression of any non-heterosexual form of behaviour, identity, relationship, or community.

One well-known example of heterosexism concerns the experience of Canadian teenager Marc Hall, who took the Durham (Ontario) Catholic District School Board to court in 2002 after his principal refused to allow Hall to bring his boyfriend to the prom. On the day of the prom, Justice Robert McKinnon issued an injunction that ordered the school to allow Hall to go to the prom with the date of his choice. Eventually, though, Hall dropped the case against the school board in 2005 when he learned it would likely take years to resolve the case (Kennedy, 2012). His story was portrayed in the television movie *Prom Queen* in 2004. But while Hall's initiative had a positive outcome, the suicide of Ottawa teenager Jamie Hubley, who took his life in 2011 after enduring years of anti-gay bullying in school, demonstrates the devastating consequences that heterosexism can have (CBC News, 2011).

Racism

Racism is a set of ideas that asserts the supremacy of one group over others based on biological or cultural characteristics, accompanied by the power to put these beliefs into practice to the exclusion of minority men and women. While scientists argue that race has no genetic basis, the concept of "race" is real enough in its consequences. Basically, it comes down to how race is socially constructed as a label attached to a system of power that has the ability to oppress or privilege, depending on the meanings and values assigned to it and on who is doing the assigning.

Fashion magazines provide excellent examples of race as a social construct. In countless instances, dark-skinned and Hispanic models and celebrities, such as Queen Latifah, Beyoncé, Tyra Banks, Rihanna, Halle Berry, and Jennifer Lopez, are "whitewashed" through digital enhancement and given lighter skin to appeal

Classism: The systematic oppression of dominant class groups on subordinate classes in order to gain advantage and strengthen their own positions.

Heterosexism: The belief in the natural superiority of heterosexuality as a way of life and therefore its logical right to dominance. Comprised of a system of ideas and institutionalized beliefs, it leads to the oppression of any non-heterosexual form of behaviour, identity, relationship, or community

Racism: An ideology that either directly or indirectly asserts that one group is superior to others, with the power to put this ideology into practice in a way that give advantages, privilege, and power to certain groups of people, and conversely, can disadvantage or limit the opportunities of racialized individuals or racialized groups.

to the dominant demographic of the fashion market (Kite, 2011). And it's not just in advertising—where there's profit to be had—that those in power manipulate individual and collective perceptions of race. In 2012, Mark Carney, then governor of the Bank of Canada, issued a formal apology for mishandling the design of the new polymer $100 bill (Isfeld, 2012). The original design showed an Asian-looking woman peering into a microscope. However, before the design was released, focus groups indicated that some people thought portraying a scientist as Asian reflected an ethnic stereotype, as well as being unfair in representing only one ethnic group on a denomination of money. In response, the bank ordered the bill to be redrawn with the image of a woman of "neutral ethnicity"— who in fact appeared to be Caucasian (The Canadian Press, 2012). It was a public relations fiasco.

How would you define "neutral ethnicity"? Do you see the Bank of Canada's re-imaging as anything different from the "whitewashing" that's done in the advertising industry? Do you think that, in a society espousing multiculturalism and ethnic diversity, the Bank of Canada's action was narrow-minded? Does it constitute racism?

Sexism

Sexism refers to the belief that one gender is superior to the other. Sexism often results in discrimination against or the devaluation of that gender and the roles related to it.

How prevalent is sexism? Results from a recent study reveal startling information about gender bias in the sciences. Researchers sent out 127 fake resumes to male and female science professors throughout the United States. The resumes contained application materials from a fictional undergraduate student applying for the position of lab manager. On 63 applications, the student's name was John; on 64 applications, the student's name was Jennifer. Every other piece of the application was identical—the resume,

GPA, references, and portfolio. The researchers also made sure to match the groups of professors receiving the resumes in terms of age distribution, scientific fields, proportion of males and females, and tenure status. Both male and female professors consistently regarded the female student applicant as less competent and less hireable than the identical male student. On a scale of 1 to 5, the average competency rating for the male applicant was 4.05, compared to 3.33 for the female applicant. The average salary offered to the male was $30 238.10, while the female was offered $26 507.94 (Moss-Racussin, Dovidio, Brescoll, Graham, & Handelsman, 2012). Could these discrepancies offer a reasonable explanation for differences between the numbers of men and women who practise in the physical sciences such as biology, chemistry, physics, and astronomy?

Sizeism

Sizeism refers to prejudice against individuals based on their body size, including height and weight; it is generally seen as leading to discrimination against those who are overweight. Social consequences experienced by persons with obesity include inequities in employment, barriers in education, compromised healthcare, and negative portrayals in the media, to the extent that the Canadian Obesity Network reports weight bias and discrimination are more prevalent than discrimination based on race, gender, or sexual orientation (Canadian Obesity Network, 2011). But our willingness to act against sizeism is a recent development: it was only in 2011 that doctors, public health policymakers, government representatives, educators, and activists gathered for Canada's first-ever conference on weight discrimination.

Jennifer Portnick knows all about sizeism. Seeking both a fitness instruction job and the purchase of a franchise with Jazzercise, a Carlsbad, California-based company that markets what it calls "the world's leading dance-fitness program," she was denied both applications. In rejecting Portnick, Jazzercise officials cited the company's policy that instructors possess a "fit appearance" (Ackman, 2002). "Jazzercise sells fitness," the company's director of franchise programs wrote to Portnick. "Consequently, a Jazzercise applicant must have a high muscle-to-fat ratio and look leaner than the public" (Brown, 2002). Portnick, a 109-kg, 172-cm (240-pound, 5-foot-8) aerobics teacher, had been taking aerobics classes for 15 years without having her skills questioned. She brought her case before the San Francisco Human Rights Commission, which enforces that city's ordinance barring discrimination based on height or weight, and won. On May 6, 2002, Jazzercise announced that it would stop requiring applicants to be fit in appearance.

Sexism: The belief that one sex is superior to the other, often resulting in the discrimination or devaluation of that gender and the roles related to it.

Sizeism: Prejudice against individuals based on their body size, including height and weight.

Discrimination: The unequal treatment of individuals or groups based on their characteristics or behaviours. It involves actions or practices of dominant group members that have a harmful impact on members of a subordinate group.

Human rights: Define how we are to be treated as human beings and what we are all entitled to, including a life of equality, dignity, and respect, free from discrimination. Human rights in Canada are protected by international, federal, provincial, and territorial laws (Canadian Human Rights Commission, 2013).

In Canada, there is currently no legal recourse for acts of discrimination based on size. The Canadian Human Rights Act does not include protection against sizeism or discrimination based on weight, nor has anyone fought an employment discrimination case based solely on the issue of weight (Immen, 2012). Similarly, in the United States, no federal legislation exists to protect individuals with obesity from discrimination based on weight. Michigan and a few localities, including San Francisco, prohibit weight-based discrimination.

Discrimination

Prejudice is an attitude that often leads to **discrimination**—the unequal treatment of individuals or groups based on their characteristics or behaviours. It involves actions or practices of dominant group members that have a harmful impact on members of a subordinate group.

Human Rights Legislation in Canada

In Canada, **human rights**, including the right to be free of discrimination, are protected by federal, provincial, and territorial laws. The **Canadian Charter of Rights and Freedoms** (1982) promises equity for all Canadians by ensuring that "[e]very individual ... has the right to the equal protection and equal benefit of the law without discrimination and, in particular, without discrimination based on race, national or ethnic origin, colour, religion, sex, age, or mental or physical disability" (Department of Justice, 2012). The **Canadian Human Rights Act** prohibits discrimination for any of the following **grounds of discrimination:** race, national or ethnic origin, colour, religion, age, sex, sexual orientation, marital status, family status, disability, or a conviction for which a pardon has been granted or a record suspended (Canadian Human Rights Commission, 2013). The Canadian Human Rights Act outlines seven **discriminatory practices** that are prohibited when based on one or more grounds of discrimination: denying goods, services, facilities or accommodation; providing goods, services, facilities or accommodation in a way that treats someone adversely and differently; refusing to hire, terminating employment or treating someone unfairly in the workplace; using policies or practices that deprive people of a chance to get a job; paying men and women differently when they do work of the same value; retaliating against a person who has made a complaint to a Human Rights Commission; and harassing someone (Canadian Human Rights Commission, 2013). The **Canadian Human Rights Commission** (2013) provides the following examples of discrimination based on the different grounds of discrimination:

- A bank has lending rules that make it unreasonably difficult for new immigrants to get loans. This is an example of discrimination based on two grounds—*race and national or ethnic origin.*
- A person is systematically referred to secondary screening at airports due to the colour of their skin. This may be a case of discrimination based on the ground of *colour.*
- An employer assigns her employees to weekend shifts without recognizing that some employees observe the Sabbath and cannot work on those days. This may be a case of discrimination based on the ground of *religion.*
- An employer's physical fitness requirements are based on the capabilities of an average 25 year old instead of being based on the actual requirements of the job. This may be a case of discrimination based on the ground of *age.*
- A female employee with an excellent performance record announces that she is pregnant. Immediately, her employer begins to identify performance issues that lead to her dismissal. This may be a case of discrimination based on the ground of *sex.*
- A policy provides benefits to some married couples but not to others. This

Canadian Charter of Rights and Freedoms: Referred to as the Charter, it is a bill of rights and the first part of the 1982 Constitution Act; it is considered the highest law of Canada and, as such, supersedes any other federal or provincial law that conflicts with it.

Canadian Human Rights Act: A federal law that protects all people who are legally in Canada from discrimination by federally regulated employers and service providers (Canadian Human Rights Commission, 2013).

Grounds of discrimination: Reasons a person may experience discrimination. Human rights legislation specifies specific reasons why employers and service providers cannot discriminate against people, such as race, religion, age, sexual orientation, marital status, family status, disability (these are only a few of many grounds used in human rights legislation).

Discriminatory practice(s): Actions that are discriminatory, and in a legal context are based upon one or more grounds of discrimination. The Canadian Human Rights Act includes seven discriminatory practices that are prohibited by law.

Canadian Human Rights Commission: An organization that was created by the *Canadian Human Rights Act* and is separate and independent from the Government of Canada and the Canadian Human Rights Tribunal (Canadian Human Rights Commission, 2013).

may be a case of discrimination based on two grounds—*sexual orientation* and *marital status*.

- After having a child, a woman cannot find childcare to continue working overnight shifts, and her employer does not allow flexibility by scheduling her on day shifts. This may be a case of discrimination based on the ground of *family status*.
- An employer requires all employees to have a valid driver's licence. People who cannot drive due to a disability are not given an opportunity to show how they could still perform the job by, for example, using public transit. This may be a case of discrimination based on the ground of *disability*.
- A person is denied a job because of a previous conviction for which a pardon has been granted or a record has been suspended. This may be a case of discrimination based on the ground of *pardoned conviction.**

The Canadian Human Rights Act requires employers and service providers to treat you equitably, and this can involve accommodating needs to be able to participate fully in work or use services. This is referred to as the **duty to accommodate**. The duty to accommodate only applies to needs that are based on one or more grounds of discrimination that are named in relevant human rights legislation (Canadian Human Rights Commission, 2013). For example, if an employer requires all job applicants to pass a written test, they have a duty to accommodate those applicants with visual disabilities with an alternate format (e.g., oral exam, Braille). There are some circumstances, such as prohibitive costs and health and safety issues, where an employer or service provider may not be legally bound by a duty to accommodate, even when this has a negative effect on a person (Canadian Human Rights Commission, 2013). This is referred to as **undue hardship**. The threshold of as to how far your employer or service provider has to go to accommodate your needs is generally high, particularly with respect to cost. For example, in the case of *Howard v. University of British Columbia,* the university was ordered to pay the interpretation costs of $40 000 for the complainant, who was deaf, so that he could attend teacher's college. The onus of proof of undue hardship rests with the service provider or employer and, in this case, while the costs were significant, the university was not able to prove that this cost would "alter the essential nature or substantially affect the viability of the educational institution" (Ontario Human Rights Commission, 2003).

Provinces and territories have their own laws, polices, and agencies to deal with acts of discrimination. "They protect people from discrimination in areas of provincial and territorial jurisdiction, such as restaurants, stores, schools, housing and most workplaces" (Canadian Human Rights Commission, 2013). In Canada these include Alberta Human Rights Commission, British Columbia Human Rights Tribunal, B.C. Human Rights Coalition, Manitoba Human Rights Commission, New Brunswick Human Rights Commission, Newfoundland and Labrador Human Rights Commission, Northwest Territories Human Rights Commission, Nova Scotia Human Rights Commission, Nunavut Human Rights Tribunal, Ontario Human Rights Commission, Human Rights Tribunal of Ontario, Ontario Human Rights Legal Support Centre, Prince Edward Island Human Rights Commission, (Québec) *Commission des droits de la personne et des droits de la jeunesse*, Saskatchewan Human Rights Commission, and Yukon Human Rights Commission. Human rights legislation generally supersedes other legislation; it is usually not a defence for an employer or service provider to say they were complying with other legislation.

Forms of Discrimination

Stereotypes, bias, and prejudice can lead to discrimination. Discrimination is defined as the unequal treatment of individuals or groups based on their characteristics or behaviours. It can include age discrimination, disability discrimination, employment discrimination, housing discrimination, racial discrimination, religious discrimination, sexual harassment, and more. Most discriminatory actions include failure to individually assess a person's merits, capacities and circumstances; making stereotypical assumptions; and excluding persons, denying benefits or imposing burdens (Ontario Human Rights Commission, 2008). Consider the example of a person refused an apartment because they have children. This action could be considered discriminatory based on family status: it may be based on stereotypical assumptions that children are noisy, messy, and destructive, and therefore not good tenants; it fails to

---

Duty to accommodate:
Human rights legislation requires employers and service providers to accommodate peoples' needs, when those needs relate to one or more grounds of discrimination.

Undue hardship:
Circumstances involving cost, or health or safety issues that would make it impossible or very difficult for an employer or service provider to meet the duty to accommodate (Canadian Human Rights Commission, 2013).

---

*Source: Canadian Human Rights Commission, "What is discrimination?" Found at: https://www.chrc-ccdp.ca/eng/content/what-discrimination.

assess the family's merits and capacities (the family could be clean, responsible, pay rent on time—all desired qualities of a good tenant); and this action results in the denial of housing. Discrimination is often the result of "a tendency to build society as though everyone is the same as the people in power—all young, one gender, one race, one religion or one level of ability" (Ontario Human Rights Commission, 2008). Failing to consider and plan for diversity can result in barriers that, even if unintended, can be considered discrimination.

Intentional versus Unintentional Discrimination

Discrimination can often occur without a deliberate intention to harm a person or group. To say "it was not my intention to harm that person or discriminate against that group" does not excuse a person under the law. Federal, provincial, and territorial human rights law applies even if the discrimination is not intentional. Intent or motive to discriminate is not a necessary element for a legal finding of discrimination—it is sufficient if the action or behaviour has a discriminatory effect (Ontario Human Rights Commission, 2008). In the case of *Ontario (Human Rights Comm.) and O'Malley v. Simpsons-Sears Ltd.,* the Supreme Court of Canada ruled that the employer's termination of Theresa O'Malley, after she refused to work on days of religious observance, was discrimination on the ground of creed. In its decision, the court found that it was "not necessary to prove that discrimination was intentional to find that a violation of human rights legislation has occurred. An employment rule, neutral on its face and honestly made, can have discriminatory effects" (Canadian Human Rights Reporter, 2013). Therefore, in determining whether discrimination has occurred, it is the "effect" of the action or behaviour that needs to be considered and not the "intent."

One prominent example of unintentional discrimination is older buildings that were built without a ramp for people who use wheelchairs. Another example might be an employer administering the same, long-standing standardized test to a new immigrant to Canada, particularly if the test has a cultural bias that does not reflect essential skills needed to perform the job. In the above examples, the service provider or employer did not necessarily intend to discriminate. Yet, according to the law, discrimination has taken place because of the effect on the protected groups. In these examples, the individuals were faced with an unfair disadvantage due to the protected characteristics of disability and ethnic or national origin.

Direct versus Indirect Discrimination

Direct discrimination is the unfair treatment of individuals or groups based on one or more of their protected characteristics, compared to other individuals or groups who do not have these characteristics in similar circumstances (Helly, 2004). This is one of the more prevalent forms of discrimination. An example of this type of discrimination would include a property owner who tells an Indigenous person there is no space available to rent, but then tells a Caucasian person there is an apartment available to rent. In this example of direct discrimination, the Indigenous person is intentionally treated less favourably based on their protected characteristic—race. Other examples of direct discrimination might include being harassed at work on the basis of sex; not hiring a person because of a disability; terminating a woman from her job because she is pregnant; denying access to a service because of sexual orientation; refusing to take reasonable steps to accommodate an employee who needs time off for a religious event; or receiving a warning from your workplace because you needed time off for a sick child.

Indirect discrimination refers to a rule, policy, practice, or requirement that applies to everyone but has the effect of creating disadvantage for people with a protected characteristic. Consider the example of an employer who creates one rule for all employees in the workplace, thinking that it will be neutral and fair for everyone, but due to lack of awareness or unconscious bias, they unintentionally create a situation that disadvantages an individual employee or group of employees with a protected characteristic. This can result in indirect discrimination. A more specific example is an employer who imposes a universal dress code for all employees that disadvantages Muslims, Sikhs, and Jews who might wear religious head coverings (Ontario Human Rights Commission, 2008).

Systemic Discrimination

Systemic (or institutionalized) discrimination refers to the policies, practices, and patterns of behaviour of social institutions that

> **Direct discrimination:** The unfair treatment of individuals or groups based on one or more of their protected characteristics, compared to other individuals or groups who do not have these characteristics in similar circumstances (Helly, 2004).
>
> **Indirect discrimination:** Refers to a rule, policy, practice, or requirement that applies to everyone, but has the effect of creating disadvantage for people with a protected characteristic.
>
> **Systemic or institutionalized discrimination:** Refers to the policies, practices, and patterns of behaviour of social institutions that can appear to be non-discriminatory and/or unintentional, but in fact have a discriminatory effect on persons based on one or more protected grounds.

can appear to be non-discriminatory and/or unintentional, but in fact have a discriminatory effect on persons based on one or more protected grounds. Systemic discrimination can be complex and difficult to identify, especially when this form of discrimination is an "unintentional secondary effect of everyday practices and thinking that are so deeply ingrained in our disciplinary and institutional cultures that we don't observe them or even think about them" (Neuman, 2003). People can also experience systemic discrimination differently based on the intersectionality of identity and grounds of discrimination. For example, a racialized woman may experience systemic discrimination differently from a woman with a disability, even when both women are part of the same organization.

Systemic discrimination can often be observed in exclusionary hiring practices. Philip Oreopoulos of the University of British Columbia was interested in finding out why immigrants who were allowed into Canada based on their skills still struggled in the labour market. He sent out 6000 fake resumes in response to online job postings across Toronto for a variety of occupations. Oreopoulos found that applicants with English-sounding names received almost 40 percent more callbacks from employers than those with Chinese-, Indian-, or Pakistani-sounding names. Further, English-named applicants with Canadian education and experience were over three times more likely to receive interview requests than applicants whose resumes gave Chinese, Indian, or Pakistani names and foreign education and experience. Oreopoulos notes that these practices and others in his findings are illegal under the Ontario Human Rights Code (Oreopoulos, 2009).

Challenging Privilege and Oppression

So, how do we end all forms of oppression? The answer to that question is worthy of its own book. Oppression is a complex and multidimensional social phenomenon that has no singular universal solution nor cookie-cutter recipe. As helping professionals, anti-oppressive practice needs to challenge oppression on all levels—personal, cultural, and structural (Mullaly, 2010). This approach needs to adopt a dialectical perspective whereby interventions at all levels are carried out at the same time, each informing and influencing the other (Mullaly, 2010). An example of this is found in the work done by this chapter's Agent of Change, Charlene Heckman. She simultaneously recognizes how knowledge of social structures that shape our everyday lives can help create social change, while also recognizing the power of human

agency on a personal and collective level in creating that change.

We can challenge oppression by acknowledging that we all have socially constructed privilege and that we all have moments when we are both oppressor and oppressed (Bishop, 2015). To perpetuate privilege and oppression, all that is needed is for people to remain silent about it (Johnson, 2005). Anne Bishop in her book, *Becoming an Ally: Breaking the Cycle of Oppression in People,* talks about the importance of people being involved in their own liberation. She describes human liberation as a spiralling process:

> It begins with breaking the silence, ending the shame and sharing our concerns and feelings. Storytelling leads to analysis, where we figure out together what is happening to us and why, and who benefits. Analysis leads to strategy, when we decide what to do about it. Strategy leads to action, together, to change the injustices we suffer. Action leads to another round of reflection, analysis, strategy, action. This is the process of liberation. (Bishop, 2015, p. 100)

The process of human liberation from oppression is premised on the belief that we are socially responsible for being a part of creating solutions that envision a just and equitable world for all people. If we choose to do nothing or remain silent, then oppression can continue to exist. If we choose to be part of a solution, "that's where our power lies, and also our responsibility" (Johnson, 2005). Transformative change requires that we begin to reflect critically upon how inclusive our thoughts and actions are; that we increase awareness of our stereotypical assumptions and prejudices; that we unmask inaccurate information; and that we become allies in actively challenging privilege and oppression in its many forms.

Edmonton urban-pop vocalist ESMA released a music single and video, "Fall Back," that was inspired by Amanda Todd's story and addresses bullying. Parts of the music video were recorded at J. Percy Page High School, where ESMA was bullied herself as a teen. She encourages people to be agents of their own liberation and to find strength in allies: "I got my army; nothing can harm me ... fall back." So what does it mean to be an ally? On her website, *Becoming an Ally,* Anne Bishop defines an ally as a person "who recognize(s) the unearned privilege they receive from society's patterns of injustice and take(s) responsibility for changing these patterns." Sir Patrick Stewart is an example of an ally who uses his privilege (what he defines as the privilege of being an old white man) to create social change to end violence against women.

AGENT OF CHANGE

Charlene Heckman

Courtesy of Charlene Heckman

Everyday Canadians, like Charlene Heckman, are allies who change our neighbourhoods, our communities, our nation, and our world.

Have you ever met someone whose everyday way of life is helping people in need—not because they are paid to do so, but because their moral fibre is woven with altruism, empathy, generosity, and compassion? Have you ever met someone who sees their daily acts of kindness as part of a collective responsibility toward achieving justice and equity for everyone? Most of us are lucky enough to meet one or two such persons throughout our lifetime; the impressions they leave on our lives are indelible. In my own life journey, I have been fortunate to have met several, one being Charlene Heckman. When teaching about "anti-oppressive practice" and what it means to be an "ally," she is the exemplar I most often refer to. It doesn't matter what day it is, you will find Charlene helping someone in the community. It might be driving someone to a support group, helping a woman who is new to Canada, taking a friend with dementia for lunch, hosting an educational workshop in her living room, supporting a woman who is a victim of abuse, building a butterfly garden, encouraging someone with a disability to attend an exercise group with her, helping someone with a mental health challenge, talking to her MP or MPP about an issue, or participating in walks to support healthcare, to support those with brain tumours, or to support Ronald McDonald House (in memory of her granddaughter Carleigh).

Charlene is a loving partner to her husband Bob; she is also a loving and devoted mother, grandmother,

sister, and aunt. She is one of the truest and most loyal friends you could ever have. A graduate of D'Youville and Canisius Colleges in Buffalo, New York, she earned a Master Degree in Deaf Education and taught at St. Mary's School for the deaf in Buffalo. Later, she became an elementary school teacher in Ontario where she continued to transform the lives of children and families as a gifted teacher. If you look closely at her picture, you will see a twinkle in her eye. It is the twinkle of a person who loves life, can make you laugh until your belly hurts, and is the life of any party.

Charlene has worked tirelessly to create a just and sustainable world for future generations through her participation in social movements and organizations that promote peace, assist newcomers, protect the environment, empower women, and support those with health issues. She has spent much of her life creating a community that is inclusive of all persons. Charlene helped start BET, a weekly support group that originally ran out of her home to provide education about neurological impairments and other health issues (you will read more about this group at the end of this chapter). Charlene and her husband Bob hosted refugee families in their home when they first arrived in Canada, and later helped start a refugee shelter known as Casa el Norte. She was an ally in the creation of Naomi's House for survivors of childhood sexual abuse. She was a member of the Yellow Shirt Brigade, a citizen's group that fought for fair access to quality hospital care for residents of the Niagara Region in Ontario. She is a long-time volunteer tutor with the Adult Literacy Council of Fort Erie, and a volunteer with the Arthritis Society and Maple Park Lodge's recreation committee.

While Charlene is the embodiment of Anne Bishop's guidelines on being an ally, found in this chapter's Reading, there are three qualities that are uniquely outstanding. The first is the way in which Charlene marries education with social action for informed social change. The second is that she advocates against micro and macro levels of oppression and makes connections between both. For example, Charlene supports large organizations dedicated to protecting the environment, and in her own community and neighbourhood she facilitates awareness about monarch butterflies and builds butterfly gardens. Finally, the hospitality of Charlene and Bob's large living room has for many years brought people together; many grassroots social movements have been born here. As an ally, Charlene has created space for people to gather and space for the voices of people who have experienced struggle to speak and be heard.

As an ally, Sir Patrick Stewart uses his privilege to stop violence against women. In what ways might you use your privilege as an ally to those experiencing oppression?

As an example of what it means to be an ally, Sir Patrick Stewart has acquired insight into violence against women through his own childhood experience of witnessing his father abuse his mother. Reflecting upon this experience, he believes men have a collective responsibility to end violence and regularly engages in action to bring about change. He has a clear understanding of his own privilege and how oppression is rooted in society's structure and power relationships.

Another example of what it means to be an ally is the Brain Empowerment Team (discussed in KWIP at the end of this chapter). BET, as it is called, is made up of members who are alternately abled as a result of neurological or other health issues who work together with allies who do not have first-person experience of these issues, but are united in a goal to end ableism. The members of this group have a remarkable understanding of power relationships, most particularly "power with" rather than "power over." Allies are people who are able to recognize their own privilege and understand the power relationships in a society that maintain this privilege. They can therefore see oppression as a structural issue without taking on individual guilt for historic injustices (Bishop, 2015). Some of the other distinguishing characteristics of allies are outlined in Figure 2.3.

FIGURE 2.3

Are You an Ally?

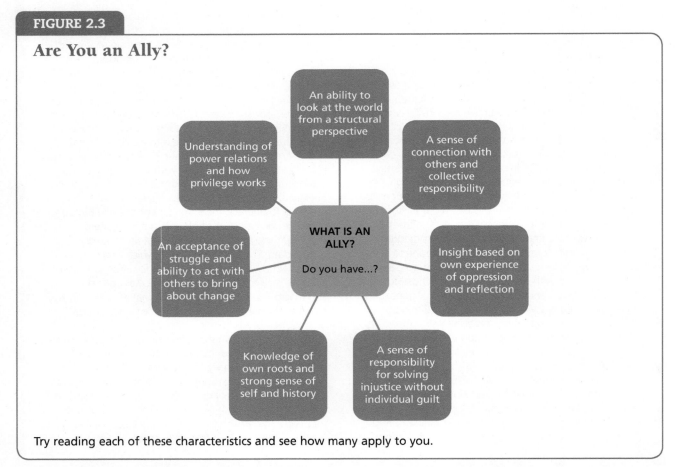

- An ability to look at the world from a structural perspective
- A sense of connection with others and collective responsibility
- Understanding of power relations and how privilege works
- **WHAT IS AN ALLY?** Do you have...?
- Insight based on own experience of oppression and reflection
- An acceptance of struggle and ability to act with others to bring about change
- Knowledge of own roots and strong sense of self and history
- A sense of responsibility for solving injustice without individual guilt

Try reading each of these characteristics and see how many apply to you.

Source: Adapted from Anne Bishop (2015), *On Becoming an Ally*, 110–111.

ENDING THOUGHTS

We know that oppression is dynamic and multidimensional; it exists in many different forms and can vary according to historical and social contexts. It is important to recognize that oppression is rooted in societal structures and power relationships. The consequences of power are manifest in every interaction in society, at all levels. Stereotypes, prejudice, and discrimination grow from the unequal distribution of power, creating dominant and subordinate groups in society and allowing the subjugation of one in favour of the other. The process of human liberation from oppression is premised on the belief that we are socially responsible for being a part of creating solutions that envision a just and equitable world for all people. It is incumbent upon all of us to assume responsibility of actively challenging privilege and oppression in its many forms.

READING

"HOW TO"—BECOMING AN ALLY

By Anne Bishop

Having written that title, I must now admit that I cannot tell anyone exactly how to become an ally. I can, however, use my growing analysis of the process and my experience to offer some guidelines. Most people in our society do not yet see the connections between different forms of oppression or even have a general sense of how oppression works. Therefore, we still find ourselves dealing in most instances with one form of oppression at a time, and in a given setting, we are either in the role of oppressed or ally. I hope these observations will be as useful to you as they have been to me when I find myself in the ally role.

1. It is important to be a worker in your own liberation struggle, whatever it is. Learn, reflect on, and understand the patterns and effects of oppression, take action with others, take risks, walk toward your fear to find your power.
2. Try to help members of your own group understand oppression and make the links among different forms of oppression.
3. I cannot overstress the need to listen. Listen and reflect.
4. Remember that everyone in the oppressor group is part of the oppression. It is ridiculous to claim that you are not sexist if you are a man, or not racist if you are white, and so on. No matter how much work you have done on that area of yourself, there is more to be done. All members of this society grow up surrounded by oppressive attitudes; we are marinated in it. It runs in our veins; it is as invisible to us as the air we breathe. I do not believe that anyone raised in Western Society can ever claim to have finished ridding themselves completely of their oppressive attitudes. It is an ongoing task, like keeping the dishes clean. In fact, the minute I hear someone claim to be free of the attitudes and actions of a certain oppression (as in "I'm not a racist"), I know they have barely begun the process. Humility is the mark of someone who has gone a ways down the read and has caught a glimpse of just how long the road is.

   There is a parallel here with the principles of the twelve-step addiction recovery process. Just as the twelve-step programs teach that the process of healing from addiction is never finished, so it is with the process of unlearning oppression. A white person never becomes non-racist but is always a "recovering racist," more often referred to as "anti-racist".

   There is another reason members of an oppressor group are always oppressors, no matter how much individual learning we have done: until we change the politics and economics of oppression, we are still "living off the avails." We would not be here, doing what we are doing, with the skills and access we have, if we did not have the colour, gender, sexual orientation, appearance, age, class, or physical abilities we have. Resources and power continue to come to us because we are members of the dominant group in relation to the particular form of oppression where we seek to be allies. So, until we succeed in making a more humane world, yes, we are racist (or ageist, or classist, or heterosexist, and so forth). Understanding this is part of the learning to think structurally rather than individually. It is part of avoiding overpersonalization of the issues.

5. Having accepted that every member of an oppressor group is an oppressor, try not to feel like this makes you a "bad" person. Self-esteem does not have to mean distancing yourself from the oppressor role; it can come instead from taking a proud part in the struggle to end oppression. This involves learning to separate guilt from responsibility. Guilt means taking on the weight of history as an individual; responsibility means accepting your share of the challenge of changing the situation. Members of oppressor groups spend a great deal of energy in denying responsibility for oppression. What would happen if all that energy could be put to work figuring out how to end it?
6. Remember that in the oppressor role, you cannot see the oppression as clearly as the oppressed group can. When people point out your oppressive attitudes or language to you, your first response should be to believe it. Ask questions and learn more about the oppression going on in that particular situation. Try not to leap to your own defence in one of the many ways oppressors use to deny responsibility for oppression. Self-defence is an overpersonalized response.

   It is true that you will likely meet members of the oppressed group who will want to claim that every little thing is oppressive and use it as a focus for their anger. You will also perhaps find members of the oppressed group who will try to use your efforts to unlearn oppression to manipulate you. It is all part of the process—their process. The point is not to defend yourself; it will not work anyway. If you can deal with your own defensive feelings, you can turn the situation into a discussion that you, and perhaps everyone else, can use to learn more about the oppression, and you will be less vulnerable to manipulation. The defensiveness, or guilt, is the hook for manipulation.

Also, if you can use your own experience of liberation to understand the anger of the oppressed, you will be able to accept it as a member of an oppressor group, not as an individual. Leave their process—working through their anger—to the oppressed group. Give your attention to your own process—becoming an ally. Then we can all participate in the process we share, ending the oppression.

7. Count your privileges; keep a list. Help others see them. Break the invisibility of privilege.

8. If you hear an oppressive comment or see an example of oppression at work, try to speak up first. Do not wait for a member of the oppressed group to point it out. Sometimes this draws a response of "Oh, I don't mind," "It was just a joke," or even anger directed at you from a member of an oppressed group. That person may be speaking out of their internalized oppression, or you may be off base. Just accept it, if you can; admit it is not your experience. More often you will find members of the oppressed group grateful that they did not have to raise the issue for a change.

9. You must be patient and leave lots of room for the greater experience of members of the oppressed group, but there are also limits. If it becomes clear over time that you are being used or mistreated, say something or leave the situation. Here is an example: a group is interested in having you present as an ally for reasons of their safety or your contacts, legitimacy, or resources, but is not ready to offer you any information or support. The message might be: "Just shut up and do everything we tell you and don't ask questions." It is also hardly fair for the members of the oppressed group to direct all their anger, over a long period of time, at a well-meaning would-be ally. This is not reasonable treatment for anyone. It is fair for you to ask them to decide: do they want you to leave, or will they provide you with some support in your efforts to become an ally?

10. Try to avoid the trap of "knowing what is good for them": Do not take leadership. They are the only ones who can figure out what is good for them, and developing their own leadership strengthens their organizations. It is fine to add thoughts or resources to the process by asking questions of the individuals with whom you have already built up some trust and equality, who will not take it as coming from an authority greater than themselves just because you are a member of the oppressor group. It is not all right to take time at their meeting or public gathering to present your own agenda or to suggest in any way that they do not understand or see the big picture.

11. Never take public attention or credit for an oppressed group's process of liberation. Refuse to act as a spokesperson, even when reporters gravitate to you because they are more comfortable with you or curious about you. You should speak in public only if members of the oppressed group have asked you to speak from your point of view as an ally or to take a public role on their behalf because speaking out will be too dangerous for them.

12. Do not expect every member of the oppressed group to agree; does your group agree on everything?

13. Learn everything you can about the oppression—read, ask questions, listen. Your ignorance is part of the oppression. Find people in the oppressed group who like to teach and who see value in cultivating allies in general or you in particular. Ask them questions. Do not expect every member of the oppressed group to be ready and willing to teach you. When you are in the ally role, you have privileges and comfort in your life that members of the oppressed group do not have because of their oppression; they may not want to also give you their time and energy so that you can learn about them. They may not have the time or the energy.

14. Support the process of unlearning oppression with other members of your own group. Do not usurp the role of communicating the experience of oppression that belongs only to members of the oppressed group. You can, however, share with other members of the oppressor group the journey of becoming an ally; you can help break through others' ignorance of the oppression. Members of your own group might hear you when they cannot hear a member of the oppressed group.

15. Remember that you will probably have to go out of your way to maintain your friendships and connections with members of the oppressed group. Our society is set up to separate different groups. Without a little extra effort, you will live in different parts of town and never cross paths. On the other hand, do not fall over backwards. It is not good to ignore the friends and support base you have already established because you are spending all your time working at the barriers or becoming a "hanger on" of the oppressed community in an inappropriate way.

16. Try not to look to the oppressed group for emotional support. They will likely be ambivalent about you, happy on one hand to have your support, annoyed on the other at your remaining oppressor arrogance, your privilege, the attention you get as a member of the dominant group. Their energy is needed for their own struggle. This does not mean you will not receive support from members of the oppressed group, sometimes more than is warranted. For example, look at the praise men get for doing housework when women still do the vast majority of it. Try not to expect the oppressed group to be grateful to you.

17. Be yourself. Do not try to claim the roots and sense of connection that a history of oppression can give to a community if it is not your own. Do not become what the Mi'Kmaw community calls a "Wannabe." Dig into your own roots. The oppressive history of the group you belong to is a burden you carry. Search out the history of allies from your group as well. Dig even deeper than that. Every group started out as a people with roots in the earth somewhere. Find your own connection with your people's history and the earth. If it is

absolutely untraceable, find appropriate ones and rebuild roots and connections in the present for yourself. But do not try to steal someone else's; you cannot anyway.

18. Be yourself. Be honest. Express your feelings. Do not defend your internalized oppressor attitudes; say it hurts to discover another piece of it. Do not sit on your doubts (except in public gatherings or meetings where you are an observer); ask them of someone you trust. The key word is *ask*. Assume that you are a learner; good learners are open.

"HOW TO"—WORKING WITH ALLIES WHEN YOU ARE A MEMBER OF THE OPPRESSED GROUP

When the shoe is on the other foot, that is, when you find yourself in a situation where it is your oppression under consideration, the same [principles] are in operation, but they are applied a little differently. Here are some guidelines, from my experience, for the situations where you are a member of an oppressed group dealing with allies.

1. Make a clear decision about if, why, when, and how you will work with allies. Do you want to work with allies at all? What can allies offer you that you would find useful? It is easy to know what you do not want members of the oppressor group to do; figure out what you do want them to do. Are there certain times, places, meetings, tasks, and functions where allies would be useful and others where their presence would be inappropriate? Be clear and concise about your degree of openness to allies. Make sure everyone agrees on what is appropriate or at least can live with the decisions without undermining the functions of the people who come in as allies. Working with allies brings a certain kind of struggle; be sure you are ready to enter into it.

2. Allies need support and information. Decide before you begin working with them what you can offer. There needs to be someone in your group who has the patience for teaching allies more about the oppression you are dealing with.

3. Be wise and canny about who is really an ally. If you end up with members of the oppressor group who are acting out of guilt, trying to replace lost roots, taking centre stage, or telling you what to do, you will end up with more frustration than help. Also, beware of people who have no consciousness of their status as a member of the oppressor group or who are unaware of their own oppression in other areas.

4. Do not lump members of the oppressor group together, thinking of them as all "white," "straight," or "male." Remember that everyone is or was also a member of an oppressed group and that people identify more with the parts of themselves that have been oppressed. You may see a woman as white, when she thinks of herself as Jewish; or you may think of a man as male, when he identifies himself as primarily gay.

5. You must listen too.

6. Be kind. Allies are taking a risk, exposing themselves to a situation that is bound to be painful at times.

7. Try to be clear about who is the enemy. There are lots of people who hate you and want to oppress you, punish you, and keep you in your place. There are the rich and powerful who are creating, sometimes deliberately, more of the oppression you suffer daily. Allies are usually well-meaning people without a great deal of power in the system. They are more vulnerable to your anger because they lack power and because of their desire to be an ally. Do not waste resources fighting with them.

8. Be yourself; be honest; express your opinions; be open. Working with allies is all part of a learning process for you too.

Source: Excerpt from Anne Bishop. (2015). *Becoming an Ally: Breaking the Cycle of Oppression in People* (3rd ed.). Halifax, NS: Fernwood Publishing.

DISCUSSION QUESTIONS

1. Imagine yourself being an ally to a group with (a) protected characteristic(s). Because of this association, you may yourself become the target of criticism, ridicule, alienation, or discrimination. How will you respond if this happens to you?

2. Due to past negative experiences, some members of the group whom you have chosen to become an ally with may not trust you and may question your motivations. Are there any points from the reading that might be useful in dealing with this situation?

3. Based on your own personal experience, what three observations presented by Anne Bishop in the reading resonate strongly with you, and why?

KWIP

KNOW IT AND OWN IT: WHAT DO I BRING TO THIS?

The "K" in the KWIP process involves examining aspects of your own identity and social location as the first step in becoming diversity competent.

Perhaps you remember from science class the lesson about dispersion of light through a prism of glass. As white light passed through a prism of glass, it separated into different colours: red, orange, yellow, green, blue, and violet. Using this metaphor, imagine taking an aspect of your own identity and watching it refract through a prism of power that had the ability to marginalize and oppress you based on this aspect of your identity.

1. **Your own "isms":** Have you ever experienced prejudice and/or discrimination based on a real or perceived aspects of your own identity? Describe this experience, its impact on you, and how you dealt with this.

2. **Other "isms":** Search the Internet for other forms of "isms" that can contribute to the oppression of non-dominant groups. Choose one, define it, and provide an example of how this form of prejudice affects members of non-dominant groups in society.

WALKING THE TALK: HOW CAN I LEARN FROM THIS?

The "W" in the KWIP process presents a scenario or case study that challenges you to "walk the talk" through problem-based learning.

ACTIVITY: CASE STUDY

Ontario (Human Rights Comm.) and O'Malley v. Simpsons-Sears Ltd. (1985), 7 C.H.R.R. D/3102 (S.C.C.)

Employer Must Take Reasonable Steps to Accommodate Employee

The Supreme Court of Canada allows an appeal by the Ontario Human Rights Commission and Theresa O'Malley from the Ontario Court of Appeal ruling, which found that O'Malley was not discriminated against because of her religion when her full-time employment was terminated because she refused to work Friday evenings and Saturdays. O'Malley's religion (Seventh-Day Adventist) required strict observance of the Sabbath from sundown Friday to sundown Saturday.

The Supreme Court of Canada, in a unanimous judgment, finds that O'Malley was discriminated against because of creed.

The Court finds that it is not necessary to prove that discrimination was intentional to determine that a violation of human rights legislation has occurred. An employment rule, neutral on its face and honestly made, can have discriminatory effects. It is the result or the effect of an act that is important in determining whether discrimination has occurred.

Where an employment rule has a discriminatory effect, the Court finds that an employer has a duty to take reasonable steps to accommodate the employee, unless accommodation creates an undue hardship for the employer. In O'Malley's case, the employment rule that all employees must work Friday evenings and Saturdays on a rotation basis had a discriminatory effect because of her religion. The employer did not show that accommodating O'Malley would have created an undue hardship.

The Court finds that the onus of proving that accommodation will result in undue hardship is on the employer since the information is in the employer's possession and the employee is not likely to be able to prove that there is no undue hardship.

The appeal is allowed. Simpsons-Sears Limited is ordered to pay O'Malley compensation for wages lost due to discrimination.

1. Discuss some of the ways in which you as an employer might have accommodated Theresa O'Malley's needs, rather than terminating her employment.

2. Provide an example of a "one size fits all" workplace rule that could unintentionally discriminate against an employee. Your answer should include a rationale as to why this might be considered discrimination.

IT IS WHAT IT IS: IS THIS INSIDE OR OUTSIDE MY COMFORT ZONE?

The "I" in the KWIP process requires you to honestly confront and identify ways in which our complex identities result in experiences of privilege and oppression, and to reflect on how we can learn to honour that privilege.

Source: Canadian Human Rights Reporter Inc. (2013). "Employer Must Take Reasonable Steps to Accommodate Employee," retrieved from *Canadian Human Rights Reporter:* http://www.cdn-hr-reporter.ca/hr_topics/systemic-discrimination/employer-must-take-reasonable-steps -accommodate-employee.

ACTIVITY: **GOT PRIVILEGE?**

In this exercise, we ask you to use self-reflective practice to examine the relationship between your own privilege and biases, stereotypes, and prejudices you may have.

ASKING: Do I Have Privilege?

1. I can be sure that I will not be denied membership in a club or organization because of some aspect of my identity.

2. I can be sure that my children will not be mistreated at school because of some aspect of their identity.

3. I can be sure that I will be not be denied meaningful work, adequate job training, promotions when earned, or equal pay for equal work because of some aspect of my identity.

4. I can be sure that I will see people who look like me widely represented in media (catalogues, magazines, books, television, etc.).

REFLECTING: Honouring Our Privilege

Describe two circumstances in which you feel disadvantaged because of some aspect of your identity. Then describe two circumstances in which some aspect of your identity gives you privilege over someone else. How can you use the advantages you experience in becoming an ally to combat the disadvantages experienced by others?

PUT IT IN PLAY: HOW CAN I USE THIS?

The "P" in the KWIP process involves examining how others are practising equity and how you might use this.

ACTIVITY: **CALL TO ACTION**

The Brain Empowerment Team (known as the "BET group") is a group of people who gather weekly in member's homes in the small town of Fort Erie in southern Ontario to transform their brains through education, networking, and mutual support. They are an informal group of alternately-abled and abled persons who educate themselves about the brain, neurological impairments, and other health issues through speakers, discussions, and books. In fact, the group first began in 2009 as a small group of women who met to read Dr. Jill Bolte Taylor's book, *My Stroke of Insight*. Some members of the group have experienced a stroke and others are volunteers who read aloud for the benefit of those who can no longer read, have memory loss, paralysis, aphasia, or vision loss as a result of having a stroke. Today's members are men and women who are readers and listeners and anyone interested in learning about the human brain. Weekly topics explore many health issues, including stroke, autism, Asperger syndrome, Parkinson's, brain injury, dementia, cerebral palsy, depression, and schizophrenia. Members focus on encouragement, empowerment, and pushing limits—the positive energy of the group is absolutely contagious. There are no dues, no fees, no fundraising, and no bosses. No BET member is alone on the road to health and recovery; and there are no lines of division (us versus them)—just deep, authentic, caring, and supportive friendships that are very apparent as members interact with one another. Healthcare professionals have now begun to refer patients to the group and have begun to explore BET as a model of community care. Perhaps what is so special about this group is found in a saying they often refer to: "Life isn't about how to survive the storm, but how to dance in the rain."

Source: Heckman, C. (2017, 02 24). Personal Interview. (T. Anzovino, Interviewer)

Study Tools
CHAPTER 2

Located at www.nelson.com/student

- Review Key terms with interactive **flash cards**
- Check your Comprehension by completing **chapter review quizzes**
- Gauge your understanding with ***Picture This*** and accompanying short answer questions
- Develop your critical thinking/reading skills through compelling **Readings** and accompanying short answer questions
- Apply your understanding to your own experience with **Connect A Concept** activities
- Evaluate Diversity in the Media with engaging ***Video Activities***
- Reflect on your Understanding with ***KWIP*** activities

Ackman, D. (2002, May 9). The case of the fat aerobics instructor. Retrieved from *Forbes*: http://www.forbes.com/2002/05/09/0509portnick.html

Agger, B. (1989). *Socio(ontology): A disciplinary reading*. Urbana, IL: University of Illinois Press.

Agger, B. (2006). *Critical social theories: An introduction* (2nd ed.). Boulder, CO: Westview Press.

Allport, G. (1954). *The nature of prejudice*. Reading, MA: Addison-Wesley.

Axner, M. (2012). Healing from the effects of internalized oppression. Retrieved from *The Community Tool Box*: http://ctb.ku.edu/en/tablecontents/sub_section_main_1172.aspx

Bassett, L. (2010, June 4). Disturbing job ads: 'The unemployed will not be considered.' Retrieved from *Huffington Post*: http://www.huffingtonpost.com/2010/06/04/disturbing-job-ads-the-un_n_600665.html

Berger, P., & Luckmann, T. (1966). *The social construction of reality: A treatise in the sociology of knowledge*. Garden City, NY: Doubleday.

Bishop, A. (2015). *Becoming an all:, Breaking the cycle of oppression in people* (3rd ed.). Halifax, NS: Fernwood Publishing.

Bishop, A. (n.d.). *Tools for achieving equity in people and institutions*. Retrieved 2017, from Becoming an ally: http://www.becominganally.ca/Becoming_an_Ally/Home.html

Brown, P. L. (2002, May 8). 240 pounds, persistent and Jazzercise's equal. Retrieved from *New York Times*: http://www.nytimes.com/2002/05/08/us/240-pounds-persistent-and-jazzercise-s-equal.html

Canadian Human Rights Reporter Inc. (2013). Human rights topics: Discrimination – employer must take reasonable steps to accommodate employee. Retrieved from *Canadian Human Rights Reporter*: http://www.cdn-hr-reporter.ca/hr_topics/systemic-discrimination/employer-must-take-reasonable-steps-accommodate-employee

Canadian Obesity Network. (2011, January 17). Canadian summit on weight bias and discrimination summit report. Retrieved from *Canadian Obesity Network*: http://www.obesitynetwork.ca/files/Weight_Bias_Summit_Report.pdf

Canadian Press. (2012, August 17). Asian-looking woman scientist image rejected for $100 bills. Retrieved from *CBC News*: http://www.cbc.ca/news/politics/story/2012/08/17/pol-cp-100-dollar-bills-asian-scientist-image.html

CBC News. (2011, June 18). Gay Ottawa teen who killed himself was bullied. Retrieved from *CBC News*: http://www.cbc.ca/news/canada/ottawa/story/2011/10/18/ottawa-teen-suicide-father.html

CBC News. (2012, June 17). Rodney King, symbol of 1992 L.A. riots, dies. Retrieved from *CBC News*: http://www.cbc.ca/news/world/story/2012/06/17/rodney-king-obit.html

Chung, S. (2007). Deconstructing lesbian and gay stereotypes in the media. *Journal of Art and Design Education*, 98–107.

Department of Justice. (2012, August 14). *Canadian Charter of Rights and Freedoms*. Retrieved from *Department of Justice*: http://laws-lois.justice.gc.ca/eng/Const/page-15.html#h-39

Draper, W., Hawley, C., McMahon, B., & Reid, C. (2012). Workplace discrimination and the record of employment. *Journal of Vocational Rehabilitation*, 199–206.

Ellin, A. (2012, August 24). Maryland man sues firing range for 'reverse sexism.' Retrieved from *ABC News*: http://abcnews.go.com/Business/maryland-man-sues-firing-range-reverse-sexism-women/story?id=17067299#.UDe67qPCRe4

Elshinnawi, M. (2009, January 8). American Muslim comic fights stereotypes with humor. Retrieved from *The Muslim Observer*: http://muslimmedianetwork.com/mmn/?p=3456

Eyssel, F., & Kuchenbrandt, D. (2012, March 5). Robot prejudice. Retrieved from *British Psychological Society–Research Digest*: http://bps-research-digest.blogspot.ca/2012/03/robot-prejudice.html

Hammer, T. (2009). Controlling images, media, and women's development: A review of the literature. *Journal of Creativity in Mental Health*, 202–216.

Heckman, C. (2017, 02 24). Personal Interview. (T. Anzovino, Interviewer)

Helly, D. (2004). Are Muslims discriminated against in Canada since September 2001? *Canadian Ethnic Studies*, 24–48.

Henrietta Muir Edwards and others v. The Attorney General of Canada, UKPC86 (Privy Council October 18, 1929).

Hensen, M. (2012, August 20). Fired 'soccer mom' seniors launch human rights battle. Retrieved from *The Daily Observer*: http://www.thedailyobserver.ca/2012/08/20/fired-soccer-mom-seniors-launch-human-rights-battle

Howard v. University of British Columbia (No. 1), (1993), 18 C.H.R.R. D/353 (B.C.C.H.R.)

Immen, W. (2012, August 23). The skinny on weight discrimination. Retrieved from *The Globe and Mail*: http://www.theglobeandmail.com/report-on-business/careers/career-advice/the-skinny-on-weight-discrimination/article4311089/

Infantry, A. (2011, August 22). 23 years later, jail guard compensated for racial taunts. Retrieved from *Toronto Star*: http://www.thestar.com/news/gta/2011/08/22/23_years_later_jail_guard_compensated_for_racial_taunts.html

Isfeld, G. (2012, August 20). Bank of Canada's Mark Carney apologizes for way removal of Asian woman on $100 bill was handled. Retrieved from *National*

Post: http://business.financialpost.com/2012/08/20/bank-of-canadas-mark-carney-apologizes-for-way-removal-of-asian-woman-on-100-bill-was-handled/

Johnson, A. G. (2005, February). Who Me? Retrieved from http:www.agjohnson.us/essays/whome/

Kan, M. (2004, April 12). Stereotypes in comedy: Harm or humor? Retrieved from *The Michigan Daily:* http://www.michigandaily.com/content/stereotypes-comedy-harm-or-humor

Keele University. (2012, May 1). Can children resist the subconscious advertising allure of Cheryl Cole? Retrieved from *Keele University:* http://www.keele.ac.uk/press releases/canchildrenresistthesubconsciousadvertisingal lureofcherylcole.php

Kellner, D. (1989). *Critical theory, Marxism, and Modernity.* Baltimore, MD: John Hopkins University Press.

Kennedy, J. R. (2012, September 25). 10 years later, Marc Hall is much more than 'the prom guy.' Retrieved from *Global Toronto:* http://www.globaltoronto.com/10+years+later+marc+hall+is+much+more+than+the+prom+guy/6442721676/story.html

Kilbourne, J. (1999). *Can't buy my love: How advertising changes the way we think and feel.* New York: Simon & Schuster.

Kite, L. (2011, February 28). Beauty whitewashed: How white ideals exclude women of color. Retrieved from *Beauty Redefined:* http://www.beautyredefined.net/beauty-white washed-how-white-ideals-exclude-women-of-color/

Leonard, S. T. (1990). *Critical Theory in Political Practice.* Princeton, NJ: Princeton University Press.

Moorman, D., & Wicks-Smith, D. (2012). Poverty discrimination revealed through student evaluations. *College Student Journal,* 141–150.

Moss-Racussin, C., Dovidio, J., Brescoll, V., Graham, M., & Handelsman, J. (2012, September 17). Science faculty's subtle gender biases favor male students. Retrieved from *Proceedings of the National Academy of Sciences:* http://www.pnas.org/content/early/2012/09/14/1211286109

Mullaly, B. (2010). *Challenging Oppression and Confronting Privilege* (2nd ed.). Don Mills, ON: Oxford University Press.

Nachbar, J., & Lause, K. (1992). Breaking the mold: The meaning and significance of stereotypes in popular culture. In J. Nachbar & K. Lause, *Popular culture: An introductory text.* Bowling Green, OH: Bowling Green State Popular Press, 236–256.

Neuman, S. (2003). Systemic discrimination and the Canada Research Chairs: Diagnosis and treatment. *Clinical & Investigative Medicine,* 35–37.

Ontario (Human Rights Comm.) and O'Malley v. Simpsons-Sears Ltd. (1985), 7 C.H.R.R. D/3102 (S.C.C.).

Ontario Human Rights Commission. (2008). Principles and concepts: What is discrimination? Retrieved from *Human Rights at Work 2008:* http://www.ohrc.on.ca/en/iii-principles-and-concepts/2-what-discrimination

Ontario Human Rights Commission. (2003). *Opportunities to succeed: Achieving barrier-free education for students with disabilities.* Retrieved from *Ontario Human Rights Commission:* http://www.ohrc.on.ca/sites/default/files/attachments/The_opportunity_to_succeed%3A_Achieving_barrier-free_education_for_students_with_disabilities.pdf

Ontario Public Interest Research Group (OPIRG). (2012). Anti-oppression. Retrieved from *OPIRG:* http://opirg.ca/ao/Op_Classism.html

Oreopoulos, P. (2009). Why do skilled immigrants struggle in the labor market? A field experiment with six thousand resumes. Retrieved from *Ideas:* http://www.nber.org/papers/w15036.pdf

Pettigrew, T., & Tropp, L. (2006). A meta-analytic test of intergroup contact theory. *Journal of Personality and Social Psychology,* 751–783.

Pharr, S. (1997). *Homophobia: A weapon of sexism.* Inverness, CA: Chardin Press.

Pincus, F. (1996). Discrimination comes in many forms. *American Behavioral Scientist,* 186–195.

Pub.L. 106-26. 106th Congress Public Law 26. May 4, 1999. http://www.gpo.gov/fdsys/pkg/PLAW-106publ26/html/PLAW-106publ26.htm

Sadler, P., & Good, E. (2006). The impact of self and peer grading on student learning. *Educational Assessment,* 1–31.

Schiller, B. (1971). Class discrimination vs. racial discrimination. *The Review of Economics and Statistics,* 263–269.

Sensoy, O., & Diangelo, R. (2009). Developing social justice literacy. *Phi Delta Kappan,* 345–352.

Seuss, D. (1961). *The Sneetches and other stories.* New York: Random House.

Shipp, E. R. (2005, October 25). Rosa Parks, 92, founding symbol of civil rights movement, dies. Retrieved from *The New York Times:* http://www.nytimes.com/2005/10/25/us/25parks.html?_r=1

Status of Women Canada. (2009, October 9). Governor General's awards in commemoration of the Persons Case. Retrieved from *Status of Women Canada:* http://www.swc-cfc.gc.ca/dates/gg/case-affaire-eng.html

Tannock, S. (2011). Points of prejudice: Education-based discrimination in Canada's immigration system. *Antipode,* 1330–1356.

Thomas, W. (1927). Situational analysis: The behaviour pattern and the situation. *Publications of the American Sociological Society,* 1–13.

Thompson, N. (1997). *Anti-discriminatory practice* (2nd ed.). London: Macmillan.

Thompson, N. (2002). Developing anti-discriminatory practice. In D. R. Tomlinson, & W. Trew (Eds.), *Equalizing opportunities, minimizing oppression: A critical review ofaAnti-discriminatory policies in health and social welfare.* London: Routledge 41–55.

Towbin, M., Haddock, S., Zimmerman, T., Lund, L., & Tanner, L. (2003). Images of gender, race, age, and sexual orientation in Disney feature-length animated films. *Journal of Feminist Family Therapy*, 19–44.

Volunteer Canada. (2012, February 29). Canadian youth perceive age discrimination while volunteering. Retrieved from *Volunteer Canada*: http://volunteer.ca/media-centre/news-releases/canadian-youth-perceive-age-discrimination-while-volunteering

Webster, R., Saucier, D., & Harris, R. (2010). Before the measurement of prejudice: Early psychological and sociological papers on prejudice. *Journal of the History of the Behavioral Sciences*, 300–313.

Women's March Canada. (2017). *Canada sister marches.* Retrieved from Women's March Canada: https://www.womensmarchcanada.com/sister-marches

Women's March Global. (2017). *Sister Marches.* Retrieved from Women's March Global: https://www.womensmarch.com/global/

Young, I. (2004). Five faces of oppression. In L. Heldke, & P. O'Connor, *Oppression, privilege, and resistance.* Boston, (MA: McGraw-Hill, 39–65).

Young, I. (1990). *Justice and the Politics of Difference.* Princeton, NJ: Princeton University Press.

Social Inequality

*"When everyone else is more comfortable remaining voiceless
Rather than fighting for humans that have had their rights stolen"*

(Macklemore & Ryan Lewis, 2012)

LEARNING OUTCOMES

By mastering this unit, students will gain the skills and ability to:

- explain how Canada is socially stratified based on income and wealth

- demonstrate problem-solving skills around the issues that cause social inequality

- analyze the differences between various measures of poverty

- identify the various types of poverty and understand the effects poverty has on Canada

- examine the various types of homelessness and investigate its impact on Canadian society

- discuss how global inequality is measured and its impact on the world we live in

As the largest city in Canada, Toronto is now being called "Canada's Inequality Capital" because the gap between the rich and the poor in the city is growing at a faster rate than in any other major Canadian city (Mojtehedzadeh, 2015). Between 1980 and 2015, the income gap among households in Toronto grew by 31 percent, more than doubling the national pace of 14 percent (United Way, 2015). When we compare Toronto to other major Canadian cities, Toronto's income inequality among households over the 25-year period surpassed that of Calgary at 28 percent, Vancouver at 17 percent, and Montreal at 15 percent (United Way, 2015). President and CEO of United Way Toronto, Susan McIsaac, argues that the "opportunity equation"—where success equals hard work plus access to opportunity—is no longer accurate (United Way, 2015). 73 percent of Torontonians agree that hard work is not enough to get ahead, and that background and life circumstance, like race, gender, and family status, often present barriers when it comes to access and opportunity (United Way, 2015). As the income gap grows, so do the social and economic consequences of income inequality. Have you ever been hungry and wondered where your next meal was coming from? In 2015, 795 million people worldwide were hungry every day (World Hunger, 2015). Have you ever experienced sleepless nights worrying about paying your rent? RentSeeker.ca, one of Canada's largest online apartment search sites, reports that a studio apartment in Vancouver rents for $917 a month, and a one-bedroom apartment goes for $1079 (Davidson, 2015). Why are so many post-secondary students, billed as the future of Canada, living below the poverty line? Perhaps it is because undergraduate students paid an average of $6191 in tuition fees in 2015–2016 (Statistics Canada, 2015a). Have you or has someone in your immediate family ever used a food bank? In 2015, food banks were used by 852 137 individuals in Canada (Hunger Count, 2015).

If you have ever grappled with these issues, you are not alone. **Social inequality** exists in the unequal distribution of tangible or intangible goods or services to individuals or groups in society. About a tenth of Canadians—more than three million people—are living in poverty (Shapcott, 2010), and that situation is having a devastating impact on their physical and mental health. Single individuals living on social assistance in Canada receive roughly $600 a month to cover all of their expenses—rent, food, utilities, and entertainment. Students receiving funding through the

Social inequality: The unequal distribution of tangible or intangible goods or services to individuals or groups in society.

Ontario Student Assistance Program (OSAP) are allocated roughly $7.50 a day for food. Could you stay within these financial guidelines and still follow the nutritional requirements of Canada's Food Guide? Not only is it a matter of eating healthy foods, but think about what those monetary restrictions would do to your social life—we often take the social processes that surround mealtime for granted. Without money to eat out or entertain at home, mealtime can be an isolating process.

In 2010, six prominent Canadians took up a one-week challenge through the "Do the Math Campaign" to live on food that came only from the local food bank (Bielski, 2010). The purpose of the challenge was to show what it is like to eat at the end of the month when all of your monthly social-assistance cheque has gone toward rent, transportation, and clothing expenses. Michael MacMillan, a former CEO, ran out of his food rations after four days, so his meals for the rest of the week consisted of hotdog/wiener rice pilaf and SpaghettiOs—meals that were not what he was used to. This campaign is just one of many (Ontario Coalition Against Poverty, MakePovertyHistory.ca, Campaign Against Child Poverty) that exist to heighten the understanding of poverty in Canada and to bring increased awareness to the government and policymakers that we are a nation in need of change.

When comparing overall child poverty rates, a 2016 UNICEF report (UNICEF Canada, 2016) ranked Canada at 26 out of 35 industrialized countries—indicating that 25 countries fare better when it comes to providing for the well-being of their children and, essentially, the future of their nations. While the gap between the rich and poor in Canada continues to grow (See Figure 3.1), the disparity between nations also increases. Of the total world income in 2014, the richest 20 percent of the population received 84 percent of the resources, while the remaining 80 percent received just 5.5 percent (OXFAM, 2015).

In this chapter, you will learn about social inequality—in all its various forms—as you discover the differences between absolute, relative, transitional, marginal, and residual poverty and the impacts they have on those who experience poverty. Once you investigate the relationships between the rich and poor, mental illness and homelessness, and inequality among nations, your thoughts on inequality may or may not change; but anything that sparks a discussion surrounding the inequalities that exist in our society is a very good thing. It becomes a starting point for social action that begins in the mind of one person—a person just like you—and that is all it takes to create change.

FIGURE 3.1

Income, Wealth, and Power

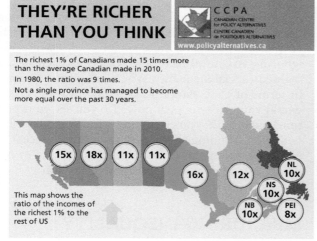

Source: Canadian Centre for Policy Alternatives (CCPA). (2016). Income inequality on the rise, especially in large cities. Retrieved from CCPA: https://www.policyalternatives.ca/newsroom/updates/income-inequality-rise-especially-large-cities. Date Accessed: November 14, 2016.

SOCIAL STRATIFICATION

Although Canada is one of the wealthiest countries in the world, with an advanced economy, universal access to healthcare, and publically funded primary and secondary education, it is also a place with great social inequality, ranging from the very poor to the extremely wealthy. Social inequality occurs when individuals or groups have unequal access to varying resources. It is present between individuals, like when your friend has more than you do; it is evident nationally in Canada, when we see homeless people living on the streets; and it is real on a global scale, as evidenced by the famine that exists in underfunded nations.

In the mid-1800s, Horatio Alger wrote "rags to riches" stories about young men who rose to the top through hard work and determination, and today, the term "rags to riches" is still commonly used. Consider, for example, Guy Laliberté—he is the founder of Cirque de Soleil, a serious poker player, a humanitarian, and a space traveller. Though he was born to middle-class parents, he began his career by busking on the streets. Even though he barely finished high school, he travelled throughout Europe and ended up studying with street artists there and in Quebec. In 1984, with the help of a Canadian government contract, he staged a grand provincial tour to celebrate the 450th anniversary of the discovery of Canada. In 1987, he took a big risk and hauled his entire troupe to a Los

Angeles Arts Festival—he had his hopes, dreams, and all his money invested in this event. In fact, if it failed, he would have no money to bring everyone with the show back home. Cirque de Soleil was a great success, and the rest is history (Wong, 2009). How often do you think of "rags to riches" stories? Do you think if you work hard enough, you will be able to claim the same kinds of victories that Guy Laliberté has? Can everyone realize their dreams in Canada? Is our society structured so that individuals have the same opportunities that will lead them down the road to success, or do some people have certain advantages over others?

The existence of classes in Canada means that we live in a socially stratified society. **Social stratification** is a common feature in systems of shared social inequality; it is common when its members are divided into categories or strata that are rewarded unequally in terms of power, property, and prestige (Berreman, 1972). In Canada, we have a stratified system that is divided into social classes. A **social class** is any group of people who share the same situations in a common social structure.

> **Social stratification:** Division of people into categories or strata that are rewarded unequally in terms of power, property, and prestige.
>
> **Social class:** Any group of people who share the same situations in a common social structure.

We usually measure social class in economic terms—according to individuals' annual income, occupation, or combined resources. The social class rankings themselves can become quite complex and might also include an individual's level of educational attainment. Sociologist Daniel Rossides (1997) uses a five-levelled model to distinguish the classes that exist in a capitalist society. They include the upper class, the upper middle class, the lower middle class, the working class, and the lower class (Rossides, 1997). The distinctions between the classes are not clear-cut, but there are marked differences between the divisions. Membership to the upper class in Canada is limited to the extremely wealthy. These individuals often belong to exclusive clubs and social circles, and generally associate with people of the same social standing. Conversely, members of the lower class find it difficult to find regular work, or must make do with low-paying employment. The difference between the classes is stark when you compare them in quantitative terms. The median net worth of the top 20 percent of income earners in Canada in 2012 was just over $1.3 million, while the median net worth of the lowest 10 percent of income earners that same year was $1100 (Statistics Canada, 2014).

According to Rossides (1997), the upper middle class consists mainly of professionals such as doctors and lawyers, while those with less affluent occupations, such as nurses and teachers, reside in the lower middle class. Though not everyone in the lower middle class may hold a university degree, they often share the common goal of sending their children to university. The working class are individuals who earn their living in blue-collar jobs, generally involving manual labour. The lower class is comprised mainly of individuals who are precariously employed (in part-time or contract labour), and who are continually looking for permanent full-time work that pays a decent salary.

As discussed in Chapter 2, social stratification or the division of society into social classes based on income, occupation, and acquired resources can lead to classism, or the systemic oppression of subordinate classes by dominant class groups in order to gain advantage and strengthen their own positions. Like racism, sexism, and ageism, classism reinforces the notion that a person's worth is partially determined by their socio-economic status. Classist attitudes are rooted in the misconception that we all have the same chances at success in society, and those who have material wealth and/or a prestigious occupation have earned that and therefore deserve what they have, while those who struggle to get by or live in low-income are often blamed for their failure. This ideology does not recognize the systemic barriers that certain vulnerable groups within society face, while also ignoring structural factors in society that create and maintain social inequality. When you identify yourself in terms of class standing, in which of these classes would you place yourself now? Do you see your position changing 20 years from now? To what extent does your class standing affect your future life chances? Do you think that your class standing has helped or hindered your current social position in any way?

MEASURING INEQUALITY

By 12:00 noon on January 3, 2012, the first official working day of the year, Canada's Elite 100 CEOs had already pocketed $44 366—what it takes the average person an entire year, working full-time, to earn (based on 2010 figures) (Hennessy, 2012). Can you imagine having earned that much money by lunchtime? How many days or weeks would you work before you would have made enough money to survive on for the year? Your answer will likely depend on what you have been used to spending and what kind of lifestyle you hope to have in the future. The bridge that spans upper and lower classes is one that you can travel in any class-based society; but in Canada, crossing that bridge is becoming a more difficult journey each year.

Canada does not have an official definition for *poverty*, so when we measure inequality in Canada, we use three different measures. The **low-income cut-off** (LICO) is established by Statistics Canada annually and refers generally to what people call a poverty line. It represents the income level at which a family may face hardship because it has to spend a greater proportion of its after-tax income on food, shelter, and clothing than the average family of similar size. There are separate cut-offs for seven sizes of family (from unattached individuals to families of seven or more persons) and for five community sizes (from rural areas to urban areas with a population of more than 500 000). For example, single individuals living in a large city, like Toronto, would be considered to have a low income if their 2014 after-tax income was below $20 160, whereas a single individual living in a smaller city (with a population of 100 000 to 499 999 people) would be considered to have a low income if their

Low-income cut-off (LICO): Measure established by Statistics Canada annually that refers generally to what people call a poverty line; it represents the income level at which a family may face hardship because it has to spend a greater proportion of its after-tax income on food, shelter, and clothing than the average family of similar size.

TABLE 3.1

Low Income Cut-Offs, After Tax, 2014

Community Size				
Census Agglomeration (CA)			**Census Metropolitan Area (CMA)**	
Rural areas outside CMA or CA[1]	Less than 30 000 inhabitants[2]	Between 30 000 and 99 999 inhabitants	Between 100 000 and 499 999 inhabitants	500 000 inhabitants or more
Current dollars				
2014				
1 person		13 188	15 093	16 836	17 050	20 160
2 persons		16 051	18 370	20 493	20 750	24 536
3 persons		19 987	22 873	25 517	25 839	30 553
4 persons		24 934	28 537	31 835	32 236	38 117
5 persons		28 394	32 495	36 251	36 707	43 404
6 persons		31 489	36 038	40 204	40 709	48 136
7 or more persons		34 585	39 581	44 155	44 711	52 869

[1]Source: Income Statistics Division, Statistics Canada

[2]The low income cut-offs after tax (LICO-AT) are income thresholds below which a family will likely devote a larger share of its after-tax income on the necessities of food, shelter, and clothing than the average family. The approach is essentially to estimate an income threshold at which families are expected to spend 20 percentage points more than the average family on food, shelter, and clothing, based on the 1992 Family Expenditures Survey. LICOs are calculated in this manner for seven family sizes and five community sizes.

Source: Statistics Canada, CANSIM Table 206-0094: Low-income cut-offs (LICOs) before and after tax by community and family size in current dollars, annual. Reproduced and distributed on an "as is" basis with the permission of Statistics Canada.

2014 after-tax income was below $17 050 (Statistics Canada, 2015b). (See Table 3.1).

Created for making international comparisons, the **low-income measure** (LIM) is another commonly used measure. The LIM compares a household's income against the median of an equivalent family. Any family whose income is less than half that median is defined as poor. The LIM adjustment for family sizes reflects the fact that a family's needs increase as the number of members increases. Most would agree that a family of five has greater needs than a family of two. Similarly, the LIM allows for the fact that it costs more to feed a family of five adults than a family of two adults and three children (Employment and Social Development, 2013).

Created in 2001, the **Market Basket Measure** (MBM) is a measure of low income based on the cost of a specified basket of goods and services representing a modest, basic standard of living in comparison to the standards of its community. It includes the costs of food, clothing, footwear, transportation, shelter, and other expenses for a reference family of two adults (aged 25 to 49) and two children (aged 9 and 13) (Employment and Social Development,

2013). It was designed to complement the LICO and LIM measures and allows for different costs for rural areas in the different provinces. The LICO and LIM are relative measures, while the MBM is an absolute measure of poverty. It is important, however, to understand that there is no perfect measure. The three measures produce different results. In 2009, according to each measure, the following numbers of Canadians were living in low income:

- LICO—3.2 million (9.6 percent of the population)
- MBM—3.5 million (10.6 percent)
- LIM—4.4 million (13.3 percent) (Conference Board of Canada, 2011)

Measuring poverty becomes increasingly important when we consider that there are varying kinds of poverty.

Low-income measure (LIM): A measure of poverty that is commonly used for making international comparisons.

Market basket measure (MBM): A measure of low income based on the cost of a specified basket of goods and services representing a modest, basic standard of living in comparison to the standards of its community.

Usually, we think of poverty in terms of rich and poor, and only in terms of the individuals who affect us in our cities or towns. Society as a whole encounters poverty on different levels and in different ways that further serve to divide it. When we talk about poverty, we generally refer to a lack of resources. Economic poverty is the most obvious kind of poverty, but people can also experience spiritual, mental, and cultural poverty. For example, people who live in cultural poverty feel marginal, helpless, or dependent, and experience a sense of not belonging. They feel isolated while living in their own country, and are often convinced that societal institutions do little to help them. They feel powerless, and it might not have anything to do with economics (Lewis, 1998).

Although they are not the only types of poverty, discussions around poverty often classify it into two major types—absolute and relative poverty. **Absolute poverty** refers to a situation where an individual lacks even the basic resources necessary for survival. People who live in absolute poverty live without food, clothing, or a roof over their heads. This is most common in developing countries, where finding necessities such as clean water for drinking and food to stave off hunger are daily struggles. **Relative poverty** refers to an individual's or group's lack of basic resources for survival when compared with other people in society as a whole. In other words, it refers to an individual's standard of living relative to someone else's. We see this in many forms in the cities that we live in. The disparities that exist highlight the difference between those families who can afford to eat

three nutritious meals a day and dress their children in the latest fashions, and those families who struggle to eat one meal a day and hope that their children have winter coats.

Transitional poverty can occur when one finds oneself living in poverty for a limited period of time. This type of poverty is usually temporary and often results from an event or circumstance, such as the unexpected loss of employment. Many people who experience transitional poverty are able to improve their circumstances within a short period of time, but if adequate supports or opportunities are not available, transitional poverty can lead to **marginal poverty**, which occurs when a person lacks stable employment over an extended period of time, or it can eventually become **chronic poverty**. Chronic poverty occurs when an individual is in a state of poverty over an extended period of time and barriers to well-being become cyclical in nature, where one circumstance can have a "snowball effect" and lead to other issues. For example, if a person drops out of high school before graduation, their employment opportunities are often restricted to low-paying jobs, forcing them to rely on low-cost housing, thereby increasing their likelihood of living in a high-poverty neighbourhood with limited access to services (Saskatoon Poverty Reduction Partnership (SPRP), 2011). Consequently, chronic poverty can lead to **intergenerational poverty** when children and youth grow up in households experiencing chronic poverty. Limited access to services and amenities leads to fewer developmental opportunities for children, and fewer developmental opportunities can result in lower educational achievement, thereby starting another cycle of poverty for the next generation (SPRP, 2011).

POVERTY IN CANADA

To understand poverty wholly in all of its forms, we must go beyond the sterile definitions and apply them to an everyday situation. According to the TVO Campaign "Why Poverty," poverty affects over 3 million Canadians and we do not all have equal chances of becoming impoverished—seniors, people with disabilities, recent immigrants, Indigenous Canadians, lone parents, singles, and children are more likely to live in poverty.

Child Poverty

Consider the responses in the following example, where a teacher asked a class of Grades 4 and 5 students

---

Absolute poverty: A situation where an individual lacks even the basic resources that are necessary for survival; people who live in absolute poverty live without food, clothing, or a roof over their heads.

Relative poverty: A situation where an individual or group lacks basic resources for survival when compared with other people in the society as a whole; relative standard of living when measured to others.

Transitional poverty: Occurs when a person is living in poverty for a limited period of time; usually results from an event or life circumstance, such as the unexpected loss of employment.

Marginal poverty: Occurs when a person lacks stable employment over an extended period of time.

Chronic poverty: Occurs when a person is in a state of poverty over an extended period of time and barriers to well-being become cyclical in nature.

Intergenerational poverty: Often occurs when children and youth grow up in households experiencing chronic poverty, where limited access to opportunities can start another cycle of poverty for the next generation.

in North Bay, Ontario, what poverty meant to them (Interfaith Social Assistance Reform Coalition, 1998):

Poverty Is...

Not being able to go to McDonald's

Getting a basket from the Santa Fund

Feeling ashamed when my dad can't get a job

Not buying books at the book fair

Not getting to go to birthday parties

Hearing my mom and dad fight over money

Not ever getting a pet because it costs too much

Wishing you had a nice house

Not being able to go camping

Not getting a hot dog on hot dog day

Not getting pizza on pizza day

Not being able to have your friends sleep over

Pretending that you forgot your lunch

Being afraid to tell your mom that you need gym shoes

Not having breakfast sometimes

Not being able to play hockey

Sometime crying really hard because my mom gets scared and she cries

Not being able to go to Cubs or play soccer

Not being able to take swimming lessons

Not being able to afford a holiday

Not having pretty barrettes for your hair

Not having your own private backyard

Being teased for the way you are dressed

Not getting to go on school trips*

Can you tell which experiences of poverty on the list are examples of absolute, relative, transitional, marginal, chronic, or intergenerational poverty? Could you identify with any of the experiences on that list? The number of children who continue to live in impoverished conditions continues to be a growing concern both nationally and internationally. How can we expect children to grow and become successful,

*Source: Interfaith Social Assistance Reform Coalition, *Our Neighbours' Voices: Who Will Listen?* (Toronto: Lorimer, 1998).

productive adults, when they begin their lives in poverty? In 2009, about 1 in 10 or 639 000 Canadian children were living in poverty (Family Service Toronto, 2011). Growing up without basic necessities affects all areas of life, but it is especially linked to mental and physical health. Poverty is a key determinant of good or poor health because we know that children growing up in poverty

- are more likely to have low birth weights, asthma, and type 2 diabetes, and suffer from malnutrition;
- are unlikely to have family benefit plans for prescription drugs or vision or dental care;
- are more likely to have learning disabilities, emotional disabilities, and behavioural problems; and
- are 2.5 times more likely to have a disability than children from wealthier families (Family Service Toronto, 2010).

New research links childhood poverty to poor performance in school and later on in life. Evans and Schamberg (2009) found strong connections between childhood poverty, physiological stress, and adult memory. The findings of their research can explain, in part, why impoverished children consistently perform worse than their middle-class peers in school, and eventually in adult life (Evans & Schamberg, 2009). Impaired health, poorer health, and school achievement are just some of the consequences children living in poverty experience.

A 2016 UNICEF report compared the state of child well-being in various countries and showed that some of the world's wealthiest countries were more successful in raising children out of poverty, despite having similar economic performances, even in challenging times. Although one would think that all wealthy countries would be successful in diminishing child poverty, UNICEF's analysis shows that the risk of poor child development is affected by government policy and spending priorities. Scandinavian countries and Switzerland have the lowest rates of child inequality, while Turkey and Israel are among the highest (UNICEF Canada, 2016). The index of child inequality looks at the gap between children at the bottom and children in the middle of each country, with a focus on inequalities in income, health, education, and overall life satisfaction (UNICEF Canada, 2016). Canada ranks in the bottom third at 26 out of 35 countries, where the poorest children have family incomes 53 percent lower than the average child (UNICEF Canada, 2016). To improve its standing, Canada must make some important changes. First, it must improve the incomes of households with children through income benefits, transfers, and taxation (UNICEF

Canada, 2016). The proposed federal Canada Child Benefit is a step in the right direction and could help to reduce child inequality by 25 percent (UNICEF Canada, 2016). Additionally, Canada must make children a priority in budget spending and allocations, especially in early years. Investing in an early-years framework, from prenatal to early child health, learning and child care could improve child inequality from the bottom up, causing all levels to rise (UNICEF Canada, 2016). Another necessary change is to improve educational outcomes and promote and support health for all children. The UNICEF report suggests one way to improve educational outcomes is to connect holistic child and youth services with educational institutions to enable the lowest achievers and disengaged children to reach academic, emotional, and social outcomes (2016). According to the report, children in low-income families and adolescent girls tend to have poorer health, so UNICEF recommends that all levels of government should invest funding in healthy food programs around children to improve access and affordability (2016). Furthermore, measurements of child well-being need to be improved and should incorporate the voices of young people. UNICEF believes that we need to "listen to young people to better understand the roles that parents, teachers, community members, employers, and policymakers could play to support their sense of well-being" (UNICEF Canada, 2016).

Student Poverty

Student poverty does not receive much attention in the media and is often overlooked when discussing poverty reduction strategies in Canada. The stereotype of the "starving student" is widely accepted, and there seems to be a misconception that student poverty is acceptable as a means to an end—with an assumption that students who struggle in school can work their way out of it once they graduate and obtain adequate employment. The reality is that student poverty is growing and researchers argue that the rise in tuition costs, the decrease in government funding, youth underemployment, and insurmountable student debt is to blame (Canadian Federation of Students (CFS), 2016). Have you ever had to prioritize housing over food in order to cover the cost of tuition and living expenses? Have you ever relied on the food banks on campus or in the surrounding community for help when you struggled to feed yourself?

Precarious employment: Employment that includes, but is not limited to, part-time, temporary, or contract work with uncertain hours, low wages, and limited to no benefits.

If you have ever found yourself struggling to balance the financial obligations of your school and home life, you are not alone. For some students, postsecondary is simply not an option because the risk of incurring insurmountable debt and not obtaining adequate employment after is just too great. According to the Canadian Federation of Students (CFS), government funding has decreased 50 percent between 1992/1993 and 2013/2014, which directly impacts the rising cost of tuition (CFS, 2016). In 2015, public funding accounted for approximately 57 percent of postsecondary operating funding, down from 80 percent just two decades before, while tuition fees have grown from 14 percent of operating funding to over 35 percent (CFS, 2016). Average tuition fees have more than tripled over the past two decades, with costs averaging $1706 in 1991–1992 and $6191 in 2015–2016 (CFS, 2016). The fundamental shift in the ratio between public funding and user fees has resulted in postsecondary education becoming unaffordable to many Canadians living in low-income households. Levels of student debt are higher than they have ever been before, with over 497 000 students forced to borrow from the Canada Student Loans Program (CSLP) in 2013, not including those who borrowed from private sources to fund their education (CFS, 2016). The amount owed to CSLP increases by nearly $1 million per day, and in the spring of 2015, the federal government increased the limit on federal student loan lending to $24 billion, resulting in even more student debt, rather than addressing the issue of unaffordable education (CFS, 2016).

The Working Poor

Precarious employment includes, but is not limited to, part-time, temporary, or contract work with uncertain hours, low wages, and limited to no benefits. Those who work in precarious employment have very little job security and often do not know how long periods of employment will last. Often referred to as Canada's "Working Poor," people in precarious employment are working but not earning enough to get by. Many take on a second or third job in order to supplement their income, but even those who obtain full-time employment, 35 hours per week at minimum wage, earn an income that falls below the low-income cut-off. For example, minimum wage in 2014 in Ontario was $11 per hour—a person working 35 hours per week, 52 weeks of the year would earn $18 480 before taxes—and the low income cut-off for a single person living in a large city in Ontario in 2014 was $20 160 (Statistics Canada, 2015b). Often associated with only service-sector jobs, precarious employment is spreading to areas of employment that were once deemed stable with benefits and pension opportunities (Sagan, 2016). Wayne Lewchuk, Economics and Labour Studies professor at McMaster University and co-author of *The Precarity Penalty: The Impact of Employment Precarity on Individuals, Households, and Communities – and What to Do About It*, identifies that precarious employment is becoming the norm with more than 40 percent of those employed in the knowledge and creative sectors in precarious or vulnerable work (Sagan, 2016). The report indicates that while precarious employment has always been most prevalent with immigrants, racialized groups, and women, it is now found at all income levels and in all demographic groups, impacting those in low income the worst (Poverty and Employment Precarity in Southern Ontario (PEPSO), 2015).

The report investigated the social impact of precarious employment, including household and community well-being, along with discrimination and health, and found that precarious workers are less likely to receive adequate training, more likely to receive negative consequences for asserting their rights related to occupational health and safety, and less likely to be paid in full for completed work (PEPSO, 2015). Workers in precarious employment are more likely to experience discrimination in the workplace, with racial discrimination as the most frequently reported, followed by discrimination based on age and gender (PEPSO, 2015). According to the report, precarious employment is also associated with a higher prevalence of poorer mental health; workers in precarious employment reported high levels of anxiety and depression, low self-esteem, and difficulty sleeping as a result of the mental stress associated with insecure employment (PEPSO, 2015). Many reported delaying starting a relationship or having children as a result of unstable employment and indicated that uncertainty with their work schedule negatively affected their home and family life (PEPSO, 2015). Finally, those who were working in precarious employment were less likely to interact socially, less likely to become involved in their communities, and less likely to participate in voting (PEPSO, 2015).

HOMELESSNESS

Before the 1980s, the word *homelessness* did not exist—really (Hulchanski, 2009). Before then, we referred to individuals who lacked homes as "the homeless"; but the term *homelessness* as an abstract concept is now used to refer not to a subset of individuals in society, but rather to the multiple blights on the social, economic, and political systems in society. The cutbacks in social housing and related housing programs and policies began in 1984. In the 1990s, the government ended federal and provincial housing programs, decreased social assistance rates, and reduced social spending. Forced onto the streets and into makeshift shelters, at-risk individuals had few, if any, social services or supports available to them. Diseases like tuberculosis returned, and the number of homeless deaths began to increase (Crowe, 2012). With a failure to provide adequate housing, income, and support services came a need to shift the focus on "the homeless" to the concept of homelessness (Hulchanski, 2009). Now, in a very broad sense, **homelessness** refers to a social category that includes anyone who cannot obtain and sustain long-term, adequate, and risk-free shelter—for any reason.

How would you define homelessness? When you think of a homeless person, who automatically comes to mind? Is it as simple as someone who does not have a place to stay, or is it far more complicated than that? Is it the stereotypical **panhandler**, asking passersby for change? Is it a squeegee kid washing windshields? Or is it the person in the seat next to you, who sleeps in his car and showers at the gym? Just as in Canada there is no formal definition for poverty, neither is there one way of categorizing those who live in less-than-desirable housing conditions. No single definition of homelessness is official in Canada. Most officials take into account the specific housing situation and the duration and/or frequency of homeless episodes (Echenberg & Jensen, 2012). If we look at homelessness on a continuum based specifically on types of shelter, at one end there is **absolute homelessness**, which refers to the situation of individuals who live either in emergency shelters or on the street. These individuals are typically the people that you see on the streets, often panhandling for change. **Hidden or concealed homelessness** refers to those people without a place of their own, who live in a car, with family or friends, or in a long-term institution such as a prison. Increasingly, these are new immigrants to Canada, who stay with family until they can afford

Homelessness: Social category that includes anyone who cannot obtain and sustain long-term, adequate, and risk-free shelter, for any reason.

Panhandler: person who approaches strangers on the street or in a public place and asks for money.

Absolute homelessness: Situation of individuals who live either in emergency shelters or on the street.

Hidden or concealed homelessness: State of those without a place of their own who live in a car, with family or friends, or in a long-term institution such as a prison.

a place of their own. These individuals fall in the middle of the continuum. At the other end of the continuum is **relative homelessness**, which refers to those individuals who have housing, but who live in substandard or undesirable shelter and/or who may be at risk of losing their homes (Girard, 2006). This might include a woman who lives with an abusive husband but stays because she and maybe her children have nowhere else to go.

The individuals who live on the streets have typically become the face of the homeless, but some suggest that for every one individual living in absolute homelessness, there are between 4 and 23 individuals whose homelessness is hidden (Condon & Newton, 2007; Hwang, 2010). Counting the homeless is a difficult process because of methodological issues such as defining the population and sample size; even conservative estimates suggest that between 200 000 and 300 000 Canadians spend their nights in shelters, on the streets, or couch surfing (Bramham, 2008). Risk factors that contribute to homelessness for the mainstream population include the lack of affordable housing, poverty and low income, mental illness, domestic violence, and drug abuse/addiction. Not all of these factors affect each individual, and some factors affect some individuals more than others. Homeless people and the issue of homelessness itself are very complex issues. Studying them as multi-layered identities and events is critical to their understanding and ultimately to their solution.

Homelessness and Mental Health

The deinstitutionalization movement started in the 1960s. With the discovery of new medications, coupled with the use of psychotherapy to help treat mental disorders, the general belief was that people with mental illnesses would fare better in the community, or back with their families, than in government-funded institutions. Consequently, the government decided to start removing psychiatric beds from mental health institutions and hospitals. Between 1960 and 1976, 27 630 beds were eliminated, reducing the number of available beds by 57 percent (Casavant, 1999). Unfortunately, in many cases, families were not prepared to care for their loved ones, social services were inadequate, and the community supports just fell through, leaving an extremely vulnerable part of the population to fend for themselves. Left with no long-term mental health treatment programs and little family or community support, many people with mental illnesses were unable to control their symptoms and ended up homeless. Recent statistics show that an estimated 25–50 percent of the homeless population have a mental illness and that up to 70 percent of those with a severe mental illness have substance abuse issues (CMHA, 2009).

People living with mental illness are disproportionately affected by homelessness. It is estimated that more than 500 000 Canadians living with mental illness are inadequately housed (Mental Health Commission of Canada (MHCC), 2013). People living with mental illness are more likely to encounter barriers to employment and more likely to remain homeless for longer periods of time (CMHA, 2016). Mental illness can lead to clouded thinking and impaired judgment, causing individuals to withdraw from family and friends, resulting in disaffiliation and further isolation. Mental illness can also impair a person's resiliency and resourcefulness, which is why homeless individuals living with mental illness tend to be in poorer health than other homeless people. Homelessness, in turn, can magnify poor mental health; high levels of stress, anxiety, fear, depression, and sleeplessness can exacerbate previous mental illness (Munn-Rivard, 2014).

The At Home/Chez Soi project was the first "Housing First" intervention to take place in Canada with over 2000 participants who were homeless and living with mental illness in Vancouver, Toronto, Montreal, Winnipeg, and Moncton (MHCC, 2014). The Housing First model created a recovery-oriented culture that emphasizes participant choice first and foremost (MHCC, 2014). The program offers people who are homeless access to various types of housing, services, and supports to aid in their recovery. A key feature of the program mandates that all participants would pay less than 30 percent of their income on rent, and the availability of housing is not contingent on participation in treatment (MHCC, 2014). Individuals are empowered to achieve full independence, and services that provide support and treatment for mental illness are voluntary (MHCC, 2014). Support services are individualized and centralized off-site but are available in the home or in the community (MHCC, 2014). You will have a chance to read more about Housing First initiatives in the Reading selection for this chapter.

After the five-year At Home/Chez Soi project concluded in 2014, the research indicated that Housing First is a model that is highly adaptable, extremely effective, and economically viable (MHCC, 2014). The research found dramatic improvements in housing stability and quality of life for participants who were experiencing chronic homelessness and mental illness (MHCC, 2014). Statistically, 61 percent of participants "described a positive life course since the study began, 31 percent reported a mixed life course, and

Relative homelessness: State of those who have housing, but who live in substandard or undesirable shelter and/or who may be at risk of losing their homes.

8 percent reported a negative life course" (MHCC, 2014, 8). From a cost perspective, Housing First initiatives are initially costly, but Housing First services end up replacing other costlier services, like hospitalization, emergency shelters, jail/prison, and home or office visits to different providers (MHCC, 2014). The Canadian government showed strong support of Housing First initiatives by announcing a five-year $600 million extension of the Homeless Partnering Strategy, with a focus on Housing First (MHCC, 2014).

GLOBAL INEQUALITY

The most commonly used measure of income inequality is the **Gini index** (also known as the Gini coefficient or ratio), which is measured on a scale of 0 to 1. A Gini index of 0 represents exact equality (that is, every person in a certain society or nation has the same amount of income), while a Gini index of 1 represents total inequality (that is, one person has all the income and the rest of the society has none) (Conference Board of Canada, 2011). In the 1980s, Canada reduced its degree of inequality, reaching a low Gini index of 0.281 in 1989. Income inequality rose in the 1990s and has remained around 0.32 into the 2000s (Conference Board of Canada, 2011). You might think that sounds good, but when you consider that number in relation to the Gini index numbers of other countries, it gives you a better picture of the level of inequality in Canada. Out of 17 peer countries, Canada ranks twelfth in terms of inequality, which means income inequality is higher in Canada than in 11 other countries. The countries with the lowest Gini index scores are Denmark (0.232) and Sweden (0.234). Canada's score (0.317) is not as high as that of the United States (0.381), but it is not nearly as low as the scores for France (0.270), Norway (0.276), or Finland (0.269) (Conference Board of Canada, 2011).

With a goal to eradicate poverty and reduce global inequalities and exclusion, the United Nations Development Programme (UNDP) works with some 170 countries and territories to support the 2030 Agenda for Sustainable Development. In effect as of January 2016, the agenda identifies 17 Sustainable Development Goals (SDGs) that build off the progress that was made between 2000 and 2015 with the 8 Millennium Development Goals (MDGs). Key achievements of the MDGs include the following:

- More than 1 billion people have been lifted out of extreme poverty (since 1990).
- Child mortality dropped by more than half (since 1990).
- The number of out-of-school children has dropped by more than half (since 1990).

- HIV/AIDS infections fell by almost 40 percent (since 2000). (United Nations Development Programme, 2016)*

The current global goals provide clear guidelines and targets for all countries to continue to work over the next 15 years to reduce global inequalities. These goals include the following:

- End poverty in all its forms everywhere.
- End hunger, achieve food security and improved nutrition, and promote sustainable agriculture.
- ensure healthy lives and promote well-being for everyone at all ages
- Ensure inclusive and equitable quality education and promote life-long learning opportunities for all.
- Achieve gender equality and empower all women and girls.
- Ensure availability and sustainable management of water and sanitation for all.
- Ensure access to affordable, reliable, sustainable, and modern energy for all.
- Promote sustained, inclusive, and sustainable economic growth, full and productive employment, and decent work for all.
- Build resilient infrastructure, promote inclusive and sustainable industrialization, and foster innovation.
- Reduce income inequalities.
- Make cities and human settlements inclusive, safe, resilient, and sustainable.
- Ensure sustainable consumption and production patterns.
- Take urgent action to combat climate change and its impacts.
- Conserve and use the oceans, seas, and marine resources for sustainable development.
- Protect, restore, and promote sustainable use of terrestrial ecosystems, sustainably manage forests, combat desertification, halt and reverse land degradation, and halt biodiversity loss.
- Promote peaceful and inclusive societies for sustainable development, provide access to justice for all, and build effective, accountable, and inclusive institutions at all levels (United Nations Development Programme, 2016)*

> **Gini index:** A commonly used measure of income inequality, which measures inequality on a scale of 0 to 1; a Gini index of 0 represents exact equality (i.e., every person in a certain society or nation has the same amount of income), while a Gini index of 1 represents total inequality (i.e., one person has all the income and the rest of the society has none).

*United Nations Development Programme (2016), "Background on the Goals," http://www.undp.org/content/undp/en/home/sustainable-development-goals/background.html.

AGENT OF CHANGE

Craig and Marc Kielburger

Craig Kielburger and Marc Kielburger

Over 20 years ago, at the age of 12, Craig Kielburger spoke out against child labour and the WE movement began. With his brother Marc, Craig founded the charity Free the Children (rebranded in July 2016 as the WE Charity). Since its creation, the WE Charity has grown into three organizations: the WE charity itself, the youth empowerment movement called WE Day, and the social enterprise called ME to WE. Working to empower people to change the world locally, globally, and through our consumer choices, the WE movement is founded on five beliefs: "Me into WE" illustrates how much stronger people are together than they are on their own; "WE is everyone" indicates that anyone can make a difference and everyone can be a part of the collective through individual action in their everyday lives; "WE are the change" reflects the tidal wave impact that takes place when our individual efforts inspire others to take action; "WE are a global community" in that we are connected at all levels to our friends, families, communities, and the world; and "I am WE," illustrating how our lives take on new meaning when we positively impact the lives of others around the world (WE Movement, 2016). The ME to WE social enterprise encourages a

shift from "me" thinking to "we" acting—inspiring people to become leaders and agents of change through their everyday experiences and purchases. It offers volunteer trips that allow young people and adults to participate in international health, education, and economic development, as well as ME to WE products for purchase that empower people and transform lives. As an annual series of events, WE Day brings together thousands of people to celebrate youth and the difference they are making in their local and global communities. WE Day features notable speakers and performers, and rather than purchase tickets, youth can earn their ticket through service in a local or global cause (WE Movement, 2016). WE Day is part of the year-long initiative, WE Schools, which provides educators and students with curriculum and educational resources to learn about the root causes of hunger, poverty, and lack of education, and participate in action campaigns to learn what they can do to help (WE Movement, 2016). Another component of the WE Charity is WE Villages, international development projects that partner with community leaders to improve education in communities abroad. WE Villages works build schools, but to build water wells near schools so that children do not miss school as a result of having to travel long distances to collect water (WE Movement, 2016). WE Villages works to improve health care programming and provide opportunities for alternative income and livelihood programs to empower mothers with financial independence (WE Movement, 2016). The Kielburger brothers are unique in that they challenge the ageism that exists within social development by recognizing the potential of young people to become agents of social change. One of the core values of this movement is that youth are "shameless idealists" who will stop at nothing to change the world, and it is this belief in change that is inspiring so many people around the world. To learn more about the WE Movement, visit https://www.we.org/

Source: Kielburger, Craig, & Kielburger, Marc. (2016). Craig and Marc Kielburger. Retrieved from MetoWe: https://www.metowe.com/craig-and-marc-kielburger/

INTERSECTIONALITY

Gender and Social Inequality

In 2010, 9.3 percent of females in Canada—over 1.5 million in total—were living on a low income (Statistics Canada, 2012). Women experience poverty in

greater numbers than men for many reasons, but two of the main reasons are related to work—both paid and unpaid.

First, women spend more time doing unpaid work such as child care and domestic labour, which leaves them less time to engage in the paid work force. Women are still expected to perform the majority

Mike Windle/Getty

If If a picture can say a thousand words, imagine the stories your shoes could tell! Try this student story on for size – have you walked in this student's shoes?

DEAR PUBLIC AUTHORITY:

Recently, we had a public meeting where you wanted to hear from different sectors of the community about poverty, the obstacles faced, and how we could deal with them. There were many people from different social agencies speaking on behalf of others, but you also wanted to hear from a single parent. I was chosen as that single parent. I apologize for having to bring my son with me, but this was first-hand experience for you on how difficult it is to find child care, especially on the spur of the moment (as was my case). I explained that as a single mom on Ontario Works, it wasn't often that I could do things on my own. Child care is expensive and a luxury I could not afford.

I believe that when I started to speak, you were shocked because you had already judged me as being lazy and looking for a handout. I also believe that you were shocked because my son sat quietly in the corner for over an hour. I would like to explain to you that not all mothers on social assistance are drug addicts

or addicts of any sort and that even though I don't work, I love my son very much and I have raised him well. I would also like you to know that I am highly intelligent, have passed on that intelligence to my son, and that I have decided to further my education in the social services field.

I want you to know that when I am actively employed in the field, I will take the lesson you taught me to every job. That lesson is that you don't prejudge someone just by the label that is placed on them. I may be a single mom, but I am every bit as ambitious and capable as you. I hope that with our meeting, you too learned a valuable lesson. Most people in my position are only asking for a hand up, not a handout.

I wish you luck as you move through the different ministries and I hope that together we can make a difference.

Yours truly,
A Single Mom

Source: Patti Pringle.

of household chores and child care. Results of the General Social Survey of 2010 indicate the stark difference between the time that men and women spend on these responsibilities. When respondents were asked to report the number of hours spent on unpaid child care in the household, women generally reported a higher number of hours per week than men: men reported spending, on average, 8.3 hours on unpaid domestic work, while women spent more than one and a half times this amount—13.8 hours (Milan, Keown, & Covadonga, 2012).

Women also face a wage gap. In 2011, women's average earnings ($32 100) were only 66.7 percent of their male counterparts' earnings ($48 100) (Statistics Canada, 2013). Since they do not earn as much as men, women are at risk of falling into poverty, especially if they have children and then find themselves single through separation, divorce, or widowhood. Women are also often involved in part-time or precarious employment. They account for 70 percent of part-time employees and two-thirds of the Canadians who work for minimum wage. Even though there are

more working mothers now than at any other time in history, 36 percent of mother-led families still live below the poverty line and 43 percent of children who live in a low-income family live with a single mother (YWCA, 2009). In the In Their Shoes feature, a single mother discusses her experience with poverty.

Race and Social Inequality

Recent immigrants are increasingly expanding the number of individuals who belong to the ranks of the hidden homeless. The increasing cost of rent, discrimination in the workplace, and language barriers all contribute to the fact that many newcomers who enter Canada live in shared, overcrowded housing, often for extended periods. In Toronto, where the average income is $69 000 and the monthly rent on a bachelor apartment is close to $800, the income of most newcomers is under $20 000. Most newcomers spend more than 50 percent of their income on housing, with 15 percent spending 75 percent or more (Preston et al., 2011).

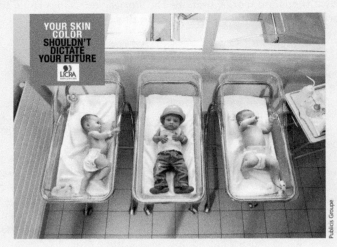

In 2010, Ligue Internationale Contre le Racisme et l'Antisémitisme (LICRA) or International League Against Racism and Antisemitism created a campaign titled "Your Skin Colour Shouldn't Dictate Your Future" to illustrate the intersectionality of social inequality and race. What do you think? In what ways does race influence socio-economic status?

The Canadian Indigenous population is also overrepresented among the homeless in Canada. They make up less than 4 percent of Canada's total population but represent 10 percent of the homeless (Sider, 2005). Indigenous people living in Vancouver represent about 2 percent of the city's people, but 30 percent of the homeless population (Native Women's Association, 2007). A history of disenfranchisement, oppression, and colonization has led to disparities in healthcare, income, and education among the Indigenous population in Canada, and this has no doubt contributed to the increased numbers in homelessness (United Native Nations Society, 2001).

Food shortages and increasing cost of food necessities in northern parts of Canada, like Nunavut, are only adding to inequalities faced by Indigenous populations. Advocacy groups like The Nunavut Food Security Coalition and Feeding Nunavut report that nearly 70 percent of Inuit homes in Nunavut are food insecure, and food staples can cost up to three times as much in these communities as they do in the rest of the country (Newman, 2015; Nunavut Food Security Coalition, 2016). How can we tackle issues of poverty in Inuit communities in Nunavut when a case of bottled water costs over $100 and a head of cabbage costs $28.54? (BBC News, 14 June 2012, http://www.bbc.com/news/world-us-canada-18413043)

Age and Social Inequality

Almost a third of Canada's homeless are youths aged 16–24 (CMHA, 2009). Public Health Canada (Government of Canada, 2012) conducted a major longitudinal study (1993–2003) on homeless youth in Canada that likely provides the most comprehensive picture of that portion of the Canadian homeless population. Highlights from this study of some 5000 homeless youths in seven urban centres found the following information:

- Most homeless youth had histories of family poverty and unstable housing.
- Fifteen percent came from families that had been homeless at some point.
- Close to half had been in foster care (45 percent), and almost 50 percent had lived in a group home at some point.
- Many had experienced violence at home (Eva's Initiatives, 2012).

Recent work shows that street youth are exposed to a number of factors that may detrimentally affect their health, including unsafe sexual practices, drug use, poor diet, inadequate shelter, exposure to violence, low levels of social support, and limited access to medical care (Boivin, Roy, Haley, & du Fort, 2005).

ENDING THOUGHTS

When you think about social inequality, it is important to remember the distinction between various types of poverty, including absolute, relative, transitional, marginal, chronic, and intergenerational poverty, and the perceptions that surround each one. The stereotypes that surround those who live in poverty lead to the misconception that poverty and homelessness are choices. We are led to believe that there are enough shelter beds to sleep in that alleviate the need to sleep on the street, that there are enough food banks to eat from and there is no need to starve, that people go hungry because they cannot or do not budget their money properly—but that is not the reality. A study conducted in 2002 defined panhandlers as people who were soliciting donations of money for personal use from passersby without providing any goods or services in return, and the authors found that 70 percent of the panhandlers in Toronto would prefer a minimum-wage job in order to have a steady income or to get off the street (Bose & Hwang, 2002). The average monthly income that a panhandler received on the street was $300; the average total monthly income was just over $600—including loans from friends and family.

The reality is that no one (or very few people) chooses a life of poverty—no one chooses mental illness, domestic violence, or addiction. Would you choose to live on the streets or to live on social assistance and bear the social stigma that is attached to it? The next time you see a street person asking for a donation, think about the others you cannot see. And then think about the millions who exist beyond the streets you live in. Remember the differences between various types of poverty and consider that everyone has a story.

READING

THE SINGER SOLUTION TO WORLD POVERTY

By Peter Singer

In the Brazilian film "Central Station," Dora is a retired schoolteacher who makes ends meet by sitting at the station writing letters for illiterate people. Suddenly she has an opportunity to pocket $1,000. All she has to do is persuade a homeless 9-year-old boy to follow her to an address she has been given. (She is told he will be adopted by wealthy foreigners.) She delivers the boy, gets the money, spends some of it on a television set and settles down to enjoy her new acquisition. Her neighbor spoils the fun, however, by telling her that the boy was too old to be adopted—he will be killed and his organs sold for transplantation. Perhaps Dora knew this all along, but after her neighbor's plain speaking, she spends a troubled night. In the morning Dora resolves to take the boy back.

Suppose Dora had told her neighbor that it is a tough world, other people have nice new TV's too, and if selling the kid is the only way she can get one, well, he was only a street kid. She would then have become, in the eyes of the audience, a monster. She redeems herself only by being prepared to bear considerable risks to save the boy.

At the end of the movie, in cinemas in the affluent nations of the world, people who would have been quick to condemn Dora if she had not rescued the boy go home to places far more comfortable than her apartment. In fact, the average family in the United States spends almost one-third of its income on things that are no more necessary to them than Dora's new TV was to her. Going out to nice restaurants, buying new clothes because the old ones are no longer stylish, vacationing at beach resorts—so much of our income is spent on things not essential to the preservation of our lives and health. Donated to one of a number of charitable agencies, that money could mean the difference between life and death for children in need.

All of which raises a question: In the end, what is the ethical distinction between a Brazilian who sells a homeless child to organ peddlers and an American who already has a TV and upgrades to a better one—knowing that the money could be donated to an organization that would use it to save the lives of kids in need?

Of course, there are several differences between the two situations that could support different moral judgments about them. For one thing, to be able to consign a child to death when he is standing right in front of you takes a chilling kind of heartlessness; it is much easier to ignore an appeal for money to help children you will never meet. Yet for a utilitarian philosopher like myself—that is, one who judges whether acts are right or wrong by their consequences—if the upshot of the American's failure to donate the money is that one more kid dies on the streets of a Brazilian city, then it is, in some sense, just as bad as selling the kid to the organ peddlers. But one doesn't need to embrace my utilitarian ethic to see that, at the very least, there is a troubling incongruity in being so quick to condemn Dora for taking the child to the organ peddlers while, at the same time, not regarding the American consumer's behavior as raising a serious moral issue.

In his 1996 book, *Living High and Letting Die*, the New York University philosopher Peter Unger presented an ingenious series of imaginary examples designed to probe our intuitions about whether it is wrong to live well without giving substantial amounts of money to help people who are hungry, malnourished or dying from easily treatable illnesses like diarrhea. Here's my paraphrase of one of these examples:

Bob is close to retirement. He has invested most of his savings in a very rare and valuable old car, a Bugatti, which he has not been able to insure. The Bugatti is his pride and joy. In addition to the pleasure he gets from driving and caring for his car, Bob knows that its rising market value means that he will always be able to sell it and live comfortably after retirement. One day when Bob is out for a drive, he parks the Bugatti near the end of a railway siding and goes for a walk up the track. As he does so, he sees that a runaway train, with no one aboard, is running down the railway track. Looking farther down the track, he sees the small figure of a child very likely to be killed by the runaway train. He can't stop the train and the child is too far away to warn of the danger, but he can throw a switch that will divert the train down the siding where his Bugatti is parked. Then nobody will be killed—but the train will destroy his Bugatti. Thinking of his joy in owning the car and the financial security it represents, Bob decides not to throw the switch. The child is killed. For many years to come, Bob enjoys owning his Bugatti and the financial security it represents.

Bob's conduct, most of us will immediately respond, was gravely wrong. Unger agrees. But then he reminds us that we, too, have opportunities to save the lives of children. We can give to organizations like UNICEF or Oxfam America. How much would we have to give one of these organizations to have a high probability of saving the life of a child threatened by easily preventable diseases? (I do not believe that children are more worth saving than adults, but since no one can argue that children have brought their poverty on themselves, focusing on them simplifies the issues.) Unger called up some experts and used the information they provided to offer some plausible estimates that include the cost of raising money, administrative expenses and the cost of delivering aid where it is most needed. By his calculation, $200 in donations would help a sickly 2-year-old transform into a healthy 6-year-old—offering safe passage through childhood's most dangerous years. To show how practical philosophical argument can be, Unger even tells his readers that they can easily donate funds by using their credit card and calling one of these toll-free numbers: (800) 367-5437 for UNICEF; (800) 693-2687 for Oxfam America. [http://supportUNICEF.org/forms/whichcountry2.html for UNICEF and http://www.oxfam.org/eng/donate.htm for Oxfam—PS]

Now you, too, have the information you need to save a child's life. How should you judge yourself if you don't do it? Think again about Bob and his Bugatti. Unlike Dora, Bob did not have to look into the eyes of the child he was sacrificing for his own material comfort. The child was a complete stranger to him and too far away to relate to in an intimate, personal way. Unlike Dora, too, he did not mislead the child or initiate the chain of events imperiling him. In all these respects, Bob's situation resembles that of people able but unwilling to donate to overseas aid and differs from Dora's situation.

If you still think that it was very wrong of Bob not to throw the switch that would have diverted the train and saved the child's life, then it is hard to see how you could deny that it is also very wrong not to send money to one of the organizations listed above. Unless, that is, there is some morally important difference between the two situations that I have overlooked.

Is it the practical uncertainties about whether aid will really reach the people who need it? Nobody who knows the world of overseas aid can doubt that such uncertainties exist. But Unger's figure of $200 to save a child's life was reached after he had made conservative assumptions about the proportion of the money donated that will actually reach its target.

One genuine difference between Bob and those who can afford to donate to overseas aid organizations but don't is that only Bob can save the child on the tracks, whereas there are hundreds of millions of people who can give $200 to overseas aid organizations. The problem is that most of them aren't doing it. Does this mean that it is all right for you not to do it?

Suppose that there were more owners of priceless vintage cars—Carol, Dave, Emma, Fred and so on, down to Ziggy—all in exactly the same situation as Bob, with their own siding and their own switch, all sacrificing the child in order to preserve their own cherished car. Would that make it all right for Bob to do the same? To answer this question affirmatively is to endorse follow-the-crowd ethics—the kind of ethics that led many Germans to look away when the Nazi atrocities were being committed. We do not excuse them because others were behaving no better.

We seem to lack a sound basis for drawing a clear moral line between Bob's situation and that of any reader of this article with $200 to spare who does not donate it to an overseas aid agency. These readers seem to be acting at least as badly as Bob was acting when he chose to let the runaway train hurtle toward the unsuspecting child. In the light of this conclusion, I trust that many readers will reach for the phone and donate that $200. Perhaps you should do it before reading further.

Now that you have distinguished yourself morally from people who put their vintage cars ahead of a child's life, how about treating yourself and your partner to dinner at your favorite restaurant? But wait. The money you will spend at the restaurant could also help save the lives of children overseas! True, you weren't planning to blow $200 tonight, but if you were to give up dining out just for one month, you would easily save that amount. And what is one month's dining out, compared to a child's life? There's the rub. Since there are a lot of desperately needy children in the world, there will always be another child whose life you could save for another $200. Are you therefore obliged to keep giving until you have nothing left? At what point can you stop?

Hypothetical examples can easily become farcical. Consider Bob. How far past losing the Bugatti should he go? Imagine that Bob had got his foot stuck in the track of the siding, and if he diverted the train, then before it rammed the car it would also amputate his big toe. Should he still throw the switch? What if it would amputate his foot? His entire leg?

As absurd as the Bugatti scenario gets when pushed to extremes, the point it raises is a serious one: only when the sacrifices become very significant indeed would most people be prepared to say that Bob does nothing wrong when he decides not to throw the switch. Of course, most people could be wrong; we can't decide moral issues by taking opinion polls. But consider for yourself the level of sacrifice that you would demand of Bob, and then think about how much money you would have to give away in order to make a sacrifice that is roughly equal to that. It's almost certainly much, much more than $200. For most middle-class Americans, it could easily be more like $200 000.

Isn't it counterproductive to ask people to do so much? Don't we run the risk that many will shrug their shoulders and say that morality, so conceived, is fine for saints but not for them? I accept that we are unlikely to see, in the near or even medium-term future, a world in which it is normal for wealthy Americans to give the bulk of their wealth to strangers. When it comes to praising or blaming people for what they do, we tend to use a standard that is relative to some conception of normal behavior. Comfortably off Americans who give, say, 10 percent of their income to overseas aid organizations are so far ahead of most of their equally comfortable fellow citizens that I wouldn't go out of my way to chastise them for not doing more. Nevertheless, they should be doing much more, and they are in no position to criticize Bob for failing to make the much greater sacrifice of his Bugatti.

At this point various objections may crop up. Someone may say: "If every citizen living in the affluent nations contributed his or her share I wouldn't have to make such a drastic sacrifice, because long before such levels were reached, the resources would have been there to save the lives of all those children dying from lack of food or medical care. So why should I give more than my fair share?" Another, related, objection is that the Government ought to increase its overseas aid allocations, since that would spread the burden more equitably across all taxpayers.

Yet the question of how much we ought to give is a matter to be decided in the real world—and that, sadly, is a world in which we know that most people do not, and in the immediate future will not, give substantial amounts to overseas aid agencies. We know, too, that at least in the next year, the United States Government is not going to meet even the very modest United Nations-recommended target of 0.7 percent of gross national product; at the moment it lags far below that, at 0.09 percent, not even half of Japan's 0.22 percent or a tenth of Denmark's 0.97 percent. Thus, we know that the money we can give beyond that theoretical "fair share" is still going to save lives that would otherwise be lost. While the idea that no one need do more than his or her fair share is a powerful one, should it prevail if we know that others are not doing their fair share and that children will die preventable deaths unless we do more than our fair share? That would be taking fairness too far.

Thus, this ground for limiting how much we ought to give also fails. In the world as it is now, I can see no escape from the conclusion that each one of us with wealth surplus to his or her essential needs should be giving most of it to help people suffering from poverty so dire as to be life-threatening. That's right: I'm saying that you shouldn't buy that new car, take that cruise, redecorate the house or get that pricey new suit. After all, a $1000 suit could save five children's lives.

So how does my philosophy break down in dollars and cents? An American household with an income of $50 000 spends around $30 000 annually on necessities, according to the Conference Board, a non-profit economic research organization. Therefore, for a household bringing in $50 000 a year, donations to help the world's poor should be as close as possible to $20 000. The $30 000 required for necessities holds for higher incomes as well. So a household making $100 000 could cut a yearly check for $70 000. Again, the formula is simple: whatever money you're spending on luxuries, not necessities, should be given away.

Now, evolutionary psychologists tell us that human nature just isn't sufficiently altruistic to make it plausible that many people will sacrifice so much for strangers. On the facts of human nature, they might be right, but they would be wrong to draw a moral conclusion from those facts. If it is the case that we ought to do things that, predictably, most of us won't do, then let's face that fact head-on. Then, if we value the life of a child more than going to fancy restaurants, the next time we dine out we will know that we could have done something better with our money. If that makes living a morally decent life extremely arduous, well, then that is the way things are. If we don't do it, then we should at least know that we are failing to live a morally decent life—not because it is good to wallow in guilt but because knowing where we should be going is the first step toward heading in that direction.

When Bob first grasped the dilemma that faced him as he stood by that railway switch, he must have thought how extraordinarily unlucky he was to be placed in a situation in which he must choose between the life of an innocent child and the sacrifice of most of his savings. But he was not unlucky at all. We are all in that situation.

Source: Peter Singer, "The Singer Solution to World Poverty," (5 September 1999).

1. What would happen if everyone took Singer's advice and donated all of their "extra" money to charitable organizations? Would this indeed create more equality among nations? What would happen to the industries that produced the luxury items that Singer suggests that individuals abandon?

2. Like many of the advertisements that we see on television, Singer suggests that individuals have a moral obligation to assist others who are less fortunate. When you compare this strategy of fundraising to different approaches that feature philanthropic efforts, like large donations made by Bill Gates and Warren Buffet or the high-profile activities of celebrities like Angelina Jolie and Brad Pitt, which initiative is more effective for you? What are the pros and cons of each approach and which strategy is likely to raise more money? Why?

3. Imagine that you had to debate the merit of Singer's ideas. How would you convince the opposing team that his solution to world poverty was sound?

KWIP

KNOW IT AND OWN IT: WHAT DO I BRING TO THIS?

The "K" in the KWIP process involves examining aspects of your own identity and social location as the first step in becoming diversity competent.

ACTIVITY: JOURNAL

Where do you fit within our socially stratified society? Do you have a firm understanding of the differences between absolute and relative poverty, as they relate to your own life? Knowing your own situation as it relates to issues of social inequality will bring a greater understanding as you consider the following:

It's your first time away from home and your life could not be better. You love your classes, your teachers are cool, and best of all, you've met a partner who you've fallen madly in love with—all in the first term! Yikes. You're from a middle class family with strong family values. Your dad works really hard and your mom does, too, since you've got four brothers and sisters—not counting the child that you sponsor. Your new partner also has four siblings; however his/her family live in the upper, upper class in society. You can't wait to meet them. They're all going away for Thanksgiving, but you'll be spending some time together at Christmas. You and your partner have decided to put a $25 limit on the amount you'll spend on gifts for each other. As a special surprise, you knit your partner a scarf. You have a lovely time meeting his/her parents, and have made them a beautiful wreath for their door. Much to your surprise, they have gifts for you too! Lots of gifts. Very expensive gifts—with labels like Lauren, Yeazy, and Apple. You're sure that your parents would not approve if you accepted the gifts.

Do you accept all of the gifts, none of the gifts or maybe just one or two?

Does your partner's parents' generous offer affect your feelings toward your partner at all?

Do you explain your own values to your partner's parents? In the end, what do your actions tell you about yourself and your new relationships?

WALKING THE TALK: HOW CAN I LEARN FROM THIS?

The "W" in the KWIP process presents a scenario or case study that challenges you to "walk the talk" through problem-based learning.

You and your friend have just seen your favourite band in concert. (Who was the band?) If you have ever been to a concert, you are aware of all the swag that is available for purchase. You see a really cool T-shirt that you want on your way into the concert, but you notice from the label that it was made by a company that has been in the news lately for exploiting child labour, as well as for health and safety issues that are threatening the lives of workers. The T-shirt was only $15. Inside the venue, you see the same T-shirt, with a different label inside, but the cost is $40. In a business course, you had learned that the company selling this T-shirt uses a business practice called "ethical supply chain management," which includes ethical sourcing of products from vendors where there is no forced or compulsory labour, fair wages, no child labour, humane treatment of animals, good health and safety standards, and environmental protection standards.

1. Which T-shirt do you purchase? You have to buy one of them, because you always buy a T-shirt from every concert you go to. Do you buy the T-shirt that is more affordable to you? Or do you purchase the $40 T-shirt because the company that produces them sources its products from socially responsible supply chains?

2. Does ethical consumerism help to alleviate poverty?

It IS WHAT IT IS: IS THIS INSIDE OR OUTSIDE MY COMFORT ZONE?

The "I" in the KWIP process requires you to honestly confront and identify ways in which our complex identities result in experiences of privilege and oppression, and to reflect on how we can learn to honour that privilege.

Social inequality is never fair, and it inevitably stems from the differences in power and privilege between groups and/or individuals in society. Perhaps you are hungry, but you have never had the nerve to ask for help because you are too afraid or embarrassed. Conversely, your parents might have thrown a huge party where there was a lot of leftover food that was tossed out, when it was still in containers and perfectly fine to donate to a shelter. Privilege comes in all forms and sizes, and in this exercise, we ask you to examine yourself in terms of privilege and power.

ASKING: Do I have privilege?

1. I am reasonably sure that I will have enough money to purchase the textbooks I need for my classes this term.

2. I will likely not have to skip meals or use the student emergency food bank to be able to eat each day.

3. I will be able to pay for a postsecondary education.

4. When I graduate from postsecondary, I will not have overwhelming debt from loans that I have to pay back.

5. I don't have to rely on public transportation to travel to school. I can afford my own vehicle.

REFLECTING: Honouring Our Privilege

In relation to social inequality, describe two traits that you possess that are disempowering to you or disadvantage you. Then describe two traits that privilege you over someone else. How might you use this social privilege to address the oppression of others?

Put IT IN PLAY: HOW CAN I USE THIS?

The "P" in the KWIP process involves examining how others are practising equity and how you might use this.

Kindness Meters are an example of a poverty initiative that has been put into play in a number of Canadian cities. Cities have revamped old parking meters by painting them bright colours and turning them into kindness meters where the money gets donated to shelters for the homeless. The strategy is to place these kindness meters in high traffic areas where there are a large number of people and panhandlers. If people want to donate but do not feel comfortable directly handing money to panhandlers, they can do so and the money will be distributed to local charities that then help those panhandlers to improve their circumstances. Do you think this is an effective poverty reduction strategy? Discuss the positive outcomes and problems associated with their use.

REFERENCES

Berreman, G. (1972). Race, caste, and other invidious distinctions in social stratification. *Race and Class*, 385–414.

Bielski, Z. (2010, April 13). Living on a welfare diet. Retrieved from *The Globe and Mail*: http://www.theglobeandmail.com/life/living-on-a-welfare-diet/article4314633/

Boivin, J., Roy, E., Haley, N., & du Fort, G. G. (2005). The health of street youth: A Canadian perspective. *Canadian Journal of Public Health*, 432–437.

Bose, R., & Hwang, S. (2002). Income and spending patterns among panhandlers. *Canadian Medical Association Journal*, 477–479.

Bramham, D. (2008, April 19). Homeless crisis grows while Canada prospers; the economy is strong, provinces run budget surpluses, yet we turn our backs on the destitute. *The Vancouver Sun*, A4.

Canadian Federation of Students (CFS). (2016). Public education for the public good. Retrieved on November 14, 2016, from Canadian Federation of Students: http://dev.cfswpnetwork.ca/wp-content/uploads/sites/71/2016/01/CFS-Public-Education.pdf

Canadian Mental Health Association (CMHA). (2016). Homelessness. Retrieved on November 2, 2016, from *CMHA*: http://www.cmha.ca/public-policy/subject/homelessness/

Casavant, L. (1999, January 1). Composition of the homeless population. Retrieved from *Parliamentary Research Branch*: http://publications.gc.ca/Collection-R/LoPBdP/modules/prb99-1-homelessness/composition-e.htm

CMHA. (2009, January 1). Out of the shadows forever: Annual Report 2008–2009. Retrieved from *Canadian Mental Health Association*: http://www.cmha.ca/public_policy/out-of-the -shadows-forever-annual-report-2008 -2009/#.T-smZMUfh40

Condon, M., & Newton, R. (2007). In the proper hands: SPARC BC research on homelessness and affordable housing. Vancouver: Social Planning and Research Council of British Columbia.

Conference Board of Canada. (2011, July 1). Canadian income inequality. Retrieved from *Conference Board of Canada*: http://www.conferenceboard.ca/hcp/hot-topics/canInequality.aspx

Crowe, C. (2012, June 8). Toronto Disaster Relief Committee: 14 years of advocacy, activism and action. Retrieved from *Toronto Disaster Relief Committee*: http://tdrc.net/uploads/tdrc14years.pdf

Davidson, S. (2015). Report reveals average rent prices across Canada. Retrieved on October 11, 2016, from *CTV News*: http://www.ctvnews.ca/canada/report-reveals-average-rent-prices-across-canada-1.2704967

Echenberg, H., & Jensen, H. (2012, May 7). Defining and enumerating homelessness in Canada. Retrieved from *Library of Parliament Research Publications*: http://www.parl.gc.ca/Content/LOP/ResearchPublications/prb0830 -e.htm

Employment and Social Development Canada. (2013, August 9). Poverty profile. Retrieved from *Employment and Social Development Canada*: http://www.hrsdc.gc.ca/eng/communities/reports/poverty_profile/2007.shtml

Evans, G., & Schamberg, M. (2009). Childhood poverty, chronic stress, and adult working memory. Proceedings of the National Academy of Sciences, Vol. 106, 13.

Eva's Initiatives. (2012, January 1). National profile – Homeless youth: Who are they and why are they on the street? Retrieved from *Eva's Phoenix Toolkit*: http://phoenixtoolkit.evasinitiatives.com/homeless-youth-background/national-profile/

Family Service Toronto. (2010, January 1). 2010 report card on child and family poverty in Canada. Retrieved from *Campaign 2000*: http://www.campaign2000.ca/reportCards/provincial/Ontario/2010OntarioReportCardEnglish.pdf

Family Service Toronto. (2011, January 1). 2011 report card on child and family poverty in Canada. Retrieved from *YWCA*: http://ywcacanada.ca/data/research_docs/00000223.pdf

Girard, M.-C. (2006). Determining the extent of the problem: The value and challenges of enumeration. *Canadian Review of Social Policy*, 104.

Government of Canada. (2012, February 2). Street youth in Canada. Retrieved from *Government of Canada Publications*: http://publications.gc.ca/collections/Collection/HP5-15-2006E.pdf

Hennessy, T. (2012, January 1). The clash for the cash: CEO vs. average Joe. Retrieved from *Canadian Centre for Policy Alternatives*: http://www.policyalternatives.ca/ceo

Hulchanski, D. J. (2009, February 18). Homelessness in Canada: Past, present, future. Retrieved from *Growing Home: Housing and Homelessness in Canada*: http://www.cprn.org/documents/51110_EN.pdf

Hunger Count. (2015). Food banks Canada. Retrieved on October 11, 2016, from *Hunger Count 2015*: https://www.foodbankscanada.ca/FoodBanks/MediaLibrary/HungerCount/HungerCount2015_singles.pdf

Hwang, S. (2010, November 1). Housing vulnerability and health: Canada's hidden emergency. Retrieved from *St. Michael's*: http://www.stmichaelshospital.com/crich/housing-vulnerability-and-health.php

Interfaith Social Assistance Reform Coalition. (1998). *Our neighbours' voices: Will we listen?* Toronto: James Lorimer and Company.

Lewis, O. (1998). The culture of poverty. *Society*, 7–9.

Mental Health Commission of Canada (MHCC). (2013). Turning the key–Assessing housing and related supports for persons living with mental health problems and illness. Retrieved on November 2, 2016, from MHCC: http://www.mentalhealthcommission.ca/sites/default/files/PrimaryCare_Turning_the_Key_Full_ENG_0_1.pdf

Mental Health Commission of Canada (MHCC). (2014). National final report: Cross-site at home/chez soi project. Retrieved on November 2, 2016, from MHCC: http://www.mentalhealthcommission.ca/sites/default/files/mhcc_at_home_report_national_cross-site_eng_2_0.pdf

Milan, A., Keown, L., & Covadonga, R. U. (2012, February 4). Families, living arrangements and unpaid work. Retrieved from *Statistics Canada*: http://www.statcan.gc.ca/pub/89-503-x/2010001/article/11546-eng.htm#a12

Mojtehedzadeh, S. (2015). Toronto is now Canada's inequality capital, United Way study shows. Retrieved on October 12, 2016, from *The Star*: https://www.thestar.com/news/gta/2015/02/27/toronto-now-canadas-inequality-capital-united-way-study-shows.html

Munn-Rivard, L. (2014). Current issues in mental health in Canada: Homelessness and access to housing. Retrieved on November 2, 2016, from *Parliament of Canada*: http://www.lop.parl.gc.ca/content/lop/ResearchPublications/2014-11-e.htm

Native Women's Association (2007, June 22). Aboriginal women and homelessness. Retrieved from *Newfoundland Labrador: Labrador Affairs Office*: http://www.laa.gov.nl.ca/laa/naws/pdf/nwac-homelessness.pdf

Newman, S. (2015). Food insecurity in Nunavut. Retrieved on November 12, 2016, from *Feeding Nunavut*: http://www.feedingnunavut.com/food-insecurity-in-nunavut/

Nunavut Food Security Coalition. (2016). Food security rates. Retrieved on November 12, 2016, from *Nunavut Food Security Coalition*: http://www.nunavutfoodsecurity.ca/Rates

OXFAM. (2015). Richest 1% will own more than all the rest by 2016. Retrieved on October 12, 2016, from OXFAM: https://www.oxfam.org/en/pressroom/pressreleases/2015-01-19/richest-1-will-own-more-all-rest-2016

Poverty and Employment Precarity in Southern Ontario (PEPSO). (2015). The precarity penalty. Retrieved on November 16, 2016, from *The United Way*: http://www.unitedwaytyr.com/document.doc?id=307

Preston, V., Murdie, R., D'Addario, S., Sibbanda, P., Murnaghan, A., Logan, J., et al. (2011, December 1). Precarious housing and hidden homelessness among refugees, asylum seekers, and immigrants in the Toronto Metropolitan Area. Retrieved from *CERIS—The Ontario Metropolis Centre*: http://www.ceris.metropolis.net/wp-content/uploads/pdf/research_publication/working_papers/wp87.pdf

Rossides, D. (1997). *Social stratification: The interplay of class, race, and gender*. Upper Saddle River, NJ: Prentice Hall.

Sagan, A. (2016). Precarious work in Canada now a white-collar problem. Retrieved on November 16, 2016, from *The Huffington Post*: http://www.huffingtonpost.ca/2016/03/28/librarians-fight-precarious-work-s-creep-into-white-collar-jobs_n_9553272.html

Saskatoon Poverty Reduction Partnership (SPRP). (2011). From poverty to possibility…to prosperity – Saskatoon, its people and poverty. Retrieved on September 16, 2016, from Vibrant Communities Canada: http://vibrantcanada.ca/files/saskatoon_its_people_and_poverty_aug_20111.pdf

Shapcott, M. (2010, November 24). More than three million Canadians forced to live in poverty: 2010 child and family poverty report card. Retrieved from *Wellesley Institute*: http://www.wellesleyinstitute.com/news/more-than-three-million-canadians-forced-to-live-in-poverty-2010-child-and-family-poverty-report-card/

Sider, D. (2005, May 1). A sociological analysis of root causes of homelessness in Sioux Lookout. Retrieved from *Canadian Race Relations Foundation*: http://www.crr.ca/divers-files/en/pub/rep/ePubRepSioLoo.pdf

Singer, P. (1999, September 5). The Singer solution to world poverty. *The New York Times Magazine*, 60–63.

Statistics Canada. (2012, June 15). Persons in low-income families. Retrieved from Table 202-0802, *Statistics Canada*: http://www5.statcan.gc.ca/cansim/a21

Statistics Canada. (2013, June 27). Average female and male earnings, and female-to-male earnings ratio, by work activity, 2010 constant dollars. Retrieved from *Statistics Canada*: http://www5.statcan.gc.ca/cansim/a05?lang=eng&id=2020102

Statistics Canada. (2014). Distribution and median net worth by quintile. Retrieved on October 11, 2016, from *The Daily*: http://www.statcan.gc.ca/daily-quotidien/140225/t140225b003-eng.htm

Statistics Canada. (2015a, September 9). University tuition fees 2015/2016. Retrieved on October 11, 2016, from *The Daily*: http://www.statcan.gc.ca/daily-quotidien/150909/dq150909b-eng.pdf

Statistics Canada. (2015b). Low-Income Lines, 2013–2014. Retrieved on October 11, 2016, from Statistics Canada: http://www.statcan.gc.ca/pub/75f0002m/75f0002m2015001-eng.pdf

UNICEF Canada. (2016). 2016 UNICEF report on child well-being. Retrieved from *UNICEF Report Card 13: Fairness For Children*: http://www.unicef.ca/sites/default/files/legacy/imce_uploads/images/advocacy/rc/irc13_canadian_companion_en_sp_new_.pdf

United Nations Development Program (UNDP). (2016). Sustainable development goals. Retrieved from *UNDP*: http://www.undp.org/content/undp/en/home/sustainable-development-goals.html

United Native Nations Society. (2001, June 1). Aboriginal homelessness in British Columbia. Retrieved from *Homeless Hub*: http://www.homelesshub.ca/%28S%280oinij454xdtat555ztjoheb%29%29/Resource/Frame.aspx?url=http%3a%2f%2fwww.urbancenter.utoronto.ca%2fpdfs%2felibrary%2fUNNS_Aboriginal_Homelessn.pdf&id=36039&title=Aboriginal+Homelessness+in+British+Columbia&owner=121

United Way. (2015). The opportunity equation: Building opportunity in the face of income inequality. Retrieved on October 12, 2016, from *The United Way*: http://www.unitedwaytyr.com/document.doc?id=285

WE Movement. (2016). Our beliefs. Retrieved on November 11, 2016, from WE: https://www.we.org/we-movement/our-beliefs/

Wong, J. (2009, October 1). Guy Laliberté; Cirque du Soleil founder: From busker to spaceman. Retrieved from *CBC News*: http://www.cbc.ca/news/arts/theatre/story/2009/10/01/f-guy-laliberte-backgrounder.html

World Hunger. (2015). 2015 World hunger and poverty facts and statistics. Retrieved from *World Hunger Education Service*: http://www.worldhunger.org/2015-world-hunger-and-poverty-facts-and-statistics/#hunger-number

YWCA. (2009, January 20). Broad investments: Counting women into the federal budget. Retrieved from *YWCA Canada*: http://ywcacanada.ca/data/publications/00000006.pdf

Gender

> *"Why should you care, what they think of you. When you're all alone, by yourself. Do you like you?"*
>
> *(Colbie Caillat, 2014)*

LEARNING OUTCOMES

By mastering this unit, students will gain the skills and ability to:

- distinguish between the biological determination of sex and the social construction of gender

- discuss the primary agents of gender socialization and how they influence dominant forms of masculinity and femininity

- analyze the various forms and causes of gender inequality in Canada

- examine how notions of gender identity and gender expression are fluid rather than fixed between male and female or placed along a continuum between these two binaries

Historically, gender and sex were rarely thought of as two separate concepts. In fact, the two terms were either used synonymously to indicate one of two possible options—male or female—or the concept of gender was ignored altogether. This rigid binary construction left little room for biological, psychological, or sociological alternatives to male or female, and the roles that women and men fulfilled in society were very clear-cut and regimented. There was a shared understanding of what were considered traditional female roles and responsibilities, and men did all the things that women would not (or supposedly could not) do. In some ways, we have come a long way since then in our understanding of gender, sex, and sexuality as distinct but interrelated concepts; for example, some men take on the role of primary caregiver for their children, while women work outside of the home. In other ways, we still have a very long way to go before the relationship between sex, gender, and sexuality is truly understood and accepted within society, evidenced by the gender pay gap that remains where women still do not earn as much as men.

Our inaccurate understanding of sex, gender, and sexuality as segmented binary groupings of male or female, masculine or feminine, and gay or straight creates rigid social groupings into which males and females are supposed to fit. Consequently, those who do not fit into these categories are sometimes labelled as deviant, criticized, or treated unfairly. Such rigid binary groupings of people based on gender, sex, and sexuality is not universal. Some cultures have different categories; for example, many Native North American subcultures (e.g., the Sioux, Lakota, and Cheyenne) recognize and revere **two-spirited** people as a **third gender** (Perkins, 1993). Additionally, the men of the Sambia tribe of Papua, New Guinea, engage in same-sex sexual behaviour as a rite of passage into becoming a warrior (McKay & McKay, 2010). The rites of the Sambia are considered "normal" and they know nothing about the binary notions of being gay or straight, even though the males in their culture experience both.

The labels that Western society places on individual practices and preferences come with all sorts of implications, innuendos, and often inhumane treatment to those who find themselves outside of the norm. We see a similar process when we reflect on stereotypical gender roles that men and women are supposed to fulfill in their lives. What are some of the stereotypical gender roles that exist within society? How much pressure do you think society puts on individuals to stay within these socially constructed gender roles? Gender stereotypes still exist in our society and they can affect us in ways that we do not often recognize. In this chapter, we will differentiate between sex, gender, and sexuality, investigate the consequences of the social construction of gender, and attempt to deconstruct binary notions of sex and gender as we explore gender and sexual diversity.

GENDER AS A SOCIAL CONSTRUCT

In recognizing the differences between male and female, let's look at how we define what it means to be a man or a woman in society today. Many people confuse the definitions of sex and gender or use them interchangeably; however, they have very different meanings. **Sex** refers to the biological components that make up who we are—the chromosomal, chemical, and anatomical components that are associated with males and females. Some of these biological differences are on the outside of your body: external genitalia and secondary sex characteristics like breast size, hair growth, and musculature; and some are on the inside of your body: DNA, gonads, sex hormones, internal reproductive organs, and brain chemistries. When we think about sex, we often think of it as an either/or concept, in that everyone is either male or female, but as we will see later in the chapter, there are ranges of male-ness and female-ness when it comes to chromosomes, chemicals, and anatomy, indicating that these are not necessarily binary constructions.

Unlike sex, **gender** is a social construct that refers to a set of social roles, attitudes, and behaviours that describe people of one sex or another. Stereotypically, we often attribute or identify different sets of roles and traits to males and females; for example, males may be expected to demonstrate strength and toughness, and to act as protectors, while females may be expected to be empathetic and caring, and take on the

Two-spirited: An umbrella term used by Native American people to recognize individuals who possess qualities or fulfill roles of both genders.

Third gender: (1) A person who does not identify with the traditional genders of "man" or "woman," but identifies with a third gender; (2) the gender category available in societies that recognize three or more genders.

Sex: The biological components—chromosomal, chemical, and anatomical—that are associated with males and females.

Gender: A social construct that refers to a set of social roles, attitudes, and behaviours that describe people of different sexes.

role of nurturer. Gender describes what it means to be a man or a woman: it is the social and cultural meanings that we attach to the biological characteristics of male and female. Similar to the male/female binary with sex, gender has been traditionally dichotomized as masculine/feminine, largely ignoring the range of differences that exist among and between each category. Gender encompasses our **gender identity**, the internal perception of an individual's gender and how they label themselves, and **gender expression**, the external display of gender that is generally measured on a scale of masculinity and femininity. Some would describe gender identity as how one feels about their gender, and gender expression as how one communicates their gender, through dress, demeanour, behaviours, and other social factors. **Sexuality** includes all the ways in which individuals express and experience themselves as sexual beings. Sexuality is discussed in more detail in Chapter 5, but traditionally in North America, like sex and gender, it has been dichotomized as gay or straight, leaving little room for those who do not fit within these rigid binary classifications.

As a social construct, our understanding of what is masculine and what is feminine varies across cultures. For example, unlike North America, the Mundugamor and Arapesh cultures of New Guinea view men and women as more similar than different (Kimmel and Holler, 2017). In the Arapesh culture, men and women share in child rearing, discourage aggressiveness among boys and girls, and are equally expected to be gentle, passive, and emotionally warm (Kimmel and Holler, 2017). In contrast, Mundugamor women and men are expected to be equally aggressive and violent, and women show little interest in childbearing and child rearing (Kimmel and Holler, 2017). For these cultures, gender roles are not polarized or defined in contrast to one another the way they are in North American societies.

Our understanding of gender also changes within cultures over time. As societal attitudes and beliefs about gender shift, so do **gender roles**. For example, child rearing was once predominantly viewed as a female role and very few women worked outside of the home, but many men are now taking parental leave and fulfilling the role of primary caregiver to their children while women spend more time in the workforce. How is gender socially constructed? Where does our understanding of masculinity and femininity come from and how are our attitudes and behaviours shaped by this shared understanding? In the next section, we examine various agents of **gender socialization** and explore the dominant notions of femininity and masculinity in Canadian society.

Agents of Gender Socialization

Research suggests that children are capable of distinguishing between boys and girls and learn to evaluate them differently as early as two years old (American Academy of Pediatrics, 2015, LoBue, 2016). As toddlers, they begin to recognize and adopt gender-stereotyped behaviours, but their understanding of gender is actually quite flexible and lacks the permanence associated with rigid classifications of male and female, masculine and feminine (LoBue, 2016). For example, children may only recognize gender by observing stereotypical boys' and girls' clothing or hairstyles—if a child had long hair and was wearing a pink dress, they would identify her as a girl, but if the same child cut their hair short and changed into jeans and a blue sweatshirt, they may identify them as a boy. Their flexible understanding of gender is also apparent when boys express a desire to be a mom when they grow up, or vice versa with a girl who wants to be a dad (LoBue, 2016). Children only begin to think of gender as a more stable trait, based on anatomical differences, between the ages of three and five (LoBue, 2016). At that time, they may start relating more to children of the same gender or start showing preference toward gender-specific toys and clothing (LoBue, 2016).

As children, we come to understand gender and gender-appropriate roles through the socialization process. We learn what it means to be a boy or girl and how to behave in masculine or feminine ways from our parents/guardians, teachers, peers, and the media. Parents are often the first and most influential when it comes to gender socialization in the early years. They pass along their own beliefs about gender and communicate gender expectations through their rearing practices—the choices that they make about children's clothes, toys, chores, discipline, and so on—that are often gendered, or different for boys and girls. Schools are also deeply gendered—children spend a great deal of time in a school environment and learn gender expectations from

Gender identity: The internal perception of an individual's gender and how they label themselves.

Gender expression: The external display of gender that is generally measured on a scale of masculinity and femininity.

Sexuality: All the ways in which individuals express and experience themselves as sexual beings.

Gender roles: A set of behaviours that are considered acceptable, appropriate, and desirable for people based on their sex or gender.

Gender socialization: The process by which males and females are informed about gendered norms and roles in a given society.

teachers and peers. Teachers often exhibit implicit or explicit bias that influences the ways in which they teach: they may model gender stereotypic behaviour, exhibit different expectations for male and female students, or use gender as a tool for organizing or grouping within the classroom environment (Bigler, Hayes, & Hamilton, 2013). In the educational setting, gender segregation is common because most children will choose to interact with same-sex peers when presented with the option, leading to more stereotypic play experiences (Bigler, Hayes, & Hamilton, 2013). Additionally, peers contribute to gender socialization by teaching their classmates stereotypes and punishing them for non-conformity (Bigler, Hayes, & Hamilton, 2013). Finally, the media—books, newspapers, magazines, television, movies, music, and social media—are powerful agents of gender socialization that shape our understanding of gender roles and expectations in Canadian society.

The media, like most agents of gender socialization, tend to reinforce the misconception that gender is inexplicably tied to sex, and males should be masculine and females should be feminine. But notions of masculinity and femininity are not universal constants; not only do they vary over time and across cultures, but also among men and women within a particular culture and over the course of their lives (Kimmel & Holler, 2017). Consequently, we should explore *femininities* and *masculinities* in a plural sense to recognize that masculinity and femininity mean different things to different people (Kimmel & Holler, 2017). Moreover, our gender intersects with other social characteristics like class, region, race, ethnicity, religion, age, and sexuality, further diversifying our experiences (Kimmel & Holler, 2017).

Although difference exists within genders as much as it exists between genders, there are dominant forms of masculinity and femininity that are used as standards or presented as ideals within Canadian society. In North America, dominant forms of masculinity and femininity are hierarchal and rooted in power, privilege, and marginalization. They exist as unrealistic and unattainable ideals that are used as standards to evaluate the degree to which men and women measure up in terms of their masculinity and femininity.

Hegemonic masculinity:
The version of masculinity that is set apart from all others and considered dominant or ideal within society; often associated with toughness, bravado, aggression, and violence.

Emphasized femininity:
The acceptance of gender inequality and a need to support the interests and desires of men; often associated with empathy, compassion, passivity, and focused on beauty and physical appearance.

Hegemonic Masculinity and Emphasized Femininity

Although there may be many forms of masculinities in Canadian society, not all are created equal. According to sociologist R.W. Connell, one version of masculinity is set apart from all others and considered dominant or ideal within society. This **hegemonic masculinity** is not only defined against femininity; it is also set in opposition to other masculinities that do not measure up to the construction of the ideal. Tenets of hegemonic masculinity can be seen in Michael Kimmel's concept of "The Guy Code"—a set of rules that men are supposed to follow in society (Kimmel, 2008). Kimmel (2008) summarizes these rules in what he calls a "Real Guys Top Ten List":

1. Boys Don't Cry
2. It's Better to be Mad than Sad
3. Don't Get Mad – Get Even
4. Take It Like a Man
5. He Who Has the Most Toys When He Dies, Wins
6. Just Do It or Ride or Die
7. Size Matters
8. I Don't Stop and Ask for Directions
9. Nice Guys Finish Last
10. It's All Good*

The rules discourage showing emotion or exemplifying vulnerability or weakness. They encourage a flippant attitude toward problems, and discourage worry or regret. Toughness, bravado, aggression, and violence are key components of hegemonic masculinity. Kindness and compassion are off-limits, and competition and winning are central to this form of masculinity. According to Connell, hegemonic masculinity is rooted in dominance of men over women and dominance of men over other men (Connell & Messerschmidt, 2005). Connell argues that this dominant form of masculinity is inexplicably tied to heterosexuality, and those who do not fully embody hegemonic masculinity are viewed as subordinate and marginalized within society (Connell & Messerschmidt, 2005). With hegemonic masculinity, there seems to be a shared understanding of what it means to be a man in society; both men and women understand what is meant by phrases like "Be a Man" or "Man Up." Connell argues that there is no "hegemonic" version of femininity, no equivalent collective understanding of the phrase "Be a Woman" (1987). Instead, gendered expectations for women come in the exaggerated ideal of femininity, which Connell terms **emphasized femininity** (2005).

*Source: "Real Guy's Top Ten List," from GUYLAND: THE PERILOUS WORLD WHERE BOYS BECOME MEN by MICHAEL KIMMEL. Copyright © 2008 by Michael Kimmel. Reprinted by permission of HarperCollins Publishers and Michael Kimmel.

Emphasized femininity is founded in the acceptance of gender inequality and a need to support the interests and desires of men (Connell, 1987). We see examples of emphasized femininity in society when women are expected to be empathetic and nurturing, emotional and complicated, vulnerable and dependent. It is evident when gender differences are exaggerated and women are depicted as fascinatingly manipulative, capable of wrapping men around their fingers by being flirtatious and sexually submissive. Emphasized femininity is also reinforced in the media through ideal representations of beauty—where images of women are edited until they depict absolute flawlessness and unattainable perfection. Attractiveness and physical appearance are central to emphasized femininity. Like hegemonic masculinity, assumptions of heterosexuality are deeply embedded in this dominant form of femininity (Connell, 1987). Emphasized femininity may not be the only form of femininity, but it is culturally valued as an ideal and there is a collective assumption that women should seek to achieve it. Connell argues that emphasized femininity and hegemonic masculinity are rooted in a long history of male domination (2016). What are the social consequences of hegemonic masculinity and emphasized femininity? How much pressure is there within society to fit within dominant forms of masculinity and femininity? Does anyone in society actually measure up to these inflated ideals? What happens to those who do not fully embody or outright reject dominant forms of masculinity and femininity? Let's explore some of the issues that arise from hegemonic masculinity and emphasized femininity.

If you have ever shopped for children's toys or clothing, you will recognize how gendered marketing and advertising reinforces hegemonic masculinity and emphasized femininity. On the one side of the aisle, there is a sea of bright pink with a focus on cosmetics and dolls; on the other side, everything is blue and focused on weapons, and gaming or action figures (Donovan, 2016). Fiona Martin, Professor of Sociology and Social Anthropology at Dalhousie University, says that research suggests that gendered marketing in children's toys has increased and become more intense over the past 15 years (Donovan, 2016). Martin argues that gender segmentation of children's toys encourages young girls to focus on their appearance and caring for others, and young boys to focus on action and aggressive behaviours (Donovan, 2016). Toys that are designed to foster imagination could actually cause children to "close their minds more than open them" (Donovan, 2016). She claims that the rise in gendered marketing in children's toys is somewhat ironic because children are naturally open to diversity, and today, people are more tolerant of gender diversity than ever before (Donovan, 2016).

Sacramento State sociologist Elizabeth Sweet believes that gendered marketing in children's toys reinforces male dominance in the fields of science, technology, engineering, and mathematics (STEM) (Vernone, 2016). She argues that when toys that develop spatial skills are only marketed to boys, and toys that cultivate empathy and verbal skills are only marketed to girls, it reinforces the stereotype that only boys are good at science and math (Vernone, 2016). Some toy manufacturers have tried to address gender inequalities by offering STEM toys in pink in order to appeal more to girls, but Sweet argues that this only further reinforces the stereotype by suggesting that "girls are so different that they need a special kind of STEM toys" (Vernone, 2016). Toy manufacturers have made some progress with offering more inclusive toys, like dolls with more realistic body types and abilities, female superhero figures, and characters with different skin tones, but Sweet argues there is still significant work to be done—"children need to see themselves in the toys and objects that they interact with" (Vernone, 2016). As we move closer to gender equality, why are toy manufacturers heading in the opposite direction? How can we as adults use our consumer power to promote social change in gendered marketing?

In the film *Killing Us Softly 4*, Jean Kilbourne examines advertising's image of women and discovers a restrictive code of femininity (Kilbourne & Jhally, 2010). After critiquing over 160 print and television ads, Kilbourne argues that the dominant message for women in advertising is that physical appearance is what is most important and that ideal female beauty is absolute flawlessness created through cosmetics, airbrushing, and digital editing (Kilbourne & Jhally, 2010). She explains that when women's bodies are sexualized, turned into objects, or only one body part is focused on in advertisements, women are dehumanized, creating a dangerous climate in which there is widespread violence against women (more on violence against women later in the chapter) (Kilbourne & Jhally, 2010). Furthermore, Kilbourne argues that the obsession with thinness and the ideal body type presented in mainstream media is teaching women to hate their bodies and to associate eating with feelings of shame and guilt (Kilbourne & Jhally, 2010). Additionally, she argues that the sexualization of women and girls in advertising sends the message that sexualized behaviour and appearance is empowering and rewarded within society (Kilbourne & Jhally, 2010). Although there is nothing wrong with wanting to be attractive and sexy, it becomes problematic when beauty is emphasized as the most important measure of success, to the exclusion of other important characteristics (Kilbourne & Jhally, 2010). These messages

© Meg Gaiger/Harpyimages

How does the media and popular culture's portrayal of what Jean Kilbourne terms "a cult of thinness" impact young girls and boys in Canadian society?

encourage unhealthy attitudes that can have serious negative impacts on physical and mental health (Kilbourne & Jhally, 2010).

Although Kilbourne claims that men's bodies are not as scrutinized and criticized in advertising, she does suggest that stereotypical advertisements that objectify men have increased, where men are increasingly presented as bigger, stronger, and presented as perpetrators of violence (Kilbourne & Jhally, 2010). She highlights that some advertisements eroticize violence, featuring women in bondage, battered, or even murdered (Kilbourne & Jhally, 2010).

Anti-violence educator and cultural theorist, Jackson Katz, supports the notion that male violence is rooted in society's inability to move away from hegemonic masculinity. In the documentaries *Tough Guise* (1999) and *Tough Guise 2* (2013), Katz examines how American popular culture (television, movies, video games, and advertisements), sports culture, and

political culture normalize violent and self-destructive forms of masculinity (Katz & Earp, 2013). He argues that when we speak about a culture of violence, we are almost always talking about a culture of violent masculinity, where boys are taught that being a man means embodying domination, power, aggression, and control (Katz & Earp, 2013). Katz suggests that we need to view this as *taught* behaviour rather than *learned* behaviour and explore the ways peer groups, fathers, coaches, and older male role models reinforce the "tough guise"—the front that boys and young men put up to shield their vulnerabilities and avoid being ridiculed (Katz & Earp, 2013). In the documentary, Katz highlights examples from mainstream media to illustrate how tenets of hegemonic masculinity are taught by male role models. In *Gran Torino* (2008), starring Clint Eastwood and Bee Vang, Eastwood's character attempts to show Vang's character how to "man up a little bit" and learn how "guys talk to each

Do images and advertisements that eroticize violence desensitize people to violence in the real world? Are people who are exposed to violent images more likely to engage in aggressive behaviour?

other" (Katz & Earp, 2013). In the children's movie *How to Train Your Dragon* (2010), the son runs into problems when he refuses to act violently, resulting in his father telling him, "you're not a Viking…you're not my son" (Katz & Earp, 2013). Finally, in *Shark Tale* (2004), the message is that violence is the way of the world, illustrated when the father shark tells his son that it is "time he learn how to be a shark … it is a fish eat fish world … you either take or get taken" (Katz & Earp, 2013). Katz concludes that we create a dangerous climate for boys and young men when "violence is taught not as a last resort but as a go-to method in resolving disputes and also as a primary means of winning respect and establishing masculine credibility" (Katz & Earp, 2013). Do you see evidence of "gender police" within society that encourage these dominant forms of masculinity and condemn or ridicule males for demonstrating qualities like compassion, caring, empathy, intellectual curiosity, and fear? What kind of impact does that have not only for young men, but also for society overall?

GENDER INEQUALITY

Living in a world divided into the distinct binary categories of male/female, masculine/feminine, and gay/straight has very real consequences. Historically and at present, most societies are **patriarchal**. This male domination has led to tangible inequities that are now systemically embedded within some of the most important social institutions in society. By looking at major sectors of society and the allocation of power, prestige, or income within them, we can see the gender inequalities that exist in Canadian society.

Education

Years ago, women in Canada did not have the same access to education as men, which resulted in men with higher educational attainment than women. Today, the situation is somewhat different. Over the past 24 years, the percentage of women aged 25–64 with a university degree more than doubled from 15 percent in 1991 to 35 percent in 2015, and the percentage of women aged 25–64 with college diplomas grew from 14 percent in 1991 to 26 percent in 2015 (Ferguson, 2016). For men, educational attainment also grew from 19 percent in 1991 to 30 percent in 2015 for university degrees, and 9 percent

Patriarchy: Historically, any social system that was based on the authority of the heads of the household, which were traditionally male; recently, the term has come to mean male domination in general.

in 1991 to 19 percent in 2015 for college diplomas (Ferguson, 2016). Although women were more likely than men to complete a university degree in 2015, women were slightly less likely than men to complete a high school diploma (23 percent and 25 percent respectively), and half as likely to complete a trades certificate (7 percent and 15 percent respectively) as their highest level of education (Ferguson, 2016). The Organisation for Economic Cooperation and Development (OECD) projects that if this trend continues in Canada, by 2025 women will outnumber men in postsecondary two to one (OECD, 2012).

At the same time, gender differences remain in chosen fields of study. According to Statistics Canada, "in 2008 women accounted for over three out of four graduates in education and in health science programs. In humanities, in visual and performing arts and communications technologies, as well as in social and behavioural sciences and law, roughly two out of three graduates were women" (Turcotte, 2012, 20). As of 2013, men were twice as likely than women to pursue STEM (science, technology, engineering, and mathematics) fields of studies (OECD, 2012; Ferguson, 2016). A 2013 study showed that even "girls with higher mathematical ability were less likely to pursue STEM fields at university than boys with lower mathematical ability" (Ferguson, 2016). Why do you think this is the case? Do negative stereotypes still persist for men and women in specific fields of study?

Employment and Government

In employment, gender disparities have significantly decreased over the years, but they are still apparent today. When comparing the number of men and women, aged 25–54, participating in the labour market, the gender participation gap in employment was 75.5 percentage points in 1950, 28.3 percentage points in 1983, and 8.9 percentage points in 2015 (Moyser, 2017). While the majority of both women and men are employed full-time, women are overrepresented when it comes to part-time or precarious employment. In 2015, three quarters (75.8 percent) of those working part-time, were women; 18.9 percent of women who were employed worked part-time, compared to only 5.5 percent of employed men (Moyser, 2017). The reasons why women and men work part-time are generally different. Research indicates that most men worked part-time involuntarily as a result of poor business conditions, whereas most women indicated a choice to work part-time in order to balance caring for children (Moyser, 2017). Women also

AGENT OF CHANGE

Feridun Hamdullahpur

President and Vice Chancellor at the University of Waterloo, Feridun Hamdullahpur, said that his personal "click" moment was when he was at his first town hall meeting and was asked how the university was going to achieve gender equality. The answer that he gave at the time was not one he was satisfied with. It was then that he made a promise to himself that the following year, he would have a much better answer.

At the University of Waterloo, Hamdullahpur wants people who identify as women, girls, and all genders to look at higher education and see equal opportunity. At the University of Waterloo, he has worked to create a comprehensive plan targeting three key areas. First, he hopes to reach gender parity in STEM outreach, experiences, and activities at the University of Waterloo

by boosting female enrolment in these outreach, experiences, and activities to 33 percent by 2020 through targeted outreach programming for young girls and women (HeForShe, 2016). Secondly, through unconscious bias training initiatives, a review of selection and recruitment hiring practices, and the creation of career opportunities for women, the University of Waterloo, under Hamdullahpur's leadership, seeks to enhance the representation of women by reaching a faculty composition of 30 percent women by 2020 (HeForShe, 2016). Lastly, the University of Waterloo plans to increase the share of women in leadership positions by reaching 29 percent representation of women in academic and senior leadership by 2020 (HeForShe, 2016). Hamdullahpur believes that, "when we achieve gender equity, women and girls will be able to see the fullness of their reflection in this University, in our programs, policies, student body, professoriate, administration, and alumni."

Feridun Hamdullahpur represents Canada as one of the HeForShe University IMPACT Champions from around the globe. Led by UN Women, the HeForShe movement works to engage men and boys to become agents of change by taking action against gender inequality. As a global initiative, one of the movements' main goals is to encourage the world to view gender inequality as a human rights issue, rather than a problem for just women, and enlist the help of all genders worldwide in the fight against gender inequality. As a HeForShe IMPACT Champion, Hamdullahpur believes that it is in solidarity and a collective partnership with IMPACT partners and all genders everywhere striving for gender equity, where we can obtain "a world's worth of gender perspectives and partnerships to advance the equity agenda" (HeForShe, 2016). To take the HeForShe pledge or for more information, visit http://www.heforshe.org/en.

---

tend to be concentrated in industries that parallel traditional gender roles. In 2015, the three industries with the greatest share of women relative to men were healthcare and social assistance (82.4 percent), educational services (69.3 percent), and accommodation and food services (58.5 percent) (Moyser, 2017). In comparison, the three industries with the greatest share of men in comparison to women were construction (88.3 percent), utilities (77.8 percent), and forestry, fishing, mining, quarrying, and oil and gas extraction (80.5 percent) (Moyser, 2017). Furthermore, the

positions that most women occupy tend to be lower-level than those of men, even in industries that are female-dominated. For example, in 2015, in accommodation and food services, 59.7 percent of chefs and cooks were men, while 71.6 percent of food counter attendants, kitchen helpers, and support personnel were women (Moyser, 2017). In 2015, women represented 25.6 percent of senior managers in the private sector, compared to 54.5 percent of legislators and senior government managers and officials in the public sector (Moyser, 2017). Some argue that the overrepresentation of women in lower-level positions is indicative of the proverbial "**glass ceiling**," an invisible barrier that prevents women and minorities from advancement in organizations. Additionally, the difference between the private and public sectors could be influenced by Canada's federal equity employment legislation and employment equity policies pertaining to provincial public servants in Nova Scotia, New Brunswick, Prince Edward Island, Quebec, Manitoba, Saskatchewan, and British Columbia (Ontario repealed this legislation in 1995) (Moyser, 2017). Furthermore, in 2016, the Government of Canada announced a commitment to gender equality, forming Canada's first gender-balanced cabinet, appointing the first woman as Government House Leader, and creating the first ever Minister fully dedicated to gender issues in Canada (Trudeau, 2016).

Despite Canada's commitment to gender equality, a gender pay gap remains between Canadian men and women. If we compare the 2014 annual earnings of full-time, full-year workers, aged 25–54, women made $0.74 for every dollar earned by men (Moyser, 2017). Since women tend to work fewer hours on average than men, it is also important to examine the gender pay gap based on hourly wages of full-time workers. In 2014, "women earned an average of $25.88 per hour, while men earned an average of $28.92, resulting in women earning $0.88 for every dollar earned by men (Moyser, 2017).

The Canadian Centre for Policy Alternatives (CCPA) issued a report in July 2015 that ranked Canada's largest 25 metropolitan areas based on a comparison between men and women in economic security, leadership, health, personal security, and education (see Figure 4.1) (CCPA, 2015). According to the report, cities in Quebec fare better than cities in any other province, which can be attributed to policies around subsidized child care, paternity leave, and more generous parental leave benefits that are unique to Quebec (CCPA, 2015). As a result of the measure of gaps that exist between men and women, Kitchener-Cambridge-Waterloo, Ontario, is the worst city for women to live in and Victoria, British Columbia, is the

FIGURE 4.1

Best and Worst Cities to be a Woman in Canada

BEST & WORST CITIES TO BE A WOMAN IN CANADA

▼ BEST
1. VICTORIA
2. GATINEAU
3. QUÉBEC CITY
4. ABBOTSFORD-MISSION
5. HALIFAX
6. LONDON
7. VANCOUVER
8. BARRIE
9. MONTRÉAL
10. OSHAWA
11. OTTAWA
12. TORONTO
13. KELOWNA
14. REGINA
15. ST. JOHN'S
16. ST. CATHARINES-NIAGARA
17. KINGSTON
18. WINNIPEG
19. SHERBROOKE
20. HAMILTON
21. SASKATOON
22. WINDSOR
23. CALGARY
24. EDMONTON
25. KITCHENER-CAMBRIDGE-WATERLOO
▲ WORST

WHERE DOES YOUR CITY RANK?
VISIT POLICYALTERNATIVES.CA/BEST-WORST2015

What are the consequences of this disparity? How would you feel knowing that employees working next to you do the same job as you but earn more or less money based solely on whether they are male or female?

Source: Canadian Centre for Policy Alternatives (CCPA). (2015). Infographic – The Best and Worst Places to Be a Woman in Canada 2015. Retrieved from CCPA: https://www.policyalternatives.ca/publications/facts-infographics/infographic-best-and-worst-places-be-woman-canada-2015. Date Accessed: January 2, 2017.

best (CCPA, 2015). The success in limiting gender gaps in Victoria is attributed to a large public sector employer where the gender pay gap is narrowed and more women are promoted into senior management positions (CCPA, 2015). The cities that ended up at the bottom of the list, like Calgary, Edmonton, and Kitchener-Cambridge-Waterloo, tend to rely on traditionally male-dominated industries, such as construction, mining, oil, and gas (CCPA, 2015).

Violence Against Women

Gender-based violence continues to be a serious problem in Canada. Both police-reported and self-reported data indicate that women are more likely than men to experience violence, such as **intimate partner violence** or spousal violence (discussed further in Chapter 13) and **sexual violence** (discussed further in Chapter 5) (Statistics Canada, 2013; Sinha, 2013).

Glass ceiling: An invisible barrier that prevents women and minorities from advancement in organizations.

Intimate partner violence: Any intentional act or series of acts by one or both partners in an intimate relationship that causes injury to either person. It can include physical assault, emotional abuse, sexual violence, and sexual harassment. Sometimes referred to as spousal violence or domestic abuse.

Sexual violence: Any sexual act or an attempt to obtain a sexual act through violence or coercion.

Furthermore, the actual incidences of violence against women and girls is estimated to be much higher than official data suggests, as a large proportion of violent acts go unreported (Sinha, 2013). The proportion of incidences that go unreported was illustrated in Antonia Zerbisias's Twitter hashtag #BeenRapedNever-Reported that went viral in October 2014, with over 8 million people across the globe taking part in the conversation to challenge social stigma surrounding sexual violence (Gallant, 2014).

R.W. Connell argues that cultural constructions of masculinity support gender-based violence, and until we tackle notions of hegemonic masculinity and emphasized femininity, we are unlikely to make progress in the fight for gender equity (Connell, 2016). She believes that family violence, collective violence, and institutional violence are all gender-based and result not from a male, biological predisposition toward violence, but from situations of social instability, particularly in societies that have models of masculinity that emphasize dominance and power (Connell, 2016).

Jackson Katz agrees, arguing in his Tedx Talk titled *Violence Against Women – It's a Men's Issue*, that violence against women is an issue for men that requires a paradigm-shifting perspective (Katz, 2013). He claims that when we call violence against women a "women's issue," we are adding to the problem by giving men an excuse not to pay attention (Katz, 2013). Additionally, Katz suggests that calling it gender-based violence might also be problematic because many people confuse the term "gender" with "women" in the same way the term "race" is mistakenly only associated with racialized communities, or the phrase "sexual orientation" is inaccurately equated with being gay, lesbian, bisexual, or other non-heterosexual identities (Katz, 2013). The confusion surrounding these terms leads to the dominant group, in this case men, never having to think about their dominance, thereby maintaining and reproducing the dominant system itself (Katz, 2013). Katz believes that if we refer to gender-based violence as a men's issue, then more men will speak up and participate in the fight against social institutions that support gender-based violence (Katz, 2013). He recognizes this shift as not only important for female victims of male violence, but also for males who have been victimized by other males (Katz, 2013). Katz encourages what he calls a "bystander approach" to gender-based violence, where instead of focusing on the perpetrator and victims, we focus on empowering bystanders, or anyone who is not a perpetrator or victim in a given situation, to challenge dominant forms of masculinity that support gender-based violence (Katz, 2013).

The HeForShe campaign (discussed in the Agent of Change section of this chapter) is an example of an initiative that specifically challenges men to participate in the fight against gender-based inequality. The HeForShe IMPACT 10x10x10 engages key male leaders in governments, corporations, and universities around the world to take on the role of Impact Champions who drive change from the top down (HeForShe, 2016). In September 2016, Canadian Prime Minister Justin Trudeau was appointed a HeForShe Champion for Youth Engagement at the UN Women HeForShe second anniversary event (HeForShe, 2016).

Another campaign challenging violence against women and promoting gender equity is *Walk a Mile in Her Shoes: The International Men's March to Stop Rape, Sexual Assault, and Gender Violence*, a playful opportunity for men to raise awareness in their communities around gender-based violence by walking one mile in women's high-heeled shoes (Walk a Mile in Her Shoes, 2017). The White Ribbon Campaign is another social movement where men and boys are working to promote gender equity, a new vision of masculinity, and end violence against women (White Ribbon, 2017). It started in 1991, when men and boys were asked to wear white ribbons as a "pledge to never commit, condone or remain silent about violence against women and girls" (White Ribbon, 2017). Since then, it has spread to over 60 countries all over the world, examining the root causes of gender-based violence and promoting a cultural shift that challenges "negative, outdated concepts of manhood and [inspires] men to understand and embrace the incredible potential they have to be a part of positive change" (White Ribbon, 2017).

THE –NESS MODEL: DECONSTRUCTING THE BINARY AND THE CONTINUUM

Now that we have critically examined how gender is socially constructed and taught through the process of gender socialization, and investigated some of the gender inequalities that emerge from the binary notions of male/female and masculine/feminine, we turn to a different model of gender in an attempt to deconstruct traditional binary and continuum models.

In his book titled A Guide to Gender: The Social Advocate's Handbook, author, artist, and comedian Sam Killermann suggests that traditional gender binaries of male/female and masculine/feminine restrict gender into only two possibilities, when gender is actually limitless (Killermann, 2013). According to Killermann, there are as many versions of gender as there are people in this world (2013). He argues that not only are sex and gender different, one does not

FIGURE 4.2

The Genderbread Person

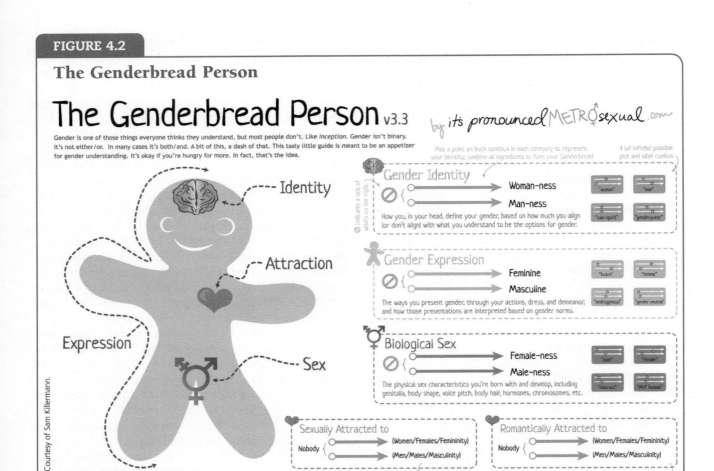

The Genderbread Person v3.3

by it's pronounced METROsexual.com

Gender is one of those things everyone thinks they understand, but most people don't. Like *inception*. Gender isn't binary. It's not either/or. In many cases it's both/and. A bit of this, a dash of that. This tasty little guide is meant to be an appetizer for gender understanding. It's okay if you're hungry for more. In fact, that's the idea.

Plot a point on both continua in each category to represent your identity; combine all ingredients to form your Genderbread

4 (of infinite) possible plot and label combos

Identity

Attraction

Expression

Sex

⊘ Indicates a lack of what's on the right.

Gender Identity
⊘ { — Woman-ness
⊘ { — Man-ness

How you, in your head, define your gender, based on how much you align (or don't align) with what you understand to be the options for gender.

"woman" "man" "two-spirit" "genderqueer"

Gender Expression
⊘ { — Feminine
⊘ { — Masculine

The ways you present gender, through your actions, dress, and demeanor, and how those presentations are interpreted based on gender norms.

"butch" "femme" "androgynous" "gender neutral"

Biological Sex
⊘ { — Female-ness
⊘ { — Male-ness

The physical sex characteristics you're born with and develop, including genitalia, body shape, voice pitch, body hair, hormones, chromosomes, etc.

"male" "female" "intersex" "MtF female"

Sexually Attracted to
Nobody { — (Women/Females/Femininity)
— (Men/Males/Masculinity)

Romantically Attracted to
Nobody { — (Women/Females/Femininity)
— (Men/Males/Masculinity)

In each grouping, circle all that apply to you and plot a point, depicting the aspects of gender toward which you experience attraction.

For a bigger bite, read more at http://bit.ly/genderbread

Courtesy of Sam Killermann.

Source: Killerman, S. (2015). The Genderbread Person v. 3.3. Retrieved from It's Pronounced Metrosexual: http://itspronouncedmetrosexual.com/2015/03/the-genderbread-person-v3/#sthash.CkL5qtqU.omteL26Q.dpbs Date accessed: January 2, 2017. Courtesy of Sam Killermann.

---

determine the other (Killermann, 2017). In other words, a person's biological sex does not determine their gender identity; a person's gender identity does not determine their gender expression; and a person's gender identity or gender expression does not determine their sexual orientation, or feelings of romantic or sexual attractions toward other people. He argues that while they are interrelated, they are not interconnected (Killermann, 2017). For some, the relationships between biological sex, gender identity, gender expression, and attraction align; for example, if someone is born with male reproductive organs and genitalia, it is likely that he will be raised a boy, identify as a man, and express his gender in masculine ways (Killermann, 2017). Individuals who are **cisgender** experience such alignment between their sex, gender, and sexuality, but a problem arises when we assume everyone who is born biologically male will identify and express themselves in the same way. Using the Genderbread Person v. 3.3 to illustrate (see Figure 4.2), Killermann introduces what he calls **"The -Ness Model,"** a unique way of exploring gender diversity in North American society (2017).

For Killerman, the –Ness Model deconstructs traditional notions of gender as binary, in that there are more options than just male or female, and it also distinguishes between sex, gender, and sexuality, further clarifying how gender identity is different from gender expression. Additionally, the –Ness model

---

Cisgender: A description of a person whose gender identity, gender expression, and biological sex align.

The –Ness Model: A gender model that differentiates between gender identity, gender expression, biological sex, sexual attraction, and romantic attraction, and presents two spectrums for each concept ranging from "0," "Null," or "Nobody" on the one side to woman-ness/man-ness, femininity/masculinity, and female-ness/male-ness on the other.

attempts to deconstruct continuum or spectrum gender models that polarize male/female, masculine/feminine, gay/straight, or place these concepts on either end of a spectrum in opposition to one another (Killermann, 2017). Killermann believes that when we place man and woman at either ends of a spectrum and then ask those who do not identify as man or woman to find themselves somewhere in between the two, we ignore the reality that "many of us aren't in-between, and many of us are a lot of both" (Killermann, 2017, 89). Instead, the –Ness Model uses two separate continua, with a "0" or "null" symbol on one side and a concept on the other, to describe biological sex, gender identity, gender expression, and sexual or romantic attraction. This model allows for someone to be "a lot of both, or neither of each" (Killermann, 2017, 90).

Killermann believes that exploring gender in this way corrects the common gender misunderstanding that to be more of one, a person has to be less of the other (2017). He uses himself as an example, illustrating that while he identifies as a man, the fact that he is sensitive, kind, familial, and likes romantic comedies (characteristics traditionally associated with "woman-ness"), does not make him any less of a man—we can be feminine and masculine or embody woman-ness and man-ness; one does not cancel the other out (Killermann, 2017). In the next sections, we will explore gender identity, gender expression, biological sex, and attraction using the –Ness Model presented in the Genderbread Person v. 3.3.

Gender Identity

As defined earlier in the chapter, gender identity refers to a person's internal perception of their gender and how they label themselves. The –Ness model asks you to plot on each of the spectrums under gender identity, how much you identify with woman-ness or man-ness, based on the social roles, attitudes, dispositions, and/or personality traits associated with those identities (Killermann, 2017). If everything you think of when you think about what it means to be a woman applies to you, you may find yourself on the far right of the spectrum for woman-ness. If everything you think of when you think about what it means to be a man applies to you, then you would fall on the far right of the man-ness spectrum. What is interesting about this model is that the two circumstances described above could apply to the same person. In the -Ness Model, there is space for infinite possibilities of gender to be visualized, allowing for one person to define themselves as a lot of woman-ness and man-ness, or by having only a little or a total lack of either

(Killermann, 2017). Someone who identifies as cisgender, can identify themselves in the model in the same way as someone who identifies as two-spirited, **transgender**, **genderqueer**, **bigender**, **genderfluid**, **genderless**, or any other gender identity.

Gender Expression

For gender expression, the –Ness Model asks you to think about the ways in which you present your gender to the world through your actions, clothing, mannerisms, grooming habits, or other personal expressions and how these presentations are interpreted based on dominant notions of masculinity and femininity (Killermann, 2017). The –Ness Model creates space for multiple forms of gender expression, allowing for individuals who express themselves in traditional gendered ways, but also for those who express themselves in non-traditional gendered ways. **Androgyny** describes a gender expression that exemplifies a combination of traditional feminine and masculine characteristics. **Agender** (also referred to as gender-neutral) refers to a gender expression that exists outside of the traditional system of gender, showing little signs of masculinity or femininity. **Gender non-conforming** refers to a gender expression that indicates a non-traditional gender presentation, such as a masculine woman or a feminine man. Additionally, there are certain social situations that often translate into extreme forms of gender expression. For example, fancy social engagements provide occasion for us to dress up in formal gowns, makeup, jewellery, tuxedos, cufflinks and tie clips, or people often exhibit extreme masculinity while

Transgender: A blanket term used to describe anyone who does not identify as cisgender.

Genderqueer: A blanket term used to describe people whose gender falls outside of the gender binary; a person who identifies as both a man and a woman, or as neither a man nor a woman; often used in exchange with transgender.

Bigender: A person who fluctuates between traditional gender-based behaviours and identities, identifying with both genders or sometimes a third gender.

Genderfluid: Describes an identity that is a fluctuating mix of the options available.

Genderless: A person who does not identify with any gender.

Androgyny: A gender expression that has characteristics of both masculinity and femininity.

Agender: A person with no (or very little) connection to traditional systems of gender; or someone who sees themselves as existing without gender; sometimes called gender neutral or genderless.

Gender non-conforming: Refers to a gender expression that indicates a non-traditional gender presentation.

If a picture can say a thousand words, imagine the stories your shoes could tell! Try this student story on for size – have you walked in this student's shoes?

The first thing I feel that everyone needs to understand, is that the LGBTQ community is more than just those specific 5 letters. Sure they may be the most prevalent aspects of the gay community, but there is so much more that the world needs to understand. There are people who are asexual; who have no sexual preference at all. The pansexual community are attracted to all humans regardless of gender identity. These are only some of the sexuality identities that need not be mixed up with romantic attractions. A person may be pansexual and identify as female, but may be heteroromantic. This would mean she is sexually attracted to all people but romantically attracted to the opposite sex. Whereas a genderfluid male may identify as an asexual but homoromantic or heteroromantic depending on the gender he identifies at the time.

I myself am a Genderfluid homosexual and homoromantic male, but sometimes a genderfluid heteroromantic and heterosexual female. It took about 6 years for me to come to the conclusion of who I am on this earth and what it was I was attracted to. At a time I had thought I was simply bisexual. I also thought I was asexual, as well as Transgendered at one point. I was so sure of myself I had gone to lengths of researching the costs of the surgeries. But like all big decisions, a mother's advice gives the clearest perspectives.

I may not have told her everything all the time, but she always just knew something was up. I remember one day, both my parents, my sister, and I were just having dinner outside on the patio when I brought up feeling like I was in the wrong body and that I was contemplating on taking either estrogen to become more feminine, or testosterone to be more male. At this time I didn't consider or even know what Genderfluid meant. I was still an identified gay male at the time but I was frustrated with how I was. I didn't have a label. I was too feminine to call myself male, but my body was too masculine to call myself female. But my mother did what all mothers do, is let me explain my reasoning behind how I felt, and then said, give it time. Which thinking back on it now, was the best advice I could have gotten. After a few months to a year from that time I found that there are so many other gender identities than what I was exposed to, or what I thought.

My first real experience with regards to my genderfluidity was back when I was in grade 11. I attended Laura Secord Secondary School, which had a reputation of being where the "gay kids" went because of its focus on the arts and its notable acceptance of the lgbtq community. It was at a Halloween dance that was themed zombies. I however didn't want to dress as a zombie, and went in drag. My mother gave me her old bridesmaids dress; did my makeup, threw on a wig from the costume store and my black Converse. Showing up at school as a "Glam-pyre" was the best thing that I did at the time. Everyone was both impressed and in subtle shock that I arrived at a school dance in drag. The attention and feeling I had with being in drag was like being in another world. It was what I was missing from my life. Now a distinguishable note I need to mention is my current life differences between my drag persona, Raven S. Klawe, and my genderfluidity. I arrive to an entertainment place in drag, with the intent of entertaining people or being a source of enjoyment. I may arrive as a woman to formal events and galas, but that's my genderfluidness. I separate my craft of drag and my gender identity through the means of the place or event I am going. People may ask me why I am doing drag at a formal event, when I'm not. I'm simply embracing my other gender at that time. There is the odd event here and there that I do combine the genders and create what is known as gender non-conforming.

A few years following that dance, when I finished my first year of college, was when I participated in my first drag pageant where I came close to winning. I was surrounded by other people like me who enjoy expressing themselves in this way and in that moment I knew I was "home." At that same time of the year I had heard of these other sexualities and gender identities that I didn't know about early in high school. So after doing some research I had found what I have been looking for; after years of wondering and searching. Genderqueer and Genderfluidity were identities that I had never known. These then lead to the discovery of being dual-spirited, agender, bigender, gender non-conforming as well as to genderfuck, which apparently means to playfully confuse others with these typical gender roles or expressions.

So now with all this in mind, don't assume I didn't have the tough childhood or teen years. All through elementary school, middle school and in the first half of my high school life I was bullied about my voice being too feminine, or that I was for lack of a better word, a fag. To say I didn't let these notions get to me would be a lie. I went through a dark stage in my life where I wanted nothing more than to not exist and just get away from everything. I kept it hidden at home; all the feelings and the bullying events I endured. There were times I tried to tell myself "You're

(Continued)

not gay. You like girls." But it was all just pointless sentiments. It wasn't until grade 10 and 11 that I had begun to do what I should have done years ago. I stopped caring. I did what I wanted to do, and loved who I wanted to, because in the end I'll be the bigger person living the life I want to, and their "verbal abuse" would have had no effect on my life.

Which leads to where I am now. April 22nd, 2016. I'm a Pride Niagara Unity award nominee for 2 awards, an emerging drag queen from the Niagara region with the title of Miss Hamilton Embassy 2016/2017. I have started my own studio business with a focus in art direction and moving to Toronto to attend a Theatre Production program. I'm finishing my second program at Niagara College, and heading to the big city where I can thrive in my life doing what I love, loving who I want and not giving a single damn about what people think about me; because I'm successful in my own eyes, and that's all that matters.

playing sports (Killermann, 2017). Gender expression can also be a conscious performance seen in the example of **cross-dressing** or in drag shows, when **drag kings** and **drag queens** present exaggerated forms of masculine or feminine expression (Killermann, 2017).

Gender expression can be a way of signifying your gender identity, or it can be an intentional way of rejecting your gender identity (Killermann, 2017). "It can align with the gender norms attached to your biological sex, or not. It can be driven by your want to conform, your want to rebel, sexual or relational desires, or something else altogether" (Killermann, 2017, 110). Gender expression is the component of gender that influences your interactions with other people the most; it can determine the adversity you might face or the privileges you might experience as a result of your gender (Killermann, 2017). It is difficult to measure because our understanding of what is masculine and feminine is always changing over time, both individually and culturally (Killermann, 2017). Additionally, one's gender expression can cause others to confuse your gender identity—because of the misconception that sex, gender, and sexuality are interconnected rather than just interrelated, some people make assumptions about gender identity or sexual orientation based on gender expression (Killermann, 2017). For example, Killermann describes having to explain that just because he embodies **metrosexual** characteristics, does not indicate that he is gay. He discusses how aspects of his gender expression that are traditionally characterized as feminine, constantly lead to circumstances where people make assumptions about his sexual orientation (Killermann, 2017). He argues that once individuals understand that biological sex, gender identity, and sexual orientation do not determine gender expression, perhaps we can get to a place where gender expression is viewed as individual expression and not "bound by assumptions about other aspects of our identity, or policed by those around us" (Killermann, 2017).

Biological Sex

As described earlier in the chapter, biological sex refers to the chromosomal, chemical, and anatomical components that are associated with males and females. For males, typical characteristics include penis, testes, scrotum, more testosterone than estrogen, XY chromosomes, thick body hair, facial hair, and a deep-pitched voice (Killermann, 2017). For females, typical sex characteristics include ovaries, vulva, vagina, uterus, more estrogen than testosterone, XX chromosomes, breasts, fine body hair, limited or no facial hair, and a high-pitched voice (Killermann, 2017). If someone has all or most of the characteristics associated with being male or female, we label them as such. Like gender identity and gender expression, with biological sex, there is and always has been enormous pressure in society to fall into one of two binary categories—male or female, which is usually assigned at birth. When sex is assigned at birth, Killermann argues that so is gender identity in that if a child is born a girl, she is likely to be raised as a girl, encouraged to exhibit femininity, and socialized to adopt traditional female gender roles (2017). Consequently, a person's assigned sex at birth strongly influences their gender identity formation, and if those two components of gender align, then there doesn't

Cross-dressing: Wearing clothing that conflicts with the traditional gender expression of your sex and gender identity.

Drag king: A person who consciously performs traditional masculinity, presenting an exaggerated form of masculine expression; often times done by a woman.

Drag queen: A person who consciously performs traditional femininity, presenting an exaggerated form of feminine expression; often times done by a man.

Metrosexual: A man with a strong aesthetic sense, who spends more time on appearance and grooming than is considered gender normative.

seem to be a problem (Killermann, 2017). However, if a child is assigned a sex at birth and then is socialized to adopt norms that do not match with their sense of self, it can create "a confusing worldview at a young age" (Killermann, 2017).

Killermann argues that our understanding of sex is much more than just the biology or the anatomy. In the –Ness Model, he suggests that there are varying degrees of male-ness and female-ness based on the physical sex characteristics you are born with and develop throughout your life, like genitalia, body shape, voice pitch, body hair, hormones, chromosomes, and so on (Killermann, 2017). He states that the more testosterone your body makes and utilizes, the more male-ness you may develop, and the more estrogen your body makes and utilizes, the more female-ness you might develop (Killermann, 2017). Again, using himself as the example, he identifies his male-ness by his wide shoulders, penis, testes, beard, and hard jaw and brow lines, and his female-ness by his lack of a protruding Adam's apple, fine body hair, wide hips, and high-pitched voice (Killermann, 2017). He argues that there are different ranges of male-ness and female-ness in everyone, and although we are born with biological sex characteristics, the way we make meaning of those characteristics is purely sociological (Killermann, 2017).

Although we tend to classify individuals into the two restrictive binary categories of male and female, the reality is that "all of the structures that make up "male" or "female" characteristics exist in people of all sexes; we often call them different names in different bodies" (Killermann, 2017, 74). For example, a "clitoris" and a "penis" are different terms used for the same anatomical structure: the clitoris is a short penis, and a penis is a long clitoris. When a baby is born with slightly ambiguous genitalia—an enlarged clitoris in an otherwise "female" body, or a small penis in an otherwise "male" body—the doctors will perform a number of tests on chromosomes, hormone levels, and ultrasounds to check for sex organs, and then they will assign a sex at birth (Killermann, 2017). If a baby is born with genitalia that are too ambiguous for a doctor to assign a sex at birth, more extensive testing is done, usually involving an endocrinologist, and then surgical interventions are used to align the child's

Intersex: A person whose physical anatomy does not fit within the traditional definitions of male or female.

Transsexual: A person who psychologically identifies with a sex/gender different from the one they were assigned at birth. Transsexuals often wish to transform their physical bodies, with puberty suppression, hormone therapy, or surgery, to align with their inner sense of sex/gender.

sex with whatever sex they decide is best for the child (Killermann, 2017). The assumption is that if a doctor assigns a sex at birth, a child can be socialized into a gender identity that corresponds with the assigned sex at birth (Killermann, 2017). Killermann argues that while this assumption highlights the right problem, it attempts to solve it with the wrong solution (2017).

The Intersex Society of North America (ISNA) defines **intersex** as "a general term used for a variety of cases where an individual is born with a reproductive or sexual anatomy that doesn't seem to fit with the typical definitions of female or male" (ISNA, 2008). It can include those who are born with ambiguous genitalia; individuals who appear to be female on the outside, but mostly have internal anatomy associated with being male; individuals who appear to be male on the outside, but have internal anatomy associated with being female; or individuals who are born with mosaic genetics, where some cells have XX chromosomes and some cells have XY chromosomes (ISNA, 2008). Furthermore, intersex anatomy does not always show up at birth; in some cases, intersex anatomy is not present until puberty, or in adulthood with infertility, or not until death when an autopsy is performed (ISNA, 2008). ISNA estimates that the total number of people whose bodies differ from the standard definitions of male or female are approximately 1 in every 100 births (Killermann, 2017). These variations challenge the notion that sex is binary and a person is either exclusively male or exclusively female—"nature doesn't decide where the category of "male" ends and the category of "intersex" begins, or where the category of "intersex" ends and the category of "female" begins...humans decide" (ISNA, 2008).

The bottom line is that sex is much more complex than what we are assigned at birth, and the biology that we use to justify the rigid categories of male and female is not as universal or objective as one might think. Some might argue that reproduction is central to our understanding of biological sex, but what about individuals who cannot have children for whatever reason, or individuals who choose not to have children, or children who have not hit puberty yet and cannot have children? Are the people in these circumstances any less "male" or "female"?

A **transsexual**, or a person who identifies psychologically with a gender/sex other than the one they were assigned at birth, sometimes undergoes a process of transition that might involve puberty suppression, hormone therapy, and/or surgery to align their bodies with their inner sense of sex or gender. Take, for example, the 2012 case of Canadian Miss Universe contestant Jenna Talackova (Raptis, 2012). Although born with a penis, Jenna knew from a very young age that she was a female. She began hormone therapy at 14 years of age and had **gender confirmation surgery** (also called sex alignment surgery, or sex reassignment surgery) when she was 19 years old. Jenna had already made it to the finals in the Miss Vancouver pageant when officials told her she could no longer continue because "the rules state that each contestant must be a 'naturally born female'" (Pullman, 2012). Jenna knew when she was four years old that she was a girl, even though her anatomical parts did not match the feelings she was experiencing. She felt she was a "naturally born female," but because her physical self did not quite fit her gendered self, she was ousted from the competition. It was only after a huge public outcry that the Miss Universe pageant changed its rules to allow transgender women to take part in all of its competitions (Reuters, 2012).

Attraction

The final component of the –Ness Model involves sexual and romantic attraction, which we discuss in greater detail in Chapter 5: Sexuality. For Killermann, "attraction is the result of your subconscious interpretation of hormonal influences on your brain chemistry, and your ability to make sense of attraction is a result of your socialization and self-awareness" (Killermann, 2017, p. 128). He argues that attraction is more about whom you are attracted to, rather than about you; people are attracted to certain kinds of people, certain expressions of masculinity and femininity, certain physical manifestations of sex and gender, and certain gender or self-identities as they relate to relationships and societal roles (Killermann, 2017). The –Ness Model divides sexual and romantic attraction into two separate components to highlight that a person can experience them in different ways. Each component is divided into two continua with "nobody" at one end of the spectrum, and "women and/or females and/or femininity" and "men and/or males and/or masculinity" at the other end of the spectrum, creating space for an infinite number of possibilities for which we experience attraction (see Chapter 5 for more detail on sexual diversity).

For Killermann, the –Ness Model is a descriptive, rather than a prescriptive, tool to use in the dialogue about gender and sexual diversity. His hope in building this tool was to illustrate that gender is more complex than the binary categories of man/woman and masculine/feminine, that biological sex is not a determinant of gender, and that gender identity and gender expression are components of gender that may or may not align with your biological sex (Killermann, 2013). What do you think? Is it possible for us to unlearn everything we have been taught to think about sex and gender and open our minds up to learning new ways of understanding the complexities of sex, gender, and sexuality?

ENDING THOUGHTS

Sex and gender are complicated subjects, even more complicated when we explore all the individuals who inhabit the spaces between and outside the traditional roles of male and female. Do you think life would be any easier if the construct of gender did not exist at all? What if you could grow up in a world where there was no pressure to conform to either role? That is exactly what is happening to a child named Storm in Toronto. Kathy Witterick gave birth to a baby named Storm in May 2011 and, along with her partner David Stocker, decided to keep the baby's sex a secret. There are only seven people who know if Storm is a boy or a girl, including the baby's two older brothers. After Storm's birth, the parents sent an email to friends and family, telling them they would not be announcing Storm's sex just yet: "A tribute to freedom and choice in place of limitation, a stand up to what the world could become in Storm's lifetime" (Poisson, 2011). The Wittericks are only one example of a family that is participating in what is being called the "gender neutrality movement," or **gender neutralism**, where there is a push for policies, language, and other social institutions to avoid distinguishing roles according to sex and gender in order to challenge gender prejudice and gender discrimination. Even Facebook has expanded its gender identifiers from traditional male and female options to include a "custom" option of over 50 more inclusive options, including agender, genderfluid, intersex, two-spirit, and other trans identities, in the United States and the United Kingdom (Goldman, 2014). The Canadian Senate is also reviewing Bill C-16, an act to amend the Canadian Human Rights Act and the Canadian Criminal Code to include gender identity and gender expression in the list of prohibited grounds of discrimination and hate propaganda legislation (Sibal, 2016). What will it take to shift traditional understandings of sex, gender, and sexuality in Canadian society? How will a shift in gender perspective influence the gender inequalities that have become rooted in some of the social institutions within our society?

> **Gender confirmation surgery:** Refers to a group of surgical options that alter a person's biological sex; also referred to as sex alignment surgery or sex reassignment surgery.
>
> **Gender neutralism:** A social movement that calls for policies, language, and other social institutions to avoid distinguishing roles according to sex and gender in order to challenge gender prejudice and gender discrimination.

FIGURE SKATING IS THE MOST MASCULINE SPORT IN THE HISTORY OF FOREVER

By Julie Mannell

In Judith Butler's novel *Gender Trouble*, she claims that masculine/feminine divisions were built upon homophobic cultural taboos that inform strict societal regulations of individual sexuality. These hetero normative acts serve the purpose of ratifying day-to-day expressions of our individuality as natural identifiers of our gender, and furthermore to marginalize and subjugate desires and expressions that fall outside of these artificial boundaries.

When I was 12 years old, I knew nothing of Judith Butler. That was, however, the first year I learned about sexual desire. It began innocently enough. I dreamed of being the first figure skater at the Pelham (Ontario) arena to make it to the Olympics. Fulfilling such a lofty ambition entailed hours on ice, practicing before and after school, special off-ice rehearsals where we jumped in shoes, and learning how to dance with a partner. For a while, my partner was an older coach from the former Soviet Union who would say "last chance" every time I went through my waltz routine. One day, I thought I'd call him out on giving more chances after saying each was the last, to which he responded, "In Russia when they say last chance, they mean last chance … and then they shoot you." He scared the shit out of me that day and I was excited when I was paired off with a younger boy instead.

His name was Isaac Molowenski, and he was the first boy to put his hand on my hip, rest his chest on mine, and evoke all the passion that could possibly transpire when you're in the sixth grade at a civic center with both of your mothers watching from the wooden bleachers. Even with his shimmering suspenders and elastic-waisted technical knit pants, I knew he was the sexiest thing. I felt it might be love.

On the ice, he was some sort of free dancing God. He'd just landed his double axel, and rumour had it he was starting work on his triple-toe-loop, an elemental feat comparable to driving a Ferrari. Yet, off the ice, the hockey players in neighboring change rooms would mutter "faggot" as they strapped cups on their crotches to protect their balls from one another. The other mothers would whisper over hot chocolate in the lobby, "How could anyone do that to their son?" And at school, he was just another kid with an elaborate collection of Pokemon cards. I'd felt I'd discovered a hidden gem in my practices. I may not have been the queen in classroom politics, but on the ice I could do a mean flying camel that would even have enticed Rudy Galindo during his sparkly-onesie "Send in the Clowns" phase.

I'd made a plan to coax Isaac's attention by coyly performing backwards spread eagles as he tried to work on his gold skills number. This merely annoyed him. Then I decided to tell my crush to Dolly Laverty who then told her mom who told Isaac's mom who giggled about it with my mom. The plan worked and I'd earned an invitation to his thirteenth birthday party—a hot tub party! I remember braiding my hair in pigtails like Britney Spears in the "Hit Me Baby One More Time" video and purchasing a brand new bikini at the local Giant Tiger. These things meant I was taking the party very seriously. I had sat up late at night dreaming about how the party would go and for some reason I'd made an unfair assumption. With all of the flak poor Isaac had received for somehow breaking with conventional notions of male-normative behaviour, I had assumed that I was the only girl who knew the truth—that he was the sexiest 12-year-old in the history of forever (or at least in Pelham in the spring of 2000). The strength of this assumption had planted a very solid image in my mind of us alone in his jacuzzi, sipping ginger ale, professing our undying passions to each other. Who else knew and could aptly appreciate the decadence of his Russian split or the suave curves to his twizzles?

I arrived ten minutes late and his mother offered me a bowl of chips. "Hurry on outside," she said, and hurry I did—through the living room, the newly refinished kitchen, the set of sliding doors, up the patio steps, and there he was, at the center of the hot tub in all of his white chested titillation. Surrounded by the entire synchro team.

That day Isaac broke my heart. The other girls' chests had filled out more and they too wanted a chance with the boy who could throw them, lift them, and death spiral them into a national championship. I would find a way to get over him, but the feeling would never leave: figure skating is the most masculine sport there is. Figure skating is the sexiest thing a man could do.

I believe the first time I wanked off, it had something to do with a fantasy of Elvis Stojko and me behind a Zamboni on our wedding night, but I digress. I do feel inclined to mention that figure skating has a bad rep. Surely there is a place for our Rudys and our Johnny Weirs and, of course, they are masculine as well. This has nothing to do with a specific aim of attraction, it's more in the essence of sexuality as a whole, as an energy you put out into the world that then draws people in. When you are skating, you are not an opinion, you are not a conversation, you are a body spinning, a body antagonizing gravity, a body alone against the elements. There are no team members, there are no supports. To watch male singles is to watch a solitary man with the force of the world on his back, a man tied to razor-thin blades, a man at war with nothing and everything, a man and physics, a man and cold air, a man as only a man—fighting himself, fighting the universe.

After Isaac, I moved on to the gutter punks at the local skate park. What amazed me was how, except for the clear disparity in musical accompaniment, what skate boys did after school was very similar to what we did after school. Their jumps had different names, their movements had a different aesthetic, and their tools were boards with wheels—but in essence they were still men jumping and twirling. The evidence is even in the lexicon: "figure skate" abbreviated to "skate," as if they wanted to exclude the bodily element, the figure, and hide what they were doing so that it would somehow be less sentimental, less attached to anything human or feeling. These boys were the cool boys and their coolness came with their blatant disregard for chivalrous

conventions—spitting on curbs, showing off scars, and smashing bones against concrete. How were their anti-establishment personae any less anti-establishment than a man putting on a sequined unitard and shamelessly shimmying to "The Samba?" I admire both for their individuality; however, I must confess that I've always been partial to any man who can wear lycra with confidence. For myself, the difference was in the fact that the skater boys were members of a collective with a similar semblance. The figure skaters, like Isaac, often stood alone—talking during ice time was frowned upon, and there was usually only one male anyhow—there wasn't necessarily a community, there wasn't any one ice-skating outfit that they all wore and fit them all. They were artists who used their body like a brush, artists who fashioned their routines and their outfits according to the colors of their movements. There is a sort of bravery in the solitude of men's singles.

In 2010, Skate Canada, basically figure skating's equivalent to the federal government, announced that they were planning to make men's singles more masculine. Essentially their aim was to "degay" the sport in order to attract hockey fans. There are several reasons such a statement is ludicrous. The first that comes to mind is the thought of my homeboys in Pelham—slightly balding, beer-bellied potato people in hockey jerseys, munching nachos on a Saturday sitting on some couch and loudly applauding the footwork of Alexei Yagudin. While the thought brings a smile to my face, it isn't realistically going to happen. These are men who grew up on hockey, men for whom it is entirely an issue of regional pride, familial alliance, and communal tradition, and not about gender performance. To assume these men would switch sports, renegotiate their fandom, simply to inhabit some ideal of their sex is a slap in the face to hockey fans, figure skating enthusiasts, and Canadian sports followers everywhere. I wouldn't want it anyhow.

Instead, this assertion reveals our own assumptions of gender. That for one reason or another we've ascribed certain cultural practices to men and women as implicit aspects of who they are as people—then gone so far as to project these assumptions onto sports. The proclamation that figure skating needs a "masculine makeover" in order to draw audiences then illustrates a society where love of sports is less of an individual autonomous act, and more of an effort to exert one's self as a certain kind of person to others. This seems absurd, and when contextualized within the framework of changing certain sports' traditions to better accommodate societal insecurities about sexuality, ultimately the message Skate Canada broadcasts to sports followers everywhere is that masculinity always wins and femininity loses. That the sports we love reveal the people we desire. As if a hockey jersey or a costume can dictate where we put our penis, non-penis, half penis, penis ambition, and so forth.

Both figure skating and hockey have an important place in our collective Canadian nostalgia—a time before we spent nights in Budget Inns, drinking cheap wine and proving some sort of point about who we are as men and women. They conjure memories of a world with twenty-five-cent hot chocolates, pay phones, wooden bleachers, smelly change-room showers and being people with people, doing what they loved because it made them feel good, because it made them a part of something, and for the pure reason that the arena was the center of most communities, a place for mothers to gossip, fathers to drink, and kids to be kids without the pressure of enacting prescribed gender codes. This is in an ideal world.

I won't ever marry Isaac Molowenski, not because he's gay (he's not), and not because I find him emasculated or defeated (he isn't). I won't ever marry Isaac Molowenski because he's now a successful computer programmer who makes good money, and I'm a lowly writer, with little to give but my words and little to gain but the (maybe) occasional twenty-five dollar cheque from some "up-and-coming" publication. He still skates, he gets many women, and I am happy for him, and would have been happy for him otherwise as well.

Sometimes I still think of Isaac as he was then, and the fact that, however light what I've written might be, I remember the real and excruciating pain he went through at a young age, simply because he preferred toe-loops to slap shots. I also remember how sexy he was to me and to other girls who hung around that rink, how my attraction revealed sexuality's multidimensional character, that it takes on many different forms, and how masculinity and femininity are never fully one or the other, and most certainly not static binaries.

Maybe you have a son who favors blades with picks. Maybe you have a daughter who'd rather hang around boys who can dance than boys who can spit, or do kick-flips in Volcom caps. Maybe, if you're lucky, they're really good at what they do and actually make it out of your small Ontario town and into the cast of Stars on Ice. Maybe. Just know that whatever any of it means, none of it is certain, and there is more to all of us than a sequined bow-tie or a goalie mask.

Source: "Figure Skating is the Most Masculine Sport in the History of Forever," Julie Mannell, previously published on The Barnstormer. (www.thebarnstormer.com).

DISCUSSION QUESTIONS

1. Male figure skaters, hockey players, and skate boarders are all extraordinary athletes, yet the labels that differentiate the groups differ significantly. Compile a list of stereotypes that identify each of the groups, and explain the formal and informal consequences they impose.
2. CBC television's *Battle of the Blades* paired female figure skaters and male hockey players, and pitted teams against each other in a figure skating competition. As suggested by the article's author, do you think this might have been an attempt to make figure skating more appealing for males?
3. Brendan Burke, the son of Toronto Maple Leafs General Manager Brian Burke, was an advocate for gays in professional sports. Openly gay himself, he tragically died in a 2010 car accident, just as his career as an aspiring social

activist started. Sexual orientation seems to be such an issue when it comes to male figure skating, yet in virtually every other sport, heterosexuality is the assumed norm. In 2009, 2 percent of the Canadian population self-reported to be either gay or bisexual (Statistics Canada, 2011). It is conceivable that some of those individuals are playing in the National Hockey League—or any other professional sports league. Analyze the reasons why we rarely read about homosexuality in professional sports, but often read about drug abuse, gambling, and womanizing.

KWIP

KNOW IT AND OWN IT: WHAT DO I BRING TO THIS?

The **"K"** in the KWIP process involves examining aspects of your own identity and social location as the first step in becoming diversity competent.

ACTIVITY: JOURNAL

How do you feel about living in a traditional "binary" world when it comes to gender? Are you aware of alternatives to the traditional dichotomies of male and female, man and woman, masculine and feminine? In your opinion, what do you think are the best and worst aspects associated with being male or female in a society that polarizes gender?

How central is your gender to your identity? When you describe yourself to someone else, does your gender identity or gender expression factor into that description? Imagine how your world would be different if your gender identity were different—what aspects of your life would change and how would you feel about those changes?

WALKING THE TALK: HOW CAN I LEARN FROM THIS?

The "W" in the KWIP process presents a scenario or case study that challenges you to "walk the talk" through problem-based learning.

ACTIVITY: CASE STUDY

Sandra has a 4-year-old daughter, named Alexa. Alexa loves to play with dolls and wear dresses, but she also loves to build Lego, play with trucks, and go outside and get really dirty. Sandra likes that Alexa doesn't restrict herself to traditional "girly" things and embraces both stereotypical boy and girl behaviours. One day, Alexa comes home from school visibly upset. When Sandra asks her what is wrong, Alexa tells her that her classmate, Charlie, came to school wearing a dress and pink boots today. She said that none of the boys seemed to care what he was wearing, but the girls got mad and wanted to know why he was wearing girl's clothes. Alexa said that she felt bad for Charlie because the girls were teasing him and you could tell that they were hurting his feelings. Alexa asked Sandra, "Mommy, why do you think Charlie came to school wearing girl's clothes? Does he want to be a girl?"

1. If you were Sandra, how would you respond to Alexa's questions?

2. If you were the teacher in the classroom that day, what would you have done (if anything) to ensure that Charlie felt comfortable in the classroom?

3. If you were Charlie's parent or guardian, how would you have responded to him wanting to wear a dress and pink boots to school?

IT IS WHAT IT IS: IS THIS INSIDE OR OUTSIDE MY COMFORT ZONE?

The "I" in the KWIP process requires you to honestly confront and identify ways in which our complex identities result in experiences of privilege and oppression, and to reflect on how we can learn to honour that privilege.

Learning about anything new is always a bit uncomfortable, and discussing our sex, gender, and sexuality can be intimidating at the best of times. Traditional binary constructions of sex, gender, and sexuality may be the norm for you, but that does not mean that they are the norm for everyone. Being open to how others choose (or do not choose) to define masculine and feminine is the key in this step.

ASKING: Do I have privilege?

1. I can use public facilities without stares, anxiety, and fear of verbal abuse, physical intimidation, or arrest.

2. I generally blend in to mainstream society, without being stared at, pointed at, laughed at, or ridiculed because of my gender expression.

3. I can get a job, rent an apartment, access healthcare, and get a loan without fear of rejection because of my gender identity or gender expression.

4. Strangers do not ask me about my biological status or genital appearance.

5. No one questions my state of mind or mental health because of my gender identity or expression.

REFLECTING: Honouring Our Privilege

Consider the complexities of gender discussed in this chapter and reflect on how you identify when it comes to your own gender identity and gender expression. Describe two circumstances in which you feel disadvantaged because of your gender. Can you also describe two circumstances where your gender may give you privilege over someone else? How can you use the advantages you experience because of your gender to combat the disadvantages experienced by others within or outside of your gender identity group?

PUT IT IN PLAY: HOW CAN I USE THIS?

The "P" in the KWIP process involves examining how others are practising equity and how you might use this.

The Prairie Valley School Division, a public school board comprised of 39 schools in 32 communities across south-eastern Saskatchewan, is taking a step forward for gender equity. The Divisional School Board has created all-gender washrooms in all of its facilities, including elementary schools. The Director of Education, Ben Grebinski, says it was particularly important to include elementary schools "because transgender children of all ages need support from educators" (Graham, 2016). There were no human rights claims to force change in this area (Graham, 2016). The Division consulted with parents and although there was some opposition, Grebinski said that once people became aware that the intent was to provide students with "an opportunity to be expressive and to be who they were on an individual basis without any kind of alienation, people were very willing and accepting" (Graham, 2016). "School boards across Alberta have since been given guidelines that say students have the right to use washroom and changeroom facilities aligned with their gender identity" (Graham, 2016). The guidelines further stipulate that "schools should provide a non-gendered, single-stall washroom for use by any student for any reason" (Graham, 2017). A number of other institutions are implementing gender-neutral, all-gender, or unisex washrooms, including Parliament Hill, where 37 of the 188 washrooms are unisex, and the Canadian National Exhibition in Toronto, with signage illustrating a half-female and half-male logo that says "We Don't Care" (Graham, 2016).

Located at www.nelson.com/student

Study Tools
CHAPTER 4

- Review Key terms with interactive **flash cards**
- Check your Comprehension by completing **chapter review quizzes**
- Gauge your understanding with *Picture This* and accompanying short answer questions
- Develop your critical thinking/reading skills through compelling **Readings** and accompanying short answer questions
- Apply your understanding to your own experience with **Connect A Concept** activities
- Evaluate Diversity in the Media with engaging *Video Activities*
- Reflect on your Understanding with *KWIP* activities

REFERENCES

American Academy of Pediatrics. (2015). Gender identity development in children. Retrieved on April 17, 2017 from *HealthyChildren.org*: https://www.healthychildren.org/English/ages-stages/gradeschool/Pages/Gender-Identity-and-Gender-Confusion-In-Children.aspx

Bigler, R., Roberson Hayes, A., & Hamilton, V. (2013). The role of schools in the early socialization of gender differences. Retrieved on April 17, 2017, from the *Encyclopedia on Early Childhood Development*: http://www.child-encyclopedia.com/sites/default/files/textes-experts/en/2492/the-role-of-schools-in-the-early-socialization-of-gender-differences.pdf

CCPA. (2015). The best and worst places to be a woman in Canada 2015: The gender gap in Canada's 25 biggest cities. Retrieved from *CCPA*: https://www.policyalternatives.ca/sites/default/files/uploads/publications/National%20Office/2015/07/Best_and_Worst_Places_to_Be_a_Woman2015.pdf

Connell, R. W. (1987). *Gender and power: Society, the person and sexual politics*. Crow's Nest NSW, Australia: Allen & Unwin.

Connell, R. W., & Messerschmidt, J. (2005). Hegemonic masculinity: Rethinking the concept. *Gender and Society 19*(6), 829–859.

Connell, R. W. (2016 March 5). *Masculinities: All about women* [Video File]. Retrieved on April 19, 2017, from *Sydney Opera House Talks and Ideas*: https://www.youtube.com/watch?v=B5fggib1-yw

Donovan, M. (2016). The gender divide: Gendered marketing of boys' and girls' toys on the rise, professor says. Retrieved on April 16, 2017, from *CBC News*: http://www.cbc.ca/news/canada/nova-scotia/toys-christmas-gender-marketing-blue-pink-boys-girls-1.3878977

Ferguson, S. J. (2016). Women and education. Retrieved on April 20, 2017, from *Statistics Canada*: http://www.statcan.gc.ca/pub/89-503-x/2015001/article/14640-eng.htm

Gallant, J. (2014). Twitter conversation about unreported rape goes global. Retrieved on April 21, 2017, from *The Star*: https://www.thestar.com/news/crime/2014/10/31/twitter_conversation_about_unreported_rape_goes_global.html

Goldman, R. (2014). Here's a list of 58 gender options for Facebook users. Retrieved on April 26, 2017, from *ABC News*: http://abcnews.go.com/blogs/headlines/2014/02/heres-a-list-of-58-gender-options-for-facebook-users/

Graham, J. (2016). All-gender washroom a big sign of inclusion in these Canadian schools. Retrieved on April 26, 2017, from *The Globe and Mail*: http://www.theglobeandmail.com/news/british-columbia/all-gender-washroom-a-big-step-toward-inclusion-for-these-canadian-schools/article33462989/

HeForShe. (2016). University Impact Champion: Feridun Hamdullahpur. Retrieved on April 22, 2017, from *HeForShe*: http://www.heforshe.org/en/impact/feridun-hamdullahpur

Intersex Society of North America (ISNA). (2008). What is intersex? Retrieved on April 26, 2017, from ISNA: http://www.isna.org/faq/what_is_intersex

Katz, J. (2013). Violence against women – It's a men's issue [Video File]. Retrieved on April 21, 2017, from *Tedx Talks*: https://www.youtube.com/watch?v=KTvSfeCRxe8

Katz, J. (Writer), & Earp, J. (Writer, Director, Producer). (2013). *Tough guise 2* [documentary]. Northhampton, MA: Media Education Foundation.

Kilbourne, J. (Writer), & Jhally, S. (Director). (2010). *Killing us softly 4* [documentary]. Northhampton, MA: Media Education Foundation.

Killermann, S. (2013). *Understanding the complexities of gender* [Video File]. Retrieved on April 26, 2017, from *Tedx Talks*: https://www.youtube.com/watch?v=NRcPXtqdKjE

Killermann, S. (2017). *A guide to gender: The social justice advocate's handbook*. Austin, TX: Impetus Books.

Kimmel, M. (2008). *Guyland: The perilous world where boys become men*. New York: Harper Collins.

Kimmel, M. & Holler, J. (2017). *The gendered society* (2nd ed.). Don Mills, ON: Oxford University Press.

LoBue, V. (2016). When do children develop their gender identity? Retrieved on April 17, 2017, from *The Conversation*: http://theconversation.com/when-do-children-develop-their-gender-identity-56480

Mannell, J. (2012, July 20). Figure skating is the most masculine sport in the history of forever. Retrieved from *The Barnstormer*: http://thebarnstormer.com/figure-skating-is-the-most-masculine-sport-in-the-history-of-forever-2/

McKay, B., & McKay, K. (2010, February 21). 8 interesting (and insane) male rites of passage from around the world. Retrieved from *The Art of Manliness*: http://artofmanliness.com/2010/02/21/male-rites-of-passage-from-around-the-world/

Moyser, M. (2017). Women and paid work. Retrieved on April 20, 2017, from *Statistics Canada*: http://www.statcan.gc.ca/pub/89-503-x/2015001/article/14694-eng.htm

Organisation for Economic Co-operation and Development (OECD). (2012). Closing the gender gap: Act now – Canada country notes. Retrieved on April 20, 2017, from *OECD*: http://www.oecd.org/gender/Closing%20The%20Gender%20Gap%20-%20Canada%20FINAL.pdf

Perkins, R. (1993, December). American Indian gender crossers. *Polare*. Retrieved from *gendercentre*: http://www.gendercentre.org.au/2article5.htm

Poisson, J. (2011, December 26). The 'genderless baby' who caused a Storm of controversy in 2011. Retrieved from *Toronto Star*: http://www.thestar.com/news/article/1105515--the-genderless-baby-who-caused-a-storm-of-controversy-in-2011

Pullman, L. (2012, March 24). Booted out for being born a boy: Transgender beauty queen kicked out of Miss Universe. Retrieved from *Mail Online*: http://www.dailymail.co.uk/news/article-2119786/Jenna-Talackova-Transgender-beauty-queen-kicked-Miss-Universe-Canada-pageant.html

Raptis, M. (2012, March 26). Miss Universe Canada disqualifies transsexual beauty queen Jenna Talackova. Retrieved from *National Post*: http://news.nationalpost.com/2012/03/26/jenna-talackova-miss-universe-canada-disqualified/

Reuters. (2012, April 10). Jenna Talackova forces Miss Universe transgender rule change. Retrieved from *National Post*: http://news.nationalpost.com/2012/04/10/jenna-talackova-forces-miss-universe-transgender-rule-change/

Sibal, A. (2016). Decoding Bill C-16: Does it threaten Canadians' freedoms? Retrieved on April 26, 2017, from *The McGill International Review*: http://mironline.ca/decoding-bill-c-16-threaten-canadians-freedoms/

Sinha, M. (2013). Measuring violence against women: Statistical trends. Retrieved on April 21, 2017, from: *Statistics Canada*: http://www.statcan.gc.ca/pub/85-002-x/2013001/article/11766-eng.pdf

Trudeau, J. (2016). Statement by the Prime Minister of Canada for Women's History Month. Retrieved on April 20, 2017, from *Government of Canada*: http://pm.gc.ca/eng/news/2016/10/01/statement-prime-minister-canada-womens-history-month

Turcotte, M. (2012, February 24). Women and education. Retrieved May 13, 2012, from *Statistics Canada*: http://www.statcan.gc.ca/pub/89-503-x/2010001/article/11542-eng.htm

Vernone, J. (2016). Sociologist explores how toys fuel stereotypes. Retrieved from *Sacramento State University*: http://www.csus.edu/news/articles/2016/11/28/sociologist-plays-with-the-idea-of-gendering-in-toys.shtml

Walk a Mile in Her Shoes. (2017). Welcome: Walk a Mile in Her Shoes: The International Men's March to Stop Rape, Sexual Assault, and Gender Violence. Retrieved on April 21, 2017, from *Walk a Mile in Her Shoes*: http://www.walkamileinhershoes.org/

White Ribbon. (2017). Who we are. Retrieved on April 21, 2017, from *White Ribbon*: http://www.whiteribbon.ca/who-we-are/

CHAPTER 5

Sexuality

"When kids are walking 'round the hallway plagued by pain in their heart, A world so hateful some would rather die than be who they are"

(Macklemore & Ryan Lewis, 2012)

LEARNING OUTCOMES

By mastering this unit, students will gain the skills and ability to:

- distinguish between the terms gender, sex, and sexuality

- analyze the social construction of sexuality

- investigate a broad spectrum of sexual identities, including gay and lesbian, heterosexuality, bisexuality, pansexuality, and asexuality

- compare and contrast monogamous and polyamorous sexual relationships

- examine current issues in sexuality, such as technology and intimate relationships, sexual education, and sexual violence in Canada

As discussed in the previous chapter, sex, gender, and sexuality interrelate, but are, in fact, distinct concepts. Sex refers to the biological components, such as chromosomes, hormones, internal reproductive organs, external genitalia, and secondary sex characteristics, that make up our physiological bodies. Gender, however, is a social construct that refers to a set of social roles, attitudes, and behaviours that describe people of one sex or another. Sexuality refers to all the ways in which individuals experience and express themselves as sexual beings; the components of sexuality discussed in this chapter will include the social construction of sexuality, sexual identities, intimate relationships, technology and intimate relationships, sexual education, and sexual violence.

As with gender, sexuality is socially constructed—we learn what is deemed appropriate and inappropriate through our culture and our experiences. In order to understand sexuality as a social construct then, we have to investigate how meaning is socially constructed around sexual desires, sexual identities, and sexual behaviours, and recognize that these meanings can change over time and from culture to culture. Sociologists John Gagnon and William Simon (1973) argue that we derive meaning about sexuality through shared beliefs within a particular social group. Like other forms of human behaviour, sexuality is learned as we internalize what Gagnon and Simon refer to as **sexual scripts**, or culturally created guidelines that define how one should behave as a sexual being. According to Gagnon and Simon, there are three levels of scripts that work together to help people make sense of their sexuality and sexual experiences: cultural scripts, interpersonal scripts, and intrapsychic scripts (1973). **Cultural scripts** indicate appropriate sexual roles, norms, and behaviours in a given society and are largely conveyed through mass media and other social institutions, such as government, law, education, family, and religion (Wiederman, 2015). Cultural scripts indicate which sexual behaviours are stigmatized, considered deviant, or deemed illegal, and which sexual behaviours are encouraged, accepted, and deemed appropriate, thereby providing context for sexual activity (Wiederman, 2015). **Interpersonal scripts** are created when individuals use the general guidelines they have learned from cultural scripts and adapt them to specific social situations (Wiederman, 2015). Interpersonal scripts are influenced by the individuals participating in the social encounter; when two or more people share similar scripts, the encounter can be positive, but when interpersonal scripts differ, problems can occur (Wiederman, 2015). The circumstances of each social encounter differ, causing people to build from, modify, and improvise previously adopted scripts. The ability to mentally rehearse outcomes before they occur is what is referred to as **intrapsychic scripts**. Intrapsychic scripts are internal and individual and can include fantasies, memories, and mental rehearsals of interpersonal scripts (Wiederman, 2015). Gagnon and Simon argue that all three levels of scripts work together in a fluid way, constantly changing and influencing a person's experiences with sexuality (1973). Scripts also vary with respect to social characteristics like gender and age; for example, cultural scripts for men and woman are different in a given society, and so are scripts for people of various ages. In this chapter, we will explore a range of sexual identities and examine issues pertaining to sexuality such as sexual attitudes and practices, intimate relationships, sexual health, and sexual violence.

SEXUAL IDENTITIES

Sexual identity is an all-encompassing concept that can include how we view ourselves as sexual beings (**sexual identity**), the ways in which we engage in sexual behaviours (**sexual expression**), and the romantic, emotional, and sexual attractions that we experience (**sexual orientation**). Similar to gender identity and expression, our sexual identities are not restricted to a choice between the two binary oppositions of heterosexual or homosexual, nor do they have to fall somewhere on a continuum between these two dichotomies.

Historically, in Canadian society, heterosexuality was viewed as the norm, and anything that deviated from that norm was considered

Sexual scripts: Socially created guidelines that define how one should behave as a sexual being—communicated through culture and learned through social interaction.

Cultural scripts: Indicate appropriate sexual roles, norms, and behaviours in a given society; largely conveyed through mass media and other social institutions, such as government, law, education, family, and religion.

Interpersonal scripts: Created when individuals use the general sexual guidelines they have learned from cultural scripts and adapt them to specific social situations.

Intrapsychic scripts: The ability to mentally rehearse sexual outcomes before they occur; internal and individual scripts based on previously adopted cultural and interpersonal scripts.

Sexual identity: How we view ourselves as sexual beings.

Sexual expression: The ways in which we engage in sexual behaviours.

Sexual orientation: The romantic, emotional, and sexual attractions that we experience.

abnormal. That changed when Alfred Kinsey started studying human sexuality in the 1940s and 1950s. Kinsey and his research team interviewed thousands of men and women about their most intimate sexual experiences, and they were among the first to publish the idea that sexuality could be measured on a continuum, suggesting that people were not necessarily exclusively heterosexual or exclusively homosexual (Kinsey, Pomeroy, & Martin, 1948). Kinsey and his colleagues went on to inform the world that there were many individual lifestyles that fell between the confines of strictly heterosexual and strictly homosexual. Instead of assigning people to the three categories of heterosexual, bisexual, and homosexual, the Kinsey team used a scale that ranged from 0–6, with an additional category of X–0, which included individuals who identified themselves as exclusively heterosexual; 1 included those who identified as predominantly heterosexual with incidences of homosexuality; 2 described individuals as predominantly heterosexual with higher incidences of homosexuality; 3 included those who were equally heterosexual and homosexual; 4 described individuals as predominantly homosexual with higher incidences of heterosexuality; 5 included those who identified as predominantly homosexual with incidences of heterosexuality; 6 described those who identified as exclusively homosexual; and X represented individuals who had no socio-sexual contact or relations (Kinsey, Pomeroy, & Martin, 1948).

Kinsey's work was groundbreaking at the time, but it was not without limitations. One of the criticisms was that the scale was too one-dimensional because it did not address levels of attraction, nor adequately account for those who identify as asexual, by labelling them "X" and placing them outside of the scale (Storms, 1980). As a result, Michael D. Storms proposed a two-dimensional model that placed homo-eroticism and hetero-eroticism on x and y axes, ranging from low to high levels of attraction, creating a place in the model for asexuality in addition to heterosexuality, bisexuality, and homosexuality (Storms, 1980). Another criticism of the Kinsey Scale was that it only measured and took into account behaviours at one point in a person's life. In an attempt to address this concern, Fritz Klein developed the Klein Sexual Orientation Grid in 1985, which considers a person's sexuality in the past, present, and an idealized future (Klein, 1993). Klein's grid contains seven variables that include not only sexual behaviour, but fantasies, emotions, and social behaviour, creating a far more comprehensive view of differing sexual identities. Kinsey, Storms, and Klein are only three of many scales used to measure and describe sexual orientation, but together, their research was pivotal in changing how we perceive sexual identity and sexual orientation.

In the next section, we will explore multiple sexual identities, including gay and lesbian, heterosexuality, bisexuality, pansexuality, and asexuality, and discuss how diversity not only exists between various sexual identities but also within each community itself. Additionally, it is important to recognize that some people do not identify with any of the sexual identities discussed in this chapter and reject the notion of being labelled with respect to their sexuality altogether.

Gay or Lesbian

A person who identifies as **gay** is romantically and/or sexually attracted to a person with the same gender identity and/or biological sex as themselves. The term *gay* is sometimes used to describe a man who is attracted to men, but it is also embraced by women who are attracted to other women. Some women in same-sex relationships also identify with the term **lesbian**. Often times, the term *homosexual* will be used synonymously with the terms *gay* or *lesbian*, but many people view this term as problematic because it has a long history of being used to pathologize gays and lesbians, or label them as psychologically abnormal or unhealthy in some medical fields (Subtirelu, 2015). Some gay rights activists and organizations, including The Gay and Lesbian Alliance Against Defamation (GLAAD), have included the word "homosexual" on their list of offensive terms and have encouraged news agencies to restrict the use of the word altogether (Peters, 2014). Part of the issue is that the term includes the word "sex," which puts the focus on the sexual activity rather than on the individual's identity (Subtirelu, 2015; Peters, 2014). GLAAD also argues that individuals should "avoid labelling an activity, emotion or relationship as gay, lesbian, or bisexual unless you would call the same activity, emotion or relationship 'straight' if engaged in by someone of another orientation" (GLAAD, 2009). Furthermore, the term *homosexual* can be problematic because in labelling a person's sexual orientation, it also requires a label with respect to gender identity and biological sex; consequently, it can isolate those who reject the notion of gender identity altogether or those who do not identify with the traditional sex binaries of male and female.

Gay: A person romantically and/or sexually attracted to a person with the same gender identity and/or biological sex as themselves.

Lesbian: A woman in a same-sex relationship who is romantically and/or sexually attracted to women.

IN THEIR SHOES

If a picture can say a thousand words, imagine the stories your shoes could tell! Try this student story on for size – have you walked in this student's shoes?

Do you ever wonder whether people are born straight or if they are born gay? I don't really. In the end, it doesn't matter to me if I was born this way or not. It doesn't change who I am and it doesn't make it any more or any less acceptable. All I know is that I am a woman who is in a relationship with a woman. Because of that, I have a different story than most. Growing up, I was one of the lucky ones. I was popular, outgoing, and athletic, and while definitely a tomboy, didn't endure any taunting, teasing, or even questioning about my sexuality. I dated men and didn't struggle with that. In fact, I never really felt that I had "different" feelings and didn't really feel that I was in any closet. I don't think I was fodder for rumours or speculation and, overall, had a pretty regular life.

In university, one of my friends told me that she was attracted to me. We started a relationship and decided to keep it a secret. We realized that once we were out, it would be tough to shed ourselves of that label if that was what we wanted. Neither of us knew what path we would take in life and neither of us wanted to have to decide immediately. After that relationship ended, I dated men again. I had some wonderful boyfriends and cared very deeply for them. The relationships were genuine and, once again, I didn't feel that I was shut in any closet or bound by any social norms.

I met a woman when I was 24 and knew that it was different. It was also the first time in my life that I started to struggle with the notion of "coming out." When, how, and what was going to happen? I was going through one of the most exciting times and was scared to death to tell people about it.

I was fortunate that my partner had been out for some time, so a lot of our common friends figured it out quickly. The next step was coming out to my family. I was having a tough time—I felt that I was being dishonest—excited, happy, and in love, but not sharing it. I started by coming out first to my brother, one of the most important people in my life. Easy—he was supportive and terrific. Next were my mom and my stepdad. It was tough and I was scared because I really didn't know what to expect. My mom's initial reaction was very supportive. She told my stepfather and, to my surprise, he accepted the news happily. As time went on, I saw that he was truly accepting my life and my choices. My mom, on the other hand, was not coping as well. She had a big birthday coming up and I mentioned that "we" would be happy to fly home for the celebration. She told me that I was welcome but that my partner was not. I was crushed. I was ready and excited to take the step to introduce my partner to my family and I was told that "we" were not welcome. I didn't go to the party. And, for two years, my mother and I rarely spoke.

During this time and after the experience with my mother, I was terrified to tell my father that I was gay. He grew up in a proper, affluent family, and I was worried that my lifestyle would be considered a mark on the family name. When I told him, he said he was thankful that I told him. Three days after, my dad called me to tell me that he "told the family my news." He had called everyone—aunts, uncles, grandmother—and told them that I was a lesbian. My initial reaction was shock but, ultimately, he was doing what he could to make things easier for me. I was proud of him and knew he was of me—it took a lot of pressure off of my shoulders and gave me a sense of peace knowing I wasn't hiding anymore. He flew to Ontario shortly after to meet my partner and was excited to know more about my life. When that relationship ended, my dad was very supportive and did all he could to help me get re-settled.

Since that time, I have grown significantly in terms of my own level of comfort with myself, my choices, and being secure with those decisions. I am in a relationship with a wonderful woman and have been for five years. She has two children and I am proud to be a stepparent to two wonderful boys. I own my own business and have terrific friends. My relationship with my mother has improved drastically and she is now pleased to be a "step-grandparent" and mother-in-law. We are all welcomed into the family like everyone else. My partner's family has also been very supportive and have welcomed me into their family without reservation.

It all sounds rather plain and normal? It is. We work, we go out with friends, we coach soccer, and we watch school plays. We have regular jobs, own a home, and have a dog and a cat. Overall, we really aren't that much different than any other family.

I realize how fortunate I am. I also believe that my family, friends, and others who have been there for me made a tough process a lot easier. What took me some time to realize was that I can't live my life afraid of what others may think—all I can do is live my life the way I choose. Daily, I give thanks for those who make coming out easier and safer for those around them. I truly hope that as our society becomes more understanding and accepting of diversity, others can have similar experiences, free from judgment, and full of acceptance!

In her project titled *Self-Evident Truths*, photographer iO Tillet Wright, attempts to illustrate the diversity that exists within the categories we create around sexual identity and sexual orientation. The project examines people who identify themselves as anything other than 100 percent straight, and illustrates their humanity through a simple photograph. Wright's goal with the project was to "show the humanity that exists within every one of us through the simplicity of the face" (Wright, 2012). Initially, she had a goal of photographing 2000 individuals, but the project took on a life of its own and soon she had revised her goal to 10 000 people (Wright, 2012). What she quickly realized was that her "mission to photograph 'gays' was inherently flawed, because there were a million different shades of gay" (Wright, 2012). Wright argues that not only do complexities exist between diverse sexual identities, but each individual is so complex that the problem is not that we have too many categories, it is that we have too few (Wright, 2012). She argues that individuals who identify as anything other than 100 percent straight share a common narrative that often includes struggle and experience with prejudice, but simply identifying as something other than straight does not necessarily mean that individuals have anything else in common (Wright, 2012). Wright's hope is that once we start seeing people in all their multiplicities, the binaries of gay and straight and the oversimplified boxes that we use to label our identities will cease to exist because they will in essence become meaningless and useless in describing what we see, who we know, and who we are (Wright, 2012).

Both in Canada and worldwide, there are a number of events that celebrate the diversity that exists within lesbian, gay, bisexual, transgender, questioning, and other sexually diverse communities, and promote awareness of issues affecting these groups. Most events take place in June to commemorate the 1969 Stonewall Riots that took place in New York City, considered to be a pivotal point in the start of the gay liberation movement. Events include cultural activities, festivals, marches, and parades organized by various not-for-profit organizations. Pride Toronto hosted Toronto's first ever Pride Month in June 2016, which culminated with the 36th annual Pride Parade (Pride Toronto, 2017). InterPride is an international organization that works to unite pride organizations on an international level, organizing World Pride events across the globe (InterPride, 2017). Canada hosted World Pride for the first time in 2014 in Toronto, with over 12 000 people participating in the parade at the end of the 10-day celebration (Mathieu, 2014).

Heterosexuality

A person who identifies as **heterosexual (or straight)** is romantically and/or sexually attracted to someone with the opposite gender identity and/or biological sex than they have. As with other sexual identities, heterosexuality can take on many forms and can mean different things to different people, often creating hierarchies within heterosexual groups.

In Canadian society, heterosexuality is much more than an individual sexual identity—it also represents an institutionalized ideology that can have significant social consequences for both heterosexuals and non-heterosexuals. First used by Adrienne Rich in her essay *Compulsory Heterosexuality and Lesbian Experience*, **compulsory heterosexuality** is the assumption that men and women are innately attracted to one another both emotionally and sexually, and that heterosexuality is natural and normal (Rich, 1980). An example of compulsory heterosexuality can be seen in Canadian society today when individuals are presumed to be heterosexual unless otherwise identified through the process of "coming out" as a non-heterosexual. The "coming out" concept itself reinforces heterosexuality as the norm and places the responsibility of identifying as something "other than" on the individual who deviates from that norm. Another example of compulsory heterosexuality lies in the controversial debate about whether sexual orientation is something you are born with or something you learn. The debate tends to focus only on the social conditioning of those who identify as non-heterosexuals, rather than questioning the social conditioning of those who identify as heterosexual. When we accept heterosexuality as a normal, expected outcome of development, we never question how one develops an attraction to the opposite sex, yet we question how one develops an attraction to same-sex partners and regard this behavior as abnormal (Habarth, 2008). It is this double-sided social regulation around sexual orientation that Rich challenges, and her conceptualization of compulsory heterosexuality was an

Heterosexual (or straight): A person romantically and/or sexually attracted to someone with the opposite gender identity and/or biological sex than they have.

Compulsory heterosexuality: The assumption that men and women are innately attracted to one another both emotionally and sexually and that heterosexuality is natural and normal.

important foundation for the concept of **heteronormativity** (Jackson, 2006; Habarth, 2008).

Heteronormativity builds off of Rich's work with compulsory heterosexuality and examines the ways in which heterosexuality is produced as a natural, unproblematic, taken-for-granted phenomenon that is maintained and reinforced through the everyday actions of individuals and through dominant social institutions (Habarth, 2008). Jackson argues that "normative heterosexuality regulates those kept within its boundaries as well as marginalizing and sanctioning those outside of them" (Jackson, 2006). Like compulsory heterosexuality, heteronormativity encompasses the idea that people are assumed to be heterosexual unless otherwise indicated, but it also examines the beliefs and attitudes about the social benefits of heterosexuality and how those beliefs and attitudes are socially constructed and reinforced by social institutions. When heteronormativity is institutionalized as a system of beliefs within society, it can lead to **heterosexism**, or the belief in the natural superiority of heterosexuality as a way of life and its logical right to social dominance. Comprised of a system of ideas and institutionalized beliefs, heterosexism leads to the oppression of any non-heterosexual form of behavior, identity, relationship, or community. As a systemic bias, heterosexism often leads to, intersects with, and can fuel **homophobia**, or the irrational fear, dislike, hatred, intolerance, and ignorance toward and the marginalization of non-heterosexuals, reinforcing the belief in the natural superiority of heterosexuality as a way of life.

Bisexuality

A person who identifies as **bisexual** is romantically and/or sexually attracted to people of their own gender as well as another gender. Bisexuality challenges the binary construction of heterosexuality and homosexuality as mutually exclusive and oppositional categories and, as such, can lead to the misconception that individuals who identify as bisexual are unsure of their sexual identity or simply move from heterosexual relationships to same-sex relationships in a fluid way throughout their lives. These misconceptions can lead to **bisexual invisibility** or the erasure and silencing of bisexual experiences, identities, and communities by presuming that individuals who identify as bisexual are in a temporary phase on their way to mature heterosexual or gay/lesbian identities (Barker et al., 2012). The denial of bisexuality as a genuine sexual orientation is a form of **biphobia** that is often reinforced by mainstream media. The article titled "Straight, Gay or Lying: Bisexuality Revisited" in the July 2005 edition of the *New York Times* is an example of such biphobia (Hutchins, 2005). The

article was criticized for insinuating that men who are attracted to more than one gender are actually repressing their true gay identities and fabricating their interest in women (Hutchins, 2005). Biphobia can also take the form of negative stereotypes about people who are bisexual; for example, the misconception that individuals who identify as bisexual are highly sexualized, promiscuous, and incapable of monogamous relationships. Furthermore, biphobia is often characterized as double discrimination because negative attitudes and assumptions about bisexual identities can come from both heterosexual and lesbian/gay communities (Barker et al., 2012).

Similar to other sexual identities, bisexuality is not a "one size fits all" sexual identity—diversity exists within the lived experiences of those who identify as bisexual, and there are often different understandings of the term *bisexuality* itself. Some individuals who identify as bisexual describe an attraction to both men and women, some identify as attracted to individuals regardless of gender, and some challenge the dichotomy and labelling of men and women altogether (Barker et al., 2012). What is true for one individual, group, or community is not true of all, and it is also important to recognize that people who are attracted to more than one gender might not use a bisexual label at all, or may use other labels, like pansexual or polysexual, to describe their sexual identity (Barker et al., 2012).

Pansexuality

Individuals who identify as **pansexual** are romantically and/or sexually attracted to members of all sexes and gender

---

Heteronormativity: Examines the ways in which heterosexuality is produced as a natural, unproblematic, taken-for-granted phenomenon that is maintained and reinforced through the everyday actions of individuals and through dominant social institutions.

Heterosexism: The belief in the natural superiority of heterosexuality as a way of life and its logical right to social dominance.

Homophobia: The irrational fear, dislike, hatred, intolerance, and ignorance toward and the marginalization of non-heterosexuals.

Bisexual: A person romantically and/or sexually attracted to people of their own gender as well as another gender.

Bisexual invisibility: The erasure and silencing of bisexual experiences, identities, and communities by presuming that individuals who identify as bisexual are in a temporary phase on their way to mature heterosexual or gay/lesbian identities.

Biphobia: The denial of bisexuality as a genuine sexual orientation, often reinforced by mainstream media.

Pansexual: A person romantically and/or sexually attracted to members of all sexes and gender identities and/or expressions.

identities and/or expressions. The prefix *pan-* is a Greek word meaning "all" or "every," and although the word *pansexual* was originally coined by Sigmund Freud in the early- to mid-1900s, its modern-day definition has little to do with Freud's psycho-sexuality theories (Jakubowski, 2014). Like all sexual identities, individuals who identify as pansexual may define their sexuality in unique ways—but the term itself eliminates the need to know a person's gender identity in order to understand their sexual identity (Jakubowski, 2014). Consequently, someone who identifies as pansexual, no matter what their gender identity/expression or biological sex is, can be attracted to any person of any gender identity/expression or biological sex (Jakubowski, 2014). For some individuals who identify as pansexual, gender distinctions that include cis men/women, intersex men/women, trans men/women, agender people, genderqueer, non-binary individuals, and any other combination of biological sex and gender identity, are essential to their understanding of their pansexual identity (Jakubowski, 2014). For others who identify as pansexual, these gender distinctions are irrelevant and unnecessary to their understanding of their pansexual identity (Jakubowski, 2014).

Pansexuality is different from bisexuality in that bisexuality refers to an attraction to the sex or gender identity that one identifies with personally, and also an attraction to other sexes or gender identities different from one's own; whereas, pansexuality refers to attraction to *any* and *all* sexes and gender identities (Jakubowski, 2014). In other words, while they are not completely different, they are definitely not the same because bisexuality often requires a person to identify with a particular gender themselves, and pansexuality does not.

Finally, a common misconception about individuals who identify as pansexual is that because they can be attracted to people of all sexes and genders, they are presumed to engage in sexual activity with people of all sexes and genders, but just because a person *can* be attracted to *anybody*, does not mean they *are* attracted to *everybody* (Jakubowski, 2014). Sexual orientation and sexual behaviour are independent of one another, and identifying as pansexual does not indicate anything about a person's sexual behaviour (Jakubowski, 2014). Although very little research has been conducted on pansexuality, it continues to challenge traditional ways of viewing the complex relationship that exists between sexuality and gender.

Asexual: A person who does not experience sexual attraction to any group of people.

Asexuality

The Asexual Visibility and Education Network (AVEN) defines individuals who identify as **asexual** as people who do not experience sexual attraction to any group of people. AVEN hosts the world's largest asexual community with a primary focus on creating acceptance and fostering discussion of asexuality, while facilitating the growth of the asexual community (AVEN, 2017). According to Karli June Cerankowski and Megan Milks, authors of the essay titled *New Orientations: Asexuality and Its Implications for Theory and Practice*, AVEN is one of the most comprehensive sources of information on asexuality, and its growth is changing the ways in which asexuality is viewed as a viable sexual and social identity (2010). Cerankowski and Milks emphasize that there is a distinct difference between those who experience decrease in sex drive or lack of sexual desire and are distressed by it, and those who do not experience sexual attraction and are not distressed by it (2010). They argue that until recently, research from medical and psychological fields have regarded asexuality as pathological, or psychologically abnormal or unhealthy, and there is a need to de-pathologize asexuality and recognize it as a distinct sexual identity (Cerankowski & Milks, 2010). Cerankowski argues that when we recognize the diversity of human sexuality, we can come to understand that some people just do not experience sexual attraction, and that does not necessarily mean there is anything wrong with them (Cerankowski & Milks, 2010). Anthony F. Bogaert, psychology professor at Brock University and author of multiple works on asexuality, agrees with Cerankowski and Milks in that the term *asexuality* should not be used to describe a health-compromised state, and he acknowledges an emerging social movement that has united a diverse group of individuals who identify as asexual (Bogaert, 2006). Like other sexual identities, there is considerable diversity among the asexual community, where individuals experience relationships, attraction, and arousal in different ways (AVEN, 2017). Some individuals who identify as asexual are happy on their own, while others seek out intimate, romantic relationships. Asexual individuals are just as likely to date sexual people as they are to date those who identify as asexual, and many experience attraction but do not need to act out that attraction sexually (AVEN, 2017). Some individuals who identify as asexual regularly experience sexual arousal, but sexual arousal does not necessarily drive a desire to find a sexual partner; others who identify as asexual feel little or no sexual arousal, but rarely view the lack of sexual arousal as a problem that requires intervention (AVEN, 2017). Some people who identify as asexual describe feeling a lack of sexual attraction for their entire lives, while others describe themselves as asexual for a brief period of time as they explore aspects of their sexual identities (AVEN, 2017). As with bisexuality and pansexuality, asexuality challenges the binary constructions of heterosexuality and homosexuality and confronts assumptions about the naturalness of heterosexuality.

The Power of Language

Language is an important aspect of our culture—words are saturated with meaning, and they can hold power over people. Consequently, we need to be conscious of the language we use to label and categorize ourselves, as well as diverse groups of individuals in society. In this section, we examine how language characterizing various sexual identities has changed significantly over the years, and investigate the relationship between language and power. The term *gay* emerged in the 1940s and 1950s as a slang term that referred to homosexual men and women, but during the 1960s and 1970s feminist movements, the term *lesbian* gained traction to highlight that lesbian experiences were distinct from the experiences of gay men (Zak, 2013). Bisexual and transgender activists at the time argued for inclusion in the language, and in the late 1990s were added to the discussion and the GLBT acronym became the common term used in the early 2000s (Zak, 2013). In the late 2000s, the acronym shifts to LGBT, most likely as a way to give lesbians more visibility, but the term *LGBT* did not encompass the experiences of all gender and sexual diversities. As a result, some have used LGBTIQ to include intersex, queer, and questioning identities, while others use LGBPTTQQIIAA + to encompass the identities of lesbian, gay, bisexual, pansexual, transgender, transsexual, queer, questioning, intersex, intergender, asexual, ally and the + symbol to represent any other gender and sexual diversities (Zak, 2013). Another commonly used form of the acronym is LGBTQQIP2SAA to represent lesbian, gay, bisexual, transgender, questioning, queer, intersex, pansexual, two-spirit (2S), androgynous, and asexual, noting that there is sometimes a third 'A' used to represent allies as well (Milligan, 2014). Greg Koskovich, curator for the GLBT History Museum in San Francisco, argues that naming various sexual identities is not only a discussion about respect, visibility, and the distribution of power, but it can lead to powerful social change, creating a world where people feel that their sense of selves and their experiences are recognized and honoured (Zak, 2013).

The term "queer" has also been reclaimed by some LGBPTTQQIIAA + communities to represent all individuals who fall outside of the gender and sexuality "norms" within society, but some still feel the isolation and hurt caused by its historical use as a derogatory slang term.

Some suggest that Gender and Sexual Diversities, or GSD, be used as a more inclusive term when discussing diverse gender and sexual communities. Therapists Pamela Gawler-Wright and Dominic Davies suggest that this term eliminates the listing of names "that kind of has an innate hierarchy to it" (quoted in Brathwaite, 2013) and is more encompassing of individuals who are in non-traditional relationships. Gawler-Wright argues that as society recognizes a larger spectrum of sexual identities at a time of more collective and inclusive thinking, we need a wider term to further that conversation (Queer Voices, 2013). She questions if the labels are in some way confining, or whether they actually matter, predicting that the language will constantly evolve in this area (Brathwaite, 2013).

The terms **androsexual/androphilic**, attraction to males, men, and/or masculinity, and **gynesexual/gynophilic**, attraction to females, women, and/or femininity, can be useful for individuals with a non-binary gender identity who want to describe their sexual orientation. A common misconception is that gender identity and sexual identity are interdependent, when in fact they are separate concepts that often relate but can also exist independent of one another. In other words, one does not necessarily determine the other—your gender identity is who you go to bed *as* and your sexual identity is who you go to bed *with* (@attn, 2017 February 10). We need to challenge assumptions about the transgender community and recognize that many people who are not cisgender can still identify as straight, gay, bisexual, pansexual, or asexual, and the most important thing a person can do to be an ally to the transgender community is to recognize that every trans person is different and not make assumptions about the community as a whole (@attn, 2017 February 10). Making an assumption about a trans person's sexual identity because of their gender identity or expression further reinforces heteronormativity and is a form of transphobia.

INTIMATE RELATIONSHIPS

Monogamy

Like sexual identities, intimate relationships are socially organized—some are considered appropriate and are encouraged and valued within society, while others are deemed problematic, discouraged,

> **Androsexual/androphilic:** Attraction to males, men, and/or masculinity.
>
> **Gynesexual/gynophilic:** Attraction to females, women, and/or femininity.

Matt Boles

Courtesy of Matt Boles

Courtesy of Matt Boles

Retired Superintendent with the Canadian Border Services Agency, Matt Boles has witnessed his fair share of prejudice and discrimination throughout his career in law enforcement. He argues that one of the most prevalent forms of prejudice in Canadian society today is directed at transgender communities and believes there is a great need for educational training in law enforcement and other community agencies to foster inclusivity and deconstruct stereotypes and assumptions about transgender individuals and groups. After a family member came out and identified as a lesbian, his experiences with transphobia hit closer to home for Boles and he began reflecting on what he could do as an ally to the transgender community to challenge transphobic attitudes and behaviours in his community. It wasn't until Boles was diagnosed with throat cancer that he really began to ponder how lucky he was to have so many successes throughout his life—a wonderful marriage, three beautiful children, and a successful career. He viewed his diagnosis as a rare gift, in that it made him realize how important it is to appreciate family, friends, and individuals within one's community. It was then that he realized how he could give back to his family, friends, and overall community and make a lasting impact on not only his own community but society as a whole.

He created the Als and Anes logo (pictured above) as a way to destroy the notion of "other" and provide language that is completely inclusive for all gender and sexual diversities (M. Boles, personal communication, January 5, 2017). He provides a conceptual framework for the logo, highlighting the meanings and messages within the design itself, and outlines ways to use the logo as an educational tool to neutralize prejudice and discrimination toward gender and sexual diversities.

Boles explains that the logo is blue because blue, as a colour, appeals most to the general public and is sometimes viewed to represent knowledge, power, integrity, and trustworthiness (M. Boles, personal communication, January 5, 2017). The ring represents the world as a whole, and the logo inside the ring is on a 45-degree angle because the groups that the logo represents are not "straight" (M. Boles, personal communication, January 5, 2017). The learning curve joining both of the signatures represents the social construction of prejudice and bigotry—if it is a learned notion, then it can also be unlearned (M. Boles, personal communication, January 5, 2017). The signatures themselves represent the spectrum of gender and sexual diversities that exist in society, pulling out the "al" or "an" from each of the names—lesbian, bisexual, transgender, transsexual, pansexual, while also including heterosexual and ally (M. Boles, personal communication, January 5, 2017). Boles recognizes that while it may not work with specific terms like gay, questioning, queer, or two-spirit, it encompasses umbrella terms that include these specific communities, like transgender communities, sexual orientations, and sexual minorities (M. Boles, personal communication, January 5, 2017). Boles hopes that the logo can be used as an educational tool in training and workshops or, at the very least, it can be used to start a conversation about inclusivity and acceptance of gender and sexual diversities. He believes that the power of the logo can have a ripple effect of social change, starting with one person at a time, moving on to one household, one neighbourhood, one community, and eventually society as a whole (M. Boles, personal communication, January 5, 2017). He has trademarked the logo and plans to use any proceeds from it to give back to the community.

Courtesy of Matt Boles

and stigmatized within society. In Canada, we idealize romantic love and **monogamy**; we expect and assume that everyone wants to pursue a relationship built on romance, love, and sexual exclusivity, and we hold it as the gold standard of relationships (Bielski, 2015). Consequently, anything that deviates from the gold standard of monogamy is often regarded as abnormal, problematic, and sometimes deviant. Although love, intimacy, and sexual expression are intertwined, they are not interdependent—and perceptions about love, intimacy, and sexual expression are culturally specific and change from time to time. If we examine the sexual behaviour of kissing as an example, we recognize that kissing is culturally significant in Canadian society, often regarded as an expression of love and/or a romantic or sexual gesture (Bever, 2015). But the kiss is not a form of sexual expression cross-culturally: researchers who studied 168 cultures only found evidence of romantic or sexual kissing in 77 societies, or 46 percent (Bever, 2015). This reminds us that not all behaviours that seem so normative to us actually occur in the rest of the world, or some may be viewed in a different way by other cultures or communities (Bever, 2015).

For many Canadians, monogamy and marriage are interrelated, but marriage is not always indicative of a monogamous relationship, nor do you have to be married to be considered monogamous. Some view monogamy as sexual exclusivity in any relationship, others believe it to mean only being married to one person at one time, and some regard monogamy as one relationship that spans an entire lifetime. Regardless of how monogamy is defined by any one person within society, there are some dominant assumptions that exist about monogamy in Canadian society. One of the dominant assumptions about monogamy is that it builds trust and respect, and validates love (Conley et al., 2012). We often assume that being in a monogamous relationship signifies safe, sexual satisfaction, or is likely to improve the quality of sexual experiences (Conley et al., 2012). These dominant assumptions can often lead to the misconceptions that non-monogamous relationships are not built on trust, respect, and love, or that people cannot engage in safe, meaningful sexual encounters outside of a monogamous relationship. Moreover, many people believe that individuals in monogamous relationships are not supposed to experience attraction to people outside of that relationship, and if they do, it somehow negates the love or commitment that they have for their partners (Bielski, 2015). Sex advice columnist Dan Savage coined the term "monogamish" to characterize relationships where couples can be sexually attracted to others outside of their monogamous relationships, without stepping outside of the relationship sexually

(Bielski, 2015). In her Canadian Ted Talk, "Rethinking Infidelity…a talk for anyone who has ever loved," sex therapist Esther Perel proposes that infidelity does not have to mean the end of a relationship, but rather can teach us a great deal about what we expect, what we think we want, and what we feel entitled to (Bielski, 2015; Perel, 2015). Along with other authors, experts, and researchers, Perel is starting a new conversation about monogamy and infidelity. In her forthcoming book, *A State of Affairs: Cheating in the Age of Transparency,* set for publication in October 2017, Perel questions why we problematize cheating when it is so common, and proposes that the security and familiarity of a monogamous relationship is actually an opposing force to the passion many of us seek in that same relationship (Bielski, 2015). Savage agrees, stating that "monogamous commitments are really at war with something else we want from our relationships, which is a passionate sex life" (Bielski, 2015). Savage and Perel, along with other experts, explore the relationship of monogamy and infidelity along with other topics of sex and love in the upcoming documentary by Tao Rusoli, "Monogamish." Savage and Perel are not alone in this new conversation around monogamy and infidelity; research psychologist Christopher Ryan argues that monogamous marriage does not necessarily come naturally to human beings (Bielski, 2016). Tracing it back to our primate ancestors, Ryan investigates how our expectations of sexual exclusivity and our attitudes about sexual jealousy are socially constructed in North America (Bielski, 2016). Lucia O'Sullivan, psychology professor at the University of New Brunswick, agrees with Ryan, stating, "Monogamy fights our natural instincts" (Bielski, 2016). O'Sullivan, who has conducted some groundbreaking research on infidelity in Canada, finds that people do not necessarily cheat because they are unhappy—happy people cheat, too, making almost anyone vulnerable to infidelity (Bielski, 2016). She questions, if sexual exclusivity is so endorsed as a cultural ideal in Canada, then why is infidelity, in all of its forms, so common? Additionally, why have we culturally defined infidelity as the ultimate betrayal—the absolute worst thing that could happen in your relationship? Perel, Savage, Ryan, and O'Sullivan are participating in a new conversation that challenges traditional expectations in intimate relationships. Consequently, we are starting to see alternative forms of relationships present themselves in North America; in addition to "monogamish" relationships, another alternative to monogamy is polyamory, or intimate relationships with multiple people at one time.

Monogamy: A relationship founded on sexual exclusivity.

Polyamory

Polyamory is defined as the state of being in love, or romantically involved with more than one person at the same time—the term *poly* is a Greek word meaning "many," and *amory* is Latin for "love," translating literally to "many loves" (Klesse, 2006). Polyamory is distinct from the concepts of **polygamy**, meaning multiple wives, and **polyandry**, meaning multiple husbands. Unlike polygamy, which is illegal in Canada, polyamory is legal because no one is married to more than one person at the same time; there is nothing in the Criminal Code that prevents three or more consenting adults from forming a family relationship in whatever way they see fit (Crawford, 2016).

Polyamory can cover a wide spectrum of relationships and sexual practices—most polyamorists use the term *partner* to characterize multiple relationships of varying degrees of intimacy and commitment (Klesse, 2006). Polyamorist relationships can include one or two primary partners and any number of secondary and/or tertiary partners, each with different expectations, characterized by the amount of time spent together and level of devotion (Klesse, 2006). Unlike other sexual relationships involving multiple partners, like 'swinging' or casual sex, love is central to polyamorous relationships, and not all polyamorous relationships are sexual (Klesse, 2006). Many polyamorist families describe their relationships as "responsible non-monogamy" where honesty and transparency are key facets of the relationship (Klesse, 2006). Although polyamorous unions are not recognized legally, they tend to resemble monogamous relationships when it comes to love, long-term commitment, and security (Klesse, 2006).

John-Paul Boyd, Executive Director of the Canadian National Research Institute, conducted the first national survey on polyamorous families and found that of the 550 people who responded, most of Canada's polyamorists live in British Columbia and Ontario, followed by Alberta (Crawford, 2016). Half of the respondents reported having relationships that involved three people, most of which lived across two households, with 23 percent of respondents indicating that at least one child was living full-time in the household (Crawford, 2016). Boyd predicts that as the polyamorist population grows, Canadian Family Law may have to adapt to include this unique style of family the way it did to accommodate common-law, same-sex, and co-parenting families (Crawford, 2016). Currently, the law only recognizes families comprised of two adults plus children, so polyamorist families are not able to claim multiple partners as dependents with social services benefits and are not protected in the event of illness or death (Crawford, 2016).

Polyamory: The state of being in love, or romantically involved with, more than one person at the same time.

Polygamy: Having multiple wives.

Polyandry: Having multiple husbands.

CURRENT ISSUES IN SEXUALITY

Technology and Intimate Relationships

There is no question that technology has changed the way we connect and communicate with one another since the dawn of the digital revolution, but there is some debate about whether dating technology has changed the way we form intimate relationships, or if the way we form intimate relationships has shaped new dating technologies. On the one side of the argument, some researchers believe that dating apps do not revolutionize sociology and psychology or change what we want, they just shape logistics and give us better access to it (Speed, 2015). They suggest that dating apps and social media change the pace of romance and dating, speeding it up so much that people can miss out on opportunities if they do not act quickly enough (Speed, 2015). Business psychology professor Tomas Chamorro-Premuzic agrees, suggesting that human behaviour tends to drive technological change, rather than technological change shifting human behaviour (Chamorro-Premuzic, 2014). In other words, although dating apps and social media may make dating more accessible and efficient, they have not changed dating expectations and ideals. Instead, they may be magnifying and highlighting already existing ideals and expectations that were not so apparent before the rise in such technology. In his article, "The Tinder Effect: Psychology of Dating in the Technosexual Era," Chamorro-Premuzic argues that the process of dating has been gamified and sexualized, making the process of mobile dating more exciting than dating itself (Chamorro-Premuzic, 2014). He argues that Tinder fulfills some basic evolutionary and social needs: "it enables people to get along, albeit in a somewhat infantile, sexual and superficial way ... enables us to get ahead, nourishing our competitive instincts ... [and] enables users to satisfy their intellectual curiosity" (Chamorro-Premuzic, 2014). Chamorro-Premuzic believes that some dating apps, particularly Tinder, emulate real-world dating, in that people get to assess physical appearance and gauge levels of interest before starting preliminary conversations (Chamorro-Premuzic, 2014). He does not think mobile or online dating necessarily translates to long-term relationship success; however, he suggests that it reduces the gap between supply and demand, making the dating market more efficient and rational (Chamorro-Premuzic, 2014).

On the flip side of the debate, many researchers believe that technology has changed romance and dating, and not necessarily for the better (Hare, 2013; Hill, 2013; Strickland, 2013; Karahassan, 2015). One argument is that because online dating technologies and social media interactions eliminate the ability to read social cues like non-verbal communication and body language, it erodes the basic intuitions required for successfully meeting new people (Strickland, 2013). When we replace the actual display of emotions with carefully chosen emoticons, the authenticity of emotions comes into question (Strickland, 2013). Some critics of online dating argue that social media and dating apps are actual antisocial, creating communities of individuals that are highly connected but feel socially isolated (Strickland, 2013). They argue that the computer screen provides a shield of protection for those who are fearful of rejection (Strickland, 2013). Patricia Wallace, author of *The Psychology of the Internet*, argues that we create new rules and social norms for digital interfacing, rather than use already established rules and social norms in face-to-face meetings (Wallace, 2015). As a result, social norms are misunderstood and rules are more difficult to enforce (Wallace, 2015). In addition to new social norms, the concept of romance in online dating is evolving to include "unromantic techspeak" and less invested ways of communicating love and affection through the use of emoticons (Hare, 2013; Hill, 2013). Moreover, some argue that dating apps have fundamentally shifted how people perceive each other; the accessibility of apps has dehumanized dating, making individuals less real and more disposable to each other (Speed, 2015). Some even suggest that the activities that take place on dating apps like Match, eHarmony, Plenty of Fish, Grindr, Tinder, Her, and OkCupid illustrate a shift to a more disposable dating culture, characterized by "hookups" and short-term relationships based solely on physical and sexual attraction (Speed, 2015). What do you think? Are new forms of digital communication getting in the way of 'real' romance? Or can countless hours of instant messaging, texting, and emailing set up a solid foundation for a sustainable long-term relationship?

Sexual Education

Ontario's newly revised Health and Physical Education curriculum has sparked controversy over the use of anatomical language, the inclusion of gender and sexual diversities, and which key concepts are appropriate for which age groups (Csanady, 2016). The new curriculum includes four main sections for each grade—Living Skills, which focuses on understanding themselves, communicating and interacting positively with others, and learning to think critically and solve problems; Active Living, which focuses on active participation, physical fitness, and safety; Movement Competence, which focuses on skills for moving properly and with confidence; and Healthy Living, which focuses on healthy choices and understanding connections to everyday life (Government of Ontario, 2016). The portion that deals with sex education is titled "Human Development and Sexual Health" and falls under the Healthy Living section, comprising 10 percent of the overall curriculum (See Table 5.1).

One study indicated that as of June 2016, almost half of Ontarians polled, or 48 percent, approve of the curriculum changes, while approximately one-third, or 36 percent of Ontarians polled, disapprove of the new curriculum (Csanady, 2016). One of the major criticisms of those who disapprove of the curriculum is that certain key concepts are introduced too early and are not age-appropriate, but when you compare sex education curriculum across provinces, there is similarity and some concepts are introduced even earlier in certain provinces (Young, 2015). For example, children are expected to know the anatomical names for all body parts as early as kindergarten in British Columbia, Alberta, and Manitoba, and children are introduced to the concept of gender identity as early as Grade 1 in Saskatchewan, compared to Grade 3 in Ontario (Young, 2015). Interestingly, Quebec is the only province which includes the concept of love in its curriculum (Young, 2015).

Those who support the changes in curriculum argue that contrary to what some critics believe, the new curriculum does not sexualize children (Soh, 2016). One sex researcher suggests that the new curriculum could decrease the likelihood of sexual experimentation and "actually lead young people to choose to delay having sex, and to use condoms and other forms of contraception when they do become sexually active" (Soh, 2016). She argues that many children will turn to pornography in the absence of sex education and will likely develop warped expectations about sex if there is limited discussion about safe-sex practices, consent, and healthy relationships (Soh, 2016).

Sexual Violence

Intimate partner violence (also discussed in Chapter 13) refers to any intentional act or series of acts by one or both partners in an intimate relationship that causes injury to either person (Statistics Canada, 2016). It can include physical assault,

TABLE 5.1

Ontario Human Development and Sexual Health Curriculum

Section 4—Human Development and Sexual Health	
Grade	**Key Concepts**
Grade 1	Body parts Senses and functions Hygienic procedures
Grade 2	Stages of development Oral Health
Grade 3	Health relationships Physical and emotional development Visible/invisible differences, respect
Grade 4	Puberty—changes; emotional, social impact Puberty—personal hygiene and care
Grade 5	Reproductive system Menstruation, sperm production Emotional, interpersonal stresses in puberty
Grade 6	Development of understanding of self Understanding puberty changes, healthy relationships Decision making in relationships Stereotypes and assumptions—impacts and strategies for responding
Grade 7	Delaying sexual activity Sexually transmitted infections (STIs) and pregnancy prevention Sexual health and decision making Relationship changes at puberty
Grade 8	Decisions about sexual activity; supports Gender identity, sexual orientation, understanding of self Decision making, contraception Relationships and intimacy
Grade 9	Preventing pregnancy and STIs Factors affecting gender identity and sexual orientation; supports Relationships—skills and strategies Thinking ahead about sexual health, consent, and personal limits
Grade 10	Decision making, communication, healthy sexuality Misconceptions relating to sexuality Relationships—effects on self and others
Grade 11	Mental illness, addictions—causes, manifestations, effects on personal health and well-being Reproductive and sexual health; proactive health measures Skills for dealing with stressful situations Mental illness, reducing stigma
Grade 12	Skills and strategies for evolving relationships Identifying personal aptitudes and interests; developing life plans Maintaining health and well-being when independent Bias and stereotyping in media portrayal of relationships

Source: Adapted from Government of Ontario. (2016). *Sex Education in Ontario*. Retrieved on February 22, 2017 from Government of Ontario: https://www.ontario.ca/page/sex-education-ontario © Queen's Printer for Ontario, 2016. Reproduced with permission.

emotional abuse, sexual violence, and sexual harassment. Although women can be violent toward men and intimate partner violence takes place in same-sex relationships, research indicates that women are more likely to experience sexual violence at the hands of their male partners (Statistics Canada, 2016; Benoit et al., 2015). According to data collected in the 2011 Uniform Crime Report, sexual assault stands out as a gender-based form of violence, where "women were eleven times more likely than men to be a victim of a sexual offence … and twice as likely to report sexual victimization" (Benoit et al., 2015). Sexual violence is defined by the World Health Organization (WHO) as "any sexual act, attempt to obtain a sexual act, or other act directed against a person's sexuality using coercion, by any person regardless of their relationship to the victim, in any setting" (quoted in Benoit et al., 2015). Sexual violence can range from obscene name calling to rape and/or homicide and includes dating violence, the sexual exploitation of minors, and violence against individuals with vulnerabilities, as well as online forms of sexual violence, such as online threats or harassment (Benoit et al., 2015). According to police-reported data collected across Canada in 2012, there were over 21 900 incidences of sexual assault, 90 percent of which were female victims, and in nearly all of these incidents (99 percent), the accused perpetrator was male (Benoit et al., 2015). Research indicates that Indigenous women, women with disabilities, immigrant and refugee women, and women in the sex trade industry are particularly vulnerable to sexual violence (Benoit et al., 2015).

Additionally, of particular concern is the widespread sexual violence that is taking place on college and university campuses across Canada—national data indicates that girls and young women between the ages of 15–24 are the most likely victims of sexual violence (Benoit et al., 2015). "Surveys across colleges and universities across Canada and the United States indicate that approximately one-quarter of female students have experienced sexual assault or had someone attempt to sexually assault them" (Benoit et al., 2015). Despite the overwhelming evidence of a high prevalence of sexual violence on campus, very few colleges or universities in Canada have policies or college/university funded services that deal directly with sexual assault (Browne, 2014; Mathieu and Poisson, 2014). As of 2014, only nine Canadian colleges and universities had special policies to address sexual violence on campus: Lakehead University, The University of Guelph, Brock University, St. Mary's University, Acadia University, St. Francis Xavier University, St. Thomas University, Mount Allison University, and Western University (Mathieu and Poisson, 2014). There is speculation that some Canadian colleges and universities

avoid directly addressing sexual violence as a serious problem on campus for fear of negative publicity, but McGill University was the first to hire a harm-reduction liaison officer in 2014 whose first priority was to create a sexual-assault policy that clearly defines what sexual assault is and outlines the institutional supports available to victims (Browne, 2014). Additionally, the Bystander Initiative at the University of Windsor is regarded as one of the most respected examples of rape prevention in Canada (Browne, 2014). The university-funded project works to educate students on "how to recognize sexual assaults, intervene appropriately and support survivors" (Browne, 2014). Other Canadian postsecondary institutions are implementing sexual violence training and focusing on educating students about consent in order to combat sexual violence on campus.

Sexual violence against women is much more than a physical or emotional assault—it is rooted historically in patriarchy and gender inequality and consistently reinforced through societal attitudes and beliefs about women and men. For example, women are often sexualized more than men in mainstream media, portrayed as sexually available in television, film, and advertising, and women are often held to different sexual standards than men (further discussed in the Intersectionality section of this chapter titled Gender and Sexuality) (Benoit et al., 2015). These images and ideas, combined with dominant gender roles, can shape people's response to sexual violence and sometimes lead to **victim blaming**, or suggesting that women invite sexual violence or deserve to be sexually assaulted because they dress or behave in ways that are considered to be sexually suggestive or inviting (Benoit et al., 2015).

The systemic attitudes that minimize, ignore, and normalize sexual violence are what is referred to as **rape culture** (Kingston, 2015). The Women Against Violence Against Women (WAVAW) organization examines ways in which society creates rape culture by normalizing sexual violence so much that both men and women perceive things like sexually suggestive remarks, unwanted touching, or even rape itself as a normalized, expected fact of life (WAVAW, 2016). They suggest that sexual coercion is so normalized that we live in a society that teaches *how not to get raped* rather than *not to rape*, and the normalization of sexual violence "has saturated every

Victim blaming: Suggesting that women invite sexual violence or deserve to be sexually assaulted because they dress or behave in ways that are considered to be sexually suggestive or inviting.

Rape culture: Systemic attitudes reflected in victim blaming that minimize, ignore, and normalize sexual violence against women.

corner of our culture so thoroughly that people can't easily wrap their heads around what rape culture actually is" (Prochuk, 2013). Experts argue that rape culture can manifest itself in a number of different ways, but every time a victim of sexual assault is questioned about her sexually suggestive behaviour or provocative attire, or told that she needs to avoid getting drunk in order to avoid sexual assault, rape culture is reinforced and strengthened (Ferguson, 2016). Researchers suggest that rape culture is evident in every joke about sexual violence, every song that glorifies sexual assault, every television show or movie that portrays sexual violence as common or downplays the effects of rape or sexual assault (Wahl, 2014). Some researchers argue that rape culture is especially evident on postsecondary campuses when institutions do not prioritize education on consent, disregard the need for sexual assault policies, or neglect to release statistics on sexual assault incidences on campus to protect the reputation of the institution (Ferguson, 2016). In October 2016, the Social Sciences and Humanities Research Council of Canada (SSHRC) awarded Shaheen Shariff, Associate Professor with the Faculty of Education at McGill University, $2.5 million to address sexual violence on university campuses across Canada and internationally (Lee, 2016). The project is set for a term of seven years (2016–2023) and includes the partnership of 9 universities, 14 community partners, 24 academics, and 13 collaborators who will "engage the most comprehensive effort to date to change the culture and practices of sexual violence on Canadian university campuses" (Lee, 2016). Shariff says that this project will "propel universities into reclaiming their central role of research and education ... as it relates to deeply embedded intersecting forms of misogyny, sexism, homophobia, and related forms of discrimination—often described as 'rape culture'" (Lee, 2016).

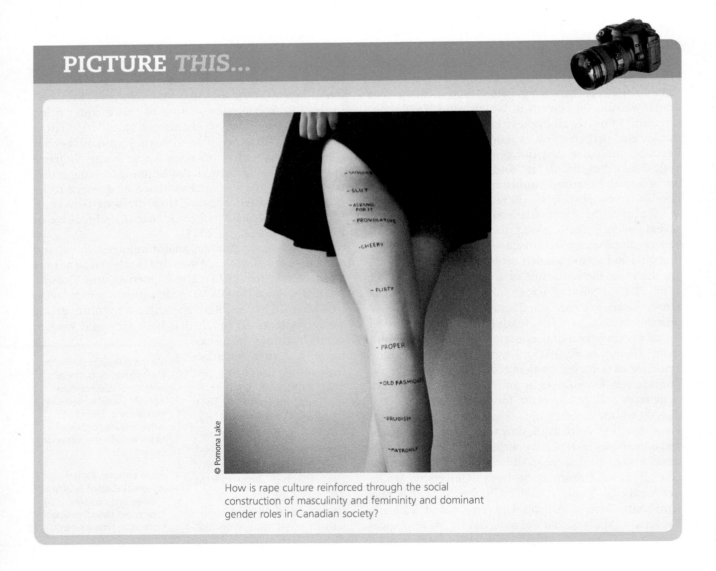

PICTURE THIS...

© Pomona Lake

How is rape culture reinforced through the social construction of masculinity and femininity and dominant gender roles in Canadian society?

INTERSECTIONALITY

Gender and Sexuality

The **sexual double standard** refers to the belief that there are different rules and standards of sexual behaviour for women and men. For example, some people believe that men are more sexual then women, and are expected to desire and seek out sexual opportunities, while women are criticized or stigmatized for doing the same thing (Bordini and Sperb, 2013). Researchers argue that the sexual double standard manifests itself in society when there are different standards for men and women when it comes to initiating sexual interaction, engaging in casual sex, or engaging in sexual activity with multiple partners (Bordini and Sperb, 2013; England and Bearak, 2014). Traditionally, the sexual double standard afforded more freedom to men when it came to sexual behaviours, and restricted the sexual behaviours of women. One study found that women face more negative judgements about engaging in casual sex or "hookup" culture, but also reported less interest in casual sex than men (England and Bearak, 2014). In the same study, researchers found that men were more likely to overestimate or exaggerate, while women were more likely to underestimate or underreport frequency of sexual intercourse or fellatio, suggesting that men see these acts as enhancing their status while women see them as diminishing theirs (England and Bearak, 2014). Mixed research indicates that there is considerable debate about the extent of the sexual double standard, and some studies argue that it no longer exists or has dramatically changed with modern views on sexuality (Bordini and Sperb, 2013; Vrangalova, 2014). Some recent studies are suggesting that there is evidence of a reverse double standard, where promiscuous men are judged more harshly than women who exhibit similar behaviour (Vrangalova, 2014; Papp et al., 2015). What do you think? Does a sexual double standard still exist in today's society? Are women judged more harshly than men when exhibiting similar levels of sexual desire or engaging in the same sexual behaviours?

ENDING THOUGHTS

The relationship between sex, gender, and sexuality is a complex one, and there is great diversity in the ways people experience and express themselves as sexual beings. Like gender, to better understand sexuality, in all of its forms, we must examine how it is socially constructed within our culture. Only then can we start to deconstruct heteronormativity and the prejudice and discrimination that result from it. When we recognize that language can hold power over people, we can predict the consequences of labels and recognize the importance of inclusive language in the naming of diverse groups. Whether we use a version of the LGBPTTQ-QIIAA + acronym, GSD, or Als and Anes to characterize ourselves and others, we must recognize that diversity exists not only between but within each community. Intimate relationships, like gender and sexuality, are socially organized with their own idealized expectations and standards that vary across time and place. As technology advances, there is great debate about whether it shapes the standards and expectations of intimate relationships or whether new technologies simply reflect existing cultural attitudes and assumptions. As our understanding of sexuality grows, so does education about human development and sexual health. At what age should we be educating our children and youth on difference and respect in gender and sexual diversities? Does sexual education sexualize children and youth or does it provide a foundation for healthy choices? Finally, as we examine sexual violence, we question the relationship between the social construction of gender and sexuality and sexual violence and ask how systemic attitudes about gender and sexuality influence concepts like the sexual double standard and rape culture in Canadian society.

> **Sexual double standard:** The belief that there are different rules and standards of sexual behaviour for women and men.

READING

RIGHTS FIGHT PROMPTS NEW TRANS-INCLUSIVE RULES FOR ONTARIO HOCKEY

By Wendy Gillis

Halfway through Grade 9, Jesse Thompson moved to a new school in a new city and faced a slew of disorienting new problems: how to make friends, how to succeed in school, how to fit in.

It was stressful—but it was also a welcome fresh start. Thompson, who is trans, was born a girl but had transitioned to a boy. No one from his new school or community in Oshawa would know him as anything other than a sports-loving teenaged guy.

Eager to find a sense of belonging, Thompson signed on to play hockey, his longtime safe haven.

"Hockey for me was a way to escape what was actually going on—school, personal stuff," Thompson, now 19, told the *Star* in an interview Wednesday. "Then that started causing me more problems, too."

What began as his amateur hockey league's staunch refusal to let Thompson change in the boys' room prompted a complaint to Ontario's Human Rights Tribunal to challenge long-held dressing-room policies within Ontario's minor hockey league—rules Thompson says ultimately outed him to his teammates and exposed him to harassment.

Thompson's complaint culminated this week in the Ontario branches of Hockey Canada posting transgender-inclusive policies for the upcoming hockey season, including a rule stating players identifying as trans can use the dressing room corresponding to their gender identity.

"I'm feeling really good," said Thompson. "It's going to take a while to see that change is actually happening, but eventually it's going to get to the point that transgendered kids are going to be playing sports more, because a lot of us quit because we start to feel uncomfortable and not accepted anymore."

Hockey Canada's inclusive policies also include ensuring trans players be addressed by their preferred pronoun and name, and that they have the confidentiality of their transgender status respected.

Hockey Canada's Ontario members have also committed to educate its more than 30 000 coaches and trainers on transgender inclusiveness by 2017.

The changes come as a result of a 2013 complaint Thompson filed against Hockey Canada, alleging discrimination based on gender identity. At the time, Hockey Canada had a dressing room policy requiring male and female players 11 and older to change in separate rooms. The policy was based on anatomical sex, not gender identity.

The Ontario Human Rights Commission intervened on the case, which resulted in a settlement in 2014. It took until this year for new policies to be developed and implemented.

By the time the complaint was filed, it was already too late for Thompson; noticing Thompson never changed in the boys' dressing room, Thompson says his teammates realized he was different.

"I wanted to quit because everyone thought of me as the little sister of the team, because they knew I was a girl because I wasn't allowed to change with them. I didn't want to play anymore."

Pushed out of frustration to lodge the human rights complaint, Thompson initially felt nervous about putting his name front and centre on the issue.

"I didn't want people to think about me being transgendered, I just wanted to be Jesse the guy. But then I realized I had to let people know that someone was standing up for them," he said.

Renu Mandhane, chief commissioner of the Ontario Human Rights Commission, praised Thompson for being "extraordinarily courageous" in taking up the fight.

"He took an experience that I think at the time was quite upsetting to him, and saw that it could really be something that could really make a change for the next generation of hockey players."

For children or teenagers only just discovering their identities, she said, it's especially important that institutions such as schools or sports teams have inclusive policies.

"Sports can be a real driver for inclusion, it can be a moment where other kids meet a trans kid or meet somebody who may be different from them and we don't want there to be barriers to that happening," she said.

The policies have yet to be adopted by Hockey Canada's other provincial branches. Both Mandhane and Thompson hope there will be a domino effect, and that other sporting associations, clubs, and teams will follow suit (Gillis, 2016).

DISCUSSION QUESTIONS

1. In the article, Thompson talks about being forced to expose his gender identity when he was "outed to his teammates and exposed to harassment." How are the policies and practices of Hockey Canada examples of systemic discrimination? How do these policies and similar policies reinforce notions of heteronormativity?

2. Can you see other examples of systemic discrimination within our society that also reinforce notions of heteronormativity? Where do they come from and how are they continually reinforced through social institutions?

3. Who should bear the most responsibility for shifting perceptions of trans communities? What do you think it will take for other provinces to adopt similar trans-inclusive policies and practices?

KWIP

KNOW IT AND OWN IT: WHAT DO I BRING TO THIS?

The "**K**" in the KWIP process involves examining aspects of your own identity and social location as the first step in becoming diversity competent.

Sexuality can be a very personal aspect of your identity and can sometimes be difficult to talk about. In this journal, we ask that you reflect on how sexuality is socially constructed and how you came to understand aspects of your own sexuality.

What do you remember most about the sexual education classes you participated in at school? Do you remember feeling confident in the knowledge these classes provided? How do your experiences with the sexual education curriculum compare to the newly revised sexual education curriculum in Ontario?

How did you come to understand aspects of your own sexuality? Did you have a traditional conversation with one or both of your parents about "the birds and the bees"? Did you feel like you had to reveal your sexual identity at some point? Or were there assumptions made about your sexual identity? As a youth, did you feel like you had a clear sense of what constituted a healthy intimate relationship? Can you identify cultural scripts that influenced your understanding of sexuality? Where were these cultural scripts most apparent?

How important is your own sexuality in terms of your identity? How does your sexual identity, sexual expression, and/or sexual orientation shape who you are?

WALKING THE TALK: HOW CAN I LEARN FROM THIS?

The "W" in the KWIP process presents a scenario or case study that challenges you to "walk the talk" through problem-based learning.

Cathy Halton recently graduated from university with a Bachelor of Arts degree in Education and has been working part-time covering a parental leave at Prince Phillip Elementary School in one of the Grade 7 classrooms. Ms. Halton loves teaching at this school, so when the principal offered her a full-time position, she gladly accepted. As a part-time faculty member, Ms. Halton never revealed that she identified as a lesbian because she didn't think her sexual orientation was relevant to her professional work. As a full-time faculty member, Ms. Halton believed it could come up, so she reluctantly mentioned to the principal that she was gay and in a relationship with a woman. The principal was surprised that she brought it up, as sexual orientation is not something he typically discusses in an interview but asked how Ms. Halton planned on handling it with her students. When Ms. Halton said that she didn't plan on revealing her sexual orientation to her students, the principal respected and supported her decision, expressing that his door was always open should she need any support in the future. Ms. Halton was relieved to not have to hide her sexual identity from her administrator and pleased that he was so supportive.

The school year continued smoothly and although Ms. Halton heard the occasional offensive comment and the usual heteronormative assumptions, she never had to directly discuss her sexual identity. She wasn't entirely comfortable with hiding who she was, but she also wasn't sure her students would be as supportive as the principal had been. Ms. Halton wondered if it ever came up how she would approach the situation and whether it would present challenges for her in the classroom.

One day in the middle of fourth period, Samuel, one of her more boisterous students, asked her in front of the class if she was gay. Ms. Halton wasn't so much surprised by the question as she was by the collective gasp of the rest of the class. She knew that she should tell Samuel that it was not an appropriate question to ask, but instead asked, "Why would you ask me that?" Samuel responded, saying that his brother had told him because he saw a photo of her and another woman on Facebook. Ms. Halton replied, "This is not an appropriate thing to discuss at school and my personal life is not up for discussion in this classroom. Let's get back to today's lesson."

Ms. Halton was disappointed with how she handled Samuel's questions. She wished she would have used this as a teaching opportunity and been more honest with her students. At the end of the day, she went in to see the principal to get his opinion on the situation. She was surprised to find out that the principal had already received a number of phone calls from parents that afternoon. Apparently, a couple of students had outed her on Facebook and parents were requesting to have their kids moved out of her class. Ms. Halton's heart sank as she asked the principal how he handled it. The principal replied, "I said that your sexual identity, whatever it may be, has absolutely nothing to do with your teaching. I told them you are an outstanding teacher and that I would not be transferring any students out of your class." Ms. Halton was relieved but was still concerned about how to approach this with her students. She felt that if she continued to avoid the subject, it would be the same thing as lying to her students and she was not comfortable with that. The principal agreed that she shouldn't have to avoid or lie about who she was, and they sat down to figure out the best way to discuss it with her students the following day.

1. In your opinion, how do you think Ms. Halton should address the rumours with her students? To what extent should she discuss aspects of her personal life in the classroom?

2. How do you feel the principal handled the concerns from parents? Should he have done more? If so, what?

3. Ms. Halton identified that she was subject to heteronormative assumptions, which take place in a context where people are assumed to be heterosexual and heterosexuality is deemed "normal." What are some of the ways in which you have witnessed heteronormativity in your experiences at school or in the larger community?

IT IS WHAT IT IS: IS THIS INSIDE OR OUTSIDE MY COMFORT ZONE?

The "I" in the KWIP process requires you to honestly confront and identify ways in which our complex identities result in experiences of privilege and oppression, and to reflect on how we can learn to honour that privilege.

ACTIVITY: GOT PRIVILEGE?

As iO Tillet suggests in her TED talk, "50 Shades of Gay," diversity exists not only between but also within various sexual identities, and we need to shift our perceptions from an "us-versus-them" dichotomy and recognize the humanity that exists within all of us. When it comes to power and privilege, how does the social construction of sexuality impact you? Does your sexual identity afford you certain privileges and power that you may or may not be aware of, or does it result in marginalization and/or sexual discrimination within society?

ASKING: Do I have privilege?

1. I can be confident that my friends, family, peers, and colleagues will be comfortable with my sexual identity.

2. I do not fear that revealing my sexual identity will result in negative economic, emotional, physical, or psychological consequences.

3. I can show signs of affection (hand-holding, kissing, etc.) toward my partner in public without being watched or stared at.

4. I am not asked to defend, justify, explain, or speak on behalf of others when it comes to my sexual identity.

5. When I read a magazine or a book, watch a television show or movie, or listen to music, I can be certain that my sexual identity will be represented.

REFLECTING: Honouring Our Privilege

Consider the various sexual identities discussed in this chapter and reflect on how you identify when it comes to your own sexuality. Describe two circumstances in which you feel disadvantaged because of your sexual identity. Can you also describe two circumstances where your sexual identity may give you privilege over someone else? How can you use the advantages you experience because of your sexual identity to combat the disadvantages experienced by others within or outside of your sexual identity group?

PUT IT IN PLAY: HOW CAN I USE THIS?

The "P" in the KWIP process involves examining how others are practising equity and how you might use this.

ACTIVITY: CALL TO ACTION

The It Gets Better Project began in September 2010 when American author and LGBT activist Dan Savage shared a video of himself and his partner Terry Miller on YouTube in an attempt to inspire hope for LGBT youth experiencing anti-gay bullying. The video was created in response to a number of students taking their own lives after being victimized by anti-gay bullying in school, and their message was that, yes, it does get better (Savage, 2010). As victims of anti-gay bullying in school themselves, Savage and Miller wanted to communicate to LGBT youth who are being bullied and feeling isolated that life does get better and that one day, they, too, will find happiness. Within days, hundreds of gay adults posted videos describing how their circumstances had improved since high school and hundreds of emails came pouring in from LGBT teens reassuring them that the videos were working and the message was being heard (Savage, 2010). Within two weeks, the YouTube channel had reached its maximum limit of 650 videos. Today, the project is now hosted on its own website at http://www.itgetsbetter.org and has become a worldwide movement with over 50 000 user-created videos from people of all sexual orientations that have been viewed more than 50 million times. Prominent Canadians, including Rick Mercer, George Smitherman, Mark Tewksbury, and the cast of MTV's *1 Girl, 5 Gays*, have submitted a video of support for alienated youth in Canada (Kupferman, 2010). Although he calls himself a very private person, Rick Mercer describes feeling a personal obligation to participate, and Canadian playwright and director Brad Fraser adds to the message saying, "no matter how hard it is now for them, it's not going to get easier … but if they want it to, it can get better" (Kupferman, 2010).

On March 22, 2011, the It Gets Better Project book was released. *It Gets Better – Coming Out, Overcoming Bullying, and Creating a Life Worth Living* is a collection of essays from various contributors, including celebrities, religious leaders, politicians, parents, educators, and youth dedicated to challenging prejudice and discrimination based on sexual orientation. The Project has become a place where young people who identify as lesbian, gay, bisexual, or transgender can receive messages of love and support, where allies can visit and support their friends and families, and where people can take the It Gets

Better pledge to create and inspire social change. Dan Savage recognizes that these videos will not solve the problem of anti-gay bullying in schools, but they can reach out to LGBT youth who are suffering, providing them with messages of hope and practical advice on how to survive what could be the toughest years of their lives (Savage, 2010). He argues that the It Gets Better Project can provide a space for people to unite and commit to the fight for safe school legislation and anti-bullying programs, and to deconstruct the prejudicial beliefs and acts of discrimination around sexual orientation (Savage, 2010).

To take the It Gets Better pledge, or for more information, visit http://www.itgetsbetter.org.

Study Tools CHAPTER 5

Located at www.nelson.com/student

- Review Key terms with interactive **flash cards**
- Check your Comprehension by completing **chapter review quizzes**
- Gauge your understanding with *Picture This* and accompanying short answer questions
- Develop your critical thinking/reading skills through compelling **Readings** and accompanying short answer questions
- Apply your understanding to your own experience with **Connect A Concept** activities
- Evaluate Diversity in the Media with engaging *Video Activities*
- Reflect on your Understanding with *KWIP* activities

REFERENCES

Asexual Visibility and Education Network (AVEN). (2017). About asexuality. Retrieved on January 12, 2017, from AVEN: http://www.asexuality.org/?q=overview.html

@attn. (2017 February 10). Fact: Not all trans people are gay. [Twitter Post]. Retrieved on February 17, 2017, from *Twitter*: twitter.com/attn/status/830093727536279552

Barker, M., Yockney, J., Richards, C., Jones, R., Bowes-Catton, H., & Plowman, T. (2012). Guidelines for researching and writing about bisexuality. *Journal of Bisexuality, 12*(3).

Benoit, C., Shumka, L., Philips, R., Kennedy, M. C., & Belle-Isle, L. (2015). Issue brief: sexual violence against women in Canada. Retrieved on February 23, 2017, from *Status of Women Canada*: http://www.swc-cfc.gc .ca//svawc-vcsfc/issue-brief-en.pdf

Bever, L. (2015). A kiss is not a kiss – In some cultures it's just gross, researchers find. Retrieved on February 12, 2017, from *The Washington Post*: https://www .washingtonpost.com/news/morning-mix/wp/2015/07/27/ romantic-kissing-is-not-a-shared-practice-across-cultures -research-shows/?utm_term=.79164d8e4a2c

Bielski, Z. (2015). A realistic look at monogamy, affairs, and the growth of "monogamishness." Retrieved on February 13, 2017, from *The Globe and Mail*: http://www .theglobeandmail.com/life/relationships/a-realistic -look-at-monogamy/article23671828/

Bielski, Z. (2016). The truth about infidelity: Why researchers say it's time to rethink cheating. Retrieved on February 13, 2017, from *The Globe and Mail*: http://www .theglobeandmail.com/life/relationships/the-truth-about -infidelity-why-researchers-say-its-time-to-rethink -cheating/article28717694/

Bogaert, A. F. (2006). Towards a conceptual understanding of asexuality. *Review of General Psychology, 10*(3), 241–250.

Boles, M. (2017, January 5). Personal Interview.

Bordini, G. S., & Sperb, T. M. (2013). Sexual double standard: A review of the literature between 2001 and 2010. *Sexuality and Culture, 17*(4), 686–704.

Brathwaite, L. F. (2013). Therapists argue to replace LGBT with more inclusive "GSD." Retrieved on February 1, 2017, from *Queerty*: https://www.queerty.com/therapists -argue-to-replace-lgbt-with-more-inclusive-gsd -20130223

Browne, R. (2014). Why don't Canadian universities want to talk about sexual assault? Retrieved on February 23, 2017, from *Maclean's*: http://www.macleans.ca/education/ unirankings/why-dont-canadian-universities-want -to-talk-about-sexual-assault/

Cerankowski, K. J., & Milks, M. (2010). New orientations: Asexuality and its implications for theory and practice. *Feminist Studies 36*(3), 650–664.

Chamorro-Premuzic, T. (2014). The Tinder effect: Psychology of dating in the technosexual era. Retrieved on February 21, 2017, from *The Guardian*: https://www.theguardian.com/media-network/media-network-blog/2014/jan/17/tinder-dating-psychology-technosexual

Conley, T. D., Ziegler, A., Moors, A. C., Matsick, J. L., & Valentine, B. (2012). A critical examination of popular assumptions about the benefits and outcomes of monogamous relationships. *Personality and Social Psychology Review 17*(2), 124–141.

Crawford, A. (2016). Canadian polyamorists face unique legal challenges, research reveals. Retrieved on February 12, 2017, from *CBC News*: http://www.cbc.ca/news/politics/polyamorous-families-legal-challenges-1.3758621

Csanady, A. (2016). One in six Ontario parents considered pulling kids from school over new sex-ed curriculum: Poll. Retrieved on February 22, 2017, from *National Post*: http://news.nationalpost.com/news/canada/canadian-politics/one-in-six-ontario-parents-considered-pulling-kids-from-school-over-new-sex-ed-curriculum-poll

England, P. & Bearak, J. (2014). The sexual double standard and gender differences in attitudes towards casual sex among U.S. university students. *Demographic Research, 30*(46), 1327–1338.

Ferguson, S. (2016). 8 appalling examples of how rape culture shows up on college campuses. Retrieved on February 23, 2017, from *Everyday Feminism*: http://everydayfeminism.com/2016/09/rape-culture-on-campuses/

Gagnon, J., & Simon, W. (1973). *Sexual conduct*. Chicago: Aldine Publishing Company.

Gillis, W. (2016). Rights fight prompts new trans-inclusive rules for Ontario hockey. Retrieved on January 4, 2017, from *The Star*: https://www.thestar.com/sports/hockey/2016/09/07/ontario-minor-hockey-moves-on-transgender-inclusiveness.html

GLAAD. (2006). GLAAD media reference guide – Terms to avoid. Retrieved on February 2, 2017, from *GLAAD*: https://www.glaad.org/reference/offensive

Government of Ontario. (2016). Sex education in Ontario. Retrieved on February 22, 2017, from Government of Ontario: https://www.ontario.ca/page/sex-education-ontario

Habarth, J. M. (2008). *Thinking 'straight': Heteronormativity and associated outcomes across sexual orientation* (unpublished doctoral dissertation). The University of Michigan, Ann Arbor, MI.

Hare, B. (2013). How technology has changed romance. Retrieved on February 20, 2017, from *CNN*: http://www.cnn.com/2013/02/12/tech/web/tech-romance-evolution/

Hill, K. (2013). Five ways technology has allegedly ruined dating. Retrieved on February 20, 2017, from *Forbes*: http://www.forbes.com/sites/kashmirhill/2013/01/14/five-ways-technology-has-allegedly-ruined-dating/#241e9fb56ad8

Hutchins, L. (2005). Sexual prejudice: The erasure of bisexuals in academia and the media. *American Sexuality Magazine 3*(4).

InterPride. (2017). About InterPride. Retrieved on January 17, 2017, from *InterPride*: http://www.interpride.org/?page=about

Jackson, S. (2006). Gender, sexuality and heterosexuality: The complexities (and limits) of heteronormativity. *Feminist Theory, 7*(1), 105–121.

Jakubowski, K. (2014). Pansexuality 101: It's more than 'just another letter.' Retrieved on January 21, 2017, from *Everyday Feminism*: http://everydayfeminism.com/2014/11/pansexuality-101/

Karahassan, P. (2015). How technology is changing dating. Retrieved on February 20, 2017, from *Psych Alive*: http://www.psychalive.org/how-technology-is-changing-dating/

Kingston, A. (2015). Thank you, Margaret Wente, for exposing rape culture. Retrieved on February 23, 2017, from *Maclean's*: http://www.macleans.ca/society/thank-you-margaret-wente-for-exposing-rape-culture/

Kinsey, A., Pomeroy, W., & Martin, C. (1948). *Sexual behaviour in the human male*. Philadelphia: W.B. Saunders.

Klein, F. (1993). *The bisexual option* (2nd ed.). Binghamton, NY: The Haworth Press.

Klesse, C. (2006). Polyamory and its 'others': Contesting the terms of non-monogamy. *Sexualities, 9*(5), 565–583. Retrieved on February 12, 2017, from *Sage Journals*: http://journals.sagepub.com/doi/abs/10.1177/1363460706069986

Kupferman, S. (2010). Canadian celebrities tell gay youth that it gets better. Retrieved on March 27, 2017, from *Torontoist*: http://torontoist.com/2010/11/canadian_celebrities_tell_gay_teens_that_it_gets_better/

Lee, C. (2016). Is there such a thing as "rape culture" on campuses in Canada? Retrieved on February 23, 2017, from McGill University: https://www.mcgill.ca/newsroom/channels/news/there-such-thing-rape-culture-campuses-canada-262861

Mathieu, E. (2014). Showing off a world of pride. Retrieved on January 18, 2017, from *The Star*: https://www.thestar.com/news/pridetoronto/2014/06/29/world_pride_12000_marchers_turn_downtown_streets_into_sea_of_colour.html

Mathieu, E., & Poisson, J. (2014). Canadian post-secondary schools failing sex assault victims. Retrieved on February 23, 2017, from *The Star*: https://www.thestar.com/news/canada/2014/11/20/canadian_postsecondary_schools_failing_sex_assault_victims.html

Milligan, K. (2014). Language matters. Retrieved on January 23, 2017, from *Citizens Project*: http://www.citizensproject.org/2014/02/20/language-matters/

Papp, L. J., Hagerman, C., Gnoleba, M. A., & Erchull, M. J. (2015). Exploring perceptions of slut-shaming on

Facebook: Evidence for a reverse sexual double standard. *Gender Issues, 32*(1), 57–76.

Perel, E. (2015, March). *Esther Perel: Rethinking infidelity... a talk for anyone who has ever loved* [Video file]. Retrieved on February 13, 2017, from Ted Talks: https://www.ted.com/talks/esther_perel_rethinking_infidelity_a_talk_for_anyone_who_has_ever_loved

Peters, J. (2014). The decline and fall of the "H" word. Retrieved on February 2, 2017, from *The New York Times*: https://www.nytimes.com/2014/03/23/fashion/gays-lesbians-the-term-homosexual.html?_r=2

Pride Toronto. (2017). About us. Retrieved on January 19, 2017, from *Pride Toronto*: http://www.pridetoronto.com/about-us/

Prochuk, A. (2013). Rape culture is real – and yes, we've had enough. Retrieved on February 23, 2017, from *WAVAW*: http://www.wavaw.ca/rape-culture-is-real-and-yes-weve-had-enough/

Queer Voices. (2013). 'Gender and sexual diversities,' or GSD, should replace 'LGBT,' say London therapists. Retrieved on February 2, 2017, from *The Huffington Post*: http://www.huffingtonpost.com/2013/02/25/gender-and-sexual-diversities-gsd-lgbt-label-_n_2758908.html

Rich, A. (1980). *Compulsory heterosexuality and lesbian experience*. Denver, CO: Antelope Publications.

Savage, D. (2010). Welcome to the It Gets Better Project. Retrieved on March 27, 2017, from ItGetsBetter.org: http://www.itgetsbetter.org/blog/entry/welcome-to-the-it-gets-better-project/

Soh, D. (2016). Let's make sure Ontario's sex-ed curriculum is here to stay. Retrieved on February 22, 2017, from *The Globe and Mail*: http://www.theglobeandmail.com/opinion/lets-make-sure-ontarios-sex-ed-curriculum-is-here-to-stay/article31605288/

Speed, B. (2015). Technology isn't ruining modern dating – humans are. Retrieved on February 21, 2017, from *New Statesman*: http://www.newstatesman.com/sci-tech/2015/08/technology-isnt-ruining-modern-dating-humans-are

Statistics Canada. (2016). Family violence in Canada: A statistical profile, 2014. Retrieved on September 16, 2016, from *Statistics Canada*: http://www.statcan.gc.ca/pub/85-002-x/2016001/article/14303-eng.pdf

Storms, M. (1980). Theories of sexual orientation. *Journal of Personality and Social Psychology, 38*(5), 783–792.

Strickland, A. (2013). The lost art of offline dating. Retrieved on February 20, 2017, from *CNN*: http://www.cnn.com/2013/02/12/living/lost-art-offline-dating/index.html

Subtirelu, N. (2015). Why the word homosexual is offensive. Retrieved on February 2, 2017, from *The Week*: http://theweek.com/articles/556341/why-word-homosexual-offensive

Vrangalova, Z. (2014). Is our sexual double standard going away? Retrieved on February 24, 2017, from *Psychology Today*: https://www.psychologytoday.com/blog/strictly-casual/201403/is-our-sexual-double-standard-going-away

Wahl, M. (2014). How rape jokes contribute to rape culture. Retrieved on February 23, 2017, from *Huffington Post*: http://www.huffingtonpost.com/madeline-wahl/how-rape-jokes-contribute_b_5240592.html

Wallace, P. (2015). *The psychology of the Internet*. Cambridge, UK: Cambridge University Press.

Wiederman, M. (2015). Sexual script theory: Past, present, and future. In J. DeLamater and R. F. Plante (eds.), *Handbook of the sociology of sexualities*. New York: Springer International Publishing, 7–22.

Women Against Violence Against Women (WAVAW). (2016). What is rape culture? Retrieved on February 23, 2017, from *WAVAW*: http://www.wavaw.ca/what-is-rape-culture/

Wright, iO. T. (2012, December). *iO Tillet Wright: Fifty shades of gay* [Video file]. Retrieved on February 17, 2017, from *Ted Talks*: https://www.ted.com/talks/io_tillet_wright_fifty_shades_of_gay/transcript?language=en

Young, L. (2015). Sexual education compared across Canada. Retrieved on February 22, 2017, from *Global News*: http://globalnews.ca/news/1847912/sexual-education-compared-across-canada/

Zak, E. (2013). LGBPTTQQIIAA + – How we got here from gay. Retrieved on February 14, 2017, from *Ms. Magazine*: http://msmagazine.com/blog/2013/10/01/lgbpttqqiiaa-how-we-got-here-from-gay/

Race and Racialization

"But if you only have love for your own race, Then you only leave space to discriminate."

("Where Is the Love," Black Eyed Peas, Elephunk, will.i.am Music Group. 2004)

LEARNING OUTCOMES

By mastering this unit, students will gain the skills and ability to:

- describe racism as a hierarchical system of power and privilege

- effectively address racism through the recognition and acknowledgment of racism as a part of Canada's past and a persistent reality in Canadian society today

- differentiate between forms of racism produced and manifest in a variety of practices, ideologies, and social relations

- reflect upon your own racialized privilege and oppression and how this can be relevant to anti-racism practices in the workplace, interpersonal encounters, and community relationships

David McGlynn/Getty

In the past, the concept of **race** was used to biologically categorize people based on physical characteristics, including skin colour. "This notion of race emerged in the context of European domination of nations and people deemed non-white" (University of Guelph, 2016). Genetic science now tells us that this biological concept of race is a myth (University of Guelph, 2016). And while the biological concept of race has been discredited, the social construction of race with a focus on difference between groups has the negative effect of marginalizing and oppressing some of these groups within a society. This process of social construction of race is referred to today as **racialization** (University of Guelph, 2016). The Ontario Human Rights Commission (2016) uses language to recognize and reflect the process of racialization by describing people as a **"racialized person"** or **"racialized group"** instead of the more out-dated and inaccurate terms "racial minority," **"visible minority,"** **"person of colour,"** or "non-White."

Racism is an ideology that one racialized group is inherently superior to others. This belief in racial superiority is associated with some of history's greatest atrocities. It is important to understand that racism operates to marginalize and oppress at a number of levels, including individual, systemic, and societal. Racism can be openly displayed as offensive jokes, emails, and hate crimes, but be more deeply rooted in prejudicial attitudes, values, and racialized stereotypes that can become embedded in systems and institutions over time, most notably in systems of criminal justice, policing, education, health, media, immigration, and employment (Roy, 2012).

As a nation, we have acknowledged and apologized to racialized groups affected by historical racism that is a part of Canada's past. Unfortunately, racism and racial discrimination remain a persistent reality in Canadian society, and this must be recognized and acknowledged as a starting point to effectively address racism (Ontario Human Rights Commission, 2016).

Consider the case of a student named Samuel, who found that as a black student studying in Canada, he was reminded every day that he was black. In fact, he described every day of his life in Canada as a personal experience in understanding racism. When asked what that meant, Samuel described daily experiences where people reminded him they didn't see the colour of his skin yet they stared at him, touched his hair, held their purses a little tighter, asked him where he came from, called upon him to speak for all black people, told him he was a role model for other students like him, and declared they were anti-racist because they had friends who were black. Samuel also described a heated discussion on racism in a class where he was one of two black students, during which time most students stared at him for his reaction. During the discussion, he listened while a student used the **reverse-discrimination** and so-called **race-card** arguments, and exasperated by their inability to recognize their own privilege, he thought to himself … try walking a day in my shoes. This is, in part, the inspiration for the title of this book.

THE CONCEPT OF RACE AS A BIOLOGICAL MYTH

So what is race? Some have used the concept of race to mean a biological division of humans based on physical attributes such as skin colour and other physical traits. And for hundreds of years, we have used these differences to draw a colour line that has categorized people into four or five groups we call races. Academics have erroneously equated biology with destiny for many years, usually with the effect of reinforcing white superiority.

During the 1820s and 1830s, American physician and scientist Samuel Morton conducted a systematic analysis of hundreds of human skulls from all over the world to confirm his hypothesis that there were differences among races not only in terms of their origin, but also in terms of their brain size. Morton

Race: A concept no longer recognized as valid, except in terms of its social consequences; in the past, the concept of race referred to biological divisions between human beings, based primarily on their skin colour.

Racialization: The process of social construction of race whereby individuals or groups are subjected to differential and/or unequal treatment based on their designation as a member of a particular "race."

Racialized person or racialized group: An individual or group of persons, other than Indigenous peoples, who are subjected to differential and/or unequal treatment based on their designation as a member of a particular "race."

Visible minority: Outdated term used primarily in Canada by Statistics Canada to refer to a category of persons who are non-Caucasian in race or non-white in colour and who do not report being Indigenous.

Person of colour: Considered to be an outdated term, which was originally intended to be more positive and inclusive of people than the terms "non-white" or "visible minorities"; was used to refer to people who may share common experiences of racism.

Reverse discrimination: Discrimination against whites, usually in the form of affirmative action, employment equity, and diversity policies; the concept of reverse discrimination, specifically reverse racism, is considered by many to be impossible because of existing power structures in society.

Race card: Term that refers to the use of race to gain an advantage.

created a hierarchy among different racial groups of their brain capacity based on skull measurements. Morton assigned the largest brain capacity to English Europeans and the smallest brain capacity to Africans and Australian Aborigines (American Anthropological Association, 2011). Neuroscience research today refutes the validity of Morton's findings, and we know that what Morton did was attempt to use science to legitimate a socially constructed inequality. After Morton, the work of early "race scientists" such as Josiah Nott, George Gliddon, and Louis Agassiz continued to attempt to prove that blacks and whites did not originate from the same species (American Anthropological Association, 2011).

In 1927, University of Toronto professor Peter Sandiford attempted to prove that racial and ethnic differences in IQ scores should be used by Canada as a selection criteria for new immigrants, encouraging the recruitment of the British, German, and Dutch, who scored well on IQ tests, and discouraging the recruitment of Poles, Italians, Greeks, and Asians who did not score as well on IQ tests (Walker, 2008). According

to Sandiford (cited in Walker, 2008, p. 197), Canada needed to protect itself from becoming a "dumping ground for misfits and defectives."

In the late 1980s, Philippe Rushton, a Canadian psychologist and professor at the University of Western Ontario, continued in the vein of earlier race scientists with theories about race and intelligence, suggesting that he had scientific evidence to prove whites are more intelligent than blacks, and that Asians are the smartest of all. In a televised debate on the University of Western Ontario's campus in 1989, geneticist David Suzuki said Rushton's ideas were "monstrous": "I did not want to be here. I do not believe that we should dignify this man and his ideas in public debate ... his claims must be denounced, his methodology discredited, his grant revoked and his position terminated at this university, this is not science" (CBC Digital Archives, 1989).

We find the theme of race as a biological myth illuminated in popular culture as well. In American comedian Dave Chappelle's parody titled "Racial Draft," Chappelle mocks the absurdity of racial classification

IN THEIR SHOES

If a picture can say a thousand words, imagine the stories your shoes could tell! Try this student story on for size – have you walked in this student's shoes?

WHAT ARE YOU?

I've been asked that question countless times. Whenever I meet someone new, they are immediately itching to find out my background. You'd think it would get annoying after a while, but in all honesty, I'd rather people ask than make assumptions.

I'm of mixed race. My father is African and my mother is European. People always comment on how unique that sounds, and I always tell them it's not unique at all. The world's mixed-race population is growing faster than ever in the 21st century. But despite all the strides it has made, things are still far from perfect.

Most people of mixed race have been fortunate enough not to have crises of identity. I, on the other hand, wasn't that lucky. All my life, I have been pushed and pulled in different directions to identify myself as either white or black. The problem with this is I'm both and neither.

Many other mixed-race individuals have fought this battle, and I will say this: it's hard to find your own identity when everyone else is trying to create one for you.

I have never wanted to be associated with any particular race. I am proud to be what I am and I don't

need to feel part of a "whole." I have never wanted to be labelled just black or just white or anything else that I am not, and I fail to understand why people can't acknowledge that fact.

I found it slightly shocking that the One Drop Rule is still referenced and relevant today. For those who may not know, the One Drop Rule stated that a person with as little as one drop of black blood in their heritage was to be considered black—a slightly racist idea, don't you think?

And so we have the rule, an echo of the United States' racist past used today, albeit unmentioned, to classify a person of mixed race as black.

People have the right to determine their racial, sexual, and national identities. No one, not even society, should have the right to tell anyone who or what they are. It is up to us to determine that for ourselves.

There are people of mixed race who feel more comfortable associating themselves with one side of their heritage, and there is really nothing wrong with that. But when all of us are expected to pick and choose, or just fall in line with illusionary expectations, then it becomes a violation of our freedom.

Source: Anonymous

as the Black delegation drafts Tiger Woods, the Jewish delegation drafts Lenny Kravitz, the Latino delegation drafts Elian Gonzalez, the White delegation drafts Colin Powell, and the Chinese delegation drafts the Wu Tang Clan. The writer of the "In Their Shoes" feature describes why the concept of race is not real in his life, other than in terms of its social consequences: "All my life I have been pushed and pulled in different directions to identify myself as either white or black. The problem with this is I'm both and neither."

Every day, as you walk the halls of your academic institution, you can see with your own eyes that human beings look different from one another. The shape of the face, the colour of the hair, the shape of the body, or the colour of the skin—all point to human physical variation. But this does not mean that we can neatly tuck everyone into a handful of groups based on these physical characteristics and then correlate these groups with attributes such as brain size, intelligence, or athletic ability. Nor does racism refer to the visceral reaction you may have to someone who has a different physical appearance. We have come to understand that the concept of race is not real—that there is no biological basis to the social categorizations we have constructed and used for the past several hundred years. Racism is really about how people assign meaning to that appearance. This is learned behaviour; and if we can learn it, we can unlearn it.

The concept of race is focused on the social construction of difference between groups, with the effect of marginalizing and oppressing some of these groups within a society. According to biological anthropologist Alan Goodman,

> To understand why the idea of race is a biological myth requires a major paradigm shift, an absolute paradigm shift, a shift in perspective. And for me, it's like seeing, you know, what it must have been like to understand that the world isn't flat ... in fact, that race is not based on biology but race is rather an idea that we ascribe to biology. (Adelman, 2003)

Race was really about the creation of a dominant group within society. According to historian Robin D.G. Kelley (PBS, 2005), "[R]ace was never just a matter of categories. It was a matter of creating hierarchies. ... or the creation of racism, was really about the invention of a dominant group." This dominant group develops the social, cultural, economic, and political power to define its own particular history and culture as representative of all. We see an example of a response to this kind of dominance in the creation of the Africentric Alternative School in Toronto. It was believed that the high dropout rate affecting students of African descent was related in part to the relevancy of school curriculum and its failure to represent anything other than the history and culture of the English and French.

If we now understand that there is no scientific basis to the concept of race, then we know racism is not about the category you belong to; rather, it is about the power of particular groups of people to construct and assign negative meaning to that category. That meaning can lead to a range of social responses—from **racial stereotyping** to hate crimes and violence. The hope for anti-racist educators and activists is rooted in this idea: if racism is socially constructed, then we might have the ability to deconstruct it as well as the desire to do so, when it is one of the root causes of human inequality (Roy, 2012).

WHAT IS RACISM?

So how can we say that race does not matter? It might not be a biological reality, but it is very real in terms of social consequences and impact on people's lived experience. As long as racism exists within a society, race will still matter. Pretending otherwise negates the experience of Samuel and other racialized persons and communities living in Canada.

Defining **racism** is difficult as its meaning is complex and multidimensional. "The most commonly accepted concept of racism in Canada is one that refers to the individual expression of overt feelings or actions" (Henry, Tator, Mattis, & Rees, 2009). For example, Canadians often identify racism as insults, name-calling, graffitti, threats of violence, or hate crimes that all target racialized people (Henry, Tator, Mattis, & Rees, 2009). Harder to identify and define is racism that is built into the systems that people live and work in, whereby racism is "the patterns of privilege and oppression themselves and anything—intentional or not—that helps to create or perpetuate those patterns" (Johnson, 2006). Fleras and Elliot (2002) use an equation to illustrate the fact that racism is pervasive and complex: *racism = prejudice + discrimination + power.*

Racism is an ideology that either directly or indirectly asserts that one

Racial stereotyping: Using the concept of race or ethnicity to attach a generalized concept that all members of a group have a particular characteristic or ability.

Racism: Racism is an ideology that either directly or indirectly asserts that one group is superior to others, with the power to put this ideology into practice in a way that gives advantages, privilege, and power to certain groups of people, and conversely, can disadvantage or limit the opportunities of racialized individuals or racialized groups.

group is superior to others, with the power to put this ideology into practice in a way that give advantages, privilege, and power to certain groups of people, and conversely, can disadvantage or limit the opportunities of racialized individuals or racialized groups. "Racism is a wider phenomenon than racial discrimination … nevertheless, racism plays a major role in fostering racial discrimination" (Ontario Human Rights Commission, 2016).

Forms of Racism

There are many different forms of racism and academics use many different terms and categorizations to classify the wide variety of practices, ideologies, and social relations that reproduce racism. It is important to remember that racism most often operates in combination with two or more forms simultaneously. Racism can be expressed through imagery, ideology, discourse, and social interaction. Racism also operates at an individual, institutional, and societal level.

Representational Racism

Representational racism communicates racist ideas through imagery that depicts racial stereotypes. Stereotypes are harmful when used as generalizations to provide an "oversimplification or an exaggerated version of the world based on preconceived and unwarranted notions that extend to all members of the devalued group" (Fleras & Elliot, 2002). Racial stereotyping is commonly used in popular culture in ways that devalue and dehumanize racialized groups. The use of blackface was used historically by white actors to stereotype, devalue, and dehumanize African Americans. Depictions of essentialized racial stereotypes for sports teams' mascots and logos are alleged to be further examples of representational

racism, including the Washington Redskins, Edmonton Eskimos, Chicago Blackhawks, and Atlanta Braves. Some broadcasters are now refusing to use racialized names or references when announcing. The Cleveland Indians and Chief Wahoo, the team's mascot, are considered by some to be the most offensive image in sports.

Other examples of representational racism in popular culture and media include the use of racialized persons and groups as criminals or "the bad guy" in film and television. Racial stereotypes have also appeared in Disney movies in the caricatures in *Dumbo* (as a reference to **Jim Crow racism**), Sunflower in *Fantasia*, Uncle Remus in *Song of the South*, Big Chief and the song "What Makes A Red Man Red?" in *Peter Pan*, Siamese cats, Si and Am, in *Lady and the Tramp*, and Pocahontas, to name a few. Representational racism is one form of racism that is often quickly dismissed as inconsequential. But representational racism can be a powerful means of perpetuating imagery in popular culture that supports racist ideology (Cole, 2016).

Ideological Racism

"Societies are ideological: they are loaded with ideas and ideals that reflect, reinforce, and advance patterns of power that empower some, disempower others: (Fleras & Elliot, 2002)." **Ideological racism** is a form of racism rooted in the ideas, beliefs, and worldviews that reflect, reinforce, and advance notions of racial superiority or inferiority. Ideological racism is often perceived as common sense that is used to justify and naturalize a hierarchy of people based on race. Historically, ideological racism has been used to justify colonialism, imperialism, and the unjust acquisition of land, people, and resources (Cole, 2016). Common ideological forms of racism are often encapsulated in racial stereotypes like those used by Donald Trump in the 2016 presidential race when he proposed a "Muslim ban," labelled Mexican immigrants "rapists" and "drug dealers," and dismissed African Americans as "inner city" dwellers with "nothing to lose."

Discursive Racism

Discursive racism is a form of racism communicated through the actual words we use. Written and spoken communication (discourse) can be a powerful tool in communicating inclusivity or discrimination. Inclusive communication purposely avoids racist or other discriminatory language, including stereotyping, loaded words, and patronizing descriptors. Examples of this kind of racism include racial slurs and hate speech, as well as words that have negative racialized meaning.

Representational racism: A form of racism that uses imagery to depict racial stereotypes, often in popular culture and media, in a manner that reinforces perceived inferiority of racialized persons or groups.

Jim Crow racism: Anti-black racism that existed in the United States during the period of 1877–1960s; Jim Crow laws enforced racial segregation and a racialized social order that resulted in the subjugation, oppression, and death (through lynching and other violence) of African Americans.

Ideological racism: A form of racism rooted in the ideas, beliefs, and worldviews that reflect, reinforce, and advance notions of racial superiority or inferiority.

Discursive racism: A form of racism that is expressed through written and spoken communication (discourse). It manifests in racial slurs and hate speech, and in words with racial meaning embedded in them.

Interactional Racism

Interactional racism is a form of racism that is expressed through social interaction with other people, more specifically how those with privilege interact with those who are oppressed. When a racialized person is verbally or physically assaulted because of their race, this is an example of interactional racism. When Samuel describes walking down the hallway and a student holds their purse a little tighter, this, too, is an example of interactional racism. Interactional racism is when someone assumes that a racialized person is a low-level employee rather than the store owner. It can also include avoiding contact or ignoring members of a racialized group.

How Racism Operates

Racism operates at a number of levels: at an individual level that may be expressed through everyday behaviour; at an institutional or systemic level that involves policies or practices that create and maintain disadvantage for racialized persons; and at a societal level where racism forms a framework for racialized thought (Ontario Human Rights Commission, 2016).

Individual Racism

Individual racism is expressed through individual discriminatory attitudes or behaviour motivated by negative evaluation of a person or group of people using a socially constructed concept of race. The expression of racism at an individual level can be overt in nature (overt racism) or can be expressed in everyday social interaction with people that can often be very subtle in nature (everyday racism). Expression of racism at the individual level could include offensive jokes, graffiti, and emails that target racialized people; insults, name-calling, verbal abuse, and threats against persons because of race; avoiding contact with, silencing, or belittling racialized individuals; and hate crimes, and violence against racialized individuals or groups (University of Guelph, 2016).

Institutional Racism

Racism that operates at the institutional or systemic level can be found in policies, practices, and procedures of organizations, businesses, or government that promotes or sustains disadvantage for racialized persons or groups (Ontario Human Rights Commission, 2016). This form of racism is often normalized as "just the way we do things" (Ontario Human Rights Commission, 2016). An example of

institutional racism is policing practices that disproportionately target specific racialized communities. Another example would be curriculum that has the effect of marginalizing, stereotyping, or making Indigenous people or racialized people invisible (University of Guelph, 2016).

Institutional racism is often more difficult to identify and address than individual racism, as it is often hidden within the implicit policies of social institutions. There are often references made to forms of racism based on intention—a categorization of racism as implicit versus explicit or conscious versus unconscious. But this kind of characterization presumes that dominant white culture in North America has not yet experienced any awareness of their privilege. Students often ask, "Isn't racism just ignorance, people simply not knowing any better?" A favourite quotation used to respond to this question comes from an interview with Robin D.G. Kelley, an American professor, who states:

> When I teach about racism the first thing I say to my students is that racism is not ignorance. Racism is knowledge. Racism in some ways is a very complicated system of knowledge, where science, religion, philosophy, is used to justify inequality and hierarchy. That is foundational. ... And that is why you can't think of racism as simply "not knowing." That is not the case at all—on the contrary. (PBS, 2005)

Societal Racism

Societal racism operates as a framework for those concepts, ideas, images, and institutions that we use to interpret and give meaning to racialized thought (Ontario Human Rights Commission, 2016). Societal racism is the basis for both individual and institutional racism, as "it is the value system which is embedded in society

Interactional racism: A form of racism that is expressed through social interaction with other people; more specifically, how those with privilege interact with those who are oppressed.

Individual racism: Individual discriminatory attitudes or behaviour motivated by negative evaluation of a person or group of people using a socially constructed concept of race.

Institutional racism: Behaviour, policies, or practices that disadvantage racialized persons; can be intentional or unintentional.

Societal racism: A framework for those concepts, ideas, images, and institutions that we use to interpret and give meaning to racialized thought.

which supports and allows discriminatory actions based on perceptions of racial difference, cultural superiority and inferiority" (Roy, 2012). Examples of societal racism might include using culturally biased assessment tests; assuming racialized Canadians were not born here; or omitting Indigenous contributions and perspectives from academic fields of study (University of Guelph, 2016).

ACKNOWLEDGING RACISM IN CANADA

By looking at the different forms in which racism is manifest, we can begin to understand some of the problems in uncovering racism in Canada. For those who do not experience racism, it is easier to deny that racism is real. The denial of the existence of racism is damaging to persons and communities affected; it is also damaging to society as a whole. In order to combat the effects of racism and racial discrimination, we must begin with the acknowledgement that racism exists. Despite progress, **racial prejudice** and **racial discrimination** are unfortunately still a real and persistent reality within Canadian society. "This fact must be acknowledged as a starting point to effectively address racism and racial discrimination" (Ontario Human Rights Commission, 2016).

In the United States, discussions of racism are often characterized as a black-and-white binary issue, with legitimate historical roots in slavery. In Canada today, discussions of racism are often truncated by the denial of its existence, or by assertions of its unintentional nature, limited by a perception of racism as something that happened in the past, something that happens in other parts of the world, or as isolated incidents against racialized persons or communities. Jennifer Roy writes for the Canadian Race Relations Foundation:

> There appears to be a refusal to recognize that racism is an issue in Canada, both presently and historically … [W]hen we think of the main examples of racism historically, the images that often spring to mind are of slavery in the United States, apartheid in South Africa or the Holocaust. … Inherent in this limited understanding is a tendency to assume both that these problems are in the past and that they do not occur in Canada. (Roy, 2012)

We would like to think of ourselves as a nation differentiated from the United States, hinged to the myth that slavery did not exist in Canada. As historian Afua Cooper (2006) states, slavery is "Canada's best kept secret, locked within the National closet." At times, we like to believe that racism is now neatly tucked away as a part of history we won't repeat. Government officials have apologized for historical racism:

- The mayor of Halifax apologized for the evictions and razing of the African-Canadian community of Africville.
- Prime Minister Stephen Harper apologized to the Indo-Canadian community for the 1914 *Komagata Maru* incident, in which immigrants were denied entry based on racist legislation. Ship passengers remained on board for months, and were denied food and water.
- The Canadian government apologized to Aboriginal peoples for Canada's role in the Indian Residential School System, where children were separated from their parents and subjected to abuse.
- The Canadian government apologized to Japanese Canadians for their internment and seizure of their property during World War II.
- The Canadian government apologized to Inuit families for their relocation from Inukjuak, Quebec, to the high Arctic.
- The Canadian government apologized to Chinese Canadians for the head tax they were forced to pay and that prevented family reunification.

So we have acknowledged and apologized to racialized groups affected by some of the racism that is a part of Canada's past. But we know the past informs both the present and the future, so how have we changed?

As Canadians we have also acknowledged that overcoming racism is of international importance. In this context, we vocalize our objections and sometimes act on them. There are many examples that attest to this practice. Many Canadians, horrified by the Holocaust, sacrificed their lives in the fight against Nazi Germany. Canada is a signatory to the United Nations International Convention on the Elimination of All Forms of Racial Discrimination, which requires us as a nation to commit to the elimination of racial discrimination, promote interracial understanding, legislate against hate speech, and criminalize membership in racist organizations (United Nations, 2011). In 2010, Canada also signed the United Nations Declaration on the Rights of Indigenous Peoples.

Racial prejudice:
Prejudgment or negative attitude based on a set of characteristics associated with the colour of a person's skin.

Racial discrimination:
Behaviour that has a discriminatory effect based on race; there does not have to be an intention to discriminate.

On the International Day for the Elimination of Racial Discrimination, held each year on March 21 to commemorate the anniversary of the 1960 Sharpeville, South Africa, massacre, Canadians renew their commitment to fight for racial equality. In 2012, that day's theme was proclaimed by the United Nations as "Racism and Conflict" to highlight ways in which racial prejudice, racial discrimination, and **xenophobia** are often at the root of conflict, **genocide, ethnic cleansing**, and **war crimes** (United Nations, 2012). As Canadians, we have been awarded the Nansen Peace Medal for our efforts in helping refugees fleeing from this form of conflict. But can we acknowledge racism exists here in Canada today?

In September 2011, students at a business school in Montreal were caught on film wearing **blackface** and chanting with Jamaican accents about smoking marijuana. Students alleged they were doing a skit to honour Jamaican sprinter Usain Bolt. A spokesperson claimed that these French-speaking Canadians were unaware of the history and significance of blackface (Wade, 2012). In November 2012, a bathroom stall at the University of Ottawa's Law School was defaced with racist graffiti (CBC News, 2012). In 2014, four students in blackface won $500 at Brock University's Student Union Halloween costume contest. In an open letter from the Department of Labour Studies at Brock University (2014), Professors Larry Savage, Kendra Coulter, and Simon Black and student Nick Ruhloff-Queiruga write:

> Blackface can never be disassociated from the vicious legacy of white supremacy and institutionalized anti-Black racism in the United States and Canada, just as redface or "playing Indian" cannot be disassociated from colonialism and the subjugation and dispossession of indigenous peoples … However, blackface is not simply a remnant of a racist historical past, but part of a broader set of cultural practices which maintain and normalize anti-Black racism and systemic oppression. Students, staff and faculty at Brock University need to understand that such costumes are not "just a joke." Regardless of the intent or motivation of the students in question, donning blackface for Halloween is never okay; it is racist, full stop.*

It is difficult to acknowledge that racism is not simply isolated incidents perpetrated by unknowing individuals. Racism is also embedded in the everyday life of our social systems and structures. Students like

*Source: Black, S., Kendra, C., Ruhloff-Queiruga, N., & Savage, L. (2014). Open Letter re: Blackface at Brock University from the Department of Labour Studies. Retrieved from Brock University: https://brocku.ca/node/27835.

Samuel can narrate the experience of everyday racism in Canada, while at the same time, many who belong to the dominant culture question whether racism still exists in Canada today. Most of us who have ever engaged in a dialogue about racism have heard the arguments of reverse discrimination, especially in relation to employment. These differing perceptions make the work of anti-racist educators and activists challenging. The first step in understanding racism in Canada is to acknowledge that it is a real, pervasive, and serious issue.

White Privilege and Forms of Its Denial

The picture featured at the beginning of this chapter is a political response to the peach-coloured crayon in the Crayola crayon box that, up until 1962, was called "flesh." Some suggest this naming is symbolic of the invisible system of privilege that has conferred dominance on those perceived as white (McIntosh, 1989). Anti-racism educator Tim Wise (2011), reflecting upon his own privilege as a white man, writes, "[T]o be white is to be born into an environment where one's legitimacy is far less likely to be questioned than would the legitimacy of a person of colour, be it in terms of where one lives, where one works, or where one goes to school."

Not surprising is the fact that any challenge to white privilege and its concomitant power is not popular among white people. One of the ways in which we reinforce white privilege in multicultural Canada is the creation of a society where white people can still assume that their social institutions, their political structures, their economy, and their neighbourhoods will

Xenophobia: Hostility to anything considered foreign.

Genocide: The intentional extermination or killing of an identifiable group.

Ethnic cleansing: The process or policy of eliminating unwanted ethnic, racial, or religious groups by deportation, forcible displacement, mass murder, or threats of such acts, with the intention of creating a homogenous population.

War crimes: Serious violations of international law applicable during armed conflict, such as ill treatment of prisoners of war, killing of prisoners of war, and so on; as of 2002, those arrested for war crimes have been tried in the International Criminal Court.

Blackface: Makeup that was used historically in minstrel shows to impersonate black people and act out racist stereotypes. Beginning in the 1800s, white American actors would rub black shoe polish or greasepaint on their faces and then perform in ways that would demean and dehumanize black people. It is associated with Jim Crow racism.

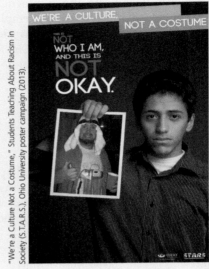

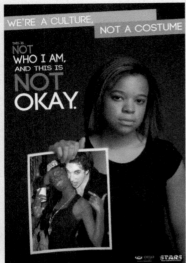

"We're a Culture Not a Costume," Students Teaching About Racism in Society (S.T.A.R.S.), Ohio University poster campaign (2013).

Students Teaching About Racism in Society (S.T.A.R.S.) is a student group at Ohio University. Their posture campaign, "We're a Culture, Not a Costume," has sparked an interesting debate over Halloween costumes that are thought to proliferate stereotypes and feed racism. Some students claim this poster campaign by students at Ohio University reveals political correctness gone too far. Other students, in rebuttal, have stated that if you are not a member of that group, then you should not be able to dictate how painful that stereotype might be to its members. Other students suggest that since stereotypes are not a person's entire identity, it is not funny or clever to suggest they are. What do you think?

still work for them as long as they play by the rules (Wise, 2010a). There still exists a pathology of isolation that creates and maintains racially and ethnically segregated neighbourhoods across Canada. The *Toronto Star*, in an article on the city of Brampton, reports that some suggest this city is experiencing a phenomenon referred to as "white flight"—where former white mainstream communities are not comfortable becoming the minority population:

> While the visible-minority segment has exploded to represent two-thirds of Brampton's population, white residents are dwindling. Their numbers went from 192 400 in 2001 to 169 230 in 2011. That's a loss of more than 23 000 people, or 12 per cent, in a decade when the city's population rose by 60 per cent. That's hardly a picture of the multicultural ideal so celebrated in this country. (Grewal, 2013)

Duke University Professor Eduardo Bonilla-Silva (2012) asserts that white isolation produces racialized perceptions and behaviour that normalizes racial inequality. Tim Wise (2010b) concurs, noting that the effects of raising children in environments where no one discusses racism can result in those children growing up to believe that racial inequalities are natural and that marginalization is not a systemic issue but one based on individual deficiencies. More generally across society, whiteness is reinforced by a kind of code that attacks equity in matters like tax exemption for First Nations people, immigration policies, **employment equity**, and inclusive language. Professor Bonilla-Silva suggests that the defence of white privilege sounds something like this: "I am all for equal opportunity; I want people to be judged by the content of their character and not by the colour of their

Employment equity:
Requirement under the Employment Equity Act of Canada that employers use proactive employment practices to increase representation of four designated groups: women, people with disabilities, Indigenous peoples, and racialized communities in the workplace. Employment equity mandates the accommodation of difference with special measures when needed.

skin; therefore, I am against affirmative action because it is discrimination in reverse" (Bonilla-Silva, 2012). He goes on to suggest that systemic white privilege is maintained by the language of liberalism: "they don't even process that because they think that they are beyond race. So they are the normative people. They are the universal people. They are the raceless people" (Bonilla-Silva, 2012).

White students often respond to Samuel's story with disbelief, blame, and certainly advice about his alleged attempts to use the so-called "race card"— as if that is something that confers real privilege. For example, a student named Ashley responded to Samuel's story by asking why he feels the need to walk around all the time thinking about race: "I don't go around thinking about my race all the time. We all belong to one race—the human race. Colour shouldn't matter." Some students in the class applauded after Ashley's statement.

Colourblindness

Racial discourse will often use words like "colour-blind" to describe the phenomenon that a person is oblivious to race and therefore not racist. Racialized persons commonly hear statements that sound something like this: "I don't care if you are black, brown, white, green, or blue." Or "When I see you, I don't see you as black (or insert any other racial category here)." Julian Bond, a civil rights legend and the former chairman of the NAACP (the National Association for the Advancement of Coloured People), said it best when he stated that colourblindness really means being blind to the consequences of colour. Colourblind policies can actually worsen the problem of racial injustice (Wise, 2010b).

So why do people say "when I see you, I don't see colour"? It is a question posed by Jane Elliott, an anti-racist activist and educator, in a workshop with college student participants that was captured in a documentary film (Elliott & Elliott Eye, 2001) titled *Angry Eye*. Elliott is the teacher who created the famous "blue-eyed/brown-eyed exercise" with her grade-school children following the assassination of Martin Luther King, Jr. In this exercise, Elliott divided the children based on their eye colour and then set out a system of discrimination based on eye colour. The purpose of Elliot's exercise was to expose the children to the experience of prejudice and discrimination based on a socially constructed hierarchy. Years later, having divided a college classroom in a blue-eyed and brown-eyed exercise, Elliott taught participants that colourblindness is a form of racism. "When you say

to a person of colour, 'when I see you I don't see you black, I just see everybody the same' ... you don't have the right to say to a person, 'I don't see you as you are, I want to see you as I would be more comfortable seeing you' ... you are denying their reality" (Elliott & Elliott Eye, 2001).

Some suggest that the insistence on colourblindness is rooted in a fear of being labelled a racist. But the point that Elliott makes is that noticing a person's race doesn't make you racist. What does make you racist is adopting attitudes and behaviours that make you believe and act upon the belief that certain characteristics are related to the race you think that person is. It is hard to overcome this behaviour when we are bombarded with racist messages through agents of socialization in our society, including families, the media, and so on.

RACISM AND THE LAW

White Supremacy and Racially Motivated Hate Crimes

Hate crimes are a focus of concern around the world. Canada works with other countries to monitor and combat hate crime (Allen & Boyce, 2013). A **hate crime** is a criminal act motivated by hate. Hate crimes can affect not only individual victims of the crime, but also the groups targeted (Allen & Boyce, 2013). Hate crimes can be either violent in nature, such as assault, or they can be non-violent in nature, such as mischief (Allen & Boyce, 2013). Hate crimes can involve intimidation, harassment, threat or use of physical force against a person, group or a property. The Criminal Code of Canada (Sections 318 and 319) makes it a crime to incite hatred against an identifiable group or to call for, support, encourage or argue for the killing of members of a group based on colour, race, religion, ethnic origin or sexual orientation. The Criminal Code of Canada (Section 718.2) asks that courts consider in sentencing any evidence that the offence was motivated by bias, prejudice or hate based on race, national or ethnic origin, language, colour, religion, sex, age, mental or physical disability, sexual orientation, or any other similar factor.

> **Hate crimes:** Crimes that are committed against people or property that are motivated by hate or prejudice against a victim's racial, ethnic, religious, or sexual identity; they can involve intimidation, harassment, destruction of property, vandalism, physical force or threat of physical force, or inciting hatred in other people.

Some of the most notable Canadian hate crime cases include verdicts against Don Andrews, James Keegstra, and Ernst Zundel. In cases such as these relating to hate, Section 2 of the Canadian Charter of Rights and Freedoms is often cited as justification for hate speech. Section 2 of the Charter guarantees the fundamental freedom of thought, belief, opinion, and expression, including freedom of the press and other media of communication. In the case of hate crimes, courts have ruled that although section 319(2) of the Canadian Criminal Code *does* limit free speech by preventing people from expressing their opinions, Section 1 of the Charter of Rights and Freedoms can restrict those freedoms granted in the Charter by making them subject "only to such reasonable limits prescribed by law as can be demonstrably justified in a free and democratic society." Canadian courts have ruled that in a democratic society like Canada, it is reasonable that we limit speech that may incite hatred against others. An example is the recent unanimous decision of the Supreme Court of Canada in the case of William Whatcott of Saskatchewan, which reaffirms the Canadian approach to freedom of speech: "that it can be limited by law to address the problem of hate speech, unlike the American approach, in which speech cannot be limited except in the most extreme circumstances" (Brean, 2013).

While the number of hate crime incidents in Canada varies from year to year, what is consistent is the fact that race or ethnicity remains the most common motivation for hate crime (Allen & Boyce, 2013). In fact, hate crimes motivated by race or ethnicity represented half (52 percent) of all police-reported hate crime incidents in 2011 (Allen & Boyce, 2013). Black populations are the most frequent targets of racially motivated hate crimes, representing 21 percent of *all* hate crimes in Canada (Allen & Boyce, 2013).

White supremacy is an ideology that supports the superiority of whites over all others and is embraced by members of white supremacist organizations. The beliefs of the white supremacist movement are examples of overt racism, and they cultivate hate (Canadian Race Relations Foundation, 2012). The ideology is premised on the myth that whites are racially superior to people of colour, Jews, and other minority populations (Canadian Race Relations Foundation, 2012). While hate-motivated acts based on race, ethnicity, religion, and sexual orientation are often associated with organized white supremacist groups, some racially motivated hate crimes are also carried out by individuals who are not associated with such groups (Canadian Race Relations Foundation, 2012).

The laws and institutions of Canadian society designed to protect the rights of all citizens are contradictory to the declared goals of the white supremacist movement (Canadian Race Relations Foundation, 2012). Perhaps this is why **white nationalism** has recently emerged as a more sophisticated political ideology incorporating a collective racialized identity for white people, premised on separatism. Examples of this are evidenced in the challenges to immigration policies, human rights, and racial integration by white nationalist organizations such as the Aryan Guard, the Canadian Heritage Alliance, and the National Socialist Party of Canada. **Neo-Nazism** is an ideology that promotes white racial superiority as well as anti-Semitism specifically. Examples of groups that operate under anti-Semitic ideologies in Canada include the Northern Alliance Canada, Blood and Honour, Canadian Heritage Alliance, Northern Hammerskins, the Final Solution Skins, and the Aryan Resistance Movement.

Law enforcement personnel use tattoos as one means of identifying membership in some white supremacist organizations. Tattoos will often include swastikas as representative of neo-Nazi ideology. Tattoos also use numbers to code specific references. For example, the number 8 signifies the eighth letter of the alphabet (H), so 88 becomes HH or "Heil Hitler." Another example is 311: the letter K is the 11th number of the alphabet, and using it three times would translate to "KKK." The number "100%" can signify racial purity. Tattoos with the number 4/20 are sometimes used to commemorate Hitler's birthday (however, in some cases, 4/20 can also be a reference to marijuana). Colour of shoelaces can also indicate membership in a racially motivated group: shoelace colour can indicate whether a member has been involved in acts of racial violence (e.g., red shoe laces can indicate spilled blood).

White supremacy: An ideology that supports the superiority of whites over all others and is embraced by members of white supremacist organizations.

White nationalism: A political ideology that advocates for a racialized identity for "white people."

Neo-Nazism: A post–World War II movement related to the white nationalist and white power skinhead movements, which seeks to revive elements of Nazi ideology such as racism, xenophobia, homophobia, holocaust denial, and anti-Semitism.

Race and Law Enforcement

There is perhaps no issue more controversial in racial discourse in Canada today than that of **racial profiling** within the Canadian criminal justice system, and specifically by the police. Law professor David Tanovich (2006), in the opening line of his book *The Colour of Justice: Policing Race in Canada,* states, "The colour of justice in Canada is White." Reference to this phenomenon is engrained in popular culture, rooted in historical events, and revealed in academic research.

In Canada, in 1995, during a peaceful protest by First Nations representatives at Ipperwash Provincial Park, Dudley George, an unarmed protestor, was shot dead by a member of the Ontario Provincial Police. The Ontario government called for an inquiry into the death of Dudley George, to make recommendations to avoid violence in similar circumstances in the future. Inquiry Commissioner Sidney B. Linden made the following statement:

> To many Aboriginal People, the shooting of Dudley George, the first Aboriginal person to be killed in a land-rights dispute in Canada since the 19th century, was the inevitable result of centuries of discrimination and dispossession. Many Aboriginal people also believed that the explanation for killing an unarmed Aboriginal occupier was rooted in racism. From this perspective, Ipperwash revealed a deep schism in Canada's relationship with its Aboriginal People and was symbolic of a sad history of government policies that harmed their long-term interests. (Linden, 2007)

Commissioner Linden's findings include the assertion that Ontario Premier Mike Harris made the racist comment, "I want the f---ing Indians out of the park," and found that racism among some members of the Ontario Provincial Police contributed to the lack of a timely, peaceful resolution to the Ipperwash dispute (Linden, 2007). The incident at Ipperwash is layered upon years of **historical disadvantage** represented in treaty violations, colonialism, residential schools, and apprehension and adoption of Indigenous children into non-Indigenous families.

It is important to note that Indigenous communities identify the concepts of racial profiling and racism rooted in a different dynamic from that of racialized communities, as they have a unique and historical reality that is different from ethnic minorities in Canada (Ontario Human Rights Commission, 2003).

Racial profiling in Canada is highlighted in these specific incidents and in research:

- Musician K'Naan, in an interview with Craig Keilburger, discussed his youth in Toronto and the use of the Trespass to Property Act in a manner that was discriminatory to black Somali youth (K'Naan, 2010), a practice confirmed in a report by the Toronto Community Housing Corporation (Ontario Human Rights Commission, 2003).

- On October 14, 2007, Polish immigrant Robert Dziekanski died in the Vancouver International Airport after being repeatedly tasered by four RCMP officers. A video of the event revealed persons in the airport waiting area advising the RCMP officers that the agitated Dziekanski spoke no English. Without ever having had the benefit of assistance from someone who spoke his language, Dziekanski was repeatedly tasered by the RCMP. An inquiry into Dziekanski's death found that the RCMP were not justified in using a taser against him, that the RCMP officers misrepresented their actions to investigators, and that the Canadian Border Security Agency needed to make changes in arrival procedures for sponsored immigrants (Braidwood Commission, 2010).

- The freezing deaths of three Indigenous men, Neil Stonechild, Rodney Naistus, and Lawrence Wegner, and the experience of Darrel Night in Saskatoon has created a great deal of controversy around the alleged police practice known as **starlight tours**, in which police pick up Indigenous people and take them to some location far away, leaving them there to get home on their own (Comack, 2012).

Racial Profiling: Any action undertaken for reasons of safety, security, or public protection that relies on stereotypes to treat persons differently; while it is most often used in reference to policing practices, it is not a term limited to the context of criminal justice.

Historical disadvantage: Disadvantage related to past historical discriminatory actions combined with current disadvantage to contribute to systemic discrimination.

Starlight tours: Alleged police practice of picking up Indigenous people and taking them to some location far away and leaving them there to get home on their own.

- A report on the victims of serial killer Robert Pickton identified unintentional systemic bias as partly to blame, as "The women were poor, they were addicted, vulnerable, aboriginal. They did not receive equal treatment by police" (Oppal, 2012).
- Racialized communities have long alleged profiling by police through use of stop-and-search procedures, a phenomenon often referred to in popular culture as "driving while black (DWB)." Rawle Maynard was awarded $40 000 by the Human Rights Tribunal of Ontario after it was found he was racially profiled by a Toronto Police Service officer. Stop-and-search practices of Kingston, Ontario, police officers reveal racial profiling by officers, confirmed in a study conducted by Scot Wortley, a criminology professor at the University of Toronto. The study found that black people were four times more likely to be pulled over than whites, that Aboriginal people were 1.4 times more likely to be pulled over by police than whites, and that 40 percent of black males between the ages of 15 and 24 were stopped by police during the study year, compared to 11 percent of their white counterparts (Wortley & Marshall, 2005).

So what does racial profiling really mean? While the definition of racial profiling is usually given in a law enforcement context, it is important to note that this is not the only context in which the concept can be applied. The Ontario Human Rights Commission, as one example, defines racial profiling as "any action undertaken for reasons of safety, security or public protection that relies on stereotypes about race, colour, **ethnicity**, ancestry, religion, or place of origin rather than on reasonable suspicion, to single out an individual for greater scrutiny or different treatment" (Ontario Human Rights Commission, 2012). Profiling can occur in many contexts involving safety, security, and public protection issues. Consider these examples heard at the Inquiry into Racial Profiling in Ontario:

- A law enforcement official assumes someone is more likely to have committed a crime because he is African Canadian.
- School personnel treat a Latino child's behaviour

Ethnicity: A person's or his or her ancestors' country of origin; includes material and non-material aspects associated with a culture and social identity.

as an infraction of its zero tolerance policy, while the same action by another child might be seen as normal "kids' play."
- A private security guard follows a shopper because she believes the shopper is more likely to steal from the store.
- An employer wants a stricter security clearance for a Muslim employee after September 11th.
- A bar refuses to serve Aboriginal patrons because of an assumption that they will get drunk and rowdy.
- A criminal justice system official refuses bail to a Latin American person because of a belief that people from her country are violent.
- A landlord asks a Chinese student to move out because she believes that the tenant will expose her to SARS (Severe Acute Respiratory Syndrome) even though the tenant has not been to any hospitals, facilities, or countries associated with a high risk of SARS (Ontario Human Rights Commission, 2003).*

In a law enforcement context, racial profiling is often defined more specifically as something that occurs when an officer consciously or unconsciously uses race or racial stereotypes associated with criminality in suspect selection for investigation, police stop-and-search practices, Canadian Border Security Agency searches at ports of entry, police patrols of racialized communities, gathering criminal intelligence that link racialized groups with particular forms of criminality, suspect treatment upon arrest and detention, and so on (CBC News, 2005).

Does racial profiling by the police in Canada exist? Some scholars (Tanovich, 2006; Tator & Henry, 2006) suggest that the evidence is so overwhelming that we need to move on from the discussion about whether or not it exists. In 1988, the Ontario Race Relations and Policing Task Force concluded that racialized groups believed they were policed differently. One of the findings of Stephen Lewis's 1992 Report to the Premier on Racism in Ontario was that racialized individuals and communities, particularly black Canadians, experienced discrimination in policing and the criminal justice system (Ontario Human Rights Commission, 2012). Following this finding, the Ontario government

*Source: Ontario Human Rights Commission, "Paying the Price: The Human Cost of Racial Profiling, Inquiry Report," p. 7. Found at: http://www.ohrc.on.ca/sites/default/files/attachments/Paying_the_price%3A_The_human_cost_of_racial_profiling.pdf) © Queen's Printer for Ontario, 2003. Reproduced with permission.

established the Commission on Systemic Racism in the Ontario Criminal Justice System to study all aspects of the Ontario criminal justice system.

The Ontario Human Rights Commission released its report from the Racial Profiling Inquiry, which documents the existence of racial profiling in the public sectors, including police services across the province (including the OPP and the RCMP), all levels of the criminal justice system (e.g., crown counsels, justices of the peace, judges, prison guards and officials, and those involved in parole and probation), all levels of the education system (e.g., school board officials, school administrators, principals, teachers, guidance counsellors, and Ministry of Education officials), and the Canada Revenue Agency, as well as within the private sector (Ontario Human Rights Commission, 2012).

More recently, the practice of carding by the Toronto police has led to allegations that racialized individuals and communities are being disproportionately targeted. Carding is a police practice known as street checks where an officer stops, questions, and documents a person. After allegations of racial profiling, the Ontario government established new regulations for the practice of carding that came into effect on January 1, 2017. Regulations prohibit the use of race or living in a racialized neighbourhood as grounds for street checks. Officers will also now be required to issue a receipt with their name and badge number at the completion of a carding interaction along with information about how to file a complaint.

The landscape is evolving. It has taken leadership from Indigenous communities, racialized communities, and policing agencies to make this evolution possible. Today, there are countless examples of police agencies across Canada that have adopted formal policies, frameworks, and best practices for hiring from and interfacing with Indigenous and racialized communities. One such example is the Ontario Provincial Police's Framework for Police Preparedness for Aboriginal Critical Incidents. Through this framework, the OPP explicitly acknowledges its intention to work with Indigenous relations teams and liaison officers—a great start that will require ongoing dialogue and participation with Indigenous and racialized communities, public transparency, and vigilance in practice.

INTERSECTIONALITY: THE COLOUR OF POVERTY

Poverty is also a racial issue. One of the greatest illuminators of blindness to the consequence of colour is Canadian statistics on poverty. If race doesn't matter, then why do Indigenous and racialized persons have a higher risk of living in poverty compared to non-racialized persons in Canada? One of the most vulnerable groups in Canada is Indigenous children, with 40 percent living in poverty and 60 percent of First Nations children on reserves living in poverty (Campaign 2000, 2016).

The racialization of poverty creates a relationship between income and reduced opportunities as a result of individual and systemic racism in Canada. Racialized communities experience ongoing, disproportionate levels of poverty and concomitant problems like poor health, lower education, and fewer job opportunities than those from European backgrounds (National Council of Welfare, 2012). Discrimination means that Indigenous and racialized communities are less likely to get jobs when equally qualified, less likely to receive equal pay for equal work, less likely to have good working conditions, and less likely to have access to healthcare (Block & Galabuzi, 2011).

A POST-RACIAL SOCIETY

So how, then, do we eliminate racism? How do we create a world where the colour of someone's skin has no greater importance than the colour of their eyes? Morgan Freeman in his *60 Minutes* interview with Mike Wallace proposed a solution to the problem of racism—stop talking about it:

Wallace: Black History Month you find …

Freeman: Ridiculous.

Wallace: Why?

Freeman: You're going to relegate my history to a month.

Wallace: Come on …

Freeman: What do you do with yours? Which month is White History Month?

Wallace: (stutters) … I'm Jewish.

Freeman: OK. Which month is Jewish History Month?

Wallace: There isn't one.

Freeman: Oh. Why not? Do you want one?

Wallace: No. No. No.

Freeman: I don't either. I don't want a Black History Month. Black history is American history.

Wallace: How are we going to get rid of racism?

Freeman: Stop talking about it. I'm going to stop calling you a white man and I'm going to ask you to stop calling me a black man. I know you as Mike Wallace; you know me as Morgan Freeman. (Freeman, 2005)

Freeman raises a valid criticism of the need to relegate the history of one race to a designated month known as **Black History Month**. As teachers rush to "celebrate" black history during the month of February, the abbreviated curriculum tells an incomplete story, focusing solely on slave narratives, civil rights, and the historical figures of Martin Luther King Jr., Rosa Parks, George Washington Carver, Frederick Douglass, Harriet Tubman, Eli Whitney, Malcolm X, and Nelson Mandela. Can we make the same argument in Canada as Freeman makes in the United States? Is black history Canadian history? The Government of Canada acknowledges that our history books do not reflect the history of African Canadians:

Black History Month: Observed during the month of February in Canada, the United States, and the United Kingdom to commemorate the important people and events and history of the African diaspora.

Asian Heritage Month: Observed annually in Canada (and the United States where it is called Asian-Pacific American Heritage Month) in May. The purpose of the month is to learn about and acknowledge the historical and contemporary contributions made by Canadians of Asian heritage.

National Aboriginal History Month: Observed annually in Canada during the month of June to recognize the past and present contributions of First Nations, Inuit, and Métis peoples to the birth and development of Canada.

Tokenism: The practice of including one or a small number of members of a minority group to create the appearance of representation, inclusion, and non-discrimination, without ever giving these members access to power.

The role of Blacks in Canada has not always been viewed as a key feature in Canada's historic landscape. There is little mention that some of the Loyalists who came here after the American Revolution and settled in the Maritimes were Blacks, or of the many sacrifices made in wartime by Black Canadian soldiers as far back as the War of 1812.

Few Canadians are aware of the fact that African people were once enslaved in the territory that is now Canada, or of how those who fought enslavement helped to lay the foundation of Canada's diverse and inclusive society.*

Is the solution the creation of a designated month? It can acknowledge the absence of the history of designated groups of people, but what happens when the month is over? Should teachers say to their students in March, "Well, it's back to white history again"? Should they say the same to students when **Asian Heritage Month** concludes in May or when **National Aboriginal History Month** concludes in June? Some suggest that these observances are merely a kind of **tokenism** used to mask real issues of power, equity, inclusion, and justice. Perhaps the most important question is how do we instead ensure that the history we share and the narratives we tell every day include the experiences, events, and voices of all people as subjects—and not just the objects—of our interest? This is much harder to achieve, as it requires the dominant group in a society to move beyond token change to break the cycle of oppression and to eliminate racism in all its forms. So, ultimately, it is an issue of power. Making only token changes risks trivializing, racially stereotyping, or demeaning the histories of racialized groups as "other" and "less than."

ENDING THOUGHTS

Most anti-oppression and anti-racism frameworks include elements of structural analysis, institutional change, self-awareness, and constant vigilance. Is the strategy for creating a post-racial society as simple as Morgan Freeman suggests—a kind of racial transcendence that comes when we stop talking about race, and when we stop assigning meaning to the colour of skin? Perhaps someday.

How, then, do we address the structural inequity that creates social injustice and oppression? The vision of a society free of oppression is perhaps most articulately described in Dr. Martin Luther King Jr.'s "I Have a Dream" speech, given in 1963 at the Lincoln Memorial, as he inspired the crowd: "I have a dream that my four little children will one day live in a nation where they will not be judged by the colour of their skin, but by the content of their character." Ironically, this goal becomes harder in a country like Canada that has been generally respected for its human rights record and has begun to acknowledge and redress some parts of its racist history. How do we as a nation begin to understand that this racist history is still echoed in modern policy and practice (Roy, 2012)? The conservative approach to racism has been colourblindness, and the liberal approach has been a framework of diversity that invites racialized communities to participate and be included under the same banner of existing power structures. Neither approach is about equity.

AGENT OF CHANGE

Babaluku

Silas, a child of diaspora, uses his authentic voice on both sides of the pond to empower youth.

Uganda's Luga Flow legend Babaluku, born Silas Babaluku Balabyekkubo, is one of the most visible and respected pioneering artists making an impact on Kampala Uganda's hip-hop scene and youth culture (Cunningham, 2014). "He's one of the first emcees to make rapping in the native Luganda tongue popular rather than deferring to colonial English" (Cunningham, 2014). It is this "authentic voice" that he explores in a Ted Talk from Kampala. He is considered the grandfather of "Lugaflow," which is rap music in Luganda.

Babaluku was born in Uganda in 1979, the eldest in a family of eight children born to the late Pastor Deo Balabyekubo and his wife Christine Balabyekubo. He came to Canada at age 12 and today splits his time between Kampala and Vancouver, BC. His life journey as a child of diaspora, sharing two cultures and two heritages, Babaluku walks with lingering questions—a search for truth and justice born from struggle that he uses as a tool for empowering youth on both sides of the pond.

He is the founder of the Bavubuka Foundation, which facilitates leadership among youth; they only need to look to the founder to find inspiration and to dare to dream of the possible. The Bavubuka Foundation provides safe spaces for Ugandan youth to develop skills through the art of self-expression via hip-hop music and culture. Silas has further developed that concept by engaging the youth he mentors in the visual arts, sports, dance, film, community advocacy, and music. "I wanted to give back and empower the youth to [become] better leaders for the continent, for the countries they represent and the communities they live in."

Babaluku's early music days in Uganda to festivals in the United States are captured in the 2008 documentary, *Diamonds in the Rough: A Ugandan Hip-hop Revolution,* which won the Audience Choice Award at the Dances with Films Festival, and Best Feature Documentary at the Peace on Earth Film Festival. This rapper, musician, producer, community youth activist, and social entrepreneur has earned a number of awards, including a Pearl of Africa Award for Best Hip Hop Single, Buzz Teen Awards for Best Hip Hop Artist, Pioneer of the Year by the Words Beats & Life Remix Teach-In Awards.

When we no longer socially construct power differences among racialized groups, then Samuel will be free to live in a post-racial society, and he will be judged by the content of his character and not the colour of his skin—where racism will no longer be a part of his everyday life.

READING

BROKEN CIRCLE: THE DARK LEGACY OF INDIAN RESIDENTIAL SCHOOLS

By Theodore Fontaine

THE MÉNAGE

"Tee-adore." The gruff voice of Sister S. shatters the silence of the study room. I cringe and slouch down at my desk—my fears are realized. I am 9 years old and a resident at the Fort Alexander Indian R esidential School, 90 miles north of Winnipeg, Manitoba. The school is run by the Oblates, a religious order of the Roman Catholic Church, and is about 2 miles from my home on the Fort Alexander Reserve.

C. has just returned and handed Sister S. a note with my name on it. I have an uncontrollable urge to yell and scream

---

and disappear into nothingness. I do not have to respond to her calling my name. I know where I am expected to go.

Every evening before we go to bed, we are in our classrooms for study periods, and every night in happens. Four or five different boys are called into a room for a weekly ritual exercise known as ménage. We have learned that ménage is a French word for cleaning.

Can I escape? If I turn right instead of left when I leave the classroom, I can be outside without anyone seeing me for at least five minutes. Then I can stick to the ditches or the shoreline until I reach home. What then? Dad is probably working night shift at the paper mill. Mom will be furious and threaten to take me right back, but then decide to wait for Dad to come home so that they can both take me back.

I decide I can't make it anyhow. Is anyone watching me? I can feel eyes on me, especially from the girls, who are segregated on the opposite side of the room in all our study sessions, as they are during regular classes. They know. It's my time for ménage—that weekly ritual, the washing of the genitals by a man in a black robe.

D. has a pained expression on his face as I catch his eye. He knows he's next or next or next. R. looks as if he enjoys my predicament. I feel the eyes of the other side of the room undressing me and imagining how I look, and what the little priest will do in a couple of minutes. That little kookoosh (pig). I feel a twinge of guilt for thinking that, as he is the priest who baptized me.

A. almost has a look of pity on her face as she realizes that I will be bared, washed, and dried, and not one voice of protest will be heard. If there was, the promise of hell or purgatory would shut the protester up immediately if the slap on the back of the head or the pulling of the ear didn't.

I wait a few seconds, hoping the class will resume studying and doing their homework and forget about me. Sister S. peers above her glasses and I realize that if she calls my name again, everyone will refocus on my dilemma. Ugly witch! I'm sorry, God, I didn't mean that. But she is not pretty. If I go, the devil will not hang around me and give me bad thoughts, because clean crotches drive him away. He likes dirty crotches.

I wonder if they wash the girls, too. M. sitting over there looks clean enough, but S. must need washing every day! I should tell her that the next time she's mean to my little sister.

The horn-rimmed glasses turn my way again, so I'd better move. I close my book and fidget to find some way of marking my place. That will make Sister S. think I am almost ready to go. There's R. looking at me again. I've seen him going to see Father more than once a week. I think he likes it. They should move him in with the priests or brothers. D. looks resigned, so I had better get it over with. It's probably been two or three seconds since my name was called. Someday I'll be away from here and everyone will forget what's happening to me at this very moment. I try to make as little noise as possible, and I crouch and attempt to become smaller as I struggle up from my desk.

I can feel my organ in the wrong place—it's supposed to be in my right pant leg, not my left! I wonder how many girls have noticed. I feel bowlegged because I wore my jeans for a couple of weeks and the stiff canvas-like denim has sagged outward at the knee. It feels like my crotch is more prominently exposed. I feel nauseated, my muscles tighten, my jaw feels rigid, and I wonder if I can walk. There is something dreadfully wrong with my paraphernalia. I hope my fly is up. I hope my shorts don't ride up my arse, and I hope my shoes don't squeak. I also hope the girls don't know that in a couple of minutes I will have an erection.

As I slither along the aisle, I can feel 60 eyes on my back, my crimson face, and my crotch. I don't turn around as I try to close the door softly behind me. I feel a breath of fresh air as I emerge from the classroom into the eerie, quiet, dimly lit hallway. I look left, down the long hall to my destination, and then turn right; I see the big doors that lead to the veranda and the fresh, crisp air and sparkling sky and fall breeze gently urging a soft, pillowy white cloud on its southerly journey. The geese and the chickens must be abundant. The muskrats must be setting up for the winter. The deer must be fat. My dog must be wondering where I have gone.

Last time I ran away, Mom and I would have to answer to God for saying those things about Father P., and I got less of the fruit from Sister C. that Mom and Dad had brought me. I also didn't get to play hockey for five days. I think the priest is scared of Dad and that's why I didn't get the strap that time. He doesn't know that Dad would never harm anyone. Besides, I might tell Dad and he might believe that what Father P. does is not natural. My grandpa had a reputation in earlier encounters with priests when Dad and his siblings were first at the school. His protective nature regarding his family triggered very strong reactions.

I wish Dad were not so busy with his work and everything else. I think he'd like to take me out of school to teach me how to work, but he also wants me to learn and to finish school in order to get a better job.

I turn left, realizing I must go and finish this. It will be over soon.

As I slink down the hallway, I hear coughing and shuffling of feet as I pass the classrooms with older boys and girls in them. I wonder why older boys don't have to go for ménage anymore. You must stay cleaner as you get older, I will be in one of those classrooms soon, and I'll be one of the big boys. Then I can laugh and make fun of the younger boys when they go for their ménage.

Passing the chapel, I think that I must go to confession tomorrow morning, because I couldn't take my eyes off Miss M.'s ankles the other day, and the outline of her legs on her dress. Miss M. is our teacher, and her dress hangs six inches off the floor. And that new sister—actually a novice, not quite a full-fledged nun and still without full nun regalia—has nice clean hands that felt warm and loving when she washed mine to clean a cut. My hands actually felt caressed and loved in the warm, soapy water.

I shuffle quietly down the hall. I think about the shop teacher H. and wonder if he confesses to Father P. I wonder if he sinned when he pulled out his penis in front of me and commanded me to sit on the chair in front of him as he tried to pull it out by its roots. Maybe they are allowed. It looked like big blood sausage, like the kind they make from a cow's blood when it's drained after slaughter.

In my mind, seeing Miss M.'s ankles, feeling good at the gentle touch of the novice nun, and watching the shop teacher are all sins to be reported in the confessional. I wonder why there are so many questions from the confessional, and only from certain priests.

I wonder if the principal knows where I am going. The light casts a shadow on the door to his office; the door is closed again. I've heard that nuns go in there at night, probably to report on our bad behaviour during the day.

I see Father P.'s door and the light peeking out from under it. My imagination takes hold and I picture what I would do if in a split second I were to become a big man! I hesitate. Then, as my hand grasps the door handle, I catch an aroma of soap in a steamy, slimy soap holder. I wonder if the water will be warm.

There are many stories about Indian residential schools in Canada, stories about physical, sexual, spiritual, and mental abuse. Some church officials and other Canadians accuse Indian residential school survivors of "telling stories," implying that their stories are untrue or exaggerated. The episode just described represents only a few minutes of my 12 years in the residential school system, but it is a scene etched into my memory. This ritual of "staying clean" happened every week or two over the years for many of the younger boys. It stopped when we became older and bigger, and our determination to threaten, maim, hurt, or even kill our tormentors gave us the power to refuse the treatment.

My shop teacher, Mr. H., was a friendly and jovial guy who befriended young boys in order to expose himself and talk dirty to them, I realize now. I recall that he forced me to unzip my trousers and expose myself to him, although I do not remember him touching me. But in exchanging stories with former classmates years later, I learned he forced himself on other boys.

I do remember sitting, as a boy, with Mr. H. and his wife at the kitchen table in their residence, situated within the school grounds, with a bottle of beer, hating the taste but feeling excited as the beer slipped down my throat and warmed my belly. The survivor friend remembers drinking with Mr. H. and having Mr. H. force him to lie on top of his wife as he stood aside and masturbated. I don't know if he ever made any other boys do that, but he certainly underwent a physical workout in front of me and others. This is one of the most belittling, embarrassing, and hurtful memories I have had since then.

Recently, such recollections have become more clear and part of my present reality. There are many like these, and they often come at awkward and unexpected moments. There are many gaps, but I am working to ensure that they become filled with missing memories, unpleasant as they may be. Each time I remember and talk about an experience, more and more comes back to me, helping me to see patterns of behaviour and understand the effects of my residential school experiences on my life.

My parents dropped me off at Fort Alexander Indian Residential School just days after I celebrated my seventh birthday, believing I would be cared for by the priests and nuns. Little did they know that the experience I was about to undergo for the next 12 years would shape and control my life for the next 40 or 50. From this point on, my life would not be my own. I would no longer be a son with a family structure. I would be parented by people who'd never known the joy of parenthood, and in some cases hadn't been parented themselves.

The system was designed by the federal government to eliminate First Nations people from the face of our land and country, to rob the world of a people simply because our values and beliefs did not fit theirs. The system was racist and based on the assumption that we were not human but rather part animal, to be desavaged and moulded into something we could never become—white.

Source: Fontaine, T. (2010). *Broken Circle: The Dark Legacy of Indian Residential Schools*. Toronto: Heritage House Publishing Company Ltd.

DISCUSSION QUESTIONS

1. Author Theodore Fontaine states that the Indian Residential School system was racist, attempting to mould children into something they would never become—white. Discuss the ways in which Indian Residential Schools in Canada were an exemplar of systemic racism.

2. What did First Nations societies lose as a result of Indian Residential Schools? What are the lasting impacts of the Indian Residential School system, and how does this narrative further your understanding of these impacts?

3. How will we know when healing and reconciliation from Indian Residential Schools has taken place in Canada?

KWIP

One of the fundamental truths about diversity is that it involves as much learning about oneself as it does learning about others. Let's use the KWIP process to look at how our identities are shaped through social interaction.

KNOW IT AND OWN IT: WHAT DO I BRING TO THIS?

The "K" in the KWIP process involves examining aspects of your own identity and social location as the first step in becoming diversity competent.

How open are you to hearing about claims of racism and listening to both narratives and scholarly evidence of racial inequality and injustices? Historical events prove that racism is not something new. The Indian Act and residential schools, Africville, the internment of Japanese Canadians, and the Chinese head tax provide evidence of systemic racism in Canada. So how do we know that racism still exists today? What can we do about it? How can you create a post-racial society where the colour of a person's skin is no longer correlated with yearly earnings, employment status, social class, incarceration rates, graduation from high school, and matriculation in postsecondary education?

WALKING THE TALK: HOW CAN I LEARN FROM THIS?

The "W" in the KWIP process presents a scenario or case study that challenges you to "walk the talk" through problem-based learning.

ACTIVITY: **CASE STUDIES**

Case One: *Rita*

Rita and her family moved to the city from a remote community in the middle of the school year. Within a week, Rita was registered at the local high school and began attending classes. She travelled to and from school by school bus.

After two weeks at the new school, Rita was just beginning to settle into her classes. However, she was somewhat nervous about her history course. After her first class, the teacher made it clear that Rita had a lot of "catching up" to do, if she were to pass the course. The following week, some students gave a presentation on Columbus's voyage in 1492 to the "New World." There was lively discussion, and readings and prints were circulated depicting Columbus's arrival in various territories. There were several references made to "Indians and savages" that the colonists "had to defeat" to settle the New World.

As a member of the Cree Band, Rita was dismayed by the way the teacher portrayed Indigenous persons in the presentation. She approached her teacher before class the next day to discuss the issue. As the class began, the teacher announced that Rita had concerns with the Columbus presentation. She then turned to Rita and asked her to give her version of the "Columbus discovery" from an Indigenous point of view.

Caught off guard, Rita haltingly made several points, and then sat down quickly when several of the students began to snicker. Later that day on the bus ride home, some of the other students jeered at her, saying if she didn't like history the way it was taught, then she should drop out. She turned away and ignored them. The next day, the jeering continued in the hallway. When she went to her locker at lunch, someone had scrawled the words "gone hunting" on her locker door. Again, she ignored the curious students around her.

Rita told her parents about the incidents. They called the principal, who said she would give "hell" to the offenders. She also suggested that Rita should make more of an effort to fit in and get along with others.

1. How should the teacher have handled Rita's concern over the Columbus presentation?

2. Should the principal deal with the situation in a different way?

Source: Ontario Human Rights Commission. (2013). Teaching Human Rights in Ontario: A Guide for Ontario Schools. Found at: http://www.ohrc.on.ca/sites/default/files/Teaching%20Human%20Rights%20in%20Ontario_2013.pdf. © Queen's Printer for Ontario, 2013. Reproduced with permission.

IT IS WHAT IT IS: IS THIS INSIDE OR OUTSIDE MY COMFORT ZONE?

The "I" in the KWIP process requires you to honestly confront and identify ways in which our complex identities result in experiences of privilege and oppression, and to reflect on how we can learn to honour that privilege.

ACTIVITY: **GOT PRIVILEGE?**

As Tim Wise (2006) suggests, the beginning of honest dialogue is the acknowledgment of the way things actually are, rather than the way we would like them to be. So where does your race land you in the hierarchy of power and privilege in our society? Does the social construction of race mean you are marginalized and discriminated against? Or does your race afford you certain privileges and power that you may not even be fully conscious of?

ASKING: Do I have privilege?

1. The colour of my skin does not negatively impact how people assess my job performance, financial credibility, intelligence, or personal character.

2. I see people of my race widely represented in my school textbooks and other teaching materials.

3. I see people of my race widely represented among my school's teaching faculty and administration.

PUT IT IN PLAY: HOW CAN I USE THIS?

The "P" in the KWIP process involves examining how others are practising equity, and how you might use this. To this end, you are invited to read a commentary about anti-black racism on

4. When I experience success, people will not tell me I am a credit to people of my race.

REFLECTING: Honouring Our Privilege

Describe two circumstances in which you feel disadvantaged because of your racialized identity. Then describe two circumstances in which your race gives you privilege over someone else. How can you use the advantages you experience to combat the disadvantages experienced by others?

Canadian campuses in an article written by Sefanit Habtemariam and Sandy Hudson, the founding members of a movement called the Black Liberation Collective-Canada that is dedicated to eliminating anti-black racism on college and university campuses.

ACTIVITY: CASE STUDIES

Canadian campuses have a racism problem

By Sefanit Habtemariam and Sandy Hudson

When we hear stories about anti-black racism and the struggle against it in the United States, we like to imagine that Canada is a superior place; a place that, with our different history, has escaped such problems. Sadly, that is not the case.

Take the recent controversy at the University of Missouri. The stories of anti-black racism at that school were all too familiar to us. One of us is a graduate of the University of British Columbia and we are both current students at the University of Toronto, and we recognized the plight of black Mizzou students from our own experience. Blackface, a lack of representation, the slow death or complete erasure of programs that focus on black people and thought; these are issues above the 49th parallel, and it's time we recognize it and do something to fix it.

At our own institution, there are no courses where you can study black people at the graduate level. The School of Global Affairs is completely devoid of programs and courses that focus on the continent of Africa. This is the largest school in Canada, often touted as the best. It's difficult to imagine such an omission with regard to, say, Europe or Asia. Canada itself has a long and vibrant black history. Should we not be able to study it?

Black students and faculty at the University of Toronto, the University of British Columbia, and other schools across southern and eastern Ontario recognize this, and have begun working with students at institutions across the United States on efforts to resist the racism we experience on our campuses. The success of our movement—and the backlash we have experienced—is telling.

On Nov. 18, black students across the United States and Canada marched, protested, and gathered on our respective campuses to share our experiences of what it means to be "black on campus." From Ottawa to Guelph to Toronto, hundreds of students and faculty participated in the actions. At the University of Toronto, we released a comprehensive list of demands focused on tackling anti-black racism at the

school's core. As a result, the university has agreed that it needs to work on the ways anti-black racism manifests on campus and has committed to begin collecting census data on the representation of black students, faculty, staff, and administration.

But the response from some in the academic community was reprehensible. At the University of Guelph, dozens of anonymous attacks on black students were made online in the form of discriminatory remarks, threats, and harassment. These responses lay bare the similarities between the black experience in Canada and the United States.

Although the University of Toronto agreeing to collect race-based census data is a great first step, it is only that. It has been two months since we met with administration to discuss our detailed demands. After promising a thorough response to each one, the administration finally responded last week with a disappointing email that did not address our issues in concrete or actionable ways. Our work continues, as it does for the black students on campuses across Canada continuing to organize events and further action.

As a follow-up to their demonstration, black organizers at the University of Guelph held an anti-black racism "teach-in," exploring the issues black students on Canadian campuses face and how to address them. At Ryerson University, black students are currently working on an anti-black racism poster campaign and a petition for their demands.

It's time we stop thinking of ourselves as superior to the United States when it comes to racism. We should be looking within and recognizing the racism that exists throughout our society—whether it's in policing, education, media representation, housing—the evidence is there. Black students know this, and live it. We were encouraged by the University of Toronto's commitment to collect data, but our demands are far from finished. Working together, black student organizers in both Canada and the U.S. know the changes we're seeing now are only the beginning.

Source: Sefanit Habtemariam and Sandy Hudson. Published in thestar.com on March 1, 2016, https://www.thestar.com/opinion/commentary/2016/03/01/canadian-campuses-have-a-racism-problem.html.

Located at *www.nelson.com/student*

- Review Key terms with interactive **flash cards**
- Check your Comprehension by completing **chapter review quizzes**
- Gauge your understanding with *Picture This* and accompanying short answer questions
- Develop your critical thinking/reading skills through compelling **Readings** and accompanying short answer questions
- Apply your understanding to your own experience with **Connect A Concept** activities
- Evaluate Diversity in the Media with engaging *Video Activities*
- Reflect on your Understanding with *KWIP* activities

REFERENCES

Adelman, L. (Director). (2003). *Race: The power of an illusion* [Motion Picture]. United States: California Newsreel.

Allen, M., & Boyce, J. (2013, July 11). *Police-reported hate crime in Canada, 2011.* Retrieved from *Statistics Canada*: http://www.statcan.gc.ca/pub/85-002-x/2013001/article/11822-eng.pdf

American Anthropological Association. (2011). One race or several species. Retrieved from *Race: Are We So Different?* http://www.understandingrace.org/history/science/one_race.html

Benjamin, R. (2009). *Searching for Whitopia: An improbable journey to the heart of white America.* Hyperion Books.

Black, S., Kendra, C., Ruhloff-Queiruga, N., & Savage, L. (2014). *Open letter re: blackface at Brock University from the Department of Labour Studies.* Retrieved from Brock University: https://brocku.ca/node/27835

Block, S., & Galabuzi, G. (2011). *Canada's colour coded labour market: The gap for racialized workers.* Ottawa: Canadian Centre for Policy Alternatives.

Bonilla-Silva, E. (2012). *Race 2012.* PBS. Retrieved from http://video.pbs.org/video/2289501021/

Braidwood Commission on the Death of Robert Dziekanski. (2010, May). *WHY? The Robert Dziekanski tragedy.* British Columbia: Provincial Government of British Columbia.

Brean, J. (2013, February 27). Supreme Court upholds Canada's hate speech laws in case involving anti-gay crusader. Retrieved from *National Post*: http://news.nationalpost.com/2013/02/27/supreme-court-upholds-canadas-hate-speech-laws-in-case-involving-anti-gay-crusader/

Campaign 2000 Family Services Toronto. (2016, November 24). 2016 *Report card on child and family poverty in Canada.* Toronto, Ontario.

Canadian Charter of Rights and Freedoms. (1982). Part I of the *Constitution Act, 1982,* RSC 1985, app. II, no. 44.

Canadian Race Relations Foundation. (2012, July 30). *CRRF facts about ... Facing hate in Canada.* Retrieved from http://www.crr.ca/divers-files/en/pub/faSh/ePubFaShFacHateCan.pdf

CBC Digital Archives. (1989). The Rushton–Suzuki debate. London, Ontario. Retrieved from http://www.cbc.ca/archives/categories/arts-entertainment/media/david-suzuki-scientist-activist-broadcaster-1/the-rushton-suzuki-debate.html

CBC News. (2005, May 26). In depth: Racial profiling. Retrieved from http://www.cbc.ca/news/background/racial_profiling/

CBC News. (2011, Jan 11). Cross-burning brother gets 2 months in jail. Retrieved from *CBC Canada*: http://www.cbc.ca/news/canada/nova-scotia/story/2011/01/11/ns-justin-rehberg-sentencing.html

CBC News. (2012, November 9). Racist bathroom graffiti angers students. Ottawa, Ontario. Retrieved from http://www.cbc.ca/news/canada/ottawa/story/2012/11/08/ottawa-racist-bathroom-graffiti-shock-university-ottawa-students.html

CBC News. (2014, November 04). Brock University students in blackface win Halloween contest. Retrieved from *CBC News Hamilton*: http://www.cbc.ca/news/canada/hamilton/news/brock-university-students-in-blackface-win-halloween-contest-1.2822958

Citizenship and Immigration Canada. (2009, August 31). About black history. Retrieved from *Government of Canada*: http://www.cic.gc.ca/english/multiculturalism/black/background.asp

Cole, N. (2016, August 5). What is racism exactly? Retrieved from *About Education*: http://sociology.about.com/od/R_Index/fl/Racism.htm

Comack, E. (2012). *Racialized policing: Aboriginal people's encounters with the police.* Winnipeg: Fernwood Publishing.

Commission on Systemic Racism in the Ontario Criminal Justice System. (1995). *Report of the Commission on Systemic Racism in the Ontario Criminal Justice System.* Toronto: Queen's Printer for Ontario.

Cooper, A. (2006). *The untold story of Canadian slavery and the burning of old Montreal*. Toronto, Ontario, Canada: Harper Perennial.

Criminal Code, R.S.C. 1985, c. C-46, s. 318, s. 319. S. 718.2. Retrieved from *Department of Justice Canada*: http://laws.justice.gc.ca

Cunningham, J. (2014, Jan 04). Uganda's Luga Flow Legend Babaluka. Retrieved from *okayafrica*: http://www.okayafrica.com/video/babaluku-uganda-luga-flow-legend/

Dunbar, G. (2012, July 30). London 2012: Swiss expel soccer player, Michel Morganella posted offensive tweet after losing to South Korea. Retrieved from *Thestar.com*.

Elliot & Elliott Eye (producer), & Golenbock, S. A. (Director). (2001). *The Angry Eye* [Motion Picture].

Fleras, A., & Elliott, J. L. (1996). *Unequal relations: An introduction to race, ethnic and Aboriginal dynamics in Canada*. Scarborough, ON: Prentice Hall Canada.

Fleras, A., & Elliot, J. L. (2002). *Engaging diversity: Multiculturalism in Canada* (2nd ed.). Toronto: Nelson Thompson Learning.

Food Banks Canada. (2012). *Hungercount 2012: A comprehensive report on hunger and food bank use in Canada and recommendations for change*. Toronto: Food Banks Canada.

Fontaine, T. (2010). *Broken circle: The dark legacy of Indian residential schools*. Toronto: Heritage House Publishing Company Ltd.

Freeman, M. (2005, December 15). *60 Minutes*. (M. Wallace, Interviewer).

Globe and Mail. (2010, November 04). Not a racist, KKK costume a mistake, says former cop who wore blackface. Campbellford, Ontario. Retrieved from http://www.theglobeandmail.com/news/national/not-a-racist-kkk-costume-a-mistake-says-former-cop-who-wore-blackface/article1241264/

Grewal, S. (2013, May 30). Brampton suffers identity crisis as newcomers swell city's population. Retrieved from *The Star*: http://www.thestar.com/news/gta/2013/05/24/brampton_suffers_identity_crisis_as_newcomers_swell_citys_population.html

Habtemariam, S., & Hudson, S. (2016, March 1). Canadian campuses have a racism problem. Retrieved from *thestar.com*: https://www.thestar.com/opinion/commentary/2016/03/01/canadian-campuses-have-a-racism-problem.html

Henry, F., Tator, C., Mattis, W., & Rees, T. (2009). The ideology of racism. In M. Wallis, & A. Fleras, *The Politics of Race in Canada*. Don Mills: Oxford University Press, 108–118.

Hill, L. (2001). *Black berry, sweet juice: On being black and white in Canada*. Toronto: Harper Flamingo Canada.

Hinkson, K. (2013, May 20). High school students put spotlight on shadeism. *Toronto Star*, GT1 & 4.

International Olympic Committee. (2011). *Olympic charter*. Switzerland: International Olympic Committee.

Johnson, A. (2006). *Privilege, power and difference* (2nd ed.). New York: McGraw-Hill.

K'Naan. (2010, October 11). *Shameless idealists*. (C. Keilburger, Interviewer).

Linden, S. B. (May 31, 2007). Commissioner's statement. Forest: Ontario Ministry of the Attorney General Ipperwash Inquiry. Retrieved from http://www.attorneygeneral.jus.gov.on.ca/inquiries/ipperwash/index.html

McIntosh, P. (1989, July/August). White privilege: Unpacking the invisible knapsack. *Peace and Freedom*, 9–10.

The Nation. (2012, October 12). *The hunted and the hated: An inside look at the NYPD's stop-and-frisk policy*. Retrieved October 20, 2013, from http://www.youtube.com/watch?v=7rWtDMPaRD8

National Council of Welfare. (2012). *Poverty profile: A snapshot of racialized poverty in Canada*. Ottawa: Government of Canada.

Ontario Human Rights Commission. (2003). *Inquiry report: Paying the price: The human cost of racial profiling*. Toronto: Ontario Human Rights Commission.

Ontario Human Rights Commission. (2012). What is racial profiling? Retrieved from *Ontario Human Rights Commission*: http://www.ohrc.on.ca/en/paying-price-human-cost-racial-profiling/what-racial-profiling

Ontario Human Rights Commission. (2013). Teaching human rights in Ontario: A guide for Ontario schools. Retrieved from *OHCR*: http://www.ohrc.on.ca/en/students%E2%80%99-handouts/case-study-1-darlene

Ontario Human Rights Commission. (2016). Racial discrimination, race and racism: Fact sheet. Retrieved from *OHCR*: http://www.ohrc.on.ca/en/racial=discrimination-race-and-racism-fact-sheet

Oppal, T. H. (2012, November 19). *Forsaken: Missing women commission of inquiry*. British Columbia: The Missing Women Commission of Inquiry.

PBS. (2005, February). *Race: The power of illusion: Interview with Robin D. G. Kelley*. Retrieved from http://www.pbs.org/race/000_About/002_04-background-02-05.htm

R. v. *Andrews*, [1990] 3 S.C.R. 870

R. v. *Keegstra*, [1996] 1 S.C.R. 458

R. v. *Zundel*, [1992] 2 S.C.R. 731

Roy, J. (2012). Acknowledging racism. Retrieved from *Canadian Race Relations Foundation*: http://www.crr.ca

Statistics Canada. (2012, June 18). Income of Canadians 2010. Retrieved from *The Daily*, http://www.statcan.gc.ca/daily-quotidien/120618/dq120618b-eng.htm

Tanovich, D. M. (2006). *The colour of justice: Policing race in Canada*. Toronto: Irwin Law.

Tator, C., & Henry, F. (2006). *Racial profiling in Canada: Challenging the myth of a few bad apples*. Toronto: University of Toronto Press.

Timberlake, J. (2003). *Where is the love.* [Recorded by Black Eyed Peas]. United States of America.

Toronto Star. (2012, July 25). Greece expels Olympic athlete over racist tweets about immigrants. Toronto, Ontario: *Thestar.com*

United Nations. (2011, January 12). International Convention on the Elimination of All Forms of Racial Discrimination. Retrieved from *Canadian Heritage*: http://www.pch.gc.ca/pgm/pdp-hrp/docs/cerd/rpprts_17_18/index-eng.cfm

United Nations. (2012, March 21). International Day for the Elimination of Racial Discrimination March 21. Retrieved from *United Nations*: http://www.un.org/en/events/racialdiscriminationday/

University of Guelph. (2016). Understanding racialization: Creating a racially equitable university. Retrieved from *Human Rights and Equity Office*: http://www.uofguelph.ca/diversity-human-rights/sytsem/files/UnderstandingRacialization.pdf

Wade, L. (2012, October 12). Race themed events at colleges. Retrieved from *Sociological Images: Inspiring Sociological Imaginations Everywhere*: http://thesocietypages.org/socimages/2012/10/20/individual-racism-alive-and-well/

Walker, B. (2008). *The history of immigration and racism in Canada.* Toronto: Canadian Scholars' Press.

Wise, T. (2006). *Beyond diversity: The hidden curriculum of privilege.* Retrieved from *YouTube*: http://www.youtube.com/watch?v=D30GOWsnVuA&list=PL64A999D2C184E259

Wise, T. (2010a). *Colorblind: The rise of post-racial politics and the retreat from racial equity.* San Francisco: City Lights Open Media.

Wise, T. (2010b, June 18). Colourblind ambition. Retrieved from *Tim Wise*: http://www.timwise.org/2010/06/colorblind-ambition-the-rise-of-post-racial-politics-and-the-retreat-from-racial-equity/

Wise, T. (2011). *White like me: Reflections on race from a privileged son.* Berkeley, CA: Soft Skull Press.

Wong, V. (24, January 2016). The problem of increasing racialized poverty in Canada. Toronto, Ontario. Retrieved from http://vincewong.ca/the-problem-of-increasing-racialized-poverty-in-canada/#respond

Wortley, S., & Marshall. L. (2005). *Race and police stops in Kingston, Ontario: Results of a pilot project.* Kingston: Kingston Police Services Board.

Indigenous Peoples

"*Stranded in the wasteland*
Set my spirit free"

(Robbie Robertson, 1998)

LEARNING OUTCOMES

By mastering this unit, students will gain the skills and ability to:

- explain the significance of Indigenous intellectual traditions in relation to Western intellectual traditions

- summarize the three treaty periods and compare the advantages and disadvantages for the colonizing agents and Aboriginal peoples involved

- identify the overt and hidden acts of assimilation and oppression Aboriginal peoples continue to confront, and compare them in terms of ongoing impacts to Aboriginal peoples' ways of knowing

- distinguish between specific and comprehensive land claims, and discuss the implications of the differences between the two as they relate to present-day ideals of territorial ownership and economic and social development

- identify reasons for the substandard living conditions among Aboriginal peoples in comparison to non-Aboriginal Canadians

- assess the impact of economic development on Aboriginal peoples, and how it impacts intra-community relations

© Tammy Luciow

Will you be celebrating June 21? Not only is it the summer solstice (the day with the most daylight), it is also National Aboriginal Day. Certain employees in the Northwest Territories will enjoy a paid holiday. Some Canadians will eat fry bread and moose stew while delighting in the summer solstice festivals and attending drumming and dancing events and ceremonies. Most, however, will let the day pass unnoticed, unaware of its existence. How much do you know about Canada's First Peoples? Scientists now have evidence that First Nations people lived in what is now Canada over 12 000 years ago, whereas Indigenous creation stories identify Turtle Island residency dating to time immemorial (Canada's First Peoples, 2007).

Without doubt, Canada is one of the most beautiful countries in the world; non-Indigenous Canadians can thank Aboriginal peoples for their historic and ongoing land stewardship. Though other interpretations exist, some believe that Canada is a word taken from the Cree language (Cardinal, 1977). Stemming from the Cree word *Ka-Kanata*, the full Cree term to describe the country is *Ka-Kanata-Aski*—"the land that is clean" (Cardinal, 1977).

The history that accompanies the rustic imagery of the heritage of European settlement, however, is anything but peaceful. Rife with ongoing and at times contentious negotiations, misunderstandings, willful ignorance, and confrontations, Aboriginal peoples (First Nations, Inuit, and Métis) in Canada have, since first contact with Europeans, constantly been challenged with new ideas and processes that undermine attempts to maintain a sense of peace between their own communities and the larger Canadian society. Through these examples, this chapter will investigate the past, present, and future relationships between Aboriginal and non-Aboriginal people in Canada.

Indigenous: The descendants of groups of people living in the territory at the time when other groups of different cultures or ethnic origin arrived there.

Aboriginal peoples: A legal/administrative category that comprises the First Nations, Inuit, and Métis; often used to describe the Indigenous peoples within Canada's boundaries.

First Nations: A legal term used to refer to those Aboriginal people who are of neither Inuit nor Métis descent. First Nations refers to the ethnicity, while a band may be a grouping of individuals within that ethnicity (e.g., Seneca or Oneida).

Inuit: The Aboriginal peoples who live in the far north or Arctic regions of Canada.

Métis: People indigenous to North America whose background is Aboriginal and European ancestry.

Indian Act: Canadian federal legislation, first passed in 1876, and subsequently amended, which details certain federal government obligations and regulates the management of reserve lands, money, and other resources. It was initiated to compel Indian assimilation into Canadian society.

THE LANGUAGE
Using the Right Words

There are many terms that are related to **Indigenous** people, and the terminology associated with **Aboriginal** peoples—**First Nations**, **Inuit**, and **Métis**—can often lead to confusion, anxiety, and misinformation. Using the right word in the right context is important in any conversation, but it becomes especially so when the chosen words belie a history where the terminology may not have been chosen by the very groups involved. Consider, for example, "status" and "non-status Indians," which are the legally defined terms under the **Indian Act**. It is important to be mindful of correct terminology and to be respectful of appropriate representations; but what becomes of paramount importance is that we continue to have many conversations, in every conceivable form, and carry on discussing the relations that exist between and among Aboriginal peoples, non-Aboriginal society, and our collective Canada.

The Role of Storytelling in Aboriginal Communities

The importance of language cannot be understated for Aboriginal peoples. There are over 60 different Aboriginal languages, grouped into 12 distinct language families, currently spoken in Canada. According to the 2011 census, nearly 231 400 people reported speaking an Aboriginal language most often, or regularly at home (Statistics Canada, 2012). Language is as much about communication as it is about identity, offering a way for Aboriginal communities to talk among themselves, but to also communicate with the earth and share their history.

What stories did your parents read to you? What lessons did they teach you? For the children of First Nations, Métis, or Inuit descent, the transmission of knowledge vis-à-vis the spoken word was, and in certain respects remains, prevalent among First Nations (Hanson, 2009a), indeed among all Aboriginal peoples. As an element of Indigenous intellectual traditions, creation stories provide people with a sense of identity, emphasizing their philosophy of life and the importance of values and historic and comtemporary beliefs (Sinquin, 2009). These stories were once told as evening family entertainment, for instance, but they are still used to convey local or family knowledge. They might include titles such as *Raven Steals the Light* (Reid & Bringhurst, 1988) and *Path with No Moccasins* (Cheechoo, 1991).

Métis stories that speak of customs while recounting details from the life and leadership of

Louis Riel have become a part of the understanding of how the Métis define and record the events of their history (Lombard, 2009). Métis children might hear stories like *The Flower Beadwork People* (Racette, 1991) or *I Knew Two Métis Women* (Scofield, 1999). Likewise, Inuit children have listened to their parents tell stories and sing songs in the Inuktitut language, and then have gone on to share those same songs and lessons with their own children (Silou, 2009). Inuit stories also teach children about their ancestors and how they lived, and often address larger themes, such as death and respect. Stories from the Inuit culture include *Qasiagssaq, The Great Liar* (Norman, 1990) and *The Sea Goddess Sedna* (Kennedy & Moss, 1997).

Though each group of Aboriginal people has their own set of cultural values and traditions, they share common beliefs that are passed on through oral traditions. Many of the creation and recreation stories that are told reflect the interdependent relationship between humans and Creation—that is, all of the relations that occupy the territory a specific people call home. Though their stories are filled with spirits, animals, and sea creatures (all or none of which may come of life), they are often deeply spiritual.

This holistic view of human and natural life, emphasizing that everything and everyone is interconnected and considered a relation, is seen in many elements of Indigenous society. For example, the **medicine wheel** demonstrates how all life travels on a circular journey. The small, inner-circle part of the wheel represents Mother Earth and the Creator, and the wheel is divided into four sections: the four cardinal directions (north, south, east, and west); the four colours of humans (red, yellow, black, and white); the four faces of man (physical, mental, emotional, and spiritual); the four seasons (fall, winter, summer, and spring); the four sacred medicines (sweet grass, tobacco, sage, and cedar); and the four stages of life (child, teen, adult, elder). There are many variations of the medicine wheel. Take, for example, the province of Alberta, where a large structure for the medicine wheel, built and arranged on the land, dates back 4500 years.

Similarly, **totem poles** are often used to document certain stories and histories that are familiar to community members or particular family or clan members. Usually carved from red cedar and painted with symbolic figures, totem poles are erected by Indigenous people of the northwest coast of North America. For example, some Kwakwaka'wakw families of northern Vancouver Island belonging to the Thunderbird clan will feature a Thunderbird crest and familial legends on their poles. Other common crests among coastal First Nations include the wolf, eagle, grizzly bear, killer whale, frog, raven, and salmon (Malin, 1986). You

Among other things, the medicine wheel teaches the importance of balance in our lives—mentally, emotionally, spiritually, and physically. Would you visit practitioners who use the medicine wheel as part of their diagnosis, treatment, or learning plans? Why, or why not?

can find them in various shapes and sizes at trading posts, on reserves, and in small tourist towns all over North America (Ramsey, 2011). Demonstrating their continued vitality, in 2017 the University of British Columbia raised a reconciliation pole to ensure that the history of Canada's Indian residential schools is not forgotten.

Medicine wheel: Ceremonial tool that symbolizes the interconnected, circular journey of all living things.

Totem pole: A pole or post usually carved from red cedar and painted with symbolic figures, erected by Indigenous people of the northwest coast of North America.

To the unaware, the totem pole designs may appear to be confusing. But on closer examination, they are very intricate and complex. Each object on the pole represents an animal, a human, or a mythological creature; the skill lies in identifying them. Every item on the pole has meaning—the choice of colour, animal, and crest all mean something special to the family or community that is represented on the pole. Generally, totem poles serve one of four purposes:

1. Crest poles give the ancestry of a particular family.
2. History poles record the history of a clan.
3. Legend poles illustrate folklore or real-life experiences.
4. Memorial poles commemorate a particular individual. (National Park Service, 2013)

The "raising" of a totem pole is surrounded with much ceremony and gaiety. Author Hilary Stewart describes a Haida pole-raising ceremony on the coast of British Columbia:

> Different groups had (and still practice) varying traditions for the pole-raising ceremonies: a popular Haida one is for the carver to dance with his tools tied around his person. Among all groups, the owner of the pole (or a speaker representing him) explains in detail the stories and meaning behind all the carved figures, and those assembled to witness the event are expected to remember what they see and hear. A particularly fine pole calls for praise, criticism and comparison, enhancing the status of the owner and the reputation of the carver. Feasting and potlatching follow in celebration, as one more carved monument stands tall and splendid against the sky. (Stewart, 1993, p. 28)

Tools like the medicine wheel and the totem pole, combined with ceremonies and rituals, contribute to a rich history that is part of the Aboriginal peoples' intellectual traditions. It is critical to recognize that their ways of knowing differ from other systems of knowledge within a Western framework, but that they are no less valid or reliable.

Totems are a distinct and important piece of Aboriginal heritage. Can you identify any figures in this Haida totem? What markers or crests signify your family's lineage?

These intellectual traditions are being recognized more frequently. For example, the Supreme Court of Canada in the landmark case of *Delgamuukw v. British Columbia*, acknowledged oral histories as legitimate forms of evidence. In this case, the Gitksan and Wet'suwet'en peoples argued that they had Aboriginal title to the lands in British Columbia that make up their traditional territories. To prove they had title to the land, evidence was needed to show that they had been living there for thousands of years. Unfortunately, no written documentation confirming this claim existed. To do so, Gitksan and Wet'suwet'en hereditary chiefs presented their histories—by telling stories, performing dances, giving speeches, and singing songs in court (Hanson, 2009a). Though the Gitskan and Wet'suwet'en nations lost the initial decision, on appeal they won a key precedent—oral histories were to be acknowledged as legal evidence. Since then, a number of cases (*Squamish Indian Band v. Canada*; *R. v. Ironeagle*; FJA, 2013) have included oral histories as legal evidence in a court of law—often with certain stipulations.

The legal right to use oral tradition in a court of law is but one battle won in a long-standing struggle. Fighting to keep their land, traditions, and culture alive has been a decades-long battle for those who have overcome colonial government policies and practices dating to the 1700s seeking the elimination of Aboriginal rights and land title. As will be discussed in greater detail, proof of this approach is found in "Act to Encourage the Gradual Civilization of Indian Tribes in this Province, and to Amend the Laws Relating to Indians" (Gradual Civilization Act, 1857). By 1867, when Canada became an independent country, the quest to integrate Aboriginal communities into "civilized" society evolved into a growing public and political concern. Canada's first prime minister, John A. Macdonald, would comment in 1887, "The great aim of our legislation has been to do away with the tribal system and assimilate the Indian people in all respects with the other inhabitants of the Dominion as speedily as they are fit to change" (Montgomery, 1965, p. 25).

What early prime ministers and lawmakers tended to ignore—and this remains an issue confronting contemporary politicians—is this: Aboriginal peoples have every right to maintain their own identity, an identity forged within the lands prior to and fortified after contact with European peoples, and one that importantly provides Canada with a sense of its own identity. Will Canada remain a land that is, ironically, unsettled by those who initially settled it? For all of our multiculturalism, industry, and pride, our Canada is not at peace. How in our separate but collective worlds did this happen?

TREATIES

When the Europeans came to the Americas, Indigenous peoples had already inhabited the land and permitted the Europeans full access to it, in accordance with their own intellectual traditions. The Indigenous peoples' first contacts with Europeans were mainly through the participation and partnerships they had in the fisheries and in the fur trade (Miller, 2017); however, they believed that "The Creator placed Aboriginal people upon this land first for a reason, and that, as the first ones on the land, they were placed in a special relationship to it" (Government of Manitoba, 2013). In keeping with this practice, Indigenous peoples framed their relationship to the land in broad conceptual terms, in that they marked time by seasons rather than by the clock, for example. Consequently, this approach supported their belief that they were inherently and innately bound to Creation. The land and their identity—as a people—were intricately woven as one, and since they had inhabited Turtle Island since time immemorial, most were adamant (and still are) that they possessed certain inalienable rights.

Reflecting various Indigenous intellectual traditions, Europeans were considered visitors and thus were expected to respect the obligations and responsibilities that came with this status. During the initial encounters between the Aboriginal peoples and the Europeans, however, evident cultural barriers led to misunderstandings, which would come to characterize the Indigenous–Canada relationship (Borrows, 2010). These misinterpretations most often centred on land concerns. In particular, Europeans had an established understanding of private ownership of land, based on capitalism, while Indigenous peoples believed that no one could own the land. Confusion was also linked to language differences. Although there were interpreters, the clash of differing intellectual traditions frequently meant that the two sides were not in sync with one another. Referring back to the land, Europeans sought to purchase Indigenous territories for settlement and to access territories for economic purposes. Indigenous peoples, on the other hand, did not believe they were entitled to sell land that was a gift from their Creator, which entrusted them to act as stewards (Roberts, 2006).

This did not stop the two sides from attempting to communicate and establish peaceful relationships. The Two Row Wampum (*gus wen ta*) was an early agreement forged between the Dutch and the Five Nations of the Iroquois in or around 1613. It was developed by Haudenosauee (Iroquois) leaders who sought peaceful coexistence with the European settlers, and in many ways reflects how Indigenous peoples in North America interacted with others: they sought to establish relationships that emphasized their political sovereignty and economic independence.

The **two-row wampum belt** has three parallel rows of white beads separated by two rows of purple beads (**wampum**). These symbolize two paths, or two vessels travelling together down the same river of life (the white beads reflecting Creation). On one side is a birchbark canoe representing Indigenous interests; on the other side is a ship carrying Europeans. The two vessels represent the **worldviews**—the laws, customs, and beliefs—of each group. In this scenario, it is possible for both peoples to travel side by side down the river, each in their own boat, without directly or negatively influencing the other. This suggests that working and living together in the same environment is possible, provided that neither side attempts to steer the other's vessel—to do so might result in a collision of cultures, thus causing each vessel to capsize, its occupants possibly left to perish. The two rows have been described as symbolizing the nation-to-nation relationship between Indigenous people and the Crown, autonomous parties that are at the same time linked by a common environment (the bed of white beads).

While many European nations agreed to the principles of the *gus wen tah*, most upon arriving on Turtle Island began to establish laws mentioning Indigenous peoples. Colonial officials created these laws to help secure their political authority that, in turn, subverted Indigenous governance structures, and social and political philosophies. Knowing about the *gus wen tah* helps us better understand why Aboriginal leaders continue to identify **treaties** as nation-to-nation agreements: they consider them to be formal agreements between independent parties. As negotiations continue between Aboriginal and non-Aboriginal parties, the two-row wampum belt symbolizes Aboriginal sovereignty and economic agency— one that includes the principles of sharing, mutual recognition, respect, and partnership (Borrows, 2002). This early treaty format eventually fell into disuse as small pox decimated Indigenous populations and they slowly

Two-row wampum belt: A treaty of respect for the dignity and integrity of the parties involved. In particular, it stresses the importance of Indigenous independence and mutual non-interference.

Wampum: White or purple shells that come from whelk (white ones) or quahog clams (purple ones). Wampum has a multitude of meanings and uses in Aboriginal cultures, including jewellery, healing, and decoration.

Worldview: The ways in which a group of people perceive and understand their place in the universe, often in relation to their interconnectedness with others.

Treaty: A formal agreement between two parties that has been negotiated, concluded, and ratified.

lost their economic and military dominance. As the ideology of the two-row wampum dissipated, three major treaty eras would emerge: pre-Confederation treaties, the Numbered Treaties, and land claims.

The Royal Proclamation of 1763 and the Pre-Confederation Treaties

The Peace and Friendship Treaties occurred between 1725 and 1779 when the British and French were vying for control of Indigenous land in Turtle Island. Each respective power formed alliances with various First Nations to help advance their own interests (Miller, 2017). One of the very first Peace and Friendship Treaties was reached in present-day Nova Scotia between the British and the Mi'kmaq Nation. In return for the Mi'kmaq's neutrality in any conflicts with the French, the British agreed to help with the advancement of trade in the area, while also promising to prevent any European interference with traditional Mi'kmaq hunting, trapping, or fishing practices (Roberts, 2006).

After the Seven Years War ended, Britain's King George III enacted the **Royal Proclamation of 1763**, a commanding piece of legislation that remains in effect and is legally binding to this day (it was formally integrated as Section 25 of the Constitution Act, 1982). One of the Royal Proclamation's main outcomes was that it established a formula enabling the British Crown to exclusively purchase Aboriginal land, after which it could be sold to settlers and newcomers. Notably, the Crown names Aboriginal peoples as nations holding title to their land, what we today describe as Aboriginal title. That is, Aboriginal title to the land existed prior to extended contact and will continue to exist until **ceded**, surrendered, or forfeited (Miller, 2017).

From these requirements emerged a treaty process binding the British government to specific protocols. Indigenous peoples prior to extended contact practiced treaties, which helps to explain why some of the earliest Indigenous–European interactions were treaties—it offered a common practice the two groups could understand. After the Royal Proclamation of 1763, treaty negotiations would assume a general pattern. It consisted of three parties—the Crown (or its representative), the First Nations party, and Creator as witness. Upon coming together, introductions were followed by gift giving and time spent renewing relationships by forgiving past transgressions through condolence ceremonies.

Royal Proclamation of 1763: One of the most important documents pertaining to Aboriginal land claims.

Cede: Surrender or forfeit possession of something, usually by treaty.

As a signal of agreement and to formalize the meeting, a pipe (calumet) would be smoked. Many First Nations' cultures considered this process as essential for building relationships with European traders, settlers, and treaty commissioners. Indeed, it was used to establish protocols between First Nations and non-First Nations parties (Miller, 2009).

Other treaties negotiated during this time included the Upper Canada Treaties (1764–1862) and the Vancouver Island Treaties (1850–1854). Under these treaties, which set the stage for the post-Confederation treaties, officials claim that First Nations peoples exchanged their interests in present-day Ontario and British Columbia in exchange for benefits that might have included land reserves, annual payments or other types of payment, and certain hunting and fishing rights (AANDC, 2010d). The British became enormously wealthy from this venture—in some of the earlier treaties in Ontario, for example, the Crown would purchase land for three pence (six cents) an acre from a First Nations band, and sell it for as much as 15 pence to a private investor (Roberts, Boyington, & Kazarian, 2008). Figure 7.1 presents a map representing some of the early treaties.

The Numbered Treaties

Since they were actually numbered 1 to 11, the treaties made between 1871 and 1921 are referred to as the Numbered Treaties. Covering Northern Ontario, Manitoba, Saskatchewan, Alberta, and parts of Yukon, Northwest Territories, and British Columbia, these treaties were signed between the British Crown, with Canadian representatives on hand, and the First Nations peoples (AANDC, 2010d).

Although there has been much written about the good or bad intentions of the British and Canadian governments leading to the victimization of the First Nations, the reality is that all parties had their own reasons for entering the treaties and saw them as necessary elements in achieving their unique goals (Applied History Research Group, 2001). Driven by the desire to build Canada as a nation, federal officials of the government sought to obtain land and resources to construct a railroad and create arable areas for settlers. The extension and expansion of Canada from 'sea-to-sea' was another goal driving Canada's desire to seek out and acquire natural resources. For members of the First Nations, and later and to a much lesser degree, for the Métis, treaties promised protected reserve lands, hunting and fishing rights, money, help with transitioning to new economic processes, annual payments, economic partnerships, and help with medical care

FIGURE 7.1

Map of Canada during Confederation

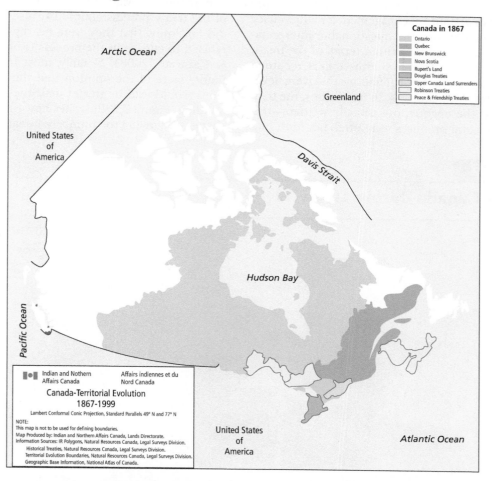

This is Canada at the time of Confederation in 1867. By the time Nova Scotia, New Brunswick, Ontario, and Quebec form the Dominion of Canada, the Robinson Treaties, the Upper Canada Land Surrenders, and the Peace and Friendship Treaties are already in place.

Source: Map of Canada during Confederation 1867, Aboriginal Affairs and Northern Development Canada, 2010. © Indigenous and Northern Affairs Canada. Used with permission.

and education. Others promised specific rights or privileges (Government of Manitoba, 2008). For example, among the exchange of land in the present-day prairies for cash, farm animals, and tools, Treaty 6 also guaranteed that a medicine chest would be kept in the home of the "Indian agent" (government official) for the use of the First Nations peoples (Roberts, 2006). This remains a contentious provision, as it stands as the basis for the provision of healthcare for all Aboriginal peoples in Treaty 6 territories, and serves to remind us that not all promises have been respected.

Since one of the main reasons for the treaty agreements was the building of the railroad, between 1881 and 1885, many Aboriginal people were uprooted from their traditional migratory lifestyle and moved onto settlements or reserves in exchange for land, farming supplies, and an annuity. With the railroad came immigrants, including additional Europeans, and a host of diseases that spread throughout the First Nations populations. Consigned to living in restricted reserves located frequently on infertile lands, many First Nations people found survival difficult in regions that were losing their natural wildlife to the "industrialization" of Canadian society.

The historic treaty processes were problematic for a number of reasons. Of particular note was the difference in communication and translation. It was not just a matter of mistranslations—Indigenous and European ways of political engagement and diplomacy differed. There was also the questionable interpretation of the oral versus the written terms of the treaty. Research shows that treaty commissioners recorded promises made during the negotiations that were never written into treaty documents. In most cases, the treaties arrived at the negotiations already written—this meant that the oral promises were often not built into the written texts. Furthermore, the treaties were most often signed by individuals who couldn't read and whose 'x' was appended to the document by the treaty commissioners. Many Indigenous nations were left out of the treaty process simply because the government did not know that they were nearby, or because they refused to engage in the process (Roberts, Boyington, & Kazarian, 2008). Notably, most Indigenous negotiators were of the opinion that these initial discussions would lead to greater dialogue and relationship building. Figure 7.2 illustrates how the map of Canada changes following the numbered treaties.

FIGURE 7.2

Map of Canada during Treaty 11

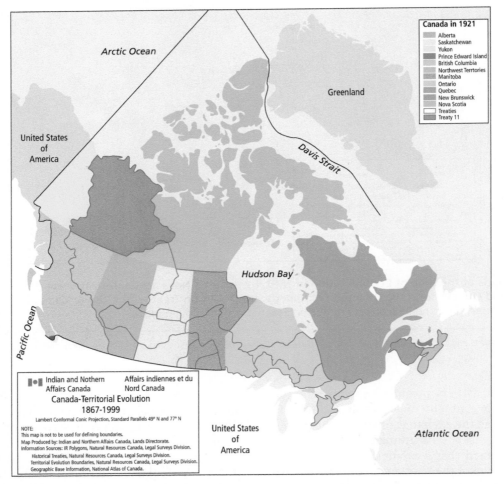

Treaty No. 11, the last Numbered Treaty, covers most of the Mackenzie District. The land in the area was deemed unsuitable for agriculture, so the federal government was reluctant to conclude treaties. Immediately following the discovery of oil at Fort Norman in 1920, however, the government moved to begin treaty negotiations.

Source: Map of Canada in 1921, Aboriginal Affairs and Northern Development Canada, 2010. © Indigenous and Northern Affairs Canada. Used with permission.

Reflecting on the complexity and history of treaties, the question remains: When the Aboriginal peoples bargained in what they thought was good faith, did they realize that they bargained away the rights to their own identities, along with the land?

LAND CLAIMS

The issues surrounding Aboriginal title, treaty rights, and land claims remain complex, and many claims are still being negotiated. Generally, there are two kinds of Aboriginal claims in Canada, commonly called land claims. Comprehensive claims are modern day treaties to resolve conflict over title to land in territories where Aboriginal land rights have not been dealt with by past treaties or through other legal means. They always involve matters related to the land, whereas specific claims are not necessarily land related (AANDC, 2010c). **Specific claims** are related to Canada's obligations under the historic treaties—a specific claim might address the failure to provide the agreed-upon amount of land in a historic treaty; or it might address the Crown's mishandling of First Nations money. Whereas treaty agreements (or other formal agreements) cannot be opened for renegotiation, fulfilling the terms surrounding the agreements enables negotiating a specific land claim. For example, a 2010 settlement involved the Bigstone Cree Nation. This First Nations band adhered to Treaty 8 in 1899, which entitled it to an area of land proportional to its population. However, the promised amount of land was not delivered at the time. In return, the band received $259.4 million in both cash and land as a settlement.

The time it takes to negotiate a specific claim varies, but it is much faster than pursuing a comprehensive claim. In fact, in 2007, new legislation was established seeking to resolve specific claims within three years. This was due to the backlog of claims. For example, from 1973 to 2016, 1713 specific land claims were filed: 133 were stuck in the assessment phase, 1222 had been concluded (412 were settled vis-à-vis negotiations; no legal obligation was found for 427, whereas 32 were resolved through administrative remedy, and for 351 the file was closed), and 138 were in active litigation or awaiting the Specific Claims Tribunal decision (AANDC, 2016a). The Canadian government, led by Prime Minister Stephen Harper's Conservatives, believed that a new and accelerated process was required to keep the process from coming to a standstill.

Comprehensive claims deal with the unfinished business of treaty-making in Canada, which is also framed as the unfinished business of Confederation.

These claims arise in areas of Canada where Aboriginal land rights have not been addressed by past treaties or through other legal means; they are used to define how traditional lands and resources can be used today. Unlike specific claims, these claims can take decades to negotiate, and on average, are settled within 15 years. Since 1973, 26 comprehensive land claims covering roughly 40 percent of Canada's land base have been completed. Of these, 18 included provisions related to self-government (AANDC, 2012b). Though the claims are lengthy in process, most are resolved without incident, although how long it will take to resolve the remaining 75 comprehensive land claims currently under negotiations is anyone's guess.

Many people mistake the emergence of Aboriginal self-government as part of the land claims process. And while claims and self-government are often negotiated simultaneously, the Canadian government acknowledges Aboriginal self-government in legislation as a separate process. Specifically, the Inherent Rights Policy of 1995 recognizes the inherent right to Aboriginal self-government. Canada's definition of self-government, however, is at odds with what Aboriginal people instead claim to be their inherent rights to self-determination. As the United Nations Declaration on the Rights of Indigenous Peoples' (UNDRIP, 2007) stated, and a conclusion that largely reflects Aboriginal leadership's vision, "Indigenous peoples have the right of self-determination [to] freely determine their political status and freely pursue their economic, social and cultural development." From the perspective of Canada's Aboriginal people then, the existing self-government process enables their communities to establish self-administration as opposed to true self-determination, which would be reflected by less federal and provincial intrusion into local decision making.

One of the problems arising when resolving land claims or negotiating Aboriginal self-government pertains to the number and kind of parties involved in the process. Often, both federal and provincial governments are involved, and, depending on the nature of the negotiations itself, other parties might include regional or municipal governments, territorial governments, private corporations, or even private individuals (Roberts, 2006).

> **Specific claim:** Process to deal with past grievances, mainly of First Nations, related to unfairly distributed treaty lands or mismanagement of First Nations funds by the Crown.
>
> **Comprehensive claim:** Modern-day treaties to resolve claims of Aboriginal land rights not yet dealt with by past treaties or through other legal means.

Sometimes, individuals change the course of a claim (and history) in ways they never thought possible. Take Dudley George, for example, who died in a confrontation with police after a protest erupted that was related to a land-claims dispute in his community that dated to the 1940s. During World War II, the government attempted to buy land from the Chippewas of Kettle and Stoney Point First Nation, with the intent to build a military camp. The land in question was a burial ground and not up for sale. In response, Canada implemented the **War Measures Act** to expropriate the required land, to which they compensated the people $15 an acre. They promised that the land (referred to as Camp Ipperwash) would be returned to Stoney Point people when it was deemed appropriate to decommission the base. At the end of the war, the Stoney Point people asked to enter into negotiations for the return of their land, but the Canadian Forces was still using it as a training camp. In 1986, the federal government agreed to pay $2.4 million to the Stoney Point band as compensation for the use of the land, and once again agreed to return it upon completion of an environmental assessment.

Cleaning the land after decades of military use proved to be a costly and time consuming venture for the Department of National Defence, and once again, the Stoney Point people were denied access to their land. It would operate as a training camp under tri-annual review and returned to the First Nation when the base was no longer needed. In 1992, the Standing Committee on Aboriginal People recommended to the federal government that the land be returned. Again, the recommendation was ignored. On September 4, 1995, unarmed Stoney Point protestors moved onto the property known as Camp Ipperwash. About 30 people planned a peaceful occupation of the land, but Ontario Premier Mike Harris insisted that they be removed from the park. Two days later, under the cover of darkness, the Ontario Provincial Police (OPP) attempted to forcefully remove the protestors during a nighttime raid during which George was shot—he would later die in hospital. (Salomons, 2009; Roberts, Boyington, & Kazarian, 2008).

A formal inquiry into the Ipperwash crisis didn't start until 2003, eight years after the confrontation, and it wrapped up in 2006. Key findings and evidence from the inquiry included surveillance tapes and audio evidence, which show certain members of the OPP and Ontario government to be racist.

War Measures Act: 1914 statute that gives emergency powers to the federal government, which allows it to govern by orders when there's a real or perceived threat of war or invasion. Changed to the Emergencies Act in 1988.

Former Attorney General Charles Harnick testified before the inquiry that Premier Harris had stated, "I want the f---ing Indians out of the park" only hours before George was shot (Harris denies making that statement). The Inquiry concluded that the OPP and the provincial and federal governments all bore some responsibility (Salomons, 2009). In June 2007, an Ontario Ministry of Aboriginal Affairs was established, and in 2010, the Ontario government took the final legislative step in relinquishing control of Ipperwash Provincial Park and putting it back in Aboriginal hands. In April 2014, the federal government and the Chippewas of Kettle and Stoney Point First Nation signed an agreement to return Camp Ipperwash and to provide $95 million "to invest in a brighter future" (AANDC, 2016b).

The convoluted nature of Canada's ongoing claims to historic and contemporary Indigenous lands means that rarely does clarity develop. For instance, Ipperwash was not the first time that violence erupted in relation to a territorial disagreement. In 1990, members of the Kanehsatà:ke, a Kanien'kéha:ka Mohawk settlement known to Canadian officials as the Kanehsatà:ke reserve, barricaded their community from construction workers seeking to expand a nine-hole golf course from Oka Township onto a burial ground surrounded by sacred pine trees. On July 11, 1990, Mayor Jean Ouellette asked the Quebec Provincial Police (Sûreté du Québec) to tear down the barricades—which they did, using tear gas and flash bang grenades. This confrontation led to the death of Corporal Marcel Lemay, which was followed by a 78-day armed standoff between the Mohawk community of Kanasetake, the Quebec police, and the Canadian army.

Since the events of 1990, negotiations to address the historical grievances of the Mohawks of Kanesatake have occurred in an effort to bring greater certainty to the Kanesatake/Oka area and the surrounding municipalities. These negotiations have led to fruition in many cases—for example, the construction of a Mohawk youth centre, elder home, and police station. Importantly, on April 14, 2008, Canada accepted the claim of the Mohawk Council of Kanesatake under the Specific Claims Policy and formally offered to negotiate the claim, restating its willingness to move forward toward resolving this historical grievance through negotiations (AANDC, 2010b; Miller, 1991; Canada History, 2012). This has been a tough struggle, however. Whereas the golf course expansion was cancelled, in 2010, tensions once again temporarily flared over a developer's plan to build three homes on a site across the street from the pine forest where the standoff began. And even though the Canadian

A Canadian soldier and Mohawk warrior confront each other during the 1990 Oka Crisis. Is this confrontation about more than a piece of land? What's really at stake here?

Shaney Komulainen/The Canadian Press

government purchased the contested lands, they have yet to be transferred to the community.

LEGISLATION

Aboriginal Identity

Up until about 1850, virtually all of the dealings between Aboriginal and non-Aboriginal peoples were connected to land issues. Colonial and, later on, Canadian officials had another covert and ever-present objective on their minds: assimilating Indigenous peoples. In an attempt to foster land acquisitions and encourage 'Indian' **assimilation**, colonial officials put into place two important acts to help protect Indigenous lands. In doing so, they had to define who an Indian was to ensure officials knew who was fully entitled to protection, and who was not. In 1850, two important acts were put in place that would attempt to determine who was of Aboriginal descent and what constituted Indian lands. Both of these acts—"An Act for the Better Protection of the Lands and Property of the Indians in Lower Canada" and "An Act for the Protection of the Indians in Upper Canada from Imposition and the Property Occupied or Enjoyed by Them from Trespass or Injury"—assigned virtually all responsibility for leasing Aboriginal land and collecting rents to the Commission of Indian Land. This also represented the first attempts at formally defining who was "Indian" and which rights and responsibilities

went with that status (Belanger, 2018). The first legislative definition of Aboriginal peoples included the following stipulations:

- all persons of "Indian" blood who were known to belong to a specific band, living on specific land, with their descendants
- all persons intermarried with any such "Indians" (and their descendants) who resided among them
- all children of mixed marriages residing among such "Indians"
- all persons adopted in infancy by such "Indians" (Frideres & Gadacz, 2008)

These stipulations set the stage for an assimilation process that continues today in the form of the Indian Act of 1876 (discussed below). This agenda was formally articulated in the 1857 "Act to Encourage the Gradual Civilization of Indian Tribes in this Province, and to Amend the Laws Relating to Indians." With this act, through the voluntary process of enfranchisement, "Indian" men of good moral character, who were literate, educated, and debt-free, could relinquish their reserve rights and the right to live with their

> **Assimilation:** A long-standing government policy, Indigenous assimilation was based on the premise that they would give up their own culture, languages, and beliefs, and live and act just like the British settlers/Canadian citizens.

families in exchange for **enfranchisement**—the right to vote and all other rights afforded other British subjects. Ironically, most British subjects would not meet these criteria and would have been denied citizenship.

Only one man applied for enfranchisement, however; and in 1869, the Civilization Act of 1857 was altered and named "An Act for the Gradual Enfranchisement of Indians, the Better Management of Indian Affairs, and To Extend the Provisions of the Act 31st Victoria." This new act developed the first system of Aboriginal self-government in the form of elective band councils, which remains in effect today. It also granted the Superintendent General of Indian Affairs virtually total control over the status of 'Indians.' For example, it forbade the sale of alcohol to Aboriginal people (for their own good) and stripped Aboriginal women (and subsequently their children) of their cultural identity and legal status if they married non-Aboriginal men (or Aboriginal men who had lost their status)—a blatantly discriminatory process, based on gender (Makarenko, 2008). These two acts formed the foundation for the Indian Act of 1876.

The Indian Act

In the Constitution Act of 1867 (originally named the British North America Act), which created Confederation, the federal government accepted responsibility for Aboriginal peoples and, as seen in the Enfranchisement Acts, lobbied heavily for the integration and civilization of all Aboriginal people into mainstream society (Roberts, 2006). In 1876, with the aim of merging all previous acts into one piece of legislation, the Government of Canada (now an independent nation) created the Indian Act (Makarenko, 2008). Essentially clumping all Aboriginal peoples into one group of Indians for ease of management and bureaucratic containment, not only did this act serve to disregard any distinct cultural differences between bands of Aboriginal people, but it also established them as wards of the government—effectively forming a **fiduciary** (or parent-like) relationship between them and the government. The problems with the Indian Act (besides the obvious moral and ethical ones) were logistical. As early as 1850, the colonial government began to keep records that identified individual Aboriginal people and the bands to which they belonged. These records formed the **Indian Register**, which helped agents of the Crown keep track of who was eligible for treaty and interest benefits and the treaties they were involved in. Working from these records and definitions set out in the previous acts, the Indian Act defined First Nations people according to an assigned status:

- Status Indians: Those individuals who are registered under the Indian Act on the Indian Register. They are entitled to certain rights and benefits under the law.
- Non-Status Indians: Those individuals who consider themselves to be Aboriginal or First Nation, but are not recognized by the Government of Canada as "Indians" under the Indian Act—either because they can't prove their status or because they've lost it. Non-Status Indians don't share the same rights and privileges as Status Indians.
- Treaty Indian: An Indian who belongs to a First Nation band that signed one of the 11 Numbered Treaties with the Crown (AANDC, 2012a).

Assigning First Nations people into categories was discriminatory and exclusionary. In fact, the Indian Act did not recognize Aboriginal peoples as individuals at all: until 1951, the term "person," as defined by the Indian Act, meant "an individual other than an Indian, unless the context clearly requires another construction" (AANDC, 2010d). The Indian Act did not (and still does not) recognize Inuit or Métis groups, resulting in the further marginalization of these peoples. The act itself and several amendments over the years only served to further oppress and discriminate against all Aboriginal peoples. Some of the discriminatory inclusions/amendments are as follows:

- Until 1960, Indians were not allowed to vote in federal Canadian elections, unless they gave up their status and the rights and privileges that went with it.
- In 1884, First Nations people were prohibited from buying arms and alcohol.
- In 1884, First Nations religious ceremonies were made illegal. The banning of **potlatches** and the

Enfranchisement: The process whereby an individual gets the right to vote or become a citizen.

Fiduciary: A person who holds power or property in trust for another person—usually for the benefit of the other person.

Indian Register: Series of documents established in the 1850s that kept track of all of the existing records of people recognized by the federal government as members of an "Indian" band.

Potlatch: Organized meeting for special ceremonies, such as name-giving, birth, rites of passage, treaties, and weddings; practised mainly by First Nations of the west coast.

sundances clearly interfered with Aboriginal traditions.

- From 1914–1951, Aboriginal people were required to get official permission before appearing in public in traditional clothing.
- In 1920, attendance at Residential Schools for First Nations children was made mandatory.
- In 1927, band members were forbidden from pursuing land claim violations using band funds, effectively making legal changes to treaty violations impossible.
- In 1930, First Nations people were banned from pool halls.
- From 1951–1985, status was stripped from Aboriginal women who married non-status men (Aboriginal or non-Aboriginal) (Kunin, 2011). Of note, a non-Aboriginal woman could acquire status by marrying a man who had status.

Although many of the practices originally written into the Indian Act have been repealed or amended, one cannot go back and rewrite the effect that this legislation has had on the peoples' lives. Overtly racist and unapologetic, the Indian Act stands as proof of the government's aim to strip all Aboriginal people of what was rightfully theirs—the land, their identity, and any desire to maintain a formal but independent relationship with government agents.

Residential Schools

In its ongoing quest to assimilate Aboriginal peoples into mainstream society, the federal government decided that the best way to "civilize" Aboriginal peoples was to get them while they were impressionable—that is, while they were children—and remove them from their family and communities. Youngsters, unlike their parents, were more susceptible to learning the values, language, and identity of a new culture.

To this end, the government institutionalized an education system that ripped young Aboriginal children away from their families and cultures to attend newly created residential schools that were located far from their homes. Beginning in 1883–1884, schools were built in every province except New Brunswick, Prince Edward Island, and Newfoundland (Belanger, 2018) that were operated by the Anglican, Presbyterian, United, and Roman Catholic churches. As wards of the state, the children and their parents were virtually powerless to stop the process of building a "civilized nation." Under the Indian Act, in 1920, it became mandatory for Aboriginal children to attend a residential school and illegal for them to attend any other educational institution. They had to attend until they were 18, but they could not obtain an education further than Grade 8, which was quite manageable, since they barely received any academic instruction at all. Religious studies took most of the morning, and chores took much of the afternoon. The goal was to assimilate them into mainstream society. The few children who did show academic promise were kept in residential schools to learn the skills that were "appropriate" for them. Consequently, the boys were taught farm work, shoemaking, or other manual-skilled trades, and girls were taught sewing, bread-baking, and household tasks (Long & Dickason, 2000).

Financed by the government, but managed by the various churches, all the schools adhered to similar common guidelines:

- Children were forbidden to speak in their native language and were punished if caught doing so.
- All references to and aspects of their customary ways of life were eliminated from school curricula.
- Boys and girls were segregated—brothers and sisters especially so, in an effort to weaken family ties.
- Children were required to cut their hair, eat European food, and wear school uniforms.
- Christian holidays were celebrated and children learned European sports, like soccer and cricket.
- School days were divided between religious classes and training for manual labour. The children were taught practical skills rather than academic skills like reading or writing. (Roberts, 2006; Hanson, 2009b)

The devastating individual and social consequences of the residential schools are immeasurable, including the loss of language and the erosion of family values, traditions, and parenting skills. The most negative effects, however, resulted from the various abuses suffered by the children (Long & Dickason, 2000). The physical and sexual abuse— the beatings and whippings—meted out for punishment or pleasure, contributed to the real or perceived "dysfunction" of Aboriginal families and their communities today. By undermining any semblance of a family structure, the children grew up lacking proper socialization and did not develop proper parenting skills. Absent any hopes and dreams to make something of

> **Sundance:** Known by the Niitsitapi people as the Okan, it is a communal year-renewal event held annually in mid-summer to reinforce intra-community bonds and interpersonal relationships.

their lives, most children confronted internal feelings of worthlessness. As George Manuel, a powerful Aboriginal political leader from the 1960s and 1970s wrote, "[T]he residential school system (not just the one that I went to—they were the common form of Indian education all across Canada) was the perfect system for instilling a strong sense of inferiority" (Manuel & Posluns, 1974, p. 67).

The 60s Scoop

The last residential school closed in 1996, but the government started phasing them out in the late 1950s and early 1960s, which continued into the 1980s. The government belief during this period suggested that the children would be better off in the Child Welfare System and integrated into the public school system where they would receive a better education (First Nations Study Program, 2009). To this end, in what has become known as the "60s Scoop," was an extension of the residential school ethos. By the mid-1960s, Canadian authorities sought to adopt reserve children into non-Aboriginal families considered superior caregivers. Thousands of children were literally scooped from their homes, often without the knowledge or consent of their families and bands, and put up for adoption. Statistics from the then-named Department of Indian Affairs and Northern Development (DIAND) reveal a total of 11 132 Indian children adopted between the years of 1960 and 1990, but it is believed that the actual number is much higher—closer to 20 000. Seventy percent of these children went to non-Aboriginal homes, and today many of those children are looking for their birth parents (Sinclair, 2011). In 2017, an Ontario Superior Court judge, responding to class-action suit seeking $1.3 billion in damages for roughly 16 000 Ontario '60s scoop' survivors, determined that the Canadian government failed to protect the identity of reserve children who were forcibly taken from their homes. Responding to the eight-year court battle, the federal government agreed that it would not appeal the ruling and would begin negotiations to resolve the grievance.

In 2008, Prime Minister Stephen Harper issued a formal apology to the former students of residential schools, admitting that the assimilation policy at the time was a "sad chapter" in Canadian history: "The government now recognizes that the consequences of the Indian Residential Schools policy were profoundly negative and that this policy has had a lasting and damaging impact on Aboriginal culture, heritage and language" (AANDC, 2008). Denouncing the residential schools program as racist, in an emotional speech that day, NDP Leader Jack Layton said, "It is the moment where we as a Parliament and as a country assume the responsibility for one of the most shameful eras of our history ... it is the moment to finally say we are sorry and it is the moment where we start to begin a shared future on equal footing through mutual respect and truth" (Canadian Press, 2008). Did two of Canada's most powerful leaders say the same thing that day? Did their words matter? Does an apology for the residential schools exonerate all that has been done to the Aboriginal peoples as a whole?

Aboriginal comedian Ryan MacMahon remembers the time he spent with his grandmother, when she told him of the abuses she suffered in the residential school system. She told him, too, that there were good things—like learning to bead, and three square meals a day. She died before she heard Stephen Harper's apology, and five years later, MacMahon thinks the apology has done little to change the reality confronting Indigenous peoples in Canada. Reconciliation is not a reality and he wonders why all the talk about Aboriginal matters centres around the notion of "moving forward": "As a father ... it is my job to break that cycle and free [my children] of the burden of the past and to teach them," he said. "On Remembrance Day [November 11] ... we say, 'lest we forget.' But in Canada [when discussing residential schools], we are always saying, can't we just all move on? Can't we just forget it already, can't you let it go? But, if we are never supposed to forget those other traumas, why should we forget these ones. ... It is our responsibility to remind people that it is an ugly past and we have to be willing to put it on the line because of that past" (Barrera, 2013).

In an effort to ensure that the travesty of the residential schools is not forgotten, so that all Canadians can learn from the injustices done, the Canadian government formed the Truth and Reconciliation Commission (TRC) in 2008. This was part of the court-approved Residential Schools Settlement Agreement that was negotiated between legal counsel for former students and legal counsel for the churches, the Canadian government, the Assembly of First Nations, and other Aboriginal organizations (CBC News, 2010). With a suggested timeline of five years and a budget of $60 million, the TRC's final report, released in 2015, assessed the residential school system (1880s–1996). The final report—the completion of which was frustrated by Ottawa's frequent refusal to provide the needed historical documents—identified more than 3000 recorded student deaths, various systemic abuses, the use of a homemade electric chair at one school to punish children, and government-sanctioned health experiments in five different institutions involving more than 1300 students. It also proposed 94 'calls to action' urging all levels of government to work together to change policies and programs to improve the lives of Aboriginal peoples nationally by dealing with the residential school legacy and contemporary issues.

The TRC report is an action-oriented response to the horrors of the residential schools that led Supreme Court of Canada Chief Justice Beverley McLachlin to describe the schooling experiment as an act of **cultural genocide** (a description that 70 percent of Canadians agree with). Will it make the whole issue disappear? No. It will, however, give Aboriginal peoples their own stake in providing their own solutions in the healing processes—the opportunity to ensure that these times are not forgotten and swept under the rug, but rather used as learning tools for the future, and healing times for those in need.

The White Paper

In 1969, Prime Minster Pierre Trudeau and then Minister of Indian Affairs Jean Chrétien set about to right the wrongs of the Indian Act, by way of dismantling it and doing away with the notion of status as it pertained to Aboriginal peoples. Essentially, their belief was that the legal status of Aboriginal people kept them from fully participating in Canadian society. With this new proposal, which has been accurately described as a policy of termination, Aboriginal peoples would no longer have special status as defined by the Indian Act, treaties would be taken at face value, and the federal responsibility for Aboriginal peoples would be transferred to the provinces (Belanger, 2018).

Many Aboriginal people were incensed by the idea of scrapping much of what had been accomplished to date, but still making them assimilate into the status quo. Others agreed, and also saw it as passing the responsibilities of the federal government on to the provinces. Dismantling the Indian Act would, in essence, rescind any legal protections helping to define who Aboriginal people were in law and policy. Aboriginal peoples wanted to be recognized as First Peoples—equal to every other Canadian, but with historic Aboriginal and treaty rights (what came to be known as

> **Cultural genocide:** The deliberate destruction of the cultural heritage and traditions of a group of people or a nation.

IN THEIR SHOES

If a picture can say a thousand words, imagine the stories your shoes could tell! Try this student story on for size – have you walked in this student's shoes?

A CONVERSATION WITH KOKUM

Two weeks ago in my Social Welfare of Canada class, we discussed issues around Aboriginals in residential schools. This wasn't a new topic for myself to discuss, I had heard of residential schools before. I didn't, however, know the horrible things that occurred in residential schools. When I was in junior high school, I remember my grandmother (or as I call her, Kokum) talking about how some of the students who attended the residential school with my grandmother were going to receive money for what had happened to them, and for sharing their stories. I remember asking, "Why wouldn't you do the same?" She said, "It wasn't that bad." Just like that—a simple answer and yet so complicated and confusing. I wondered … if it wasn't so bad, why was my family so traumatized, hurt, holding on to so many secrets?

I knew not to ask any more questions. Even though she said it wasn't that bad, I had a feeling it was. She made me know that this wasn't a topic I should bring up again. For 26 years of my life, I chose

not to talk about residential schools again within my family. That all changed a couple weeks ago. There was an assignment due about intergenerational trauma. Who better to write about this topic than myself? I knew this topic so well, and yet not at all. I decided to ask my mother if she could tell me how my grandmother raised her. She ended the conversation before I had even asked one question. There was too much pain and hurt for my mother to talk about what her childhood was like with my grandmother. I decided to call my cousin who had already graduated from the Social Service Worker Program, in Alberta. She told me she tried to ask our grandmother about the school she attended, and got very basic information. She was third generation to attend a residential school system, and she attended from the age of six for a total of eleven years. This left me with more questions than answers.

I decided one last time to try and ask my Kokum. I started my conversation off by telling her the horrible things I had seen on the movies we had to

(Continued)

CHAPTER 7 Indigenous Peoples

watch in class. I knew when I would ask her a question, I would have to manipulate an answer from her. So I did. I would start off every question by telling her something horrible I had seen. I think this method of asking helped make her realize I wasn't too young to know. I already had heard the worst of the worst. She was able to relax a little at these thoughts, and open up a little more than she would have.

Me: How did you feel going to the school at such a young age?

Kokum: It wasn't a secret. Everyone around me went. We all knew from a young age this is what happens.

Me: What was it like for you in your first year of being there?

Kokum: For the first couple of months, I cried myself to sleep. One of my cousins would sneak over and hold me and try to get me to be quiet because she knew if we got caught, or I was heard crying, we would be in trouble.

Me: Did you get to see your siblings or cousins during the day?

Kokum: I would see them, but we were trained not to talk to each other. Only when we were alone could we.

Me: Were you allowed to speak your language?

Kokum: No. But when the kids were outside in the fields, they would do it anyways. They all knew not to get caught though.

Me: Did you get to see your parents during the year while at school?

Kokum: Yes. But it hurt. I missed them before they even left. The time never felt long enough.

Me: Did you ever see anyone get abused, or were you abused?

Kokum: Of course. (*Her voice said don't ask any more about that. So I didn't*).

Kokum: After high school, I had a choice to go back to my reserve or go to Edmonton. I decided to go work in Edmonton. I never returned to live on the reserve again.

Me: What did it feel like to go back to your reserve? Did you feel like you abandoned your family? Did you feel like you would be welcomed if you wanted to go back?

Kokum: Once I knew what it was like to live in the city, I didn't ever want to go back. If I had wanted to, I could have though. My family wasn't too happy but they got over it.

Me: What was it like for your family when you married a white man?

Kokum: They were mad. My mother especially. We didn't talk for years until she found out I was pregnant. (*Once my kokum had her children, this mended her relationship with her mother a little*).

Me: How did your children feel when they went to the reserve to visit?

Kokum: They were bullied and called names for being whiter than their cousins. My children always felt like they didn't fit in and were not accepted anywhere they went. In the cities, they were different colour than the whites, and on reserve, they were too white.

Me: How do you think you were as a parent?

Kokum: I used to drink a lot of alcohol when my children were young. Once I became a grandmother, I stopped.

Me: What were your parents like as parents? Did you notice how the residential schools had affected them?

Kokum: Yes, in some ways. My father was silent and cold at times. I never got a reaction from him. My mother was loving and caring. She always hugged me and my siblings as children.

Me: Did your parents tell you they loved you?

Kokum: No, but they showed it at times, so we knew they did.

Me: Did you tell your children you loved them?

Kokum: Not at first. It took a lot of years before I could get the words out.

Me: Were the residential schools the reason for the depression over the reserve you were from? Were they to blame for the cycle of abuse, neglect, and addictions our family is still facing today?

Kokum: I don't doubt about that. Yes, they are to blame.

No one in Kokum's family was ever ready to share their stories. Everyone kept that piece of their life to themselves. No way of healing. I often wonder what would happen to my family if these horrible secrets were let out? Would it create more hurt and pain, or would it help everyone heal? I'm not sure I'll ever really know. I may not have been able to get my grandmother to get too detailed with myself, but I was able to get her to open up to myself in a way she had never done with another family member. My grandmother is 77 years old now. She can still recall the feelings and emotions she had when she was six years old. Although she wasn't perfect as a parent and her children still hold a lot of pain and resentment toward her, she was the most amazing grandmother. I wouldn't be who I am today without her. She took care of me for most of my life. I've had the honour of not only knowing she loves me, but also hearing the words.

Courtesy of Mia Bakker

Citizen's Plus). Though Trudeau believed the proposed changes would culminate in what he described as 'a just society,' the prevailing belief among Aboriginal leaders at the time reflected the opposite. Harold Cardinal's biting and satirical response to the White Paper indicated that he equated it to cultural genocide, and what he believed Indian Affairs Minister Jean Chrétien's belief was at the time: "The only good Indian is a non-Indian" (Cardinal, 1969, p. 1). The White Paper provoked wide and organized public outcry from various Aboriginal groups and was withdrawn in 1971.

Bill C-31

Also known as "A Bill to Amend the Indian Act," Bill C-31 was passed into law in 1985 largely in order to address the gender discrimination that resulted from the Indian Act and to better align the Indian Act with the Canadian Charter of Rights and Freedoms. The complex issues of Aboriginal identity were magnified in the case of Aboriginal women, who lost their Indian status and thus community connections, and if they married outside their band. Bill C-31 ended this discrimination by reinstating status to more than 127 000 individuals, and by allowing them to retain their status no matter whom they married. In addition, as a move toward self-government, Bill C-31 made changes in order to allow bands greater control over their membership (First Nations Study Program, 2009).

The Royal Commission on Aboriginal Peoples (RCAP)

The issue of self-government has always been controversial, but for many Aboriginal people, the right to govern themselves equals the right to self-determination (Roberts, 2006). With the ability to manage their own affairs, Aboriginal people would have greater autonomy and decrease their dependence on government supports, with the goals of self-sufficiency and improved living conditions. To this end, in 1991, after the events at Oka demonstrated the resolve of Aboriginal peoples fighting for their rights, the Canadian government led by Brian Mulroney's Progressive Conservatives established the Royal Commission on Aboriginal Peoples to examine the relationships between Aboriginal peoples, the government, and the larger society as a whole. In 1996, the five-volume, 4000-page report was published, with some 440 recommendations, which included the following:

- restructuring the Indian Act
- self-determination through self-government
- the creation of an Aboriginal parliament
- dual citizenship as Aboriginal nationals and Canadians
- Aboriginal economic initiatives through the provision of more land
- the establishment of an Aboriginal national bank (Roberts, 2006)

Although negotiations on many of the recommendations are ongoing, over ten years after the report, how much progress has been made? According to Métis writer Chelsea Vowel, " ... since the RCAP was released, almost nothing has been accomplished" (Vowel, 2012). The Assembly of First Nations released a Report Card 10 years after the RCAP, pointing out the lack of progress in these areas:

- The Department of Indian Affairs (now AANDC) had yet to be abolished.
- There has been no commitment to train 10 000 Aboriginal professionals in health and social services over 10 years.
- There is no First Nations jurisdiction over housing, no independent administrative tribunal for lands and treaties, and no sustained investment in meeting basic needs in First Nation communities. (Auditor General of Canada, 2006; Vowel, 2012).

Murdered and Missing Indigenous Women (MMIW)

As the above-mentioned sections show, legislative changes and the ongoing study of the Indigenous–Canada relationship are common. Yet Aboriginal living conditions only slowly improve. In 2016, yet another inquiry was announced, this time to draw the public's attention to what has been termed murdered and missing Indigenous women. Between 1980 and 2014, an estimated 1181 Aboriginal women in Canada were murdered or went missing, a low estimate according to Indigenous and Northern Affairs Canada (INAC) minister Carolyn Bennett. Ongoing calls for a national inquiry supported by the United Nations that were previously ignored by Prime Minister Harper (2006–2015) were acknowledged by Prime Minister Justin Trudeau in 2015 with the support of the provincial premiers. On September 1, 2016 the two-year, $53.86 million MMIW inquiry was officially launched with the intention of releasing its report with recommendations by the end of 2018.

This is, however, but one of several issues that demonstrate Canada's reluctance to instituting the types of changes identified in the RCAP's final report. And changes are needed. Take, for example, the

AGENT OF CHANGE

Cindy Blackstock

Cindy Blackstock, Executive Director, First Nations Child and Family Caring Society of Canada, and Professor, School of Social Work, McGill University

As a newly minted child protection worker working for the BC government in the 1980s, Dr. Cindy Blackstock (Gitksan First Nation) worked closely with Aboriginal families. In doing so, she concluded that Ottawa spent poorly on Aboriginal child welfare services. Specific details were lacking, but Blackstock also knew that more Aboriginal children were being sent to foster homes, often with non-Aboriginal families. After she formed the non-profit advocacy group First Nations Child and Family Caring Society in 1998, she began to collect data that would prove 20–30 percent less was being spent on Aboriginal children.

In 2007, after 20 years of advocacy work, Blackstock forced the Canadian government to account for its pitiable treatment of Aboriginal children. As the executive director of the First Nations Child and Family Caring Society, Blackstock, along with her colleagues at the First Nations Child and Family Caring Society, and the Assembly of First Nations, filed a human rights complaint against the Attorney General of Canada, representing the department today named Indigenous and Northern Affairs.

It took almost one full decade, but the Canadian Human Rights Tribunal in 2016 ruled that reserve children were adversely impacted by the services provided by the Canadian government. The message was simple: Indigenous and Northern Affairs had discriminated against these children. The long delays in proceedings were attributable to Canada's attempts to have the case thrown out by the Federal Court and the Federal Court of Appeal, and its refusal to disclose 100 000 documents related to the case—documents that appeared only after Blackstock filed an access to information request.

The federal files tell a disturbing story, in particular that the 163 000 children in foster care spent 66 million nights away from their families. Indigenous children are also three times more likely to be in foster care than they were during the residential school period. Yet, as of this writing, the federal government had yet to respond to the Tribunal's initial compliance order demanding the end to discriminatory practices. Dissatisfied with the Canadian government's failure to respond, the Tribunal in September 2016 again delivered an order demanding Indigenous Affairs take immediate action and provide clear information about how it is implementing its original order.

It was in these files that Blackstock learned that the government spied on her illegally—there was a 2500-page file cataloguing her actions. In 2013, the Federal Privacy Commissioner Jennifer Stoddart ruled that having 189 federal government officials from the Justice Department and Aboriginal Affairs collecting data on and monitoring Blackstock violated the spirit of the Privacy Act.

Blackstock continues her fight against government discrimination against Aboriginal children, efforts that have been recognized by among others the Nobel Women's Initiative, the Aboriginal Achievement Foundation, and Frontline Defenders. She publishes her work to assist academics and field practitioners, and she is widely sought after as a public speaker. Her collaborations have come to include working with the United Nations Committee on the Rights of the Child, UNICEF, and the United Nations Permanent Forum on Indigenous Issues to produce a youth-friendly version of the United Nations Declaration on the Rights of the Child.

Millennium Scoop. Similar to the '60s Scoop, currently more Aboriginal children are placed in childcare in Canada than attended residential schools in any one year (roughly one in every three Aboriginal children). As an example, at the peak of residential school operations in 1953, there were 11 000 students being housed. Today, there are roughly 30 000 Aboriginal children in care in Canada, which amounts to 15 percent of children in care nationally—despite making up 3 percent of the overall population.

As has been discussed by others, "the *Indian Act* and, just as importantly, the attitudes that informed its development, continued to affect how we in Canada formulate [Aboriginal] policies and the related interventions" (Belanger, 2018, p. 117). For many, the White Paper philosophy advocating the legal termination of Indians, reserves, and treaties, remains feasible.

Lessons

How do Aboriginal communities fare in Canada today? In an effort to debunk the myths surrounding Aboriginal peoples, a 2011 study from TD Economics reported that the personal income of Aboriginal people has grown 7.5 percent a year for the preceding decade, and has increased from $6.9 billion in 2001 to $14.2 billion in 2011 (TD Economics, 2011). Using labour market data from Statistics Canada, the report's authors suggest that there has been a shift toward occupations based in the natural resource sector (oil and gas mining), along with construction and development that accounts for the higher wages, often found in these areas. The report predicted that by 2016, the combined total income of Aboriginal households, business, and government sectors could reach $32 billion. Though the shift toward gas and oil mining foretells income increases, at what cost will this economic uplift come?

The costs to the land will be detrimental and irreversible, according to the manifesto of the grassroots movement Idle No More: "The taking of resources has left many lands and waters poisoned—the animals and plants are dying in many areas in Canada" (Idle No More, 2012). Idle No More started in December 2012 in response to Prime Minister Harper's tabling Omnibus Budget Bill C-45, a 450-page document that buried provisions that would lead to troubling legislative alterations, including proposed changes to the Indian Act, the Navigation Protection Act (the former Navigable Waters Protection Act), and the Environmental Assessment Act.

The hashtag #idlenomore was create to help publicize the issue. The original resistance to and dialogue exploring the **omnibus bill's** meaning that began in Saskatchewan soon evolved into the mobilization of Aboriginal peoples nationally. By early December, Idle No More leaders requested and then witnessed Assembly of First Nations (AFN) leaders demand Canada remove the proposed Bill. Within two weeks, a national day of action had been planned and occurred on December 10 as flash mobs and round dances came to dominate the national news until April 2013.

Idle No More spoke in universal terms to Aboriginal concerns about federal Indian policies that helped to spark regional activists to heightened levels of mobilization. Social media anchored by Twitter, Facebook, YouTube, and other platforms helped drive mobilization that led to increased interface. This, in turn, served to motivate "social forces, values, community passions, and historical grievances" to "ignite and deliver exponential impact." As Ken Coates (2015) has written, the level of social media engagement was incredible considering that Idle No More was "a largely uncontrolled social movement, with a minimal budget, no paid staff, and little concerted effort to deliberately build audience size." In terms of overall social media activity, Idle No More garnered the following response:

- Overall Mentions: 1 366 156
- Blog Mentions: 11 296
- Facebook Mentions: 100 011
- Forum Mentions: 13 669
- New Mentions: 19 189
- Twitter Mentions: 1 215 569
- Twitter Participants: 143 173
- YouTube Mentions: 6422

Do the Idle No More protestors make sound arguments? Especially when we factor in that not all Aboriginal peoples in Canada have the same social, political, and economic beliefs? Take, for example, recent events as they relate to oil and gas development. Though the media would have you believe that all Aboriginal people oppose such ventures, this is not the case. Aboriginal development is much more dynamic: roughly 8.3 percent of all Native employment is directly related to natural resource projects, and 19 of 634 Fist Nations in Canada are oil-producing communities (with more seeking industry entry). Yet, in certain cases, Aboriginal peoples support natural resource development, and for various reasons. In British Columbia, 300 benefit- and revenue-sharing agreements worth $6 billion have been negotiated with First Nations by the provincial government and private sector firms. And a corridor coalition of 28 First Nations and Métis communities in British Columbia negotiated equity ownership in the now-defunct expansion of the Northern Gateway Pipeline project, which could have benefited those communities upward of an $800 million stake (10 percent). Interestingly, Aboriginal leaders seek to establish their own projects rather than awaiting corporate outreach. The

Omnibus bills: Proposed laws or legislature that can cover a number of different subjects, but are packaged together in one bill.

© Tammy Luciow

Developing gas and oil resources on Aboriginal lands can lead to environmental damage, poverty, and the loss of traditional lifestyles. However, what are the potential benefits (both Aboriginal and non-Aboriginal) from development?

Eagle Spirit pipeline plan, for example, is a $14 billion Aboriginal-led project between Fort McMurray, Alberta, and Prince Rupert, B.C., that took three years to finalize due to the need to secure buy-in of the Aboriginal communities located along the proposed transportation corridor. The second pipeline plan is an Assembly of First Nations–Alaska Native oil-by-rail accord to transport 1.5 million barrels daily along a $10.4 billion oil railway from Fort McMurray, Alberta, to Valdez, Alaska. To reiterate, Aboriginal development (or the resistance to it) is not, nor should it be, portrayed as uniform: not all Aboriginal communities refuse to pursue oil and gas and other extraction industries. There are several northern Alberta First Nations, such as Fort McKay, that encourage oil and gas development to aid with local development initiatives. This has pitted them against the Beaver Lake Cree,

National Day of Action: Day devoted to raising the awareness of serious issues facing Aboriginal people in Canada; first organized on June 29, 2007.

who have sued the federal government to slow development.

If the Idle No More campaign and other such protests brought media attention to the potential harms of development projects, this media attention was insignificant in comparison to that of Chief Theresa Spence, whose simultaneous hunger strike brought national attention to Canada's Aboriginal peoples—whether they wanted it or not. The Idle No More campaign and Chief Spence's intent to initiate a hunger strike (with the mission to meet with Prime Minister Harper in order to discuss the dismal living conditions of Canada's Aboriginal people) both occurred on the **National Day of Action**. The two events were not part of a coordinated plan, but Spence's 43-day hunger strike raised media attention for both the substandard living conditions in the Attawapiskat First Nation community and for the Idle No More campaign (CBC News, 2013). Although Chief Spence's actions brought media attention once again to the crisis in Attawapiskat (it had been declared in a state of emergency in 2011), they brought forward as well the state of Aboriginal affairs in general.

According to a new study from the Canadian Centre for Policy Alternatives, Canada's Indigenous children fall behind other Canadian children in virtually every measure of well-being: family income, educational attainment, water quality, infant mortality, health, suicide, crowding, and homelessness. The authors of this study found that in the first tier, children who experience the lowest rate of 12 percent are *not* Indigenous, racialized, or immigrant. The second tier includes racialized children, with a poverty rate of 22 percent; first-generation immigrant children, whose poverty rate is 33 percent; and Métis, Inuit, and non-Status First Nations children, who suffer with a poverty rate of 27 percent. Most disturbing is the third tier, where fully half—50 percent—of Canada's Status First Nations children live below the poverty line—a number that grows to 62 percent in Manitoba and 64 percent in Saskatchewan (see Figure 7.3; Macdonald & Wilson, 2013).

Adult Aboriginal people—regardless of their band membership—do not fare much better than the children. Consider the following (see Belanger 2018, unless otherwise noted):

- With a First Nation suicide rate twice that of other Canadians (6–11 times higher for Inuit), over a third of Aboriginal deaths are self-induced (CMHA, 2013).
- According to the 2006 census, which provides the latest Canadian statistics on the topic, the projected life expectancy for most Canadians in 2017 is 79 years for men and 83 years for

FIGURE 7.3

From Bad to Worse—Child Poverty Rates in Canada

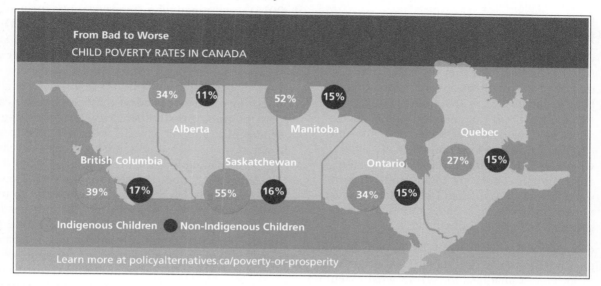

From Bad to Worse
CHILD POVERTY RATES IN CANADA

British Columbia — 39% / 17%
Alberta — 34% / 11%
Saskatchewan — 55% / 16%
Manitoba — 52% / 15%
Ontario — 34% / 15%
Quebec — 27% / 15%

Indigenous Children ● Non-Indigenous Children

Learn more at policyalternatives.ca/poverty-or-prosperity

In some cases, child poverty rates between Indigenous and non-Indigenous children are tripled. Sometimes, money isn't the answer. Besides financial assistance, what would you do to address the poverty that Indigenous children suffer?

women. The numbers are dramatically different for Aboriginal peoples in 2017—at 73–74 for men and 78–80 years for First Nations and Métis women. It is higher than the projected ages of 64 for Inuit men and 73 years for Inuit women.

- In 2015, the Aboriginal employment rate stood at 52.1 percent compared to non-Aboriginal people at 61.2 percent, whereas the 15 percent Aboriginal unemployment rate was double that of their non-Aboriginal counterparts.
- In 2006, fully 25 percent of non-Aboriginal adults had a university degree, compared to 9 percent of Métis, 7 percent of First Nations people, and 4 percent of Inuit.
- In 2014, despite accounting for roughly 4 percent of the Canadian population, Native people account for 22.8 percent of the total incarcerated population. More specifically, 25 percent of men and 36 percent of women sentenced to provincial and territorial custody are Native.
- In 2013–14, despite accounting for just 7 percent of the youth in the nine jurisdictions reporting,

Native youth accounted for 41 percent of all admissions. Once again, gender disparities were evident: girls accounted for 53 percent and boys for 38 percent of admissions.

- Nationally, Aboriginal people have a median income of $20 701 compared to $30 195 for non-Aboriginal people.
- In 2011, Aboriginal Affairs officials estimated that 20 000–35 000 new houses were needed to meet reserve community demands, 16 900 needed repairs, and 5200 needed replacing—out of an estimated national reserve housing stock of around 110 000 units. In Nunavut (population 31 695), more than 3000 households (roughly 11 000 people) are homeless and waiting for public housing. Between 2004 and 2014, more than 400 out of 614 First Nations had some kind of water problem.

The substandard living conditions of Aboriginal peoples have existed and continue to exist in Canada, across all measurable indicators, indicative of the systemic discrimination that has kept them from achieving equality with the larger Canadian society. Earlier queries in this chapter have asked you how this could have happened. Now, we ask you—when and how will this change?

In 1969, Harold Cardinal, a Cree leader, lawyer, and political activist, wrote:

> Canadians worry about their identity. Are they too English? Are they too American? Are they French Canadians or some other kind of hybrid? Indians worry about their identity too. For the most part they like to think of themselves as Canadians. But there are towns and cities in Canada, in every province of Canada, where an Indian dares not forget his identity as an Indian. There are towns and cities in Canada where a Canadian Indian simply does not go ... where simply being an Indian means getting a beating. (Cardinal, 1969, p. 18)

Can Harold Cardinal's words be, in part, an explanation for rifts that exist between Aboriginal and non-Aboriginal Canadians today? To what extent do you think Cardinal's words are still relevant? Is the link between the land and identity the weakest or the strongest link in the chain of forgiveness and acceptance? Only time will tell.

Eagle feather: A great honour among Aboriginal peoples; represents a mark of distinction.

ENDING THOUGHTS

During the writing of this chapter, Canada lost a great Aboriginal leader and activist. Elijah Harper (1949–2013) was "a man who stood for unity, inclusiveness and equality for all Canadians" (Rigden, 2013). Best known for putting a halt to the controversial Meech Lake Accord in 1987, by raising an **eagle feather** when it came time for him to vote, Harper was a wise man. In Aboriginal belief systems, the law of nature calls for one order of life to depend on another—like two canoes perhaps guiding each other along, but respecting each other's places in the river. Taking the time to make a new history—separately but together—re-learning the reasons and rewriting the future.

Language. Land. Legislation. Lessons. These are four quadrants that can serve as an attempt to understand the identity of a people, like the four quadrants in the medicine wheel. But to do that would be to impose our intellectual traditions on those of Canada's First Peoples, and that has already been done before—hasn't it? Perhaps the metaphor of the medicine wheel is best left to those who use it best. Maybe the time has come to share our experiences and respect our distinct intellectual traditions, as equal, in an arena built on attitudes of mutual respect and a collective desire to learn—regardless of the physicality of the structures in which the lessons take place. Together, yet separate, we can make this happen, don't you agree?

READING

A MOTHER TO A TEACHER

Before you take charge of the classroom that contains my child, please ask yourself why you are going to teach Indian children. What are your expectations? What rewards do you anticipate? What ego-needs will our children have to meet?

Write down and examine all the information and opinions you possess about Indians. What are the stereotypes and untested assumptions that you bring with you into the classroom? How many negative attitudes toward Indians will you put before my child?

What values, class prejudices, and moral principles do you take for granted as universal? Please remember that "different from" is not the same as "worse than" or "better than," and the yardstick you use to measure your own life satisfactorily may not be appropriate for their lives.

The term "culturally deprived" was invented by well-meaning middle-class Canadians to describe something they could not understand.

Too many teachers, unfortunately, seem to see their role as rescuer. My child does not need to be rescued; he does not consider being an Indian a misfortune. He has a culture, probably older than yours; he has meaningful values and a rich and varied experiential background. However strange or incomprehensible it may seem to you, you have no right to do or say anything that implies to him that it is less than satisfactory.

Our children's experiences have been different from those of the "typical" white middle-class child for whom most school curricula seems to have been designed (I suspect that this "typical" child does not exist except in the minds of curriculum writers). Nonetheless, my child's experiences have been as intense and meaningful to him as any child's.

Like most Indian children his age, he is competent. He can dress himself, prepare a meal for himself, clean up afterward, care for a younger child. He knows his Reserve, all of which is his home, like the back of his hand.

He is not accustomed to having to ask permission to do the ordinary things that are part of normal living. He is seldom forbidden to do anything; more usually the consequences of an action are explained to him and he is allowed to decide for himself whether or not to act. His entire existence since he has been old enough to see and hear has been an experiential learning situation, arranged to provide him with the opportunity to develop his skills and confidence in his own capacities. Didactic teaching will be an alien experience for him.

He is not self-conscious in the way many white children are. Nobody has ever told him his efforts toward independence are cute. He is a young human being energetically doing his job, which is to get on with the process of learning to function as an adult human being. He will respect you as a person, but he will expect you to do likewise to him.

He has been taught, by precept, that courtesy is an essential part of human conduct and rudeness is any action that makes another person feel stupid or foolish. Do not mistake his patient courtesy for indifference or passivity.

He does not speak Standard English, but he is no way "linguistically handicapped." If you will take the time and courtesy to listen and observe carefully, you will see that he and the other Indian children communicate very well, both among themselves and with other Indians. They speak "functional" English, very effectively augmented by their fluency in the silent language, the subtle, unspoken communication of facial expressions, gestures, body movements, and the use of personal space.

You will be well advised to remember that our children are skillful interpreters of the silent language. They will know your feelings and attitudes with unerring precision, no matter how carefully you arrange your smile or modulate your voice. They will learn in your classroom, because children learn involuntarily. What they learn will depend on you.

Will you help my child to learn to read or will you teach him that he has a reading problem? Will you help him develop his problem-solving skills, or will you teach him that school is where you try to guess what answer the teacher wants?

Will he learn that his sense of his own value and dignity is valid, or will he learn that he must forever be apologetic and "trying harder" because he isn't white? Can you help him acquire the intellectual skills he needs without at the same time imposing your values on top of those he already has?

Respect my child. He is a person. He has a right to be himself.

Yours very sincerely,
His Mother

Source: Anonymous, "A Mother to a Teacher: Respect My Child—He Has A Right To Be Himself," *Saskatchewan Indian Magazine:* Vol. 12 no.7, 1982.

DISCUSSION QUESTIONS

1. Imagine you are the teacher to whom this letter is directed. How would you respond to this student's mother? What are some effective teaching strategies that you might employ?
2. Given what you've read in this chapter, do you think that this mother's letter was justified in any way? Will she be helping or harming her child's chances in that teacher's class?
3. How easily does this letter transcend boundaries? Could it and does it apply to any child that is different from the norm? What is an example?

KWIP

One of the fundamental truths about diversity is that it involves as much learning about you as it does learning about others. Let's use the KWIP process to look at how our identities are shaped through social interaction.

KNOW IT AND OWN IT: WHAT DO I BRING TO THIS?

The "K" in the KWIP process involves examining aspects of your own identity and social location as the first step in becoming diversity competent.

ACTIVITY: JOURNAL

In Canada, from an early age we are taught about our British and French heritage, and how these peoples settled untamed and often wild lands. Have you given thought to narratives that counter these stories? That is, we know that Indigenous peoples occupied what is Canada prior to Europeans arriving—are you able to name the Indigenous place names of your community? For example, did these people always live here? If not, where did they come from? What is their Creation story? Who are the Indigenous peoples who presently reside in the community you are examining? How do the Indigenous and settler inhabitants see each other? How do they interact with each other? What issues do the non-Aboriginal inhabitants see? What issues do the Indigenous people see? What do you see?

WALKING THE TALK: HOW CAN I LEARN FROM THIS?

The "W" in the KWIP process presents a scenario or case study that challenges you to "walk the talk" through problem-based learning.

ACTIVITY: CASE STUDY

Rhonda is a third-generation Aboriginal mother, born and raised in Winnipeg. She has frequently visited the home community her family has historic ties to, as well as her husband's. Their child Joseph attends Grade 3 at a local public school.

At Thanksgiving, Joseph came home in tears. During the previous classes, the teacher has been discussing the importance of the relationship between Aboriginal people and the first settlers. As part of a classroom assignment, the teacher asked that on the Friday before the holiday that the children come to class dressed as 'Indians.'

Joseph was quite confused with the assignment. In his mind, he came to school everyday dressed as an 'Indian,' so what he was being asked to do didn't make any sense. Rather than discuss this with his parents, he decided to just go to school dressed like he did every other day.

When he arrived, he noticed that some his classmates were wearing feathers cut out of cardboard in their hair. Others were gripping plastic bows and arrows. Some were running around making war whoops. Not only was Joseph confused, but also the teacher then chastised him for not following instructions and getting dressed up like an Indian—and he received a failing grade.

When Joseph tried to tell the teacher that he was Aboriginal, and of his confusion with the assignment, he was told to go to the principal's office. When he told his parents, they chose not to confront the principal or the teacher due to their fears that because of prominent evidence of racism in the city, they could escalate the situation—and that systemic beliefs such as the teacher's, that are also built into the curriculum, cannot be changed in one-on-one settings.

1. How should the teacher have handled Joseph's concerns over the Thanksgiving dress-up activity?

2. What options are available to Joseph's parents, who were concerned about this blatantly insensitive activity based on stereotypes that historicizes Indians in such a fashion?

IT IS WHAT IT IS: IS THIS INSIDE OR OUTSIDE MY COMFORT ZONE?

The "I" in the KWIP process requires you to honestly confront and identify ways in which our complex identities result in experiences of privilege and oppression, and to reflect on how we can learn to honour that privilege.

ACTIVITY: GOT PRIVILEGE?

One of this chapter's themes is that 'Aboriginal people' is a racially constructed legal category that identifies in negative terms as less civilized. As a result, it also means that Aboriginal schools receive less funding, reserve communities are constrained in their abilities to develop, and that in addition to often times looking visibly different as an Aboriginal person, you are legally distinct. This was done to help dispossess 'Indians' from their lands to aid in Canada's nation-building project. It continues to impact Aboriginal people, whose educational outcomes are lower than non-Aboriginal, as are their health status and employment opportunities. This systemic marginalization often results in higher incarceration rates, drug and alcohol use, and feelings of not belonging. So where does your race land you in the hierarchy of power and privilege in our society? Does the social construction of race mean you are marginalized and discriminated against? Or does your race afford you certain privileges and power that you may not even be fully conscious of?

ASKING: Do I have privilege?

1. Is my family's economic success or stability traced to historic policies and actions that continue to negatively influence Aboriginal people?

2. Do I fully understand the ramifications of residential schools, not only from historical perspectives but also from contemporary perspectives as well?

3. Are Aboriginal people regularly represented in the work force or among my fellow students?

4. Will I be picked for a job over another similarly qualified individual, simply because of my skin colour or my race?

REFLECTING: Honouring Our Privilege

Describe two circumstances in which you feel disadvantaged because of your racialized identity. Then describe two circumstances in which your race gives you privilege over someone else. How can you use the advantages you experience to combat the disadvantages experienced by others?

Put It In Play: How Can I Use This?

The "P" in the KWIP process involves examining how others are practising equity and how you might use this. To this end, you are invited to read a commentary about reconciliation in Canada written by former Grand Chief of the Assembly of First Nations and residential school survivor, Phil Fontaine, that is dedicated to expanding our understanding of what reconciliation means and its essential role in improving the Indigenous–Canada relationship.

ACTIVITY: CASE STUDY

There Were Three Founding Peoples of Canada

Phil Fontaine, residential school survivor, is former Chief of the Assembly of First Nations.

Reconciliation is not an easy topic to talk about. In order to achieve true reconciliation in Canada, we're going to have to take some bold steps.

Confederation is held up as a monumental achievement that formed our Canadian identity and provided the foundation of the free and democratic nation we believe ourselves to be.

But what's missing from this story? Canada would not have been formed in 1867 had the Indigenous peoples not made possible the fur trade that ensured a strong economy.

But by far the most important contribution to the formation of our nation were treaties that resulted in the take-up of the most valuable land for settlement and resource development. The treaty relationship underscored the idea of peaceful coexistence as a central theme of relations between Canada and Indigenous peoples.

[But] in 1876, shortly after Confederation, Parliament passed the Indian Act, which not only ignored our contributions but also subjugated us as wards of the state. It essentially made us non-citizens in our own country.

Ever since then, we have been on the outside looking in, not really a part of Confederation, not really full citizens. We have no protection of our languages, no protection of our laws, no protection of our cultures.

How do we tell the true origin story of Canada? It can only be told if Parliament, formally through legislation, recognizes that there are three founding peoples of Canada: the British, the French and Indigenous peoples.

What would this accomplish? It would set the record straight. It would make Canada whole. The correct and powerful narrative of Canada's origins would become part of the shared story of every Canadian for generations to come. It would open up possibilities for genuine and lasting reconciliation, not the lie that's been imposed on all Canadians.

And it's important that we do it now.

In 1996, the Royal Commission on Aboriginal Peoples wrote, "A country cannot be built on a living lie." We should celebrate Canada's 150th anniversary by officially telling the truth about Canada's history.

Recognition and reconciliation go hand in hand. For the Parliament of Canada to recognize Indigenous peoples as equal partners in the nation would be a profound gesture of reconciliation. It would be a moment for all Canadians to celebrate.

Source: Senator Murray Sinclair, Dr. Dawn Lavell-Harvard and Phil Fontaine, "Indigenous leaders issue call to action: here's what Canada must do to make amends for residential schools tragedy," *NOW Magazine*, March 31, 2017. Used with permission.

Study Tools
CHAPTER 7

Located at www.nelson.com/student

- Review Key terms with interactive **flash cards**
- Check your Comprehension by completing **chapter review quizzes**
- Gauge your understanding with *Picture This* and accompanying short answer questions
- Develop your critical thinking/reading skills through compelling **Readings** and accompanying short answer questions
- Apply your understanding to your own experience with **Connect A Concept** activities
- Evaluate Diversity in the Media with engaging *Video Activities*
- Reflect on your Understanding with *KWIP* activities

REFERENCES

AANDC. (2008). Statement of apology. Retrieved from *Aboriginal Affairs and Northern Development Canada*: http://www.aadnc-aandc.gc.ca/eng/1100100015644/1100100015649

AANDC. (2010a, September 15). Chapter 18—An act to amend and consolidate the laws respecting Indians. Retrieved from *Aboriginal Affairs and Northern*

Development Canada: http://www.aadnc-aandc.gc.ca/eng/1100100010252/1100100010254

AANDC. (2010b, September 15). Fact sheet—Progress report—Kanesatake. Retrieved from *Aboriginal Affairs and Northern Development Canada*: http://www.aadnc-aandc.gc.ca/eng/1100100016305/1100100016306

AANDC. (2010c, September 15). Land claims. Retrieved from *Aboriginal Affairs and Northern Development Canada*: http://www.aadnc-aandc.gc.ca/eng/1100100030285/1100100030289

AANDC. (2010d, September 15). Treaties with Aboriginal people in Canada. Retrieved from *Aboriginal Affairs and Northern Development Canada*: http://www.aadnc-aandc.gc.ca/eng/1100100032291/1100100032292

AANDC. (2011, September 14). Fact sheet—Settlement agreement with Bigstone Cree Nation. Retrieved from *Aboriginal Affairs and Northern Development Canada*: http://www.aadnc-aandc.gc.ca/eng/1316020893971/1316021019328

AANDC. (2012a, October 1). Words first: An evolving terminology relating to Aboriginal peoples in Canada. Retrieved from *Aboriginal Affairs and Northern Development Canada*: http://www.aadnc-aandc.gc.ca/eng/1100100014642/1100100014643

AANDC. (2012b, October 5). Minister John Duncan congratulates Sioux Valley Dakota Nation on successful self-government agreements community vote. Retrieved from *Aboriginal Affairs and Northern Development*: http://www.aadnc-aandc.gc.ca/eng/1349475217859/1349475288013

AANDC. (2012c, December 5). Aboriginal peoples and communities. Retrieved from *Aboriginal Affairs and Northern Development Canada*: http://www.aadnc-aandc.gc.ca/eng/1100100013785/1304467449155

AANDC. (2016a, 26 March). The Ipperwash final settlement agreement: A journey towards reconciliation. Retrieved from *Aboriginal Affairs and Northern Development*: http://www.aadnc-aandc.gc.ca/DAM/DAM-INTER-HQ-MR/STAGING/texte-text/ipperwash_1460388200951_eng.pdf

AANDC. (2016b). Specific claims snapshot. Retrieved from *Aboriginal Affairs and Northern Development*: https://www.aadnc-aandc.gc.ca/eng/1395939024596/1395939088362.

Anonymous. (1982). A mother to a teacher: Respect my child—he has a right to be himself. *Saskatchewan Indian*, Vol. 12, no. 7, 45–47.

Applied History Research Group. (2001). Canada's first nations. Retrieved from *University of Calgary*: http://www.ucalgary.ca/applied_history/tutor/firstnations/

Assembly of First Nations (2006, November 26). Aboriginal peoples, 10 years after the royal commission. Retrieved June 27, 2013, from *CBC News*: http://www.cbc.ca/news/background/aboriginals/pdf/afn_rcap.pdf

Auditor General of Canada. (2006, May). Management of programs for First Nations. Retrieved from *CBC News*: http://www.cbc.ca/news2/background/auditorgeneral/ag_report200605/20060505ce.pdf

Barrera, J. (2013, June 11). A grandson reflects on Harper's Indian residential school apology and the day his grandmother revealed her story. Retrieved from *APTN*: http://aptn.ca/pages/news/2013/06/11/24358/

Belanger, Y. (2018). *Ways of knowing*, 3rd Ed. Toronto: Nelson.

Borrows, J. (2002). *Recovering Canada: The resurgence of Indigenous law*. Toronto: University of Toronto Press.

Borrows, J. (2010). Canada's Indigenous constitution. Toronto: University of Toronto Press.

Canada History. (2012, January 1). Oka. Retrieved from *Canada History*: http://www.canadahistory.com/old/sections/Eras/pcsinpower/oka.htm

Canada's First Peoples (2007). The first peoples of Canada. Retrieved from *Canada's First Peoples*: http://firstpeoplesofcanada.com/fp_groups/fp_groups_origins.html

Canadian Press. (2008, June 11). PM cites 'sad chapter' in apology for residential schools. Retrieved from *CBC News*: http://www.cbc.ca/news/canada/story/2008/06/11/aboriginal-apology.html

Cardinal, H. (1969). *The unjust society: The tragedy of Canada's Indians*. Edmonton: M.G. Hurtig Publishers.

Cardinal, H. (1977). *The rebirth of Canada's Indians*. Edmonton: Hurtig Publishers.

CBC News. (2010, June 14). Federal commission FAQs: Truth and reconciliation commission. Retrieved from *CBC News*: http://www.cbc.ca/news/canada/story/2008/05/16/f-faqs-truth-reconciliation.html

CBC News. (2013, January 5). 9 questions about Idle No More. Retrieved from *CBC News*: http://www.cbc.ca/news/canada/story/2013/01/04/f-idlenomore-faq.html

Cheechoo, S. (1991). *Path with no moccasins*. West Bay.

CMHA. (2013). Suicide among Aboriginal people in Canada. Retrieved from *CMHA*: http://london.cmha.ca/mental_health/suicide-among-aboriginal-people-in-canada/#.UczgjJz3Nug

Coates, K. (2015). *#Idlenomore and the remaking of Canada*. Regina: University of Regina Press.

Directorate of Human Rights and Diversity. (2008). Religions in Canada. Retrieved from *Government of Canada Publications*: http://publications.gc.ca/collections/collection_2011/dn-nd/D2-147-2008-eng.pdf

Enbridge. (2013). Benefits for Aboriginals. Retrieved from *Enbridge Northern Gateway Pipeline*: http://www.northerngateway.ca/aboriginal-engagement/benefits-for-aboriginals/

Federal Judicial Affairs. (2013, May 17). *Squamish Indian Band v. Canada*, [1996] 3 F.C. Retrieved May 24, 2013, from *Office of the Commissioner for Federal Judicial Affairs Canada*: www.fja-cmf.gc.ca

First Nations Study Program. (2009, January 1). Bill C-31. Retrieved from *University of British Columbia*: http://indigenousfoundations.arts.ubc.ca/home/government-policy/the-indian-act/bill-c-31.html

Frideres, J., & Gadacz, R. (2008). *Aboriginal peoples in Canada*. Toronto: Pearson.

Ganondagon. (2012). The two row wampum. Retrieved from *Ganondagon: Preserving a past; providing a future*: http://www.ganondagan.org/wampum.html

Gobeil, M., & Monpetit, I. (2011, May 11). When the government fails to honour its commitments. Retrieved from *CBC News*: http://www.cbc.ca/news/canada/story/2011/05/30/f-mapping-future-specific-claims.html

Government of Manitoba. (2008, November 24). Numbered treaties. Retrieved from *Education and Literacy*: www.edu.gov.mb.ca/k12/cur/socstud/foundation_gr6/blms/6-1-4f.pdf

Government of Manitoba. (2013). The justice system and Aboriginal people. Retrieved from The *Aboriginal Justice Implementation Commission*: http://www.ajic.mb.ca/volume1/chapter5.html

Gradual Civilization Act, 1857. Statutes of Canada, 20 Vict., c. 26, 10 June 1857.

Hanson, E. (2009a). Oral traditions. Retrieved from *University of British Columbia*: http://indigenousfoundations.arts.ubc.ca/home/culture/oral-traditions.html

Hanson, E. (2009b, January 1). The residential school. Retrieved from *University of British Columbia*: http://indigenousfoundations.arts.ubc.ca/home/government-policy/the-residential-school-system.html

Idle No More. (2012, January 24). Manifesto. Retrieved from *Idle No More*: http://idlenomore.ca/manifesto

Kennedy, M., & Moss, J. (1997). *Echoing silence: Essays on Arctic narrative*. Ottawa: University of Ottawa Press.

Kunin, J. (2011, May 24). Highlights of the Indian Act. Retrieved from *Educators for Peace and Justice*: http://epjweb.org/resources/lessons/social-sciences/some-highlights-of-the-indian-act/

Lombard, A. (2009, June 18). Our voices, our stories: First Nations, Métis and Inuit stories–Voices of Métis. Retrieved from *Library and Archives Canada*: http://www.collectionscanada.gc.ca/stories/020020-2000-e.html

Long, D., & Dickason, O. (2000). *Visions of the heart*. Toronto: Harcourt.

Macdonald, D., & Wilson, D. (2013, June 22). Poverty or prosperity: Indigenous children in Canada. Retrieved from *Canadian Centre for Policy Alternatives*: http://www.policyalternatives.ca/sites/default/files/uploads/publications/National%20Office/2013/06/Poverty_or_Prosperity_Indigenous_Children.pdf

Makarenko, J. (2008, June 2). The Indian Act: Historical overview. Retrieved from *Maple Leaf Web*: http://www.mapleleafweb.com/features/the-indian-act-historical-overview

Malin, E. (1986). *Totem poles of the Pacific north coast*. Portland: Timber Press.

Manuel, G., & Posluns, M. (1974). *The fourth world: An Indian reality*. Toronto: Collier Macmillan.

Miller, J. R. (1991). Great White Father knows best: Oka and the land claims process. *Native Studies*, 23–52.

Miller, J. R. (2009). *Compact, Contract, Covenant*. University of Toronto Press.

Miller, J. R. (2017). *Skyscrapers hide the heavens*. University of Toronto Press.

Montgomery, M. (1965). The Six Nations and the MacDonald Franchise. *Ontario History*, 25.

National Park Service. (2013, May 15). Sitka National Historical Park. Retrieved from *National Park Service*: http://www.nps.gov/sitk/historyculture/totem-poles.htm

Nature Canada. (2013). Enbridge Northern Gateway Project. Retrieved from *Nature Canada*: http://naturecanada.ca/enbridge_northern_gateway.asp

Norman, H. (1990). *Northern tales: Traditional stories of Eskimo and Indian peoples*. New York: Pantheon.

Racette, S. (1991). *The flower beadwork people*. Regina: Gabriel Dumont Institute.

Ramsey, H. (2011, March 31). Totem poles: Myth and fact. Retrieved from *The Tyee*: http://thetyee.ca/Books/2011/03/31/TotemPoles/

Reid, B., & Bringhurst, R. (1988). *Raven steals the light*. Vancouver: Douglas and MacIntyre.

Rigden, M. (2013, May 21). The humble and powerful Elijah Harper will be missed. Retrieved from *APTN*: http://aptn.ca/pages/news/2013/05/21/the-humble-and-powerful-elijah-harper-will-be-missed/

Roberts, J. (2006). *First Nations, Inuit, and Métis peoples*. Toronto: Emond Montgomery Publications.

Roberts, J., Boyington, D., & Kazarian, S. (2008). *Diversity and First Nations in Canada*. Toronto: Emond Montgomery.

Rowland, R. (2013, June 24). First Nations don't have right to direct tankers, Northern Gateway lawyer says. Retrieved from *The Vancouver Sun*: http://www.vancouversun.com/news/metro/First+Nations+have+right+direct+tankers+Northern+Gateway/8571353/story.html

Royal Alberta Museum. (2005). What is a medicine wheel? Retrieved from *Royal Alberta Museum*: http://www.royalalbertamuseum.ca/human/archaeo/faq/medwhls.htm

Scofield, G. (1999). *I knew two Métis women*. Victoria, BC: Polestar Book Publishers.

Silou, S. (2009, June 18). Our voices, our stories: First Nations, Métis and Inuit stories–Inuit oral traditions: The social conscience of Inuit culture. Retrieved from *Library and Archives Canada*: http://www.collectionscanada.gc.ca/stories/020020-3000-e.html

Sinclair, R. (2011). The 60's scoop. Retrieved from *Origins Canada*: http://www.originscanada.org/the-stolen-generation/

Sinquin, A. (2009, June 18). Our voices, our stories: First Nations, Métis and Inuit stories–Voices of First Nations. Retrieved from *Library and Archives Canada*: http://www.collectionscanada.gc.ca/stories/020020-1000-e.html

Salomons, T. (2009, January 1). Ipperwash crisis. Retrieved from *University of British Columbia*: http://

indigenousfoundations.arts.ubc.ca/home/community
-politics/ipperwash-crisis.html

Statistics Canada. (2010, June 21). Aboriginal statistics at a glance. Retrieved from *Statistics Canada*: http://www .statcan.gc.ca/pub/89-645-x/89-645-x2010001-eng.htm

Statistics Canada. (2012). Census in brief: Aboriginal languages in Canada. Retrieved from *Statistics Canada*: http://www12.statcan.gc.ca/census-recensement/2011/ as-sa/98-314-x/98-314-x2011003_3-eng.pdf

Stewart, H. (1993). *Looking at totem poles*. Vancouver: Douglas & McIntyre.

TD Economics. (2011, June 17). Estimating the size of the Aboriginal market. Retrieved from *Canadian Council for Aboriginal Business*: (https://www.td.com/document/PDF/ economics/special/sg0611_aboriginal.pdf)

Treaty Relations Commission of Manitoba. (2013). Treaties in Canada. Retrieved from *Treaty Relations Commission of Manitoba*: http://www.trcm.ca/about_treaties.php

UNDRIP. (2007). United Nations Declaration on the Rights of Indigenous People. http://www.un.org/esa/socdev/unpfii/ documents/DRIPS_en.pdf

Vowel, C. (2012, December 31). Chelsea Vowel: Assimilation is not the answer to the Aboriginal 'problem.' Retrieved from *The National Post*: http://fullcomment.nationalpost .com/2012/12/31/chelsea-vowel-assimilation-is-not-the -answer-to-the-aboriginal-problem/

Immigration

"So why does it feel so wrong/To reach for something more/ To wanna live a better life"

(Sick Puppies, 2010)

LEARNING OUTCOMES

By mastering this unit, students will gain the skills and ability to:

- analyze historic and contemporary patterns of immigration and resettlement in Canada

- critique Canada's immigration policy as a regulatory mechanism that determines who can come into a country and on what terms

- examine some of the difficulties experienced by immigrant and refugees groups as they strive to resettle and integrate into Canadian society

- distinguish between some of the common myths and facts surrounding debates on immigration policy and causes of international human migration

- reflect upon your own personal migration history and locate it within the context of Canadian immigration history

Alex/Thinkstock.com

The Canadian **immigration** process has significantly contributed to our evolving character as a multicultural nation and will likely continue to contribute to significant demographic changes in the future. Successive waves of **immigrants** throughout Canada's history have made extraordinary contributions to our nation. Look at the contributions made to Canada's social, cultural, political, and economic landscape by immigrants such as Thomas Bata, Alexander Graham Bell, Olivia Chow, Adrienne Clarkson, Tommy Douglas, Michaëlle Jean, Peter Mansbridge, and Joe Schlesinger, to name only a few.

As you learn about the history of Canada's immigration policy, some of the different ways in which people come here to live, a few of the commonly held myths and facts about immigrants and **refugees**, and the contemporary issues faced in resettling in Canada, it is important to remember that behind all of these issues are human stories that are worth exploring. Unless you are an Indigenous student, then you have a personal **migration** history of how you or your ancestors came to Canada—the challenges faced and the opportunities realized.

The stories of **asylum seekers** and others forced to migrate also expose the limits of human endurance and the indomitable human spirit. Narratives of exile and quests for safe asylum are detailed in this chapter's Reading, "The Thinnest Line," and *In Their Shoes*. The drowning death of Alan Kurdi on September 2, 2015, in the Mediterranean Sea, and the picture of his lifeless body on the beach that appeared in media around the world, narrated a powerful reminder of the depth of human tragedy and suffering for those seeking asylum and cut through the refugee debate. In a world that was riddled with anti-migration rhetoric, this image mobilized everyday people and governments to move from sentiments of intolerance and indifference to act and to welcome Syrian refugees seeking asylum. These narratives give testament to the importance of the Canadian **Immigration and Refugee Protection Act's** objective in fulfilling our humanitarian obligations under international law (Immigration and Refugee Protection Act, 2001).

HISTORY OF CANADIAN IMMIGRATION POLICY

An analysis of Canada's immigration policies helps us to understand some of our nation's greatest controversies around inclusion and exclusion—essentially, who gets in and who doesn't. The history of Canadian immigration policy can best be summarized in eight periods—a model formulated by Geneviève Bouchard in a presentation by the Canadian Institute for Research on Public Policy on Canada's immigration system (Bouchard, 2007).

Period One

Period One (1867–1913) was a time when the main goal of immigration was to secure farmers and labourers to populate and settle western Canada. During this time, immigration was encouraged from source areas such as Great Britain, the United States, and northwestern Europe. The Canadian Pacific Railway was completed, and Chinese labourers were no longer required—this is important because, at this time, the government imposed a **head tax** on new Chinese immigrants. The first head tax imposed by the federal government in 1885 was $50 (Canadians for Redress, 2002). In 1900, the Canadian government doubled the head tax to $100, and then increased it once again to $500 in 1903. In 1913, over 400 000 immigrants arrived in Canada, marking the largest influx of immigrants in Canadian history (Citizenship and Immigration Canada, 2006a).

Period Two

Period Two (1919–1929) was a time when revisions to the Immigration Act resulted in more restrictive and selective procedures based on the country's "absorptive capacity," which is essentially a country's perceived saturation level—the level at which it cannot take on any more immigrants without risking peril to its own inhabitants. Immigrants had to pass a literacy test, and the government could limit the number of immigrants allowed into Canada. Source countries were officially divided into preferred and non-preferred groups.

Immigration: Entering into and becoming established in a new place of residence; usually means entering a country that one was not born in.

Immigrants: People residing in Canada who were born outside of Canada; this category excludes Canadian citizens born outside of Canada and people residing in Canada on temporary status, such as those with a student visa or temporary foreign workers.

Refugee: A person who is forced to flee from persecution and is outside of his or her country of origin.

Migration: To move from one country, place, or region to another.

Asylum seeker: A person seeking protection as a refugee in another country but has not yet been found to meet the definition of a refugee.

Immigration and Refugee Protection Act (IRPA): Legislation whose mission is "respecting immigration to Canada and the granting of refugee protection to persons who are displaced, persecuted, or in danger."

Head tax: Tax imposed by the Canadian government on anyone immigrating to Canada from China between 1885 and 1923.

Preferred source countries included Great Britain, the United States, the Irish Free State, Newfoundland, Australia, and New Zealand. The Chinese were formally excluded from immigrating to Canada from 1923 until 1947.

Period Three

Period Three (1930–1945) was during the time of the Great Depression and World War II. When the Canadian unemployment rate reached 27 percent in 1933, the door was closed to most newcomers (except those from Britain and the United States) and active immigration recruitment ended.

Period Four

Period Four (1946–1962) was significant, as Canada saw a large influx of displaced persons from Europe. Approximately 70 000 war brides (foreign wives of Canadian soldiers) and their children arrived in Canada. Canada's immigration policy now had very clear ethnic/racial and economic goals. In 1947, Prime Minister William Lyon Mackenzie King stated that the purpose of immigration was to improve Canada's standard of living, but that immigration should not change the basic character of the Canadian population.

Period Five

Period Five (1962–1973) was a time when the government abolished what it acknowledged as racist immigration policy: "Henceforth any unsponsored immigrants who had the requisite education, skill, or other qualifications were to be considered suitable for admission, irrespective of colour, race, or national origin" (Citizenship and Immigration Canada, 2006b). One of the most significant changes in immigration policy occurred in 1967 with the creation of a point system that facilitated the immigration of skilled workers based on an assessment of factors such as education, employment opportunities in Canada, adaptability, and language proficiency. The point system is still the method of selection used today for skilled workers.

Period Six

Period Six (1974–1984) was the time of a new Immigration Act (1976), which defined the three objectives of Canada's immigration policy: family reunification; humanitarian concerns; and the promotion of Canada's economic, demographic, social, and cultural goals. These same objectives have endured as part of today's immigration policy.

Period Seven

Period Seven (1985–1993) was a time when a landmark decision made by the Supreme Court of Canada changed the refugee determination system, resulting in the eventual creation of the Immigration and Refugee Board. This decision, known as the Singh Decision, required that all people seeking asylum in Canada be provided with a full oral hearing on the merits of their claim. This decision, rendered on April 4, 1985, is now commemorated annually on this date as Refugee Rights Day. During this period, immigration policy increased the inflow of economic immigration to 250 000 people, but not at the expense of humanitarian obligations.

Period Eight

Period Eight (1993–2010) was a time that saw the 1976 Immigration Act replaced with the Immigration and Refugee Protection Act of 2002. The desire to increase the number of skilled workers continued during this period, with only a few changes in the point system, which was designed to draw a younger and bilingual demographic. The definition of refugee was expanded to include a class of protected persons, honouring international obligations under the **Convention against Torture** (Danelius, 2002). There was an increased use of temporary foreign workers and a decrease in the number of family class sponsorships approved.

Period Nine

Period Nine (2010–) is added to the model to capture some of the changes with respect to refugee protection. The federal department referred to as Citizenship and Immigration Canada is renamed Immigration, Refugees and Citizenship Canada in 2016. Parliament passes two new immigration acts, namely the **Balanced Refugee Reform Act** and the **Protecting Canada's Immigration System Act**. In 2014, Canada resettled 23 286 refugees in

Convention against Torture: United Nations international human rights agreement signed on December 10, 1984, as a commitment against the use of torture in their country for any reason; signatory nations also agree not to use any evidence obtained under torture and not to deport or return people to countries where they are at risk of being tortured.

Balanced Refugee Reform Act: Immigration legislation that, together with the Protecting Canada's Immigration System Act, makes changes to the Immigration and Refugee Protection Act and the refugee claimant process in Canada.

Protecting Canada's Immigration System Act: Immigration legislation designed to make the review and determination of refugee claims faster and to expedite removals of those who do not qualify.

total. Responding to the Syrian refugee crisis, Canada is expected in 2016 to resettle approximately 44 000 refugees: 25 000 government sponsored refugees, and the rest a mix of privately sponsored and blended visa cases whereby the federal government and private groups split the cost (Friscolanti, 2016).

CANADA APOLOGIZES

Despite the outstanding contributions made to Canada by successive waves of immigrants throughout our history, Canada's immigration policy has not always been an open door. Broad claims about the Canadian government's longstanding humanitarian tradition suffer from historical amnesia. The Canadian government overtly discriminated through an exclusionary immigration policy designed to keep out certain ethnicities who were deemed "unfit" to enter—some call it a history of "whites only" immigration.

The "Undesirables"

The Canadian Immigration Act of 1910 stated that the government could exclude immigrants of any race; in 1919, language was added to the act to provide a rationale for deeming some immigrant groups undesirable "owing to their peculiar customs, habits, modes of life and methods of holding property and because of their probable inability to become readily assimilated" (Matas, 1985). This legislation was used at points throughout the period of 1910–1962 to prohibit, restrict, or expel immigrants from Germany, Austria, Hungary, Bulgaria, Turkey, India, Pakistan, Ceylon, China, Japan, and other nations (Matas, 1985). It was also used to exclude religious groups like Doukhobors, Hutterites, Mennonites, and Jews (Matas, 1985).

An Apology for *Komagata Maru*

The incident involving a ship named the *Komagata Maru* is one example of the way the Canadian government applied exclusionary immigration policy to bar entry to those deemed "undesirable." As Ali Kazimi (2012) suggests, the *Komagata Maru* is not just an incident, it is part of a continuum in creating Canada as a white settler state.

When the *Komagata Maru* dropped anchor in the Vancouver harbour in 1914, aboard were 376 passengers from British India, many bearing the Sikh ceremonial name of Singh (Bissoondath, 1994). Many of the men on board were veterans of the British Indian Army, who believed their military service would support their right to settle anywhere in the empire that they had fought to defend. They were wrong (Kazimi, 2012).

"The immigration restrictions experienced by some people of Indian descent mark an unfortunate period in our nation's history. This monument commemorating the *Komagata Maru* incident recognizes this past."— Citizenship, Immigration and Multiculturalism Minister Jason Kenney, July 24, 2012

Using the "continuous journey" regulation, immigration officials held passengers on board for over two months, and eventually only 24 passengers were allowed to remain in Canada. The other 352 passengers were forced to sail back to India, where upon arrival, 29 passengers were shot and 20 were killed. Others were jailed. On August 3, 2008, Prime Minister Stephen Harper issued an apology on behalf of the government of Canada for the *Komagata Maru* incident. There was some disappointment among some members of the Sikh community, who expected a formal apology to be made in Parliament.

An Apology for the Chinese Head Tax and Exclusion

Another example of Canada's racially motivated exclusionary immigration policies was the Chinese head tax levied by the Canadian government in 1885 through the Chinese Immigration Act as a means of discouraging new Chinese immigrants from entering Canada after the completion of the Canadian Pacific Railway. By today's standard, the tax would be the equivalent of $100 000 per person (Canadians for Redress, 2002). This act was replaced by the Chinese Exclusion Act, which specifically prohibited Chinese immigration to Canada from 1923–1947. In 2006, Prime Minister Stephen Harper issued a formal apology and financial redress to the Chinese Canadian community.

An Apology for Japanese Internment

Following the bombing of Pearl Harbour during World War II, the Government of Canada confined Japanese immigrants and Canadian citizens of Japanese descent in internment camps, confiscated their personal property, and forced their repatriation (deportation) from Canada. Once again, Canadian immigration policy was used to exclude a particular group of people—this time those of Japanese descent.

Famous Canadian architect Raymond Moriyama and well-known Japanese-Canadian scientist and environmental activist David Suzuki were among the 23 000 Japanese taken to internment camps in British Columbia, branded as "enemy aliens." Their personal property was confiscated; Price Waterhouse estimated the value of this property in 1986 dollars at $443 million (Roberts-Moore, 2002). Regulations passed under the authority of the War Measures Act restricted immigration from Japan and also provided for the deportation of Japanese Canadians.

In 1988, Prime Minister Brian Mulroney issued a formal government apology to the Japanese-Canadian community. Under the terms of the Japanese Canadian Redress Agreement signed in 1988 between the Government of Canada and the National Association of Japanese Canadians, the federal government agreed to create a Canadian Race Relations Foundation to help eliminate racism. The federal government proclaimed the Canadian Race Relations Foundation Act into law on October 28, 1996, and the Foundation officially opened in 1997.

No Apology for Jewish Refugees aboard the St. Louis

Another example of xenophobia manifest in Canadian immigration policy was the systematic exclusion of Jews as immigrants or refugees during the period of 1933–1948. During this period, Canada's immigration policy is best summarized in the words of an immigration agent, who, when asked how many Jews would be allowed into Canada, replied, "None is too many." This phrase has become the title of a book by Irving Abella, a Toronto history professor, who argues that from 1933–1948, Canada had the worst record of any immigration country in the world in providing asylum to Jews, despite mounting reports of Adolf Hitler's genocide (Abella, 2008). The closed-door policy on Jewish immigration was led by Frederick Blair, the head of immigration in the King administration. In one letter, Blair compared Jews clamouring to get into the country to hogs on a farm at feeding time (CBC, 2011).

In 1939, 907 Jewish refugees aboard the *St. Louis* were refused asylum by Canada, and many returned to Europe where they faced death (Thomas & Witts, 1974). A memorial for these Jewish refugees was unveiled in Halifax at Pier 21 in 2011, but no official government apology was ever issued. In 2000, a group of Canadian clergy gathered with 25 survivors to issue an apology. One of those clergymen was Douglas Blair, a Baptist minister whose great uncle was none other than Frederick Blair (CBC News, 2000).

IMMIGRATING TO CANADA

According to the most recent Canadian Census data available, immigrants made up nearly one-fifth (19.8 percent) of Canada's population, a percentage that is expected to reach at least 25 percent by 2031 (Malenfant, Lebel, & Martel, 2010). More than half (54 percent) of the adult immigrant population in Canada lives in the cities of Toronto, Montreal, or Vancouver (Ng, 2011). Canada is now home to people from more than 200 countries who speak over 200 languages (Statistics Canada, 2012).

People can come to live in Canada in a variety of different ways on either a temporary or a permanent basis by applying to **Immigration, Refugees and Citizenship Canada (IRCC).** They come to work, study, visit, set up a business, reunite with family, be adopted, or find protection. The legislation currently used to regulate immigration to Canada is called the Immigration and Refugee Protection Act (IRPA). Passed in 2001, some of the objectives of this act are to support the development of a strong and prosperous

Immigration, Refugees and Citizenship Canada (IRCC): The branch of the federal government that is responsible for immigration, settlement, and citizenship; formerly called Citizenship and Immigration Canada (CIC).

Family class: Immigration category used to describe immigrants who have been sponsored to come to Canada as a spouse, partner, dependent child, parent, or grandparent.

Permanent resident: According to the Canadian Immigration and Refugee Protection Act (2002), a person who has come to Canada and successfully applied and received immigration status to live here permanently.

Undertaking: A contract signed by a sponsor with the Minister of Citizenship and Immigration, or with the Ministère de l'Immigration, de la Diversité et de l'Inclusion [MIDI] if you live in Québec, promising to provide financial support for basic requirements, including healthcare, for those relatives sponsored to come to Canada.

Economic class: An immigration category that includes federal and Quebec-selected skilled workers, federal and Quebec-selected business immigrants, provincial and territorial nominees, the Canadian Experience Class (CEC), and caregivers, as well as spouses, partners, and dependants who accompany the principal applicants in any of these economic categories (Government of Canada, 2016a).

Canadian economy, family reunification, and fulfillment of international humanitarian obligations. Recently, Canada passed new legislation to deal with refugee migration to Canada, namely the Balanced Refugee Reform Act and Protecting Canada's Immigration System Act. Receiving Royal Assent on June 29, 2010, the Balanced Refugee Act is intended to improve the expediency and fairness of the refugee determination process. Some of the changes include the launch of a Refugee Appeal Division, the designation of countries considered safe to live in, and an expedited removal process for failed refugee claimants. The Protecting Canada's Immigration System Act, implemented on December 15, 2012, brings further reform to the refugee determination process, but it is also intended to address issues of human smuggling and adds a requirement of biometric data for temporary resident visas, work permits, and study permit applications.

Ways of Immigrating to Canada

Family Class

For people looking to come to Canada permanently, there are three basic ways to do this. First, they can reunite with family members in Canada if someone sponsors them. This is called a **family class** sponsorship. A Canadian citizen or **permanent resident** of Canada can sponsor a spouse, common-law partner, conjugal partner, dependent child, or other eligible relatives (e.g., a parent or grandparent) (Government of Canada, 2016b). Sponsors sign a document called an **undertaking**, promising that they will support the family member financially in Canada, and that they will not seek financial assistance from the government for this purpose.

Economic Class

The second way to apply to come to live in Canada permanently is through a category referred to as the **economic class**. An immigrant can apply independently to immigrate to Canada as a skilled and/or experienced worker through the Federal Skilled Worker (FSW) program, the Federal Skilled Trades Program (FSTP), the Canadian Experience Class (CEC), or the Provincial and Territorial Nominee programs (Government of Canada, 2016). As of January 1, 2015, applications to these programs are made through a new online immigration application system called **Express Entry**. This new system allows potential immigrants to Canada to submit an online profile to be considered as a skilled worker in any one of these programs, and to apply to become a permanent resident if they so choose (Government of Canada, 2016b). The **Federal Skilled Worker (FSW) Program** selects people to immigrate to Canada because of their work experience and skills. As a skilled worker, an applicant is assessed under a point system based on education, work experience, knowledge of official languages, and other criteria that prove the person can make an economic contribution to Canada (Government of Canada, 2016a). (See Table 8.1.)

The **Federal Skilled Trades Program (FSTP)** facilitates the immigration of skilled tradespersons. Eligibility for this program is based on practical training and work experience. The **Canadian Experience Class Program** allows some skilled temporary foreign workers and international student graduates with at least one year of full-time work experience to stay in Canada permanently (Government of Canada, 2016a). The **Provincial and Territorial Nominee programs** help provinces and territories meet specific labour market demands through immigration and have resulted in the resettlement of immigrants to areas that are not traditional immigrant destinations (Government of Canada, 2016a).

In recent years, the government has made some significant changes to two categories within the economic class, namely **business immigrants** and live-in caregivers. In November 2014, the Government of Canada redesigned the Live-In Caregiver Program, renaming it the **Caregiver Program** and adding two new streams: the *Caring For Children* pathway to permanent

Express Entry: An online immigration application process that came into effect as of January 1, 2015, for persons applying to come to Canada through the Federal Skilled Worker (FSW) program, the Federal Skilled Trades Program (FSTP), the Canadian Experience Class (CEC), and some provincial/territorial nominee programs.

Federal Skilled Worker (FSW) Program: Program for people who want to apply to become permanent residents of Canada based on work experience and skills that are assessed through a point system.

Federal Skilled Trades Program (FSTP): Program for people who want to apply to become permanent residents of Canada based on full-time work experience and qualifications in an eligible skilled trade.

Canadian Experience Class (CEC): Program for people who want to apply to become permanent residents of Canada based on skilled work experience acquired in Canada while having legal status to work or study, such as temporary foreign workers and international student graduates with one year of full-time work experience.

Provincial and Territorial Nominee programs: Programs for people who want to apply to become permanent residents of Canada, who are selected by participating provinces or territories to live there. Selection criteria are based on streams that target specific skills, education, and work experience needed to contribute to the economy of that province or territory.

TABLE 8.1

Canada Point System for Federal Skilled Workers

Factor	Description	Maximum Points
English and/or French Language Skills	Being able to communicate and work in one or both of Canada's official languages is very important. Must send proof through approved testing. You will be given points based on your ability to listen, speak, read, and write.	/28
Education Can include recognized equivalencies obtained outside of Canada	Different points allocated for different post-secondary degrees and their equivalencies: Doctorate (PhD) level is 25 points; Master's Degree or specific occupational/ specified professional degree is 23 points; two or more post-secondary degrees or diplomas where one is three year credential is 22 points; Canadian post-secondary degree or diploma for a program of three years or longer is 21 points; Canadian post-secondary degree or diploma for a two-year program is 19 points; Canadian post-secondary degree or diploma for a one-year program is 15 points; Canadian high school diploma is 5 points.	/25
Work Experience	You can get points for the number of years you have spent in full-time paid work (at least 30 hours per week, or an equal amount of part-time): 1 year worth 9 points; 2–3 years worth 11 points; 4–5 years worth 13 points; 6 + years worth 15 points.	/15
Age	You will get points based on your age on the day your application is received. Under 18 receives 0 points. Persons 18–35 receive maximum 12 points. Every year after 35 years of age, your points decrease by one until those 47 + receive 0 points.	/12
Arranged Employment in Canada	In some cases, you can get points if you have a full-time job offer of at least one year from a Canadian employer. The job must be arranged before you apply to come to Canada as a federal skilled worker. A valid job offer has to be for continuous, full-time work that is not seasonal and at least one year and in an occupation listed as Skill Type 0 or Skill Level A or B of the National Occupational Classification (NOC).	/10
Adaptability	Relatives in Canada, past work experience or study experience in Canada for you or partner, language skills of partner.	/10

If you score 67 points or higher (out of 100), you may qualify to immigrate to Canada as a federal skilled worker. If you score lower than the pass mark of 67 points, you will not qualify to immigrate to Canada as a federal skilled worker.

Source: Government of Canada 2016a, http://www.cic.gc.ca/english/immigrate/skilled/apply-factors.aspCanadian Immigration and Citizenship, "Six selection factors–Federal skilled workers," http://www.cic.gc.ca/english/immigrate/skilled/apply-factors.asp.

residence for caregivers who have provided child care in a home; and the *Caring for People with High Medical Needs* pathway to permanent residence for caregivers who have provided care for the elderly or for those with disabilities or chronic disease in a health facility or in a home (Government of Canada, 2016a). Those persons currently providing care through the traditional Live-In Caregiver Program will be grandfathered in to the new program. In June 2014, the government also ended the existing federal Immigrant Investor and Entrepreneur programs, as research indicated they provided limited economic benefit to Canada (Government of Canada, 2016b). The government will use pilot initiatives to

Business Immigrants: A category of people in the economic class who invest in or start a business in Canada with the expectation that it contributes to the development of a strong Canadian economy.

Caregiver Program: New immigration program replacing the Live-In Caregiver Program with two new pathways to permanent residence for caregivers: 1) Caring for Children for caregivers who have provided child care in a home; 2) Caring for People with High Medical Needs for caregivers who have provided care for the elderly or for those with disabilities or chronic disease.

Immigrant Investor Venture Capital Pilot Program: Pilot program launched by the government in 2015 for people who want to apply to become permanent residents of Canada, who have a personal net worth of $10 million CDN (acquired through lawful, private sector business or investment activities) and are willing to invest a minimum of $2 million CDN for fifteen 15 years in the Immigrant Investor Venture Capital Fund.

Convention refugee: Individual who has been granted asylum by the 1951 Geneva Convention Relating to the Status of Refugees; someone who has reason to fear persecution in his or her country of origin due to race, religion, nationality, membership in a social group, or political opinion.

Government Assisted Refugee (GAR): A government-sponsored refugee selected overseas for resettlement in Canada.

Blended Visa Office-Referred Refugees: Refugees for whom UNHCR matches a private sponsor to share income support with the Government of Canada.

Immigration and Refugee Board (IRB): An independent tribunal established by the Parliament of Canada that is responsible for hearing refugee claims and appeals in accordance with the law.

replace investment programs, like the **Immigrant Investor Venture Capital Pilot Program**, launched in January 2015. The Canadian government now selects business class immigrants through two streams, start-up visas and self-employment, both of which are based on applicants' abilities to become economically established and support the development of the Canadian economy (Government of Canada, 2016c).

Resettled Refugee and Protected Person

The third way someone can apply to come to Canada permanently is as a **Convention refugee** or person in need of protection (Government of Canada, 2016b). Canada will offer protection to those who fear persecution or who could be tortured or suffer cruel and unusual punishment and are therefore unable to return to their home country (Citizenship and Immigration Canada, 2013). There are several ways that people come to Canada as refugees. The first way is as a **Government Assisted Refugee (GAR)**—a government-sponsored refugee selected overseas for resettlement in Canada. The second method is to be privately sponsored as a refugee by a group within Canada that has been approved by the government. The third is a new category called **Blended Visa Office-Referred Refugees** who are referred by the United Nations High

Commission for Refugees and matched with a private sponsor in Canada. The Government of Canada and the private sponsor each share six months of income support. The fourth way is to make a refugee claim upon arrival in Canada. The Canadian Border Services Agency decides if you are eligible to make a refugee claim. If deemed eligible, they refer the case to the **Immigration and Refugee Board (IRB)** to decide, based on testimony and evidence presented, whether a person meets the criteria as a **protected person**, according to Canada's Immigration and Refugee Protection Act (Citizenship and Immigration Canada, 2013).

An Overview of the Number of Immigrants Coming to Canada

The majority of immigrants come to Canada as members of the economic class (see Table 8.2). The Protected Persons class, which includes all refugees, protected persons, and their dependants abroad, is a comparatively smaller group, although 2016 will see an increase as a result of sponsorships of Syrian refugees.

DECIDING WHO GETS IN

Myths and Facts about Refugees and Immigrants Coming to Canada

Unfortunately, there are many enduring myths about refugees and immigrants coming to and living in Canada. Here are some of the statements and questions people make around refugee and immigrant issues, and some of the facts to consider in response.

When Someone Asks … Why Don't Refugees Have to Line Up and Wait Like Other Immigrants?

This question refers to the practice known as queue jumping—like when someone cuts in front of you in the coffee line. Refugees need to come to Canada more quickly than ordinary immigrants because their lives are usually in danger. Many refugees come to Canada's borders to apply because they can't safely apply overseas. In many countries where there is conflict, Canadian embassy or consulate offices are not accessible. Time needed for processing applications can endanger lives and safety. There are also situations where Canadian embassies and consulates are watched by those perpetrating violence in that country; obviously, they do not want their citizens talking to the international community about human rights violations. Sometimes, waiting in line can mean the difference between life and death for the asylum seeker.

AGENT OF CHANGE

Dr. Jean Placide Rubabaza

Dr. Jean Placide Rubabaza fled Burundi during the genocide and arrived in Canada in 1994 seeking asylum. He had been a second-year medical school student and, in his exodus, left behind six orphaned siblings. Lack of credential recognition in Canada forced Dr. Rubabaza to return to Grade 13 to complete his OAC credits. He graduated as valedictorian and Ontario scholar from high school and was admitted to the University of Ottawa on scholarship. He graduated with honours in Biochemistry and was later accepted into medical school. While studying, Dr. Rubabaza also worked to support his siblings. With the help of Casa El Norte, he was eventually reunited with his family of origin in Canada.

After many years of education and residency, Dr. Rubabaza is today a practicing obstetrician/ gynecologist. His patients describe him as a brilliant, caring, and compassionate doctor and a highly skilled surgeon. He is Secretary of the Medical Staff Society of Rouge Valley Health Care System, and Minimally Invasive Gynecology Lead and Continuing Medical Education Lead of the OBGYN group at Rouge Valley Health Care, Ajax Campus.

Dr. Rubabaza is also the recipient of numerous awards for his extensive community service and humanitarian efforts. Paying it forward, he has established several scholarships for students. Dr. Rubabaza is President of the Burundi-Canadian Professional Alliance. He is also President of the Black Physicians' Association of Ontario, an organization incorporated to establish a formal network of physicians committed to excellence in healthcare and improving the health and wellness of the black population of Ontario by addressing issues of health equity and cultural competency.

In meeting Dr. Rubabaza's family, one is struck by the strength, beauty, and wisdom of his wife, the giftedness of his children, and the loving bond that exists with his family of origin—all exceptional and talented people in their own right. Dr. Rubabaza signs his emails with a quote:

"Our struggle for freedom and justice was a collective effort... It is in your hands to create a better world for all who live in it."

– Tata Madiba Mandela

They are words he lives his life by. Creating a better world for all who live in it is something that Dr. Rubabaza does on a daily basis in both his professional and personal life and why he was selected as this chapter's Agent of Change.

Courtesy of Dr. Jean Placide Rubabaza

When Someone Asks … Doesn't Canada Already Do Enough to Help Refugees?

This is really a question about what Canada's role is in terms of its international obligations. The United Nations High Commissioner for Refugees in the 2015 Global Trends Report (2016) notes some of the following facts:

- Wars and persecution have displaced a total of 65.3 million people at the end of 2015; this is the first time the threshold of 60 million people has been crossed and was the *largest number of displaced persons in UNHCR history*.
- An unprecedented one in every 113 people globally is now either an asylum seeker, **internally displaced person**, or a refugee.
- The rate at which people are fleeing war and persecution has soared from 6 per minute in 2005 to 24 per minute in 2015.
- Three countries produce half the world's refugees: Syria at 4.9 million, Afghanistan at 2.7 million, and Somalia at 1.1 million together accounted for more than half the refugees under UNHCR's mandate worldwide.

- Colombia at 6.9 million, Syria at 6.6 million, and Iraq at 4.4 million had the largest numbers of internally displaced people.

Given the UNHCR's global statistics on world refugees, to suggest Canada needs stricter limits so that we do not end up hosting millions of refugees is not valid (see Table 8.3). In 2015, the Canadian government sponsored approximately

Protected person: Someone who, according to the Immigration and Refugee Protection Act, meets the definition of a Convention refugee and can also mean a person in Canada who, if they were sent home, would be tortured or at risk of cruel and unusual treatment or punishment.

Internally Displaced Persons: Persons forced to flee their home for safety but do not cross a border or leave their country; they seek safety in another location within their own country.

TABLE 8.2

Immigration Overview 2013–2015

Category	2013	2014	2015
Canadian Experience	7 209	23 783	20 059
Caregiver	8 799	17 689	27 230
Skilled Trade	17	139	1 972
Skilled Worker	83 230	67 596	70 145
Entrepreneur	426	499	259
Investor	8 407	7 450	5 460
Self-Employed	265	399	677
Start-up Business	0	9	62
Provincial Nominee Program	39 901	47 624	44 534
Economic	**148 254**	**165 188**	**170 398**
Sponsored Spouse or Partner	45 633	45 064	46 356
Sponsored Children	3 109	3 561	3 316
Sponsored Parent or Grandparent	32 320	18 203	15 489
Sponsored Extended Family Member	319	285	329
Sponsored Family Member - H&C Consideration	1 998	534	0
Sponsored Family	**83 379**	**67 647**	**65 490**
Government-Assisted Refugee	5 726	7 626	9 488
Privately Sponsored Refugee	6 330	5 070	9 746
Blended Sponsorship Refugee	153	177	811
Protected Person in Canada	11 930	11 197	12 070
Resettled Refugee & Protected Person in Canada	**24 139**	**24 070**	**32 115**
Humanitarian & Compassionate	3 194	3 333	3 799
Public Policy	29	15	0
Other Immigrants not included elsewhere	44	29	45
All Other Immigration	**3 267**	**3 377**	**3 844**
Total	**259 039**	**260 282**	**271 847**

Source: Government of Canada, Facts and Figures, 2015. Reproduced and distributed on an "as is" basis with the permission of Statistics Canada.

9488 refugees; private individuals and organizations sponsored 9746 refugees; and 811 refugees were blended sponsorships. Canada accepted a further 12 070 asylum seekers. In total, in 2015, Canada accepted 32 115 refugees. Canada is beginning to do more in terms of its international obligations to world refugees, especially with respect to Syrian refugee resettlement. But in comparison to other top hosting

TABLE 8.3

Top Host Countries for Refugees Worldwide in 2015

No.	Country	Refugees
1	Turkey	2.5 million
2	Pakistan	1.6 million
3	Lebanon	1.1 million
4	Islamic Republic of Iran	979 400
5	Ethiopia	736 100
6	Jordan	664 100

Source: UNHCR: The UN Refugee Agency. Global Trends: Forced Displacement in 2015, p. 3. Found at: http://www.unhcr.org/576408cd7.pdf.

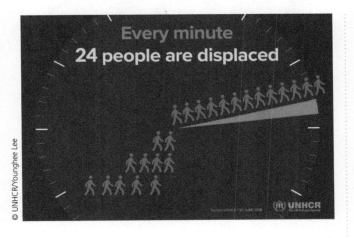

countries around the world, our numbers are small. As well, no one, including the UNHCR, is suggesting that all the world's refugees should be resettled in a third country, when **voluntary repatriation** and settlement of countries of first asylum are preferred.

When Someone Asks ... It's Nice That We Help, But Shouldn't We Look After People in Our Own Country First?

Canada has a legal obligation under international law as a signatory to the UN Convention on Refugee Protection (1951) and the 1967 Protocol. The spirit of the Convention and the Protocol suggests that all human beings have a right to seek asylum from persecution in other countries. As such, refugee protection is a universal responsibility for all of us as global citizens.

When Someone Asks ... If a Refugee Can Afford to Wear Nice Clothes and Drive a Car, Is That Person a Genuine Refugee?

Refugees are not **economic migrants**. Economic status has no bearing on refugee status. A refugee is someone who has a well-founded fear of being persecuted because of his or her race, religion, nationality, membership in a particular social group, or political opinion. It makes no difference to the granting of status whether a refugee is rich or poor—the point is that the person is at risk of, or has experienced, persecution. Many refugees are well-educated professionals in their own countries, such as doctors, nurses, judges, lawyers, professors, journalists, and so on.

When Someone Asks ... Why Can't These People Just Go to Refugee Camps?

This is not a question anyone might ask who has visited or worked in a refugee camp overseas or even participated in a simulation exercise of life for a refugee in a camp. The average length of time spent in a refugee camp is 17 years (UNHCR, 2010). Security of food and water in camps is unpredictable. Refugees living in camps are often not allowed to leave or work outside the camp.

Voluntary repatriation: The process of returning voluntarily to one's place of origin or citizenship.

Economic migrant: Person who moves to another country for employment or a better economic future.

When Someone Asks … Aren't There a Lot of Bogus Refugees Coming to Canada Abusing Our System? Shouldn't They Apply to Be a Refugee Before They Come to Canada?

Persons who seek protection onshore and are granted status are no less "genuine" than refugees who are resettled from offshore. Refugees who are resettled in Canada, regardless of whether they apply onshore or offshore, must meet the criteria for refugee status outlined in the UN Refugee Convention (UNHCR, 2010). A decision by the Supreme Court of Canada, referred to as the **Singh Decision**, requires all people seeking asylum in Canada to be provided with a full oral hearing of the merits of their claim (Immigration and Refugee Board of Canada, 2013). Unfortunately, the terms "bogus refugee," "phoney claim," and "abuser" are often used when governments are looking to make restrictive changes to the refugee determination process. Some of the reforms to this process throughout history have been positive and necessary. Unfortunately, the use of this language and rationale brands all asylum seekers indiscriminately.

When Someone Asks … Don't Refugees Bring Crime and Pose a Terrorist Threat to Canada?

The UN Refugee Convention excludes people who have committed war crimes, crimes against peace, crimes against humanity, or other serious non-political crimes from obtaining refugee status. All asylum seekers must undergo rigorous security checks before being granted protection. Applicants with a serious criminal record or considered to be a security risk because of suspected involvement in terrorism, organized crime, espionage, or human rights violations are not eligible for a refugee hearing in Canada. The Canadian Security Intelligence Service (CSIS) Security Screening program manages the *Refugee Claimant Screening Program*, providing security assessments on refugee applicants to Immigration, Refugees and Citizenship Canada (Canadian Security Intelligence Service, 2013). This program is designed to ensure that those who are inadmissible to Canada for security reasons under the Immigration and Refugee

Singh Decision: Landmark decision in refugee determination made by the Supreme Court of Canada in 1985; the decision protects the right of every refugee claimant to an oral hearing.

Lump of labour fallacy: The false belief that the amount of work available to labourers is a fixed amount and, therefore, there is no capacity to absorb more labourers into an economy.

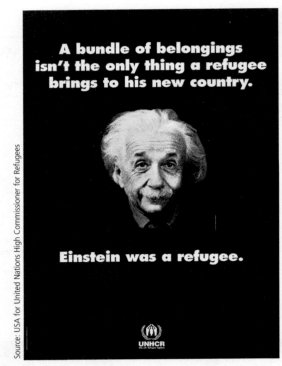

Source: USA for United Nations High Commissioner for Refugees

Can you imagine a world without the contributions of Albert Einstein, a Jewish refugee to America from Germany?

Protection Act are identified early and prevented from taking up residence in this country (Canadian Security Intelligence Service, 2013).

When Someone Asks … We Don't Have Enough Jobs for Canadians, So Why Are We Inviting Immigrants to Come In?

Whenever we experience hard times economically in Canada, the first target we generally look to are immigrants. The fear is that immigrants will come into the country and take jobs away from Canadians who have lived here all their lives. Economists refer to this as the **lump of labour fallacy**. The lump of labour theory suggests that there is only a fixed amount of labour within a country for a fixed number of workers, but economists know this is a fallacy. Immigrants make a positive contribution to the economy through the goods and services they buy and the taxes they pay. They also have an important contribution to make toward filling skilled labour shortages in Canada. The Canadian Council of Human Resources Associations explains:

> [R]esearch has shown that without immigration there would be only two ways to fuel the workforce: natural increase (more births than deaths) or movement from rural to urban

National Immigration Centre—The Conference Board of Canada

WHY IS IMMIGRATION IMPORTANT TO CANADA?

25% of Canada's population will be over 65 by 2035

5 000 000 Canadians set to retire by 2035

Canada's worker-to-retiree ratio TODAY

Canada's worker-to-retiree ratio in 2035

1.6 Canada's fertility rate, which is ranked **181st** globally, is well below Canada's replacement rate of **2.1**

Immigrants TODAY make up 65% of Canada's net annual population growth

Almost **100%** of Canada's net population growth will be through immigration by 2035

350 000 Estimated number of immigrants Canada will need annually by 2035 to meet its workforce needs

NOW **HIRING**

The Conference Board of Canada   Le Conference Board du Canada

Canada's acceptance of immigrants on humanitarian grounds demonstrates compassion, leadership and enhances Canada's global standing

IMMIGRANTS...

boost trade ties between Canada and the world

strengthen culture and diversity

are motivated, innovative and entrepreneurial

National Immigration Centre | conferenceboard.ca/NIC | @ImmigrationCBoC

areas. Both have levelled off in Canada in recent years, making the role of immigrants more important for the Canadian economy. (Canadian Council of HR Associations, 2008)

SETTLEMENT AND INTEGRATION

Defining Settlement

The need for a comprehensive, collaborative approach to the successful **settlement** of newcomers in Canada is something that private, public, and voluntary sectors agree on. The Ontario Council of Agencies Serving Immigrants (2000) defines the term *settlement* as "a long-term, dynamic, two-way process through which, ideally, immigrants would achieve full equality and freedom of participation in society and society would gain access to the full human resource potential of its immigrant communities." The Canadian Council for Refugees describes the settlement process as a continuum whereby newcomers move from **acclimatization**, to adaptation, to integration (see Figure 8.1).

Upon arrival in Canada, newcomers have a wide variety of issues to deal with. Learning or improving proficiency in English and/or French is one of the first steps in this process. The ability to communicate in one or both of Canada's official languages is integral to a newcomer's ability to navigate systems that include critical issues like finding housing and employment, enrolling children in schools, finding a doctor, and making new friends (see Figure 8.2). At the other end of this continuum is **integration**—a long-term

Settlement: The process of settling in another place or country to live; resettlement generally refers to newcomers' acclimatization and the early stages of adaptation. Settlement and resettlement are used interchangeably.

Acclimatization: The process of beginning to adapt to a new environment.

Integration: The long-term, multidimensional process through which newcomers become full and equal participants in all aspects of society. As part of this process, newcomers interact with the larger society and also maintain their own identity.

FIGURE 8.1

The Settlement/Integration Continuum

Acclimatization   Adaptation   Integration

Source: "Best Settlement Practices," Settlement Services for Refugees and Immigrants in Canada, Canadian Council for Refugees (February 1998).

FIGURE 8.2

Newcomer Needs and Challenges

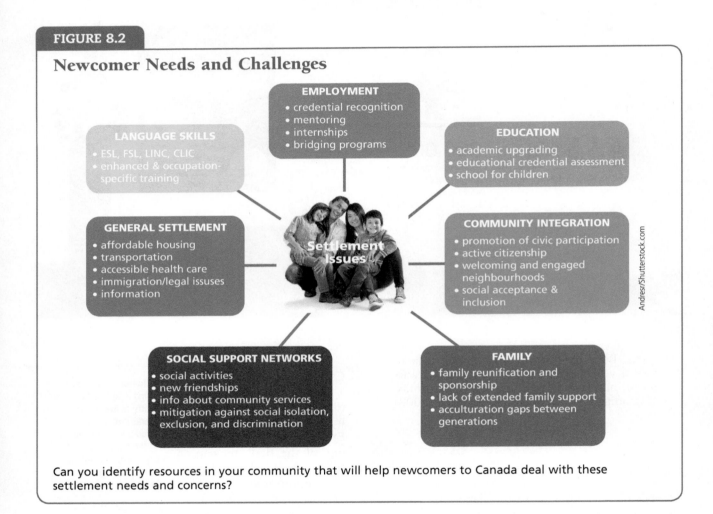

EMPLOYMENT
- credential recognition
- mentoring
- internships
- bridging programs

LANGUAGE SKILLS
- ESL, FSL, LINC, CLIC
- enhanced & occupation-specific training

EDUCATION
- academic upgrading
- educational credential assessment
- school for children

GENERAL SETTLEMENT
- affordable housing
- transportation
- accessible health care
- immigration/legal issues
- information

Settlement Issues

COMMUNITY INTEGRATION
- promotion of civic participation
- active citizenship
- welcoming and engaged neighbourhoods
- social acceptance & inclusion

SOCIAL SUPPORT NETWORKS
- social activities
- new friendships
- info about community services
- mitigation against social isolation, exclusion, and discrimination

FAMILY
- family reunification and sponsorship
- lack of extended family support
- acculturation gaps between generations

Andresr/Shutterstock.com

Can you identify resources in your community that will help newcomers to Canada deal with these settlement needs and concerns?

process whereby newcomers become full, equal, and active participants in the institutions of their new host country.

The model developed here illustrates a continuum of phases in the settlement process, but it is important to remember that settlement is a multidimensional process and each newcomer's experience is multifaceted and unique. Settlement does not occur at a similar rate across all aspects of a newcomer's life. Adaptation issues can arise long after newcomers arrive in Canada. Settlement in a new country is not a linear experience beginning with acclimatization and a guaranteed happy ending of successful integration for all newcomers.

Settlement as a Two-Way Process

Settlement is a reciprocal process; in other words, it is a two-way street in which the needs of Canadian society and the newcomer populations are intertwined. Through this two-way process, newcomers have the opportunity and freedom to actively participate in Canadian society and in return, Canada as a **host country**, gains access to the full human resource potential in its newcomer communities (Ontario Council of Agencies Serving Immigrants, 2000). Within this concept of settlement as a two-way process is the premise that the needs of individual **newcomers** correspond to the needs of our society for the mutual benefit of both, as illustrated in Table 8.4.

The goal of the settlement process is not the assimilation of newcomers. Part of the reciprocity of the settlement process involves societal changes as new groups enter and challenge the norms of mainstream culture (Ontario Council of Agencies Serving Immigrants, 2000).

Factors Affecting Settlement

Factors affecting settlement in a new country can include how receptive and welcoming a new culture is, a person's pre-migration experiences (whether that person voluntarily chose to move or was forced to move), the individual attitude and personality of the newcomers, and the support services available to help newcomers adjust.

It would appear that with some exceptions, Canada has generally become a more receptive and welcoming nation for newcomers. Perhaps we have learned from moments in our history as a nation when our policies were clearly exclusionary. At first, Canada moved ideologically from xenophobia to assimilation, telling new immigrants that they would now be welcomed but they would have to melt into the dominant culture—what was thought of as "Canadian" at that time. In 1947, for example, Prime Minister King stated in the House of Commons that "the people of Canada do not wish as a result of mass immigration to make a fundamental alteration in the character of our population" (Canadian Council for Refugees, 2000). Canada's receptivity to newcomers at that time is described by the former governor general of Canada, Adrienne Clarkson (2011, p. 32), in her recent book titled *Room for Us All*. She characterized Canada's receptivity as one of benevolent neglect—"so that people find themselves, sometime stumbling but without obstacles put in their way."

Host country: A country where representatives or organizations of another country co-exist, either because they have been invited by the government, or because an international agreement exists.

Newcomer: Blanket term used to refer inclusively to all immigrants and refugees who have recently arrived in the country to settle on a permanent basis.

Assimilation: Total re-socialization of individuals from one culture to another, such that traces of the former culture eventually deteriorate.

TABLE 8.4

Settlement: A Two-Way Process

Society Needs	Newcomer Needs
Labour Force	Jobs
Tenants/Homeowners	Housing
Skilled Workers	Accreditation/Training
Service Users	Health Care/Social Services
Stability/Harmony	Security
Leadership	Opportunities to Advance
Growth/Diversity	Growth
Contributions from Its Citizens	Participation
Congruence with Its Principles	Equality/Freedom
Peace	Self-Esteem, Happiness

Source: OCASI (2000) Immigrant Settlement Counselling: A Training Guide—Part 1.

As a result, many immigrants who arrived in Canada during the 1950s and 1960s relied upon their families and their ethnic communities to assist them with initial settlement, with the expectation that they would become self-sufficient as quickly as possible (Anzovino, 2009). In 1971, Canada implemented a new national policy called multiculturalism, and the celebration of our cultural diversity as a nation became one of the policy's underlying principles. But today, the Canadian government provides significant funding for the language training and settlement services necessary to support the settlement and integration process of newcomers to Canada. While years from now, we may have to apologize for some exclusionary aspects of recent refugee law and policy changes, the process of settlement has certainly evolved from one of benevolent neglect.

Both immigration and **emigration** are processes that often involve psychological processes of loss: family members may be separated and left behind, community ties and social networks are lost; and even language and systems of meaning used to communicate may be gone. The **push and pull factors** for migration can have implications for the settlement process. For refugees or protected persons who are forced to migrate, their settlement in Canada is often difficult. Some of their settlement challenges may stem from

> **Emigration:** Leaving one country with the intention of settling in another.
>
> **Push Factors:** Things that influence a person to emigrate.
>
> **Pull Factors:** Things that influence a person to immigrate.
>
> **Acculturation:** The process of adaptation to a new culture, whereby prolonged contact between two cultures begins to modify both cultures.

IN THEIR SHOES

If a picture can say a thousand words, imagine the stories your shoes could tell! Try this student story on for size – have you walked in this student's shoes?

Have you ever left your home, knowing it wasn't yours anymore, knowing that your identity was lost and that every action and reaction from that moment on was focused on the future that didn't exist in the place you called home—the fact that you would have to look forward, and turn away from family that you chose to leave behind to find a better life?

My story is not about feeling sympathy for me in any way. It is about relating to people who have gone through horrific events and survived not just physically but emotionally and spiritually.

When I was a little girl, I had a happy and worry-free life. Being the first-born grandchild on both sides of the family, I had been spoiled by most family members, especially my maternal grandfather, who saw me as the biggest joy in his life. We had many animals such as, pigs, chickens, and rabbits, and my grandfather even bought a horse just for me. Life had been pleasant as we enjoyed spending time with family all hours of the day—eating, having coffee, singing, and just loving life. Since my grandfather's siblings had built houses along a road in the more rural part of town with apartments for their children, I lived beside my cousins and there was not a day that I was friendless or alone.

All was well until one day when I was about 4 or 5 years old. I heard this sudden and forcefully loud siren that kept getting louder and louder each second. Because my dad and I had been in front of the house, I remember him quickly picking me up, running toward the closest window of the house, opening it, and throwing me in. He followed after. I didn't know what was going on. I had never heard a sound anything quite like that, and my dad seemed panicked, which wasn't like him at all.

Before we all knew it, we were living in silence during a brutal war between three different religious and political groups. We had to live quietly so that we would remain safe. With the electricity and water being shut off for months at a time and food being scarce, we had to eat what we could. My parents would regularly let themselves go hungry so that my sister and I could eat. They would pick insects out of small amounts of rice and boil the rice for us because it had been some of the only food we had. Sometimes, we had to share one egg and a glass of water between eight of us, or one bean of coffee just to remember the taste of coffee my family used to enjoy a few times daily when times were better. Since my grandfather was a man of many trades, he had a small butcher shop at the back of the house, which was underground. That was mostly where we hid because enemy soldiers lived in the daycare centre next to our home. The cellar in which we stayed in was rough and cold, sort of like a cave. If there was no candlelight, it was pitch black, and the feeling of being in there was somewhat unsettling. But we were safe there.

One day, when it was safe to go outside for a few minutes, I went upstairs and above the butcher

(Continued)

shop. My grandpa had a small barn at the top of the hill behind the house, and a slaughterhouse. When I opened the door, I saw the body of my horse hanging on giant hooks by its feet and with its head severed and laying on the floor. To most, this would have seemed malicious and evil, but we had to eat. Even at the age of 5, I knew it had to be done because he would have died of starvation anyway. I just remember not having any feeling at all.

My mom had worked in the town hospital as an X-ray technician before the war and especially during it. There were days when she would run to work because there would be a sniper on the roof of a tall building, shooting anyone and everyone below. It wasn't a world I was used to, and it was confusing trying to figure out what went wrong.

There was one night that I remember hearing a loud knock on the front door. I went to answer the door with my parents while my sister slept. There was a black jeep parked in front of our house with its lights shining on the house. Even though it was difficult to see who was in it, the minute one of the people got out of the jeep holding a gun, I knew that something bad would happen. They were enemy soldiers, dressed in camouflage, who had discovered that my mom worked at the local hospital, so they came for her. My mom, standing in her long, white nightgown, had to leave us. She had no choice. So my dad went with her because he knew that terrible things would happen to her if she made the wrong move, and he wanted to be there to protect her. They left me and my sister, and I had to stay in the upstairs apartment with our grandparents as we hoped and prayed that our parents would return.

They returned, but my mom would still be forced to spend days at the hospital, taking bullets out of soldiers, performing impossible surgeries, sleeping on the floor, and hoping to survive in the control of the enemy. Due to all of this chaos, my parents tried to plan an escape. They had tried hard to get out, but because the borders between cities had been blocked, it was difficult to get out, especially without being noticed.

Since my dad had been very ill and had to stay at the hospital for weeks, my mom used his hospital documents to arrange helicopter transportation to the coast of Croatia while she and my grandpa planned for my sister, my mom, and me to get out of the country with a transport truck driver. Although the plan had once failed for the three of us to leave, we tried again because my mom was determined to take us to safety. During that time, my dad left. I can never forget the moment when I had to say good-bye to him. I knew that I may not ever see him again and I remember screaming "Dad" with endless tears running down my cheeks. But I knew it had to be like that.

Eventually, my sister, my mom, and I left our town with a random truck driver. We didn't know who he was or if he was a good person or a bad person, but we had to make a decision in order to survive. This had just been another step toward possible freedom. As we travelled, my mom would play a game with us. She would say, "It's time for a game! Let's see who can keep their heads down and eyes closed for a longer time." We played this game a lot and my sister and I took pleasure in it. We liked games and my mom was always playing something with us, trying to make life more enjoyable, especially during that time. It wasn't until years later that I had realized why my mom would play games like that with us. She wanted to protect us from seeing the spew of dead bodies and limbs on the streets, which had recently been bombed and burned. I have always been grateful to her for that because she had tried to make life as innocent as she possibly could so that my sister and I could have "normal" lives after living in such circumstances.

We drove for the longest time to get to our destination. I remember the truck driver having to stop at each border of every city so that the soldiers guarding the cities could search the truck repeatedly. Most of the soldiers were not of the same religion as us and they searched for my dad. My mom had lied that he was still at home and hadn't come with us. Had he been with us, he would have been arrested and executed. That was the norm during the war.

We eventually got to Croatia in the middle of the night and the truck driver dropped us off in a random parking lot. There was one taxi cab there and luckily, we had gotten in it before anyone else, including an older lady with a colourful shawl around her head who was looking for her missing son. The driver drove us to the address where my dad was supposedly staying at, but at one moment, after driving for a while, he stopped the cab and told us that that was the farthest he could go and that we had to get out. So, we did. Unbeknownst to any of us, including the cab driver, he had dropped us off directly in front of the address my mom had given him. We then took the few bags that we had and knocked on the door. One of my dad's family members answered and invited us in. That was the day that I had been waiting for since we had absolutely no contact with my dad's family or my dad. My dad came out of a room and when I saw him, I ran up to him and jumped into his arms as he knelt down to catch me. It had been the happiest moment in my life. That had been the first time that I cried of happiness and I'll never forget it.

Soon after, we decided to move to Sweden as the country had been accepting refugees and many people from Yugoslavia headed there. We went by bus, which took quite a few days, and when we had finally arrived, we moved into a refugee camp where we met lots of amazing people who shared similar stories with us. Later on, we moved into an apartment building with roommates from the same country, and then a compound of apartments with only refugees. It was a

(*Continued*)

nice place to live and we enjoyed it there. It was clean, safe, and fun to finally be playing outside without being scared of getting shot.

Life in Sweden had been simple and pleasant. I liked going to school there with all of the other refugees. I became friends with people from my own country as well as from Albania and Somalia. The teachers there had been very kind and welcoming. They somehow knew what we had gone through and treated us with such care that my experiences from my life there will never escape my memories.

After about three years of waiting for permanent residency in Sweden, my parents and my sister and I decided to go back to Croatia and then plan our move to Canada where a lot of our relatives lived, including my dad's sister. We left Sweden in 1997, went to Croatia to arrange our move to Canada, and said our good-byes once again to family from our hometown in Bosnia that had driven all the way there to see us.

When we came to Canada, the transition for me, personally, had been difficult. It hadn't been like Sweden where I was in a class full of students who couldn't speak the language so we worked together to learn. I felt out of place here even though teachers had been compassionate and understanding. As a student here, I had a lot of anxiety when I went to school, especially when my parents would leave me. I didn't want to be separated from them. It scared me. I felt alone and overwhelmed in my classes not knowing the language.

Over time, I still hated school. I wasn't a great student as most people would assume. I sometimes studied hard only to receive a poor grade. There had been times when I would do very well and become proud of myself, but usually, moments like that were short-lived.

Although my efforts to do well in school weren't that pleasant for me because of my anxiety, which, by the way, caused me to cry almost every day in Grade 5, I had many friends who supported me and treated me nicely throughout my years here. I learned to speak English quite quickly because I had to use it in the class at all times even though I didn't have ESL classes as often as I would have liked, but teachers understood and always motivated me to try.

Throughout my life, I had started out being confident and happy, and when life had changed drastically and negatively, it was surreal and strange for me. When I had to transition to living in a war and then in a refugee camp, it had been possible for me, but yet another transition of moving to another new place had been more than enough for me and I had not known its harm at the time.

Today, I still live with moments of a lack of confidence, and sometimes a great amount of anxiety within new situations, but I believe it happens to everyone at times. We just have to learn to overcome it and when we do, it becomes easier to keep going.

I am a graduate student now and also work as an ESL teacher, and I sometimes share parts of my story with my students. Some of them find it hard to believe that I had learned English so quickly or that I had come to Canada as a result of a war, but it's what makes me who I am. I wouldn't change a thing about my life because it makes me stronger and more grateful for what I have now and for all of the education that I have had the opportunity to acquire while in Canada.

It can take a lot to make each of us truly happy and for us to realize it, but who we are individually is what makes us unique as long as we keep moving forward. We all have some type of past, but we cannot continue living in it, for it will drag us down. So, try to take every opportunity that you get because you live in one of the best countries in the world.

Martina

---

pre-migration experiences that result in **acculturative stress** and may include the effects of war, the effects of torture, the loss of loved ones, and **post-traumatic stress disorder (PTSD).** Immigrants who voluntarily migrate to Canada by choice, on the other hand, have time to prepare before their arrival. They usually have the opportunity to sell their homes, say goodbye to family members, and bring all of their documents and belongings.

Acculturative stress:
The stress experienced by immigrants when there are difficulties resulting from the acculturation process. The stress of this process increases when immigrants experience discrimination, language difficulties, and incongruences in non-material aspects of culture.

Choice and the opportunity for closure are two factors that can help to mitigate against the feelings of loss.

Many believe that the stress of the settlement process in Canada can affect the health of newcomers. There is a complex relationship between migration and health, but research generally supports the conclusion that while recent immigrants' health is generally better than that of people who are born in Canada (known as the **healthy immigrant effect**), this declines as their years in Canada increase. It is believed that the difficulties of settling and integration into Canada contribute to this phenomenon (Ng, 2011).

The community-based settlement sector in Canada is comprised of not-for-profit agencies and support services whose mandate is to help newcomers with settlement issues. Settlement counsellors provide services that may include assistance finding a job or training programs, help filling out forms and applications, interpretation and translation of documents, referrals to community services, language classes, healthcare, and so on. Most agencies have mechanisms in place to ensure newcomers participate in all aspects of the operation, including service delivery, management, and governance (Ontario Council of Agencies Serving Immigrants, 2000).

CONTEMPORARY MIGRATION ISSUES IN CANADA

Syrian Refugee Crisis and Canada's Response

The civil war in Syria, which began in 2011, has resulted in a global refugee crisis as Syrians flee their homes and country. As of August 2016, the United Nations High Commission for Refugees (2016) reported 4 808 229 registered Syrian refugees and approximately 8.7 million internally displaced persons as the crisis continues. As of June 2016, the UNHCR reports that the number of refugees seeking safety in neighbouring countries of Turkey, Lebanon, Jordan, Iraq, and Egypt continues to grow.

Canada expects to welcome in total around 44 000 refugees this year: 25 000 government-sponsored refugees and the rest a mix of privately sponsored and blended visa cases (Friscolanti, 2016). "Driven to act by a single heart-breaking image—a three-year-old boy dead on a beach—Canadians rallied in droves to #WelcomeRefugees, as the hashtag said, offering safe haven to thousands …"(Friscolanti, 2016). Resettled Syrian refugees in Canada describe overwhelming feelings of personal safety that bring relief and gratitude for the ability to sleep again, knowing they will survive, have food to eat, and a place to call home (Anzovino, personal interviews, August 22, 2016). In the face of anti-immigrant sentiments around the world, Canada has "literally redrawn the blueprint for how a western country can respond, in rapid time, to a refugee crisis halfway around the world" (Friscolanti, 2016). Warm and welcoming hearts will bolster the initial period of elation that Syrian refugees may experience during the first six months in Canada, but there will be many ongoing and difficult settlement issues in the months and years ahead, including access to language classes, income support, cultural differences, mental health issues, and family reunification (Friscolanti, 2016).

Temporary Foreign Worker Program

The use of temporary foreign workers in Canada is facilitated by two programs: the Temporary Foreign Workers (TFW) Program and the International Mobility Program. In 2014, 95 086 individuals were admitted to Canada under the TFW Program, and 197 924 under the International Mobility Program (Government of Canada, 2016b).

Some suggest that the use of temporary workers is simply a method of importing cheap labour to Canada, and that it disadvantages Canadian workers and undermines the long-term contributions made to the Canadian economy by economic class immigrants. There have been many questions about the increased use of this program in Canada. Is the increased use of temporary foreign workers the human resource strategy needed to fill skilled labour demands of the Canadian workplace? While temporary foreign workers may fill immediate skill shortages, is this a short-sighted strategy in terms of the business investment in human capital and retention of skilled employees in Canada? (Canadian Council of Human Resources Associations, 2008).

Post-traumatic stress disorder (PTSD): A psychological disorder caused by a traumatic event involving actual or threatened death or serious injury to oneself or others; symptoms may include anxiety, depression, survivor guilt, sleep disturbances, nightmares, impaired use or loss of memory, concentration difficulties, hyper arousal, hypersensitivity, suspiciousness, fear of authority, and paranoia.

Healthy immigrant effect: Immigrants' health is generally better than the health of those born in Canada, but this declines as their years spent living in Canada increase.

Prime Minister Justin Trudeau greets newly arrived Syrian refugees.

The photo of three-year-old Alan Kurdi, having drowned at sea with his brother and mother, went viral in social media and sent shock waves around the world. Alan's image became an icon of the Syrian refugee crisis, and people and governments were moved from feelings of intolerance and indifference to action. In the words of Alan's aunt, Tima Kurdi, "It is very painful to go through this tragedy, but in other ways, we are so proud of this picture [which] saved thousands of refugees," she says (Devichand, 2016). In the wake of this photo, Canada responded with a pledge to resettle 25 000 Syrian refugees. Do you think Canada is doing enough to fulfill our international obligations to Syrian refugees?

One of the ways in which the government attempted to address these concerns was through the development of the Canadian Experience Class program where work permit holders in the **Temporary Foreign Worker Program** and **International Mobility Program** could now apply for permanent residence in Canada. In 2014, there were 46 520 persons who transitioned from these two programs to become permanent residents of Canada (Government of Canada, 2016b). The other way in which concerns regarding job security for Canadian employees was addressed was through the use of a tool called the **Labour Market Impact Assessment,** which allows an employer to fill acute labour shortages on a temporary basis through the Temporary Foreign Worker Program, only when there are no qualified Canadians or permanent residents available to do so.

Credential Recognition for Immigrant Professionals

One of the biggest frustrations for immigrants remains credential recognition and employment in Canada. We invite highly trained immigrant professionals into Canada each year, only to deny them employment because their credentials are not recognized or equivalent here. This frustration is very palpable within immigrant communities.

The Canadian Council of Human Resources Associations (2008) notes that Canada's population and its workforce are undergoing demographic changes that will impact employers. We have a significant portion of our workforce made up of aging baby boomers who are getting close to retiring. There

Temporary Foreign Worker Program: A program for persons wanting to work temporarily in Canada. Work permits and Labour Market Impact Assessments are required to be allowed to work in Canada under this program (Government of Canada, 2016).

International Mobility Program: A program that allows employers to hire temporary workers without a Labour Market Impact Assessment.

Labour Market Impact Assessment: Needed for temporary foreign workers to be granted work permits; assesses impact of hiring on the Canadian labour market.

are fewer young people in the workforce because of a declining birth rate (CCHRA, 2008). The demand for more highly skilled and educated workers is increasing at the same time as we have skilled immigrants arriving in Canada who have higher levels of education and international expertise (CCHRA, 2008). Despite labour market demands, one of the most significant challenges faced by internationally trained immigrants is finding employment (OCASI, 2012).

This disconnect is highlighted in a series of television advertisements titled *Recognize Immigrant Credentials*, which dramatize this question: If Canada is a land of opportunity, then why is a doctor driving a cab, an MBA cleaning offices, and an engineer serving fast food? The individuals featured in these television advertisements have presumably received the maximum points allotted through the Federal Skilled Workers Program for their educational qualifications and employment experience. Yet the likelihood that a new immigrant will work in his or her chosen profession in Canada is a dream that generally takes many years and much effort to realize. The cliché of overqualified immigrants driving taxis is supported to some extent in a study entitled *Who Drives a Taxi in Canada?* Taxi drivers do include internationally trained physicians, architects, engineers, and other highly trained professionals (Xu, 2012).

ENDING THOUGHTS

Some of the world's greatest controversies today surround human migration and the inclusion and exclusion of migrants. These controversies are linked to security concerns around human trafficking, terrorism, and border control. What often gets lost in dialogue are the social, cultural, and economic benefits of immigration, our humanitarian ideals, and our international legal obligations.

Given Canada's demographic pressures resulting from a declining **birth rate**, an aging population, and resulting labour shortages, we will need to continue to look to immigration as a durable solution for economic prosperity. At the same time, Canada will need to continue to support solutions to resolve systemic barriers resulting in the underemployment and unemployment of immigrants. If we want to move forward as socially responsible global citizens, we will also have to critically examine the "brain drain effect" of our immigration policies. At the top of the humanitarian response agenda will be the continued support of refugees and asylum seekers, as the numbers of persons displaced and in need of protection is at an unprecedented number.

> **Birth rate:** The number of live births per year per 1000 population.

READING

THE THINNEST LINE
WHEN DOES A REFUGEE STOP BEING A REFUGEE?

By Beth Gallagher

> "The line that separates me now, an assistant professor at a university living an incredibly comfortable life, from a body on the beach is the thinnest line imaginable."
>
> VINH NGUYEN, professor of English, University of Waterloo

There's a childhood family photo of Vinh Nguyen on his seventh birthday. The party table is loaded for a celebration with spring rolls, fried noodles, jelly, and raisins— but Nguyen is not smiling.

You might assume it's because the boy in the photo is celebrating his birthday in the Ban Thad refugee camp in Thailand, but that wouldn't begin to capture the depth of the loss he felt that day. Nguyen, now an English professor at the University of Waterloo, was one of the three million people who fled Vietnam, Cambodia, and Laos between 1975 and 1995. His father had survived years of war and imprisonment in Communist "re-education" camps, only to die while fleeing in a small boat — something Nguyen learned through bits of overheard conversation as his family gathered to celebrate his seventh birthday in Ban Thad.

"There was never a single moment where I was told about my father's death," says Nguyen. "Everything I learned was through fragments of conversations that I overheard."

"I think there's a connection between that kind of activity and the kind of work I do now. It's very similar, in that literary scholarship asks you to read between the lines, to look for meaning, to fill in the gaps—to put pieces together."

Today, Nguyen, also a professor of humanities and East Asian studies at Renison University College, researches the life stories of refugees. In part, it's because his own life experiences are not always reflected in conventional refugee narratives found in novels, poetry, or film.

"So much of my work is trying to enlarge the understanding of who refugees are and, really importantly, to look at their power, even when they are in dire need," says Nguyen. "They are people who make decisions and have very complex lives, backgrounds, and desires."

One of his research projects looks at how former Southeast Asian refugees in Canada are stepping up to support refugees fleeing the war in Syria. Outside his role as professor, Nguyen is also co-founder of Southeast Asian Canadians for Refugees, an advocacy group that focuses on increasing awareness of the current refugee crisis.

While the photo last fall of Alan Kurdi, a three-year-old Syrian child whose body washed up on a Turkish beach, shocked the world, it had particular resonance for millions of former refugees, who risked their lives and lost loved ones in desperate attempts to gain asylum during war.

"That photo struck a chord in the Vietnamese-Canadian community. So many people saw themselves in that picture. I surely did," says Nguyen. "The line that separates me now, an assistant professor at a university living an incredibly comfortable life, from a body on the beach is the thinnest line imaginable."

"It could have ended so differently, and it's a difference nobody can explain. It was a ripple in the ocean that capsized the Kurdi family boat, and a ripple in the ocean that didn't happen for me."

VINH NGUYEN

Nguyen was just five years old when he, along with his mother, brother, and two sisters, fled Vietnam: "I remember being stopped by pirates, once, for sure," he recalls. "I remember at one point we had to go through what I can only describe as quicksand. Part of the journey was through the jungle. The swamp was sucking people in and I remember my mother just gave me to a random man to carry me across. Then, we hid in a shed for a long time before actually getting to a camp. I do have fragments of those memories."

Nguyen is hoping to both challenge and illuminate commonly held notions of what it means to be a refugee and to also examine how refugee narratives shape our understanding of history, nationhood, and citizenship. Ultimately, his work is an attempt to begin to answer the question: When does a refugee stop being a refugee?

The question hit home on a very personal level one day while he was researching the life stories of Southeast Asian refugees in an archive in California. Nguyen opened a random folder with yellowed newspaper clippings and there, unmistakably, was a 1988 news photo of his mother in a Thai refugee camp. She is the only one looking directly into the camera, and she is smiling broadly.

"It was literally the second file that I opened, not even an hour in on my first day in the archive. I was just searching blindly. It felt really surreal; just the shock of identifying yourself or your personal history in a larger history," says Nguyen.

"There was something powerful about finding that photo of my mother. It was also very reassuring," he says.

With more than 50 million people around the world fleeing their homes due to war and other crises, part of Nguyen's work lies in an emerging field in literary and cultural studies known as "affect theory." He is examining refugee narratives in novels, films, and other media to see how emotions go beyond the personal and the psychological to tell us something about the historical and political.

While the recent arrival of 25 000 Syrian refugees in Canada is unprecedented in some aspects, Marlene Epp, a University of Waterloo history professor who also teaches at Conrad Grebel University College on campus, says it's important to understand the current crisis in the context of other displaced persons that shape Canada and the world.

"Vinh Nguyen's work on refugees, diaspora, and affect is important, not only because his personal life history shapes his scholarship, but because it provides broader themes within which to think about current events," says Epp, who researches the history of ethnicity and immigration in Canada. Epp is teaching a new course in Conrad Grebel's Peace and Conflict Studies program called Refugees and Forced Migration.

In another research project, Nguyen asks how the "gratitude" expressed by refugees has been used throughout history as evidence for the justness of war. "Gratitude is the primary thing refugees are supposed to feel. We as a nation want to hear that; it's very easy for us to hear, but how is the nation itself being constructed through these feelings?"

Refugee gratitude in Canada can embolden our national identity, providing evidence of a compassionate country, says Nguyen. "The immigrant success story is so central to the Canadian nation."

He says refugee gratitude becomes more problematic when it's used by governments as human evidence of the appropriateness of going to war. In particular, Nguyen questions how stories of the gratitude of Southeast Asian refugees who sought asylum in the United States after the Vietnam War was used by Americans during the height of public debate about the war. "That's why gratefulness can be dangerous," says Nguyen. "It can be used by other powers for different narratives."

Society's appetite for refugee gratitude, says Nguyen, often leaves no room for other emotions like bitterness, resentment and anger.

"This compulsion to be grateful is really powerful," says Nguyen. "Anything else is off-script."

Nguyen's work also examines how refugee narratives are often restricted to trauma, when in fact, refugees, like most people, lead richly layered lives.

While trauma marks Nguyen's story, he says other forces, like his adolescence in Calgary, teaching English in Japan, and backpacking through Asia as a young adult, were, for him, equally as formative. He points out that former Vietnamese refugees helping Syrians challenge the stereotype of refugees being forever hopeless, helpless figures without rights.

Nguyen stresses while he now shares his own story and researches other refugee narratives, there are others who choose not to share their stories because it's too painful and still others who are never asked. "I think childhood protected me, in a way, not always having the information that my mother had," he says. "I look back on those days and there were some very bad things, but my mother always made me feel loved. I've always felt like I had a good childhood."

Which brings Nguyen back to the photo of his seventh birthday, the first birthday he remembers, and the loved ones not seated at the celebration table: his mother, a grieving wife hoping to distract her son from pain with a refugee camp birthday party, and his father.

"I write so that those who didn't survive aren't erased from memory. In many ways, my story is a refusal to let them go."

Source: Gallagher, Beth. (2016, Spring). "The Thinnest Line: When Does A Refugee Stop Being A Refugee?" *University of Waterloo Magazine.* Reprinted with permission.

DISCUSSION QUESTIONS

1. Why is the "refugee narrative" important according to Professor Vinh Nguyen? Why might it be important to create public space for self-representation whereby refugees can tell their own stories?

2. Do you think refugees should be grateful to Canada when they are accepted to live here permanently? According to Professor Vinh Nguyen, what are some of the problems associated with "refugee gratitude"?

3. "The line that separates me now, an assistant professor at a university living an incredibly comfortable life, from a body on the beach is the thinnest line imaginable." What does Professor Nguyen mean by the "thinnest line imaginable"? When does a person who has been a refugee ever feel like they stop being a refugee?

KWIP

KNOW IT AND OWN IT: WHAT DO I BRING TO THIS?

The "K" in the KWIP process involves examining aspects of your own identity and social location as the first step in becoming diversity competent.

ACTIVITY: JOURNAL

As a non-Indigenous student, you share a common history as an immigrant to Canada. Understanding this aspect of who you are may help you to understand and respect the migration experience of others. If you are an Indigenous student, you may want to begin by examining your ancestors' history since inviting others to live on your land. In this journal, you are asked to reflect upon your experience based on your identity as either an Indigenous student or a non-Indigenous student.

As a non-Indigenous student, you have a personal history of immigrating to Canada. From what country or countries did your family immigrate to Canada? What generation within your family immigrated here? Was it your great-grandparents, grandparents, or parents? Are you first generation Canadian?

Try to pinpoint the period in immigration history when they came to Canada. What was going on during that time? What were some of the push and pull factors that contributed to the decision to immigrate to Canada? What was the journey to Canada like? Can you describe the first day arriving in Canada? What were some of the opportunities and challenges for yourself and/or your family in resettling in Canada?

As you reflect on what it was like for you or your ancestors coming to Canada, can you begin to identify with the experiences of others as immigrants? As an Indigenous student, how can you continue to teach the non-Indigenous ancestors who came to live on your land respect for nations bound together but separate?

WALKING THE TALK: HOW CAN I LEARN FROM THIS?

The "W" in the KWIP process presents a scenario or case study that challenges you to "walk the talk" through problem-based learning.

ACTIVITY: CASE STUDY

Keicha is a young woman who arrived in Canada as a refugee from the Democratic Republic of Congo. She lives in Canada at the married students' residence at the local university. She has two children living in Canada with her whom she parents alone. Her son is 4 years old and in daycare full time. Her daughter is 10 years old and is in Grade 4 at the local elementary school. Another student living in residence has made a complaint to child protection services that Keicha's 10-year-old daughter is home alone after school every night without adult supervision.

You are one of the few people Keicha knows in Canada and she calls you to tell you she has a problem and asks if you could come over to her apartment right away. You agree. When you arrive, there is someone from child protection services already there with Keicha. The worker asks you if you are "Canadian" and if you could answer some questions for Keicha. You explain that you are her friend and have been asked by Keicha to come, and that you will do what you can to help. You know that Keicha reads, writes, and comprehends English with good proficiency, but her oral communication skills are more limited.

Throughout the interview, the worker looks at you directly and makes no eye contact with Keisha. She refers to Keisha in the third person, making statements like "she doesn't seem to understand" and "the mother must understand Canadian law." You suggest that the agency may want to call for an interpreter and are told that the agency doesn't have money for things like that. You can sense that Keisha is frightened because she comes from a country where any attention from "the authorities" is cause for grave concern, since the "law" usually acts to injure, imprison, or jail.

The worker asks Keisha a series of yes and no questions from a paper that she has and without looking at her, checks off the boxes on the form. Keisha is terrified by the warnings that her children could be taken away and therefore responds yes to all questions asked. At this time, the worker turns to you and says, "see, she speaks English just fine." The worker continually repeats the seriousness of this situation (all in the third person) but intersperses her warnings with statements of how healthy the children looked for being refugees.

The worker leaves, recommending that Keisha thank the Canadian student mother who had called the authorities and taken her child in—otherwise she would have been placed in temporary foster care. Keisha sits very still, eyes blinking but otherwise nonresponsive.

1. Can you identify some of the communication issues that were problematic in this case study?

2. Can you recommend alternative ways of dealing with the child protection issues in this case?

Create a standard of practice based on this case study. Adapted from *A Shared Experience: Bridging Cultures, Resources for Cross-Cultural Training*. London: London Cross Cultural Learner Centre.

Source: Adapted from *A Shared Experience: Bridging Cultures, Resources For Cross-Cultural Training*. London: London Cross Cultural Learner Centre.

IT IS WHAT IT IS: IS THIS INSIDE OR OUTSIDE MY COMFORT ZONE?

The "I" in the KWIP process requires you to honestly confront and identify ways in which our complex identities result in experiences of privilege and oppression, and to reflect on how we can learn to honour that privilege.

ACTIVITY: GOT PRIVILEGE?

Don't be surprised if you find that you have been challenged to move outside your comfort zone in this chapter. Immigration is about the power to decide who gets to live where. If you don't think immigration is about power, then try imagining a world without borders where you could move anywhere and live any place you wanted at any time—no visas, no applications, no hearings. The words "undocumented immigration" and "illegal migrants" would cease to exist. We ask you to examine yourself in terms of the unequal distribution of power and privilege based on immigration status.

ASKING: Do I have privilege?

1. I can travel freely to almost any country.

2. People generally assume I can communicate well in English or French.

3. When I apply for employment, my legal right to work in Canada is not questioned.

4. People do not assume I am poor because of my country of birth.

REFLECTING: Honouring Our Privilege

Describe two circumstances in which you feel disadvantaged because of your migration history. Then describe two circumstances in which your migration history gives you privilege over someone else. How can you use the advantages you experience to combat the disadvantages experienced by others?

PUT IT IN PLAY: HOW CAN I USE THIS?

The "P" in the KWIP process involves examining how others are practising equity and how you might use this.

ACTIVITY: CALL TO ACTION

Cultural anthropologist Margaret Meade is quoted (Lutkehaus, 2008) as saying, "Never doubt that a small group of thoughtful, committed citizens can change the world; indeed, it's the only thing that ever has."

In 1984, a small group of thoughtful, committed citizens in a small border town in Ontario were called to action. The ripple effect of their action has been felt by asylum seekers across Canada and beyond our borders. It has been felt for over 30 years, and its legacy continues on today in the work of an organization known as Casa El Norte.

The story begins with a boy named Isaac who arrives at the Peace Bridge in 1984 with his father. Both were from El Salvador and were seeking safety in Canada. Survivors of a massacre in their village, Isaac had been shot and then bayonetted in the back, causing a spinal cord injury that left him unable to walk. After a long and arduous journey to the border, neither father nor son had a place to stay in Canada.

Later that evening, a small faith-based group of people had gathered and were looking at ways in which their faith might inspire social change. Within the group was a very compassionate immigration officer, Vince Buklis, who had met Isaac. He shared his story. Another group member, Gerard Mindorff, himself the father of 11 children, turned to the others and suggested that this was the time and opportunity to act. That was the beginning of a volunteer social movement that today has helped tens of thousands of asylum seekers in the past 25 years.

Isaac was the first of thousands of stories yet to come. There were human stories of great tragedy, human stories of great resilience, and human stories of great sacrifice. In response, this small group of volunteers went to work—without pay. They worked closely with a very inspiring, dedicated, and caring group of Immigration officers who would call members of their "bridge committee" when refugees arrived at the border and had no place to go in Canada. The calls came at all times of the day and night. Charlene and Bob Heckman were the first members of this group to host a family in their own home. Other group members, including Lynn and Pat Hannigan, Gerard and Betty Mindorff, Pat and Patricia Anzovino, Roderick and Laurie McDowell, Doris and John Lapp, Sid and Hedy Surtel, Charlie and Theresa Jones, Michael Hamman, Lorraine Clemens, and Frank and Pauline Orendorff made room in their homes, gathered furniture and clothing, set up apartments, helped children enrol in school, and helped with legal needs. The numbers of asylum seekers needing help grew.

One day a young woman arrived at the border from Guatemala. In her hand was a crumpled-up piece of paper with a name and phone number on it that was given to her in Mexico. She was told that when she got to Canada, this woman would help her. On the paper was the name of a member of this group. Word of the group's volunteer help was spreading, and as the need grew, the group reached out to other host families in the community and across the region and province. The Sisters of St. Joseph in Peterborough donated money to open a reception home for refugees, and it still operates today as Casa el Norte. The home is run by one of the original members of the group, Lynn Hannigan, with support from the Sisters of St. Mary of Namur, the Sisters of St. Joseph, the Holy Cross Fathers, and the Sisters of the Holy Cross, as well as a large contingent of volunteers and students on placement from Niagara College.

Once upon a time … there was a boy named Isaac who came to Canada so that he and his father might live. He met a group of ordinary people who did extraordinary things, who began a social movement to help refugees that would spread and connect from Fort Erie to Niagara, throughout Ontario, Canada, and beyond.

Located at www.nelson.com/student

Study Tools
CHAPTER 8

- Review Key terms with interactive **flash cards**
- Check your Comprehension by completing **chapter review quizzes**
- Gauge your understanding with *Picture This* and accompanying short answer questions
- Develop your critical thinking/reading skills through compelling **Readings** and accompanying short answer questions
- Apply your understanding to your own experience with **Connect A Concept** activities
- Evaluate Diversity in the Media with engaging *Video Activities*
- Reflect on your Understanding with *KWIP* activities

REFERENCES

Abella, I. (2008). None is too many: Canada and the Jews of Europe 1933–1948. Toronto: Key Porter Books.

Anzovino, T. (2009). A national movement begins with one man named Isaac. World Cafe Presentation. Fort Erie, Ontario.

Beah, I. (2007). A long way gone. Vancouver: Douglas and McIntyre.

Benson, J., Haris, T. A., & Saaid, B. (2010). The meaning and the story: Reflecting on a refugee's experiences of mental health services in Australia. Mental Health in Family Medicine, 7(1), 3–8.

Bissoondath, N. (1994). Selling illusions: The cult of multiculturalism in Canada. Toronto: Penguin Books Canada.

Bouchard, G. (2007). The Canadian immigration system: An overview. Retrieved from Institute for Research on Public Policy: http://archive.irpp.org/miscpubs/archive/bouchard_immig.pdf

Canadian Council for Refugees. (1998). Best settlement practices. Retrieved from http://ccrweb.ca/bpfinal.htm

Canadian Council for Refugees. (2000). A hundred years of immigration to Canada 1900–1999. Retrieved from Canadian Council for Refugees: http://ccrweb.ca/en/hundred-years-immigration-canada-1900-1999

Canadian Council of Human Resources Associations. (2008). Integrating new Canadians into Canada and the workplace: Maximizing potential–white paper. Retrieved from National Forums: http://www.ccarh.ca/uploadedFiles/Content_-_Primary/National_Forum/CCHRA_White_Paper_FINAL.pdf

Canadian Security Intelligence Service. (2013, June 11). Security screening. Retrieved from CSIS: http://www.csis-scrs.gc.ca/prrts/scrt-scrnng-eng.asp#bm05

Canadians for Redress. (2002). History of Chinese head tax. Retrieved from Canadians for Redress: http://www.ccnc.ca/redress/history.html

CBC. (2011). Life after Auschwitz. Retrieved from CBC Digital Archives: http://archives.cbc.ca/war_conflict/second_world_war/topics/1579-10644/

CBC News. (2000, November 6). Canadian clergy apologize to 'Voyage of the Damned' survivors. Retrieved from CBC News Canada: http://www.cbc.ca/news/canada/story/2000/11/06/holocaust001106.html

Citizenship and Immigration Canada. (2006a, July 1). Forging our legacy: Canadian Citizenship and Immigration, 1900–1977. Chapter 1. Retrieved from Citizenship and Immigration Canada: http://www.cic.gc.ca/english/resources/publications/legacy/chap-1.asp

Citizenship and Immigration Canada. (2006b, July 1). Forging our legacy: Canadian Citizenship and Immigration, 1900–1977. Chapter 6. Retrieved from Citizenship and Immigration Canada: http://www.cic.gc.ca/english/resources/publications/legacy/chap-6.asp#chap6-3

Citizenship and Immigration Canada. (2009, May 6). Sponsoring your family. Retrieved from Citizenship and Immigration Canada: http://www.cic.gc.ca/english/immigrate/sponsor/index.asp

Citizenship and Immigration Canada. (2010, September 9). Immigration overview: Permanent and temporary residents. Retrieved from Citizenship and Immigration Canada: http://www.cic.gc.ca/english/resource/statistics/facts2009/permanent/01.asp#category

Citizenship and Immigration Canada. (2013, August 23). Refugees. Retrieved from Citizenship and Immigration Canada: http://www.cic.gc.ca/english/refugees/

Citizenship and Immigration Canada. (2013a, August 1). Facts and figures 2012 – Immigration overview: Permanent and temporary residents. Retrieved from Citizenship and Immigration Canada: http://www.cic.gc.ca/english/resources/statistics/facts2012/permanent/02.asp

Clarkson, A. (2011) Room for us all. Toronto: Penguin Books Canada.

Conference Board of Canada. (2011). How Canada performs: A report card on Canada. Retrieved June 23, 2011, from Conference Board of Canada: http://www.conferenceboard.ca/e-library/abstract.aspx?did=4423

Danelius, H. (2002, December 18). Convention against torture and other cruel, inhuman or degrading treatment or punishment. Retrieved June 14, 2011, from Audiovisual Library of International Law: http://untreaty.un.org/cod/avl/ha/catcidtp/catcidtp.html

Devichand, M. (2016, January 02). Alan Kurdi's aunt: 'My dead nephew's picture saved thousands of lives.' Retrieved from BBC Trending: http://www.bbc.com/news/blogs-trending-35116022

Friscolanti, M. (2016, August 22). Warm hearts, cold reality. MacLean's Magazine, 129(32 & 33), 24–27.

Gallagher, B. (2016, Spring). The thinnest line: When does a refugee stop being a refugee? University of Waterloo Magazine.

Government of Canada. (2016a). Six selection factors – Federal skilled workers. Retrieved from Citizenship and Immigration Canada: http://www.cic.gc.ca/english/immigrate/skilled/apply-factors.asp

Government of Canada. (2016b, March 08). 2015 annual report to Parliament on immigration. Retrieved from Immigration, Refugees and Citizenship Canada: http://www.cic.gc.ca/English/resources/publications/annual-report-2015/index.asp

Government of Canada. (2016c). Determine your eligibility – Immigrant investor venture capital pilot program. Retrieved from Citizenship and Immigration Canada: http://www.cic.gc.ca/english/immigrate/business/iivc/eligibility.asp

Government of Canada. (2017). Facts and figures 2015 – Immigration overview: Permanent residents. Retrieved from Citizenship and Immigration Canada: http://open.canada.ca/data/en/dataset/2fbb56bd-eae7-4582-af7d-a197d185fc93?_ga=1.244793420.83203155.1486166786

Immigration and Refugee Board of Canada. (2013, July 31). Immigration and Refugee Board of Canada's twentieth anniversary: The landmark Singh decision. Retrieved from http://www.irb-cisr.gc.ca/Eng/NewsNouv/NewNou/2009/Pages/singh.aspx

Immigration and Refugee Protection Act. (S.C. 2001, c 27). Retrieved from Justice Laws Website: http://laws-lois.justice.gc.ca/eng/acts/1-2.5/index/html

Immigration Watch Canada. (2011, April 15). The leaders' immigration debate. Retrieved June 2, 2011, from Immigration Watch Canada: http://www.immigrationwatchcanada.org/2011/04/19/april-15-2011-the-leaders-immigration-debate-how-things-went-downhill-fast/

Iroquois Indian Museum. (n.d.). What is wampum? Retrieved from Iroquois Indian Museum: http://www.iroquoismuseum.org/ve11.html

Kazimi, A. (2012). Undesirables: White Canada and the Komagata Maru. Toronto: Douglas & McIntyre.

Komagata Maru incident [Gurdit Singh with passengers]. (1914). Retrieved from http://komagatamarujourney.ca/node/14661

London Cross Cultural Learner Centre. (1983). A shared experience: Bridging cultures, resources for cross-cultural training. London: London Cross Cultural Learner Centre.

Lutkehaus, N. C. (2008). Margaret Mead: The making of an American icon . Princeton, NJ: Princeton University Press.

Malenfant E. C., Lebel A., & Martel L. (2010). *Projections of the diversity of the Canadian population, 2006–2031* (Statistics Canada, Catalogue 91-551-X). Ottawa: Statistics Canada.

Matas, D. (1985). Racism in Canadian immigration policy. Refuge, v5, n2.

National Association of Japanese Canadians. (2008). The 20th anniversary of the Japanese Canadian redress settlement celebration and conference. Retrieved May 2011 from Japanese Canadian Redress Anniversary: http://redressanniversary.najc.ca/redress/

Ng, E. (2011). Longitudinal health and administrative data research team. Insights into the healthy immigrant effect: Mortality by period of immigration and place of birth (Statistics Canada, Catalogue 82-622- X, Number 8) Ottawa: Statistics Canada.

Ontario Council of Agencies Serving Immigrants. (1991). Immigrant settlement counselling: A training guide. Toronto: OCASI.

Ontario Council of Agencies Serving Immigrants. (2000). Immigration settlement counselling: A training guide Part 1. Retrieved from http://atwork.settlement.org/downloads/atwork/Training_Guide_CHAPTER_1.pd

Ontario Council of Agencies Serving Immigrants. (2012). Making Ontario home 2012: A study of settlement and integration services. Toronto, Ontario, Canada. Retrieved from http://www.ocasi.org/downloads/OCASI_MOH_ENGLISH.pdf

Powles, J. (2004). Life history and personal narrative: Theoretical and methodological issues relevant to research and evaluation in refugee contexts. Working paper No 106, United Nations High Commission for Refugees, Evaluation and Policy Analysis Unit. Retrieved from www.unhcr.org/4147fe764.pdf

Richards, S. (2009, September 9). Close that door. Retrieved from YouTube: http://www.youtube.com/watch?v=iyxWGqFQalY&feature=related

Roberts-Moore, J. (2002). Establishing recognition of past injustices: Uses of archival records in documenting the experience of Japanese Canadians during the Second World War. Archivaria: The Journal of the Association of Canadian Archivists, 53 (2002), 64–75.

Say, A. (1993). Grandfather's journey. Boston: Houghton Mifflin Company.

Showler, P. (2006). Refugee sandwich: Stories of exile and asylum. Kingston: McGill-Queen's University Press.

Smart City News. (2012, September 12). Smart City Business: "Diversity for Growth and Innovation" feat. RBC, Season 2: Episode 1. (C. Layton, Ed.) Halifax, Nova Scotia.

Statistics Canada. (2007, September 10). Study: Canada's immigrant labour market. Retrieved April 2, 2011, from Statistics Canada: http://www.statcan.gc.ca/daily-quotidien/070910/dq070910a-eng.htm

Statistics Canada. (2007, September 10). The immigrant labour force analysis series. Retrieved June 12, 2011, from Statistics Canada: http://www.statcan.gc.ca/pub/71-606-x/2007001/4129573-eng.htm

Statistics Canada. (2012, October 24). 2011 census of population: Linguistic characteristics of Canadians. Retrieved from The Daily: http://www.statcan.gc.ca/daily-quotidien/121024/dq121024a-eng.htm

Thomas, G., & Witts, M. M. (1974). Voyage of the damned. Konecky & Konecky.

Toronto Region Immigrant Employment Council. (2011). The mentoring partnership. Retrieved from Toronto Region Immigrant Employment Council: http://triec.ca/?nr=1

United Nations High Commissioner for Refugees. (2010a). The 1951 Refugee Convention: The legislation that underpins our work. Retrieved from UNHCR: http://www.unhcr.org/pages/49da0e466.html

United Nations High Commissioner for Refugees. (2013). 2012 global trends: Refugees, asylum-seekers, returnees, internally displaced and stateless persons. Retrieved from http://www.unhcr.org/4c11f0be9.html

United Nations High Comissioner for Refugees. (2016, June). 3RP mid year report. Retrieved from Regional Refugee & Resilience Plan 2016–2017 In response to Syrian Crisis: http://data.unhcr.org/syrianrefugees/regional.php

United Nations High Commissioner for Refugees. (2016). Global trends: Forced displacement in 2015. Retrieved from UNHCR: http://www.unhcr.org/news/latest/2016/6/5763b65a4/global-forced-displacement-hits-record-high.html

Xu, L. (2012, March). Who drives a taxi in Canada? Retrieved from Citizenship and Immigration Canada: http://www.cic.gc.ca/english/pdf/research-stats/taxi.pdf

Multiculturalism

> "Hear your voice 'cross a frozen lake
> Voice like the end of a leaf"
>
> – *The Tragically Hip (2006)*

LEARNING OUTCOMES

By mastering this unit, students will gain the skills and ability to:

- trace the history and various definitions of multiculturalism

- compare and contrast multiculturalism concepts and practices between countries

- assess whether multiculturalism is working in Canada

- reflect upon multiculturalism as a unifying and inclusive national identity and policy, and as a divisive and marginalizing reality

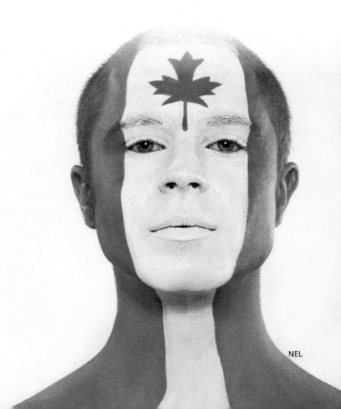

Paper Boat Creative/Getty

As a nation, we take pride in the fact that we were the first country in the world to adopt multiculturalism as an official policy that affirmed the value and dignity of all citizens of this racially, ethnically, linguistically and religiously diverse society. And while we know that Canadian multiculturalism creates a sense of belonging and is interwoven with our belief that all citizens are equal, the framework and definition of multiculturalism continues to change.

What does it mean to live in a country where hundreds of thousands of immigrants are welcomed each year? Is multiculturalism a fundamental right or freedom? Is it a policy or a law? Is it a political ideology that is mandated differently through varying government factions?

For some, multiculturalism simply boils down to traditions, customs, and costumes that are worn at folk festivals and annual celebrations—like Canadian Multiculturalism Day, celebrated on June 27 of each year. Others see it as a movement toward the inclusion of all people who are marginalized or disenfranchised—yet welcomed—into Canada every year. Still others see it as an imposition on what they view is "their" Canada. Whatever your viewpoint is, two things are certain. First, as a society, we will likely never agree on a universal definition of multiculturalism. Second, we must never stop trying to create one.

Debating what multiculturalism is and what it is not means questioning the value and rights of immigrants as they attempt to rebuild their lives in new surroundings. As the face of Canadian society continues to change, we must continue to ask ourselves if multiculturalism in Canada is working. Conversations surrounding multiculturalism in Canada must continue to be open and ongoing until inclusion and equity for all become the norm.

IN THEIR SHOES

If a picture can say a thousand words, imagine the stories your shoes could tell! Try this student story on for size – have you walked in this student's shoes?

"Where are you from?" I can't begin to count the number of times I have been asked this question. I am Canadian. I was born in Canada and have lived here all of my life. I have a Canadian passport and am proud to call myself Canadian. I know that I am privileged to be born in this great nation. I am bilingual in English and French. My parents are originally from Africa, but came to Canada before I was born. I attended elementary and secondary school in Canada. I am now attending college, studying in a Police Foundations program. I hope to one day be employed as a police officer where I will uphold the laws of Canada. So why do people frequently ask me the question, "Where are you from?" Why do they assume that my sense of belonging and my national identity are acquired elsewhere?

Since about Grade 3, I remember teachers talking about Canada as a multicultural country. They used metaphors like mosaics and mixed green salads to illustrate how as a nation, Canada, too, was made up of distinct pieces and ingredients. We were taught that with the exception of Aboriginal peoples, all Canadians are immigrants to Canada or descended from immigrants. People have labelled me as a second-generation Canadian. One person even labelled my oldest brother as 1.5-generation Canadian because he was born in Africa but came to Canada with my parents when he was 10 years old. You can only imagine how many jokes were made about

"one and a half." So why are people so eager to identify others by their migration history? Is it relevant? I have friends who are "second-generation Canadian" and the only things we have in common are that our parents immigrated to Canada and we love soccer. I don't come from the same country as them. We don't share the same religion, culture, customs or traditions. I spoke English at home with my parents growing up; they didn't. They attended ESL classes in elementary school. I didn't. In fact, I have more in common with friends who are "fourth- and fifth-generation Canadian," so why not just identify us as Canadian—period. Isn't that what multiculturalism is all about—we can be different but we still "belong" as part of the whole?

If ethnic and racial diversity and integration are really part of Canadian multiculturalism, then why do people still ask me on a regular basis where I come from? When I tell them I am Canadian, I can usually predict their next question: "But where were you born?" When I tell them Fort Erie, I can usually predict their next question: "Is that in Canada?" And you might think it would end there. But I have had people ask me where I learned to speak such good English, how long have I been in Canada, if they had McDonald's restaurants back where I came from, and where did I learn to skate?

Now I just wear a T-shirt that says, "My name is John and I am CANADIAN."

CANADIANS ARE NOT AMERICANS

It's not always easy living next door to the world's largest economic, military, and cultural giant. Canada can often feel like a satellite in America's planetary orbit. But in many ways, Canada defines itself and its culture specifically by distinguishing itself from its neighbour. In 1990, sociologist Seymour Martin Lipset proposed that one of the most important defining elements of Canadian identity was that Canadians have historically defined themselves as "not Americans" (Lipset, 1990).

Though many have contested this, it's hard not to use the United States as a point of reference against which to determine our own national culture. Nowhere is this matter better illustrated than with Canada's stance on multiculturalism. Recent polls show that Canadians consider multiculturalism a fundamental Canadian value (Mosaic Institute, 2012). Further, they are quick to point out the distinction between "our multiculturalism" and "their multiculturalism."

In looking at models of acculturation, the **mosaic model** and the **melting pot model** have emerged in North America as these binary concepts used to explain and distinguish Canada and the United States' relationship with immigration, and beliefs about individual and collective belonging and citizenship (Glendon College, York University, 2012). The concept of the cultural mosaic has been used in Canada to embody **multiculturalism** as a national ideology and suggests newcomers can maintain their unique and distinct cultures and live alongside other Canadians with other distinct cultures. The concept of the melting pot is a metaphor for assimilation where newcomers to a country are expected to dissolve that which makes them different into a "pot" mixed with everyone else. The melting pot model is often associated with how the United States thinks about integration as well as cultural and linguistic pluralism (Glendon College, York University, 2012). The important ideological difference between these models centres on how much newcomers are expected to shed their traditional cultures. The mosaic model, in theory, asserts that all Canadians have the right to maintain their own distinct cultures, as long as the values expressed do not clash with Canadian laws and values. At a roundtable discussion held at York University in 2012 entitled "*The Mosaic vs. the Melting Pot? Myths and Realities of Cultural Pluralism in Canada and the United States,*" panelists suggested that while "Many Canadians view the melting pot as the opposite of the mosaic and official multiculturalism ... the two ideologies have much in common and mask many similarities when we examine the everyday realities of cultural pluralism in North America."

IS MULTIKULTI A FAILURE?

In 2010, German Chancellor Angela Merkel announced to the world that the so-called "*Multikulti*" concept—where people would "live side-by-side" happily—did not work, and immigrants needed to do more to integrate (Evans, 2010). Her comments came alongside a rising wave of anti-immigration sentiments that were sweeping across Europe at the time, and which are continuing to do so. Here, decades after Pierre Trudeau's government enshrined multiculturalism as a national value and a law in Canada, social and political commentators claim we have gone too far with the experiment. Multiculturalism has been called a "pc" movement and a governmental money-drain, and it has been blamed for ethnic ghettos, violence against women, a lack of a cohesive national identity, and **home-grown terrorism**. Nor is it just extreme right-wing politicians blowing the whistle on immigration. According to Transatlantic Trends (an annual survey of American and European public opinion), the strongest opposition to multiculturalism in Germany and France comes from the left (Todd, 2011a).

Great Britain's commitment to multiculturalism naturally took a sharp blow after the subway bombings in July of 2005. Overnight, low-grade grumblings became full-fledged and feverish protests against the nation's perceived over-acceptance of "foreigners." There was an immediate demand for newcomers (and long-established ethnic ghettos) to start assimilating and dropping any allegiances to other countries.

Mosaic model: Immigration and settlement model that is often used in reference to Canada, wherein newcomers are encouraged to maintain their unique and distinct cultures and live alongside other Canadians with other distinct cultures.

Melting pot model: Immigration and settlement model that is often used in reference to the United States, wherein newcomers are expected to dissolve that which makes them different into a "pot" mixed with everyone else.

Multiculturalism: A concept often used in different ways. It can refer to the demographic reality that we are a multicultural society made up of ethnoculturally and racially diverse groups of people. Multiculturalism can also refer to an ideology that uses ethnocultural and racial diversity and equity as its framework and a mosaic as its metaphor. It may also refer to formal policy and initiatives.

Home-grown terrorism: Acts or plans of mass violence for political purposes that are initiated and/or carried out by residents or citizens as opposed to foreign nationals.

Months later, ethnic riots erupted in North African and Arab neighbourhoods in France after a French police chase of a group of North African teenagers ended in two fatal electrocutions. Nine thousand cars and hundreds of buildings were destroyed in the ensuing race riots. But in France's case, multicultural policy could hardly be blamed: France has always insisted on a strict **assimilationist policy** (Gregg, 2006) for immigrants.

In the Netherlands, a country which prides itself on its broad tolerance toward immigrants and which took on an aggressive multicultural policy in the 1980s, the retreat has been swift and startling. The nation has now instituted a restrictive visa system; unless immigrants come from one of the exempt (wealthy) nations, anyone wishing to live in the Netherlands now must submit to a rigorous civic-integration examination, including components that test their compatibility with liberal Dutch values in the form of films depicting homosexuality and nude beaches. Under the new policy, Dutch men and women of foreign descent would not be able to sponsor a spouse if that spouse did not also pass the exam. The government claims that it is simply responding to public opinion, where polls show 90 percent of Dutch citizens wanting a more assimilationist integration policy (Bransten, 2006).

Even intergovernmental groups have joined in the condemnation of multiculturalism, with the Council of Europe calling it "the flip side of assimilation, equally based on the assumption of an irreconcilable opposition between majority and minority" (Kymlicka, 2010) and blaming the naive acceptance of a live-and-let-live attitude for increasing ethnic **segregation** and marginalization throughout the continent.

The high unemployment rates, low education levels, and general view that immigrants are not successfully integrating in many European countries have put Canada in the spotlight. Some countries, such as Sweden, recognizing that what they've done historically has not worked, are looking westward. In their book *Kanadamodellen* ("The Canada Model"), editors Peter Hojem and Martin Adahl examine the various factors that have allowed Canada to be seen as the promised land where people of hundreds of different backgrounds are able to live and work together in relative peace and prosperity (Saunders, 2011).

Canadians themselves have become somewhat ambivalent about their own multicultural reality. Almost half of Canadians polled in 2010 believed immigrants should give up their customs and traditions and more closely assimilate to the majority culture (Angus Reid, 2010). But this view is the exact

Bhola Chauhan and Harnarayan Singh broadcast *Hockey Night in Canada* in Punjabi. They may not be as flashy as Don Cherry, but they manage to draw a lot of fans.

opposite of what the nation has been known for. Is *Multikulti* failing in Canada, too?

Because one of the perceived goals of the Canadian model of multiculturalism is to allow diverse communities to retain their various backgrounds without having to assimilate to succeed in Canada, there was always the possibility that historic ethnic tensions would find their way to Canadian soil. When a Canadian politician of Croatian descent appeared on television in 1991 and proclaimed, "I don't think I'd be able to live next door to a Serb," it was immediately taken as proof that this fear was already a reality. (Bissoondath, 1988).

Backlash against multiculturalism is not simply a matter of anti-immigrant sentiment. Opponents of multiculturalism claim that the Canadian model does not so much respect the diversity of its population as it treats its various ethnic groups as archetypes, not individuals—viewing the superficial differences as exotic, and turning various ethnic traditions into an amusement park ride (Bissoondath, 1994). There is also the danger that ethnic minorities themselves do not want recognition as such. A study of recent Colombian immigrants found that 13 percent of the sampled group responded that they rarely, if ever, listened to Colombian music, read Colombian periodicals, or ate Colombian food. That same group indicated that they spend little or no time with other

> **Assimilationist policy:** Policy that encourages immigrants to integrate into the mainstream culture.
>
> **Segregation:** Imposed separation of different groups, usually unequally.

Colombians. More importantly, over 90 percent of the study respondents felt that interacting with other ethnic groups was particularly important (Urzúa, 2012). Is multiculturalism putting Canadians in boxes they don't want to be in?

While immigrants often do better in Canada than they do elsewhere, it could be a result not of our multicultural policy, but of the fact that our immigration policy cherry-picks the best and brightest of those seeking to come. After all, refugees make up only a small fraction of Canada's immigrant numbers, especially compared to other countries who either share a land border with poorer nations or are an obvious sea voyage port such as Australia (Kymlicka, 2012). Canada does not have to worry about illegal immigration. Essentially, immigrants who come to Canada are those whom the government has allowed to come.

In 2005, Bernard Ostry, one of the original architects of Trudeau's multiculturalism policy, sounded a tepid retreat from the direction Canada had been going for the past 40 years (Kymlicka, 2012). Was multiculturalism just forcing everyone into different corners?

MULTICULTURALISM AS A DESCRIPTIVE AND PRESCRIPTIVE CONCEPT

That Canada is a culturally diverse nation is undeniable. Whether in major metropolises like Vancouver, Montreal, or Toronto, or in far-flung rural Northern communities, the near-homogeneity found at the turn of the 20th century is now almost impossible to find. Where the population once consisted of almost 90 percent British or French origin, the 2011 National Household Survey reports respondents from over 200 ethnic origins (Statistics Canada, 2013). In addition to the two official languages sanctioned and protected by name in the Constitution, there are now over 200 languages spoken in homes across Canada (Canadian Press, 2012b); and some of these languages have now escaped the confines of the home and made their way onto street signs and shop windows. One has only to walk along almost any street in Canada to find ample evidence of the diversity of this nation.

Cultural relativism: Assumption that all cultures are essentially equally valuable and should be judged according to their own standards; no cultural standpoint is privileged over another.

Anglo- or Franco-conformity: Policy that informed early Canadian immigration; immigrants were expected to conform to dominant British-based or French-based culture as opposed to retaining their own.

But diversity is a fact, not a moral choice. To return to a comparison with the United States, for every Caribana festival in Toronto, there is an equally colourful and "ethnic" festival in the United States—Carifest in Baltimore, Bayou Bacchanal in New Orleans, and dozens of other Caribbean festivals throughout the country. So what makes Canada different? When multiculturalism became an officially recognized policy in our Constitution, did we automatically become more diverse? more tolerant?

One way to understand multiculturalism and its relationship to diversity is that multiculturalism can refer to governmental policies that aim to manage a diverse population. Multiculturalism can be descriptive or prescriptive: demographics or policy. For this textbook, we will examine the issues from the prescriptive angle.

Multiculturalism is often characterized as an "anything-goes" celebration of superficial differences where **cultural relativism** allows independent subcultures to live outside the bounds of the law. But neither the architects nor the proponents of multicultural policy ever intended or allowed for a flouting of the law. Anything that violates federal law—regardless of whether it is a cultural norm elsewhere—remains illegal for anyone living in Canada.

The Architecture of Multiculturalism

Now that multiculturalism is a buzzword, it's hard to remember a time when it wasn't even mentioned. How did this happen? Where did this come from? Perhaps a historical rather than a geographic comparison would clarify matters. Though not enshrined in official federal documents, the general policy that informed early Canadian immigration and that preceded multiculturalism was known as **Anglo- or Franco-conformity.**

Immigrants were expected to conform to dominant British-based or French-based culture, as opposed to retaining their own. In the years leading up to World War II, English-speaking Canada had no identity crisis. Canada was a British outpost and proud of it. The only acceptable "ethnic" affiliation was to England and to a lesser degree, France; all other ethnic identities were to be purged, and new immigrants were encouraged to assimilate—and the sooner the better. This was relatively easily done, as the immigration policies discouraged large numbers of visible minorities ("non-Caucasian") unless they were needed as cheap labour (e.g., the building of the Canadian Pacific Railway). The Empire Settlement Scheme of the 1920s explicitly stated the preference of British immigrants to the prairies. Although not preferred, central and Eastern Europeans were considered acceptable, but only in limited number and capacity—namely, as agricultural and domestic workers (Belanger, 2006). But in addition to the hostility and xenophobia that many of these new immigrants faced (including the internment of 5000 Ukrainian men in enemy camps during World War I), official laws were enacted to enforce English-only instruction in schools (Stoddard, 2012).

The final "caste" of undesirables consisted of all visible minorities. Section 38 of the Canadian Immigration Act of 1910 gave the federal government the right to forbid entry "of immigrants belonging to any race deemed unsuited to the climate or requirements of Canada." Among those deemed unsuited to immigrate to Canada were those from "warm climates"—in other words, those who could fully assimilate but would never really be "Canadian" (Belanger, 2006).

By the 1960s, the federal government introduced the idea of the non-racist points system allowing for immigration from non-European countries to increase (Belanger, 2006). While this changed the demographics of Canada, it would still be a decade before the ideology caught up. At this point in the story, the assumption is that multicultural policy was enacted as a direct result of this demographic shift. History, however, tells a different story.

Early Multiculturalism

The Quiet Revolution

Ironically, **Quebec nationalism** is at least partly responsible for Canada as we know it today. In the 1960s, rumblings of the **separatist movement** that sought separation from Canada to form a smaller independent nation could be heard throughout the province. Campaigning under slogans of "Maîtres chez nous" ("Masters of our house") and "Il faut que ça change" ("Things must change"), the **Quiet Revolution**, a period in the 1960s, was poised to split the nation in two. This spurred Prime Minister Lester B. Pearson's government to engineer policies that would enshrine the place of both England and France as "founding nations" and woo Quebec back from the brink of secession. This included making the country officially bilingual at all levels of government with the Official Languages Act of 1969 (Stoddard, 2012). But in trying to appease one group, the Canadian government awoke another group who watched these developments with concern.

Did making Canada officially **bilingual** and **bicultural** mean that there was no room for other ethnicities? This was the message that "white -ethnics"—Ukrainians, Poles, Germans, Jews—across Canada received. Many of these groups were the ones forced to give up their own languages shortly upon arrival in Canada. Why should they now be forced to recognize French? These groups mobilized and began to put pressure on the government to ensure that they would be included as full Canadians in the "new Canada," and that French-speaking Quebec wouldn't be given higher accord at their expense. After all, they had proven themselves as model citizens, building the country in its early years, and fighting in two world wars. In response, the federal government commissioned a report clumsily entitled "The Cultural Contributions of Other Ethnic Groups in Canada," which laid out 16 recommendations to recognizing the "other ethnic groups" (Kymlicka, 2010).

The Formative Stage of Multiculturalism: Pluralism

This is how the current formulation of Canada's multicultural policy started to emerge. The government's attempts at a bilingual country expanded to include a somewhat broader multiculturalism within that framework. As Will Kymlicka put it, "The formula which gradually emerged—namely multiculturalism within a bilingual framework—was essentially a bargain to ensure white ethnic support for the more urgent task of accommodating Quebec" (Kymlicka, 2004, p. 7). What this means is that Canada's multiculturalism was not initially an attempt to be inclusive of a diverse nation of visible minorities. Perhaps it could be argued that we didn't quite know what we were in for with this bargain.

The period between 1971 and 1985 is considered the formative stage of multiculturalism (Urzúa, 2012). Prior to this period, the default policy in most cases was Anglo-conformity. But before it became a nationally recognized value, in its earliest form, official Canadian

> **Quebec nationalism:** The belief that Quebec should be recognized as a sovereign and separate nation.
>
> **Separatist movement:** A movement that seeks separation from a nation to form a smaller independent nation, as with Quebec nationalism.
>
> **Quiet Revolution:** Period during the 1960s when Quebec saw rapid secularization and an increased sense of nationalism.
>
> **Bilingual:** Able to speak two languages; in Canadian history, the official recognition of English and French as equal official languages.
>
> **Bicultural:** Retaining two distinct cultural identities; in Canadian history, it often refers to the recognition of England and France as primary cultures within Canada.

multiculturalism only existed in the context of Prime Minister Pierre Trudeau's speech to the House of Commons:

> Cultural **pluralism** is the very essence of Canadian identity. Every ethnic group has the right to preserve and develop its own culture and values within the Canadian context. To say that we have two official languages is not to say we have two official cultures, and no particular culture is more official than another. (Trudeau, 1971)

The resulting policy pledged to do the following:

1. Recognize and respect the multicultural nature of Canada's population.
2. Eliminate any barriers to full participation for all Canadians, regardless of ethnic origin.
3. Enhance the development of communities with shared origins and recognize their contributions to Canada.
4. Ensure equality of all Canadians.
5. Preserve the use of languages other than Canada's two official languages.
6. Foster multiculturalism through social, cultural, economic, and political institutions (Hyman, Meinhard, & Shields, 2011).

Much of the early support went into funding folk festivals and artistic programs. It's important to recognize that the early policy was intent on new Canadians retaining their heritage; no funding was provided for the Métis, First Nations, or long-established black Canadian communities. That was to come in the next iteration of multicultural policy in the 1980s (Hyman, Meinhard, & Shields, 2011).

Pluralism: Existence of recognized diverse groups within a single (peaceful) society.

Expansionism: A country's practice or policy of getting bigger, usually in terms of territory or currency.

Canadian Multiculturalism Act: Law passed in 1988 that aims to preserve and enhance multiculturalism in Canada.

Employment Equity Act: Law that requires employers in Canada to eliminate barriers to and increase the hiring of women, people with disabilities, Aboriginal people, and visible minorities.

The Expansionist (Institutionalization) Stage

After the formative stage, Canada entered the **expansionist** or institutionalization stage of official multiculturalism, where multiculturalism was formally recognized and institutionalized as law (Urzúa, 2012). The 1988 revamp of the policy saw Parliament's passing of the **Canadian Multiculturalism Act,** which aimed to preserve and enhance multiculturalism in Canada. In addition to the continued emphasis on the celebration of diversity, the act recognized the changing nature of Canadian demographics and placed special emphasis on eliminating the growing racial tensions and subsequent social exclusion of new visible minorities. This era saw the birth of the **Employment Equity Act** of 1986, which specifically addressed the barriers faced by women, Aboriginal peoples, persons with disabilities, and members of visible minorities. The legislation insists that it is not enough to merely treat everyone the same, but that an effort must be made to accommodate differences (Abella, 1984).

In 1996, the federal government, through the Department of Canadian Heritage, began a review of multicultural policy, and the resulting findings concluded that the "non-diverse" segments of the population had a duty to allow immigrants and minorities to more fully integrate (Gingrich, 2004). Program funding was tied to equity outcomes for ethnically diverse and racialized communities. The backlash to this was seen almost immediately in cries of reverse-discrimination, and this period witnessed the lowest satisfaction with multiculturalism since its inception. Ensuing budget cuts began chipping away at the funding for multiculturalism, and today the promotion of multiculturalism is a function of the federal department, referred to as Immigration, Refugees and Citizenship Canada.

DOES CANADIAN MULTICULTURALISM WORK?

Canada has been hailed all over the world as a successful, functioning model of multiculturalism done right. They are perhaps too polite to admit it, but Canadians—in government, in the press, in the streets—take pride in this and wonder why other Western countries seem to keep getting it wrong. A statistical outlier, Canada has managed to maintain all the markers of modern success—economic prosperity, a thriving democracy, and individual freedom—with high levels of immigration, without resorting to assimilationist policies. It is no wonder that Canadians have taken such pride in their reputation.

International leaders and organizations ranging from Nelson Mandela to the United Nations have looked to Canada as the leader in the field of multiculturalism and have set up think tanks to examine why and how it works. All the indices of success are there: high intermarriage rate, high citizenship rate, high levels of acquisition of one or both official languages, and a significantly high level of political participation at the voter level as well as in terms of elected officials.

But one of Canada's leading scholars of political philosophy argues that far from Canada's success resulting from unique policy, it is a set of fortunate circumstances that have allowed Canada's multiculturalism to work while other Western countries have faced emphatic anti-immigrant backlash. Some governments—of Australia, England, and the Netherlands, for example—are sounding the retreat from their attempts. In fact, Australia, the only country that accepts more immigrants per capita than Canada, jettisoned the term "multiculturalism" altogether from its federal vocabulary. In 2007, it replaced the Department of Multicultural Affairs with the Department of Immigration and Citizenship (Kymlicka, 2012).

As mentioned earlier, Canada's earliest iterations of multiculturalism referred to "white ethnics." For the most part, these Canadians looked and acted like the majority of the population. The differences were mostly benign—cuisine, ethnic clothing, language spoken by elders in the home—and they had already proven that they would provide no clash of civilizations. In fact, most, if not all of these groups, came from a Judeo-Christian tradition and shared many of the same values.

The face of multiculturalism today is very different. These early ethnic groups have now been absorbed as "real Canadians" to make way for the visible minorities who now take centre stage in the issue. Many of these early ethnic groups, in fact, feel a great deal of resentment toward later immigrants who they see as taking advantage of their hard-won gains in acceptance.

PICTURE THIS...

THE CANADIAN PRESS IMAGES/Adrian Wyld

In 2014 Minister of State for Multiculturalism Tim Uppal faced an apparent racist rant while leaving a tennis court with his wife when they overheard a passerby say, "are they members? Why can't they play in the day – they don't have jobs." Is this an isolated incident or is it reflective of the concern that multiculturalism in Canada is failing?

What if these early ethnic groups were seen as bringing with them ideas and practices that clashed with Canada's Western democratic ideals? This is not simply the early prejudicial xenophobia that all newcomers to Canada faced when the country was young; this is a real question with which both proponents and opponents of multiculturalism today contend. It is one thing to have a Polish perogy festival; it is another to allow cock-fighting—a sport that is widely engaged in all over the world, particularly in Latin American countries. It is one thing to build a museum to recognize the suffering of Jews during World War II; it is another to allow female genital mutilation (FGM), a practice that is common in many parts of Africa and Asia Minor. These are the kinds of issues that are simmering and have flared up repeatedly since the 1970s.

But a simple comparison between immigrants of similar backgrounds coming to Canada and the United States illustrates that—all other things being equal—immigrants to Canada are more quickly able to feel "Canadian," despite being allowed and encouraged to retain their ethnic heritage, than immigrants to the United States are able to feel "American." Irene Bloemraad, a sociology professor at the University of California at Berkeley, conducted studies comparing Vietnamese and Portuguese immigrants of similar socioeconomic and language backgrounds coming to Boston and Toronto. In both cases, the Toronto group was found to have a far higher sense of citizenship in their new country and was more actively participating in civic life. After examining all the alternative reasons for this disparity, Bloemraad concludes that it is indeed the mechanism of Canada's institutionalized multiculturalism that allows for a healthier, smoother, and faster integration into mainstream society (Bloemraad, 2010).

What Do the Numbers Tell Us?

As diverse as Canada may sometimes feel, the vast majority of the population is still of European heritage, with only 13 percent of the total population reporting non-European descent. But official ethnicity does not determine a person's ethnic identity. The Ethnic Diversity Survey found that—not surprisingly—ethnic identification varied among diverse groups and tended to decline with subsequent generations. Seventy-eight percent of Filipinos reported a strong sense of ethnic affiliation compared to 58 percent of Chinese (Statistics Canada, 2003). Presumably, this means that the remaining 42 percent of Chinese Canadians

Ethnic enclave: Focused areas of homogeneous ethnic groups, usually coupled with some business and institutional activity.

consider themselves primarily Canadian. Consider the dissonance created when someone feels Canadian, but their "Canadian-ness" is repeatedly questioned.

Statistics Canada projects that by 2031, at least one out of every four people in Canada will be foreign-born, and over half this number will likely be from Asia. Compared to the rest of the population, the foreign-born segment will be increasing at four times the rate.

As for visible minorities, the 2006 report (Statistics Canada, 2010) documented just over 5 million, and the projections by 2031 nearly double, meaning that almost one-third of the population will belong to a visible minority group. The question remains as to whether the minority continues to have any meaning in this context, especially in cities like Toronto and Vancouver where visible minorities are expected to represent 63 percent and 59 percent respectively of the cities' populations (Statistics Canada, 2010).

South Asian minorities will continue to be the most numerous visible minority, with the Chinese population likely doubling as well. However, the Chinese population will lose some of its percentage of representation of the visible minority segment due to low fertility rates among Chinese women. The second and third largest visible minority groups in 2006, black and Filipino Canadians, will also possibly double in size. The fastest growing groups of all will be Arabs and West Indians, who are predicted to triple their current numbers by 2031 (Statistics Canada, 2010).

Is "Ethnic Enclave" Just a Fancy Name for "Ghetto"?

Almost every ethnic group that has immigrated to Canada has some claim to a particular geographic concentration. According to Statistics Canada projections, by 2031, more than 71 percent of all visible minority Canadians will live in Canada's three largest census metropolitan areas: Toronto, Vancouver, and Montreal (Statistics Canada, 2010). In the Greater Toronto Area alone, there are numerous well-established ethnic neighbourhoods: Italians in Woodbridge and Vaughan, Chinese in Richmond Hill, and South Asians in northern Brampton. Sociologists call these focused areas of homogeneous ethnic groups, which are usually coupled with some business and institutional activity, **ethnic enclaves** (Keung, 2013). People don't need charts or maps to tell them they have entered into an ethnic enclave; a quick look at the storefronts, billboards, and religious buildings is usually enough.

To some, these enclaves proclaim the beauty of diversity and the triumph of capitalism, where demand meets previously non-existent supply. To others, these enclaves are the harbingers of the impending cultural wars Canada faces if it does not resume its policy of assimilation. With each ethnic group crouched behind its walls, how can we hope to have a cohesive national identity?

Ethnic enclaves serve a number of positive functions. These neighbourhoods help new immigrants acculturate slowly into their new home country. Minimizing culture shock, they allow various ethnic groups to establish stores and places of worship; they make it easier to deliver civic services in a culturally appropriate way; and, in many cases, they transform the landscape, adding numerous vibrant tourist attractions—Little Italy, Little India, Chinatown; all distinct, yet all Canadian.

In 1981, Canada had only six ethnic enclaves; today there are more than 260 throughout the country, half of which are found in metropolitan Vancouver (Hopper, 2011). But what differentiates them from the ethnic ghettos seen in the United States and Western Europe is that, by and large, these enclaves are enclaves of choice—not the result of discriminatory housing practices or poverty (Kymlicka, 2010). Another radical difference is the mobility both in and out of these enclaves. While many of the previously homogeneous neighbourhoods are now being **gentrified**, causing housing prices to rise as wealthy young urbanites populate them, many ethnic minorities are moving out of urban areas and into the suburbs.

That's the good news. The bad news is far louder—and it's the bad news that makes the actual news. Segregation, social exclusion, lack of language acquisition—all these issues are blamed on ethnic enclaves, which are viewed as the enemies of integration. Although not necessarily determined by socio-economics, some ethnic enclaves were historically indistinguishable from slums, and many still are. When ethnic neighbourhoods get large and dense enough, they are able to support their own schools, thereby eliminating one of the fastest means of civic integration and allowing minority groups to socialize almost entirely with "their own." Despite reports that Canada's immigrant population has gotten high marks on all the standard benchmarks of assimilation—home-ownership, language acquisition, and attainment of citizenship (Kymlicka, 2004)—Canada's former Minister of Immigration and Citizenship, Jason Kenney, worried that the government's commitment to preserving and enhancing Canada's multicultural heritage and allowing the continuation of nearly autonomous ethnic enclaves threatened national cohesion and did a disservice to new immigrants (MacDonald, 2013). And though some of these neighbourhoods remain prosperous, newcomers to Canada today—earning on average 61 cents to the dollar of their Canadian-born counterparts—have nowhere near the economic mobility that they had in the early 1990s (Hopper, 2011).

University of Victoria scholar Zheng Wu found that although ethnic enclaves can help recent immigrants feel more comfortable and protected, they also decrease a resident's sense of belonging to Canada (Todd, 2011b). Ethnic segregation and economic inequality are ripe conditions for exacting the kind of cultural warfare that some right-wing politicians often warn about as a result of multicultural "political correctness."

There are also serious issues of violence in ethnic communities. When a Punjabi teenager was stabbed to death in Brampton in 2007, residents voiced their concerns that "Vancouver-style" violence was spreading into their communities. What they were referring to was the hundreds of Punjabi-on-Punjabi murders in British Columbia's ethnic gang warfare (Grewal, 2007). Canadians were horrified when they learned of Pakistani-Canadian teenager Aqsa Parvez being murdered by her father for supposedly violating the family's honour. Soon after, other stories of so-called "honour killings" started appearing in the headlines. Many feared that the perpetrators would get a lighter sentence, hiding behind a cultural defence. A Canadian blogger recently told Canadians that they should "consider ourselves fortunate that Canada's Somali settlers seem more or less content with killing each other" (Shaidle, 2011). In 2011, Edmonton led the nation in homicide rates, and a significant number of those murdered were young Somali men. With Alberta pledging close to $2 million to address the issues surrounding the Somali communities' integration, many, like then-Minister Jason Kenney, wondered if it was time to restructure Canada's multicultural stance (Sun News, 2011).

CANADA'S MULTICULTURAL FUTURE: DOES IT EXIST?

One of the reasons the debate around multiculturalism is so heated stems from lack of a cohesive definition of what multiculturalism actually is.

Opponents equate multiculturalism with political correctness and a war on "real Canadian" culture. Further, they say multiculturalism divides Canada into ethnic silos and does not promote national unity. Speaking to the Trudeau Foundation conference in Nova Scotia, a European critic of the Canadian model contends that

> Multiculturalism undermines much of what is valuable about the lived experience of diversity. Diversity is important because it allows us to expand our horizons, to

> **Gentrified:** Improved so as to appear more middle class.

The Right Honourable Justin Trudeau, Prime Minister of Canada

NICHOLAS KAMM/AFP/Getty Images

As Prime Minister of Canada, the Right Honourable Justin Trudeau has championed multiculturalism and diversity in many international and national forums. In doing so, he carries on the legacy of his father, Prime Minister Pierre Trudeau, who on October 10, 1971 announced Canada would adopt official government policy on multiculturalism, making Canada the first on the world stage to do so. When Prime Minister Justin Trudeau spoke at the World Economic Forum in Davos, Switzerland (2016), he spoke about why multiculturalism needs to be an integral part of our education systems:

> When kids grow up in the schoolyard where people speak every different language, have every different background, where instead of looking at multiculturalism as a whole bunch of

a mainstream culture going to a school gym on a given day, and going to different booths and sampling samosas here, and then going over to see a Berber dance over here, we have instead, an entire school celebrating Diwali the Festival of Lights, or looking up their Chinese horoscope, or talking about how to support your friends going through Ramadan. The range of experiences become the mainstream in Canada and for me that happens within our public schools. It happens within our education. That is, for me, the answer when people say, "Oh, these folks aren't integrating into our value systems quick enough." For me, how we ensure that education gives people the tools to understand that you don't have to choose between the identity that your parents have and being a full citizen in Canada. Yes, there are behaviours and attitudes that are different. You're growing up a second generation Muslim girl in Canada means you may have to have a difficult conversation with your parents about lipstick or about that Indian boy you're dating. Or whatever. But these are things that do not weaken the fabric of who you are and the society you belong to. And it's not easy, that's why you can't do it overnight. But that's where a diverse and open and inclusive education system and open circle of friends is what we have to work toward in our communities.

think about different values and beliefs, and to engage in political dialogue and debate that can help create a more universal language of citizenship. But it is precisely such dialogue and debate that multicultural policy makes so difficult by boxing people into particular ethnic or cultural categories. (Malik, 2011)

At the federal level, in the 1990s, the Reform Party of Canada (an early branch of today's Conservative party) explicitly stated its aim to cut any and all funding to multicultural programs and to abolish the entire department. Some have even made the link between "homegrown terrorism" and multicultural policy (Wong, 2011).

Proponents, on the other hand, define multiculturalism as part of the Canadian identity and a necessary component of a modern liberal society. Tom Axworthy, one of the main architects of Prime Minister Pierre

Trudeau's original policy calls it "our Alamo, without the original war," referring to the unifying rallying cry of the war between Texas and Mexico (Fleras & Elliot, 2007). Without it, Canada's immigration drive would not be successful, and the nation would find itself with a rapidly declining population. With 84 percent of recent immigrants reporting a strong sense of belonging to Canada, it is clear that Canada must be doing something right. On the other hand, 20 percent of visible minorities report experiencing some degree of discrimination, and racialized Canadians are at least twice as likely to be poor as non-racialized Canadians (Galabuzi, 2006).

To further illustrate Canada's struggle with its cultural identity, consider the range of opinions that come in through polls. For a decade, research has continued to affirm that Canadians overwhelmingly support multiculturalism and proudly consider it one of the nation's defining features. Compare that to recent findings that as many as 30 percent of Canadians think that multiculturalism has been bad or very bad for Canada. Perhaps even more surprisingly, a strong majority (54 percent) support a "melting pot" model as opposed to the traditional "mosaic" model that Canada made famous (Angus Reid, 2010).

Social scientists are attempting to reinvigorate Canada's multicultural policies to address not only the reality of new groups coming in greater numbers to Canada, but also the accusations or fears of cultural isolation as a result of emphasizing the differences between various ethnic groups. Sociologist Lloyd Wong proposes including programs that foster inter-ethnic cohesion, manufacturing forums where different groups would regularly interact, and—it is hoped—develop bonds based on their allegiance to Canada. Instead of funding groups that simply promote entrenchment in their own heritage identity, he proposes that the federal and provincial governments begin funding groups and programs that get different groups meeting on common ground—whether that be politically, artistically, or simply socially (Wong, 2011).

Others propose a more cosmopolitan approach that essentially erases the categories of ethnicity altogether. Neil Bissoondath, author of the highly controversial and best-selling book *Selling Illusions: The Cult of Multiculturalism in Canada*, illustrates with his own example:

> I feel greater affinity for the work of Timothy Mo—a British novelist born of an English mother and a Chinese father—than I do for that of Salman Rushdie, with whom I share an ethnicity. … Ethnically, Mo and I share nothing, but imaginatively we share much. (Bissoondath, 1994)

Still others, like Phil Ryan, author of *Multicultiphobia*, propose a blending of both the mosaic and melting pot model with a few guiding principles. Among these principles, he counsels that we should avoid comparing the best of one ethnic group with the worst of another; that respecting multiculturalism does not necessitate signing a "non-interference clause," where everyone is forced to remain silent in the face of illegal or immoral activities committed under the guise of cultural relativism; and that we stop considering white Canadians the default and stop measuring everyone else by that mark (Todd, 2010). This last point has recently been brought to light in a number of headlines. For example, when the Bank of Canada commissioned an artist to portray a female Canadian scientist on the new polymer $100 bills, focus groups responded to the first draft with concerns that the scientist looked "too Asian." The Bank withdrew this version and replaced it with one depicting a more "neutral"-looking scientist (Canadian Press, 2012a).

In a study conducted on behalf of Metropolis British Columbia—entitled "Why Do Some Employers Prefer to Interview Matthew, But Not Samir?"—researchers collected evidence from Toronto, Vancouver, and Montreal, and confirmed what was perhaps already anecdotally obvious: that résumés with English-sounding names had a far better chance getting a call back—in some cases, a 40 percent greater chance (Oreopoulous & Dechief, 2011). Whether this is evidence of latent discrimination, or of recruiters' attempt to avoid a "bad hire" with language-skill deficiencies, the study echoes the Bank of Canada focus group findings. An Anglo background is still the default in Canada. The report ends with numerous suggestions for avoiding this sort of discrimination without resorting to a quota system or resulting in unqualified hiring (Oreopoulos & Dechief, 2011).

ENDING THOUGHTS

As Canadians become more self-conscious about their country's multicultural policy, it is natural to wonder if the bloom is off the rose. Evidence is coming to light that challenges former assumptions about the early days and motivations for multiculturalism, and there is a growing cynicism that we are perhaps not as tolerant as we once believed. Recent research is promising, however. Studies have shown that multiculturalism in Canada prevents national identity turning into xenophobia as it often does and has in other countries. With the Canadian policy, national identity is inextricably linked to the stance toward immigrants and visible minorities. The mythology, well founded or not, means that to be Canadian is to take pride in the country's multiculturalism:

> Multiculturalism provides a locus for the high level of mutual identification among native-born citizens and immigrants in Canada … the fact that Canada has officially defined itself as a multicultural nation means that immigrants are a constituent part of the nation that citizens feel pride in. (Kymlicka, 2010)

Reciprocally, according to the "multicultural hypothesis," immigrants integrate into mainstream society more successfully when they feel their ethnic identity or their ancestral roots are publicly acknowledged and respected (Kymlicka, 2010).

If Canadians' interactions with the federal government become increasingly limited, coupled with a lack of any other defining narrative, there are those who believe multiculturalism may be the unifying rallying point for the nation. Still others fear that years of encouraging segregated cultural identities have fundamentally weakened the fabric of national cohesion. No matter the future direction Canada takes with regard to its diverse population, it will forever retain the distinction of being the first nation to implement an official policy of multiculturalism.

READING

AN IMMIGRANT'S SPLIT PERSONALITY

By Sun-Kyung Yi

I am Korean-Canadian. But the hyphen often snaps in two, obliging me to choose to act as either a Korean or a Canadian, depending on where I am and who I'm with. After sixteen years of living in Canada, I discovered that it's very different to be both at any given time or place. When I was younger, toying with the idea of entertaining two separate identities was a real treat, like a secret game for which no one knew the rules but me.

I was known as Angela to the outside world, and as Sun-Kyung at home. I ate bologna sandwiches in the school lunchroom and rice and kimchee for dinner. I chatted about teen idols and giggled with my girlfriends during my classes, and ambitiously practised piano and studied in the evenings, planning to become a doctor when I grew up. I waved hellos and goodbyes to my teachers, but bowed to my parents' friends visiting our home.

I could also look straight in the eyes of my teachers and friends and talk frankly with them instead of staring at my feet with my mouth shut when Koreans talked to me.

Going outside the home meant I was able to relax from the constraints of my cultural conditioning, until I walked back in the door and had to return to being obedient and submissive daughter.

The game soon ended when I realized that it had become a way of life, that I couldn't change the rules without disappointing my parents and questioning all the cultural implications and consequences that came with being a hyphenated Canadian.

Many have convinced me that I am a Canadian, like all other immigrants in the country, but those same people also ask me which country I came from with great curiosity, following with questions about the type of food I ate and the language I spoke. It's difficult to feel a sense of belonging and acceptance when you are regarded as "one of them." "Those Koreans, they work hard … You must be fantastic at math and science." (No.) "Do your parents own a corner store?" (No.) Koreans and Canadians just can't seem to merge into "us" and "we."

Some people advised me that I should just take the best of both worlds and disregard the rest. That's ideal, but unrealistic when my old culture demands a complete conformity with very little room to manoeuvre for new and different ideas.

After a lifetime of practice, I thought I could change faces and become Korean on demand with grace and perfection. But working with a small Korean company in Toronto proved me wrong. I quickly became estranged from my own people.

My parents were ecstatic at the thought of their daughter finally finding her roots and having a working opportunity to speak my native tongue and absorb the culture. For me, it was the most painful and frustrating 2-1/2 months of my life.

When the president of the company boasted the he "operated little Korea," he meant it literally. A Canadian-bred Korean was not tolerated. I looked like a Korean; therefore, I had to talk, act, and think like one, too. Being accepted meant to totally surrender to ancient codes of behaviour rooted in Confucian thought, while leaving the "Canadian" part of me out in the parking lot with my '86 Buick. In the first few days at work, I was bombarded with inquiries about my marital status. When I told them I was single, they spent the following days trying to match me up with available bachelors in the company and the community.

I was expected to accept my inferior position as a woman and had to behave accordingly. It was not a place to practise my feminist views, or be an individual without being condemned. Little Korea is a place for men (who filled all the senior positions) and women don't dare speak up or disagree with their male counterparts.

The president (all employees bow to him and call him Mr. President) asked me to act more like a lady and smile. I was openly scorned by a senior employee because I spoke more fluent English than Korean. The cook in the kitchen shook her head in disbelief upon discovering that my cooking skills were limited to boiling a package of instant noodles. "You want a good husband, learn to cook," she advised me.

In less than a week, I became an outsider because I refused to conform and blindly nod my head in agreement to what my elders (which happened to be everybody else in the company) said. A month later, I was demoted because "members of the workplace and the Korean community" had complained that I just wasn't "Korean enough" and I had "too much power for a single woman." My father suggested that "when in Rome do as the Romans." But that's exactly what I was doing. I am in Canada, so I was freely acting like a Canadian, and it cost me my job.

My father also said, "It doesn't matter how Canadian you think you are, just look in the mirror and it'll tell you who you really are." But what he didn't realize is that an immigrant has to embrace the new culture to enjoy and benefit from what it has to offer. Of course, I will always be Korean by virtue of my appearance and early conditioning, but I am also happily Canadian and want to take full advantage of all that such citizenship confers.

But for now I remain slightly distant from both cultures, accepted fully by neither. The hyphenated Canadian personifies the ideal of multiculturalism, but unless the host culture and the immigrant cultures can find ways to merge their distinct identities, sharing the best of both, this cultural schizophrenia will continue.

Source: "An Immigrant's Split Personality" by Sun Kyung Yi. Reprinted from *The Globe and Mail*, April 12, 1992. Reprinted by permission of the author. Sun-Kyung Yi is a documentary film maker in Toronto.

1. Sociologist Erving Goffman's (1956) theory of impression management suggests that we divide ourselves into our "front stage" selves, which we use to wind through the realities of everyday life, and our "back stage" selves, which allow us to take off our masks and exist as who we really are, when we are not "acting." In some ways, this division becomes a coping mechanism. For those who live their lives as "hyphenated Canadians," how and when can they experience their "back stage" self?

2. As an ideal, multiculturalism was supposed to alleviate the author's notion of "cultural schizophrenia," but what can be done to merge the opposing views of her elders, employers, and the author herself?

3. The author identifies herself in dimensions that include other factors besides her ethnicity; for example, she indicates that she's a feminist, single, and not a very good cook. Her parents, however, seem to identify themselves strictly in terms of their heritage. Do you think that this is a generational phenomenon, or do you think this is because the author is "hyphenated"?

KWIP

One of the fundamental truths about diversity is that it involves as much learning about oneself as it does learning about others. Let's use the KWIP process to look at how our identities are shaped through social interaction.

KNOW IT AND OWN IT: WHAT DO I BRING TO THIS?

The "K" in the KWIP process involves examining aspects of your own identity and social location as the first step in becoming diversity competent.

ACTIVITY: JOURNAL

Being mindful of how your own identity and social location might influence your responses, explain why you agree or disagree with the following statements about multiculturalism:

- Multiculturalism as an ideal is divisive for Canadians. It is hard to construct a Canadian national identity when multiculturalism encourages everyone to identify with their different ethnic ancestries. This encourages people to identify with hyphenated identities like Japanese-Canadian or Italo-Canadian at the expense of national unity. Official multiculturalism places too much emphasis on Canada as a mosaic and not enough on the glue that binds us—our identity as Canadians.

- Multiculturalism as an official policy strengthens Canada by binding Canadians together through the idea that we do not have to be the same to belong and participate equally in society. By embracing the slogan "unity in diversity," multiculturalism provides a unifying framework for a country whose demographic reality is a population characterized by ethnocultural, racial, and linguistic diversity.

- Multiculturalism has marginalized racialized and ethnic groups in Canada. Multiculturalism's celebratory framework has eclipsed issues of social inequality by focusing on superficial aspects of difference (often referred to as the saris, samosas, and steel bands approach). In doing so, it has ignored the power differences that have contributed to the social, economic, and political marginalization of ethnic and racialized groups and individuals within Canadian society.

- Official multiculturalism is an instrument for ensuring full participation, equity, and inclusion for all Canadians by ensuring their race or ethnicity does not deny involvement or citizenship. Official multiculturalism provides both a social and legal framework for peaceful and respectful coexistence.

- Multiculturalism is essentializing and stereotypes. Using the metaphor of the mosaic, multiculturalism envisions a society made up of distinct ethnic groups that runs the risk of failing to appreciate the diversity within groups and reducing members to ethnic stereotypes. Some examples might include the following: all families are extended; all women are subordinated; and all marriages are arranged.

WALKING THE TALK: HOW CAN I LEARN FROM THIS?

The "W" in the KWIP process presents a scenario or case study that challenges you to "walk the talk" through problem-based learning.

One of the challenges for healthcare professionals in a multicultural society is delivering diversity-competent healthcare services that meet the social, cultural, spiritual, and linguistic needs of patients and their families. There are solutions to consider in each of the following cases. Remember, becoming a culturally competent professional is a continuous journey involving ongoing education and learning.

Case One: A recent Chinese immigrant had major bladder surgery. He was told by the nursing staff to force fluids. The client did not understand the forced fluid order. He refused to drink the glasses of cold water from the big pitcher left on his bedside table. Each time the nursing staff entered the client's room, they reminded him that he needed to force fluids and drink many glasses of water. They threatened that his physician would order intravenous fluids if he did not drink more water. He still refused to drink the cold water on his bedside. The staff said he was uncooperative, strange, and a noncompliant client. When the client's daughter came to see him, she told the nursing staff that he would drink hot herbal tea but not cold water. Finally, the nurses gave him the hot tea and he drank several cups. The nurses did not understand why the hot tea was culturally acceptable and why he had refused to drink tap water. A transcultural nurse came to explain the clinical hot and cold theory of the Chinese and its importance in nursing care.

Describe this hot and cold theory and its importance in nursing care. What other cultural factors and principles in this nursing situation were evident that needed to be addressed?

Source: Leininger, M., and McFarland, M. (2002b). Transcultural Nursing and Globalization of Health Care. In *Transcultural Nursing: Concepts, Theories, Research and Practice 3rd ed.* New York: McGraw-Hill.

Case Two: An elderly Bosnian woman admitted with terminal cancer presents the following challenges for healthcare staff and organizations: she and her family do not read, speak, or understand English or French; her Muslim faith requires modesty during physical examinations; and her family may have cultural reasons for not discussing end-of-life concerns or her impending death.

Discuss some of the culturally, spiritually, and linguistically appropriate strategies and interventions you would consider in this case as a culturally competent professional.

Source: Chin, J. L. (2000). Culturally Competent Health Care. *Public Health Reports.* 115(1/2):25–33.

Case Three: Hector is originally from El Salvador and only speaks Spanish. He goes to his health clinic to see a physician for HIV care. The physician needs an interpreter in order to treat Hector, so he calls in the Spanish-speaking receptionist to help. Hector is taken by surprise: The receptionist is related to his boss at work. Now, he fears that she will tell his boss that he is HIV positive and that word will spread through the small Latino community in his town. He leaves the clinic upset and afraid.

Discuss some of the do's and don'ts for using a language interpreter, and assess which of these are specifically relevant in this case study.

Source: www.hab.hrsa.gov/publications/august2002.htm

Case Four: A Cambodian refugee uses *"cao gio,"* or coin rubbing, to dispel the bad wind and restore the natural balance between hot and cold elements of the universe when her daughter is feverish. The bruise left by this remedy is reported as a sign of abuse by the provider.

What cultural factors and principles in this situation need to be addressed with the provider and family?

Source: Chin, J. L. (2000). Culturally Competent Health Care. *Public Health Reports.* 115(1/2):25–33.

IT IS WHAT IT IS: IS THIS INSIDE OR OUTSIDE MY COMFORT ZONE?

The "I" in the KWIP process requires you to honestly confront and identify ways in which our complex identities result in experiences of privilege and oppression, and to reflect on how we can learn to honour that privilege.

ACTIVITY: GOT PRIVILEGE?

In this exercise, we ask you to examine your privilege and power in relation to your national and ethnic identity. Perhaps you are a student who refers to themselves as "Canadian" without much thought as to its importance in terms of your identity and citizenship rights. Conversely, you may be a student who strongly identifies as "Canadian," and for particular reasons, you are very proud of this identity and very conscious of your citizenship rights. Maybe you are a student who self-identifies by ethnic origin first, such as Japanese. You could be a student who uses a hyphenated national identity, such as Japanese-Canadian. Perhaps you prefer to reverse the order and refer to yourself as a Canadian of Japanese descent.

ASKING: Do I have privilege?

1. I have never been told in everyday social interaction not to speak my first language.

2. People rarely ask me "Where are you from?" in reference to my country of origin.

3. People from my country are positively and visibly represented in Canadian media.

4. When I apply for a job, I am sure that I will not be discriminated against in the hiring process because of my name.

REFLECTING: Honouring Our Privilege

Describe two circumstances in which you feel disadvantaged because of your nationality. Then describe two circumstances in which your nationality gives you privilege over someone else. How can you use the advantages you experience to combat the disadvantages experienced by others?

PUT IT IN PLAY: HOW CAN I USE THIS?

The "P" in the KWIP process involves examining how others are practising equity and how you might use this.

ACTIVITY: CALL TO ACTION

While multiculturalism exists as a sociological fact and public policy, across Canada service agencies within the voluntary sector are funded to "put it in play." Historically, these agencies were funded to promote multiculturalism and often did so through a celebratory framework. Later, the Department of Canadian Heritage funded many of these organizations through their Multiculturalism Program to implement institutional and organization change based on anti-racism and anti-oppression frameworks. Today, the federal government, through the Department of Immigration, Refugees and Citizenship, funds these agencies to provide integration services for newcomers to Canada, including settlement, language training, and employment assistance programs. Most of these agencies are governed by volunteer Boards of Directors who are representative of communities served. They are dispersed across Canada, from the Affiliation of Multicultural Societies and Service Agencies of British Columbia, Moose Jaw Multicultural Council, Access Alliance Multicultural Health and Community Services in Toronto, Service d'éducation et d'intégration interculturelle de Montréal, to New Brunswick Multicultural Council, to name only a few. Try and find a service agency in your community or region that puts multiculturalism in play!

Study Tools
CHAPTER 9

Located at www.nelson.com/student

- Review Key terms with interactive **flash cards**
- Check your Comprehension by completing **chapter review quizzes**
- Gauge your understanding with *Picture This* and accompanying short answer questions
- Develop your critical thinking/reading skills through compelling **Readings** and accompanying short answer questions
- Apply your understanding to your own experience with **Connect A Concept** activities
- Evaluate Diversity in the Media with engaging *Video Activities*
- Reflect on your Understanding with *KWIP* activities

Abella, R. (1984). Equality in employment: A Royal Commission report. Retrieved from http://epe.lac-bac .gc.ca/100/200/301/pco-bcp/commissions-ef/abella1984 -eng/abella1984-eng.htm

Angus Reid Public Opinion. (2010, November 8). Canadians endorse multiculturalism, but pick melting pot over mosaic. Retrieved from *Angus Reid Global*: http://www .angus-reid.com/wp-content/uploads/2010/11/ 2010.11.08_Melting_CAN.pdf

Belanger, C. (2006, May). Canadian immigration history: Lecture plan. Retrieved from *Marianopolis College*: http:// faculty.marianopolis.edu/c.belanger/quebechistory/ readings/CanadianImmigrationPolicyLectureoutline.html

Bissoondath, N. (1988, September 5). Multiculturalism. Retrieved from *New Internationalist*: http://newint.org/ features/1998/09/05/multiculturalism/

Bissoondath, N. (1994). *Selling illusions: The cult of multicul-turalism in Canada.* Toronto: Penguin Books.

Bloemraad, I. (2010, October 28). Multiculturalism has been Canada's solution, not its problem. Retrieved from *The Globe and Mail*: http://www.theglobeandmail.com/ commentary/multiculturalism-has-been-canadas-solution -not-its-problem/article4330460/

Bransten, J. (2006, April 5). Netherlands leading trend to more stringent immigration rules. Retrieved from *Radio Free Europe*: http://www.rferl.org/content/article/1067418. html

Canadian Press (2010, March 16). Native gangs spreading across Canada. Retrieved May 2013, from *CBC News*: http://www.cbc.ca/news/canada/manitoba/ story/2010/03/16/mb-native-gangs-manitoba.html

Canadian Press (2012a, August 17). Asian-looking woman scientist image rejected for $100 bills. Retrieved from *CBC News*: http://www.cbc.ca/news/canada/ story/2012/08/17/pol-cp-100-dollar-bills-asian-scientist -image.html

Canadian Press (2012b, October 24). Census trends: Canada's many mother tongues. Retrieved from *CBC News*: http:// www.cbc.ca/news/interactives/cp-census/index-oct -mother-tongues.html

Evans, S. (2010, October 17). Merkel says German multicul-tural society has failed. Retrieved from *BBC News*: http:// www.bbc.co.uk/news/world-europe-11559451

Fleras, A., & Elliott, J. L. (2007). *Unequal relations: An intro-duction to race, ethnic, and Aboriginal dynamics in Canada.* 5th ed. Toronto: Pearson.

Fleras, Augie, and Elliott, Jean Leonard. (1992). *Multiculturalism in Canada: The challenge of diversity.* Scarborough, ON: Nelson Canada.

Galabuzi, G.-E. (2006). *Canada's economic apartheid: The social exclusion of racialized groups in the new century.* Toronto: Canadian Scholars' Press.

Gingrich, P. (2004, September). Sociology 211: History of multiculturalism in Canada. Retrieved from http:// uregina.ca/~gingrich/211s2904.htm

Glendon College, York University (2012, October 19). Borderlands, Transnationalism, and Migration in North America. *"The Mosaic vs. the Melting Pot? Myths and Realities of Cultural Pluralism in Canada and the United States."* Toronto: Glendon College, York University. Retrieved from https://borderlandsworkshop.wordpress. com/roundtable-discussion/

Goffman, E. (1956). *The presentation of self in everyday life.* Anchor Books.

Government of Canada (1988, July 21). Canadian Multi-culturalism Act. Retrieved from *Justice Laws Website*: http://laws-lois.justice.gc.ca/eng/acts/C-18.7/page-1.html

Gregg, A. (2006). Multiculturalism: A twentieth-century dream becomes a twenty-first-century conundrum. Retrieved from *The Walrus*: http://walrusmagazine .com/article .php?ref=2006.03-society-canada -multiculturalism&page=

Grewal, S. (2007, March 9). Punjabis in Peel warn of teen violence. *Toronto Star*. Retrieved from http://www.thestar .com/news/2007/03/09/punjabis_in_peel_warn_of_teen _violence.html

Hopper, T. (2011, December 2). Canada: As immigration booms, ethnic enclaves swell and segregate. Retrieved from *National Post*: http://news.nationalpost.com/ 2012/02/11/canada-as-immigration-booms-ethnic -enclaves-swell-and-segregate/

Hyman, I., Meinhard, A., & Shields, J. (2011). The role of multiculturalism policy in addressing social inclusion processes in Canada. Retrieved from *Centre for Voluntary Sector Studies*: http://www.ryerson.ca/content/dam/cvss/ files/new-WORKING-PAPERS/2011-3%20The% 20Role%20of%20Multiculturalism%20Policy%20in% 20Addressing%20Social%20Inclusion.pdf

Jovanovic, M. (2008, September 18). A thesis presented to the University of Waterloo as requirement of Master of Arts Degree in Sociology. *Cultural Competency in Hospice Care: A Case Study of Hospice Toronto.* Waterloo, ON: University of Waterloo. Retrieved from https://uwspace .uwaterloo.ca/handle/10012/3990

Keung, N. (2013, May 7). Toronto's immigrant enclaves spread to suburbs. Retrieved from *Toronto Star*: http:// www.thestar.com/news/immigration/2013/05/07/ torontos_immigrant_enclaves_spread_to_suburbs.html

Kymlicka, W. (2004). The Canadian model of diversity in a comparative perspective. University of Edinburgh: Eighth Standard Life Visiting Lecture.

Kymlicka, W. (2010, January 12). The current state of multi-culturalism in Canada and research themes on Canadian multiculturalism 2008–2010. Retrieved from *Citizenship* and *Immigration Canada*: http://www.cic.gc.ca/english/resources/publications/multi-state/section1.asp#evidence

Kymlicka, W. (2012, February). Multiculturalism: Success, failure, and the future. Retrieved from *Migration Policy Institute Europe*: http://www.migrationpolicy.org/pubs/multiculturalism.pdf

Lipset, S. M. (1990). *Continental divide: The values and institutions of the United States and Canada.* New York: Routledge.

MacDonald, A. (2013, March 20). Canada worries over 'deepening ethnic enclaves.' Retrieved from *The Wall Street Journal*: http://online.wsj.com/article/SB10001424127887324557804578372483418341140.html

Malik, K. (2011, December 6). Canada's multiculturalism is no model for Europe. Retrieved from *The Guardian*: http://www.guardian.co.uk/commentisfree/2011/dec/06/canada-multiculturalism-europe

Mosaic Institute. (2012, April 24). Younger Canadians believe multiculturalism works; older Canadians, not so sure. Retrieved from *Mosaic Institute*: http://mosaicinstitute.wordpress.com/2012/04/24/younger-canadians-believe-multiculturalism-works-older-canadians-not-so-sure

Nilsen, K. (2001). Deconstructing multiculturalism: How government policy is reflected in current education and practice. Retrieved from *Canadian Association of Information Science*: http://www.cais-acsi.ca/proceedings/2001/Nilsen_2001.pdf

Oreopoulos, P., & Dechief, D. (2011, September). Why do some employers prefer to interview Matthew, but not Samir? Retrieved from *Metropolis British Columbia*: http://mbc.metropolis.net/assets/uploads/files/wp/2011/WP11-13.pdf

Ryan, P. (2010). *Multicultiphobia.* Toronto: University of Toronto Press.

Satzewich, V., & Liodakis, N. (2010). *'Race' & ethnicity in Canada: A critical introduction* (2nd ed.). Canada: Oxford University Press.

Saunders, D. (2011, October 13). Sweden's big immigration idea: The 'Canada model.' Retrieved from *The Globe and Mail*: http://www.theglobeandmail.com/news/world/worldview/swedens-big-immigration-idea-the-canada-model/article618559/

Saunders, D. (2013, April 13). Time to make temporary foreign workers permanent. Retrieved from *The Globe and Mail*: http://www.theglobeandmail.com/commentary/time-to-make-temporary-foreign-workers-permanent/article11142634/

Shaidle, K. (2011, August 8). Canada's Somali problem. Retrieved from *Taki's Magazine*: http://takimag.com/article/canadas_somali_problem/print#axzz2UWI01PWE

Statistics Canada. (2003, September). *Ethnic diversity survey: Portrait of a multicultural society. Catalogue no. 89-593-XIE.* Retrieved from http://publications.gc.ca/Collection/Statcan/89-593-X/89-593-XIE2003001.pdf

Statistics Canada. (2010, March). Projections of the diversity of the Canadian population 2006–2031. Retrieved from

Statistics Canada: http://www.statcan.gc.ca/pub/91-551-x/91-551-x2010001-eng.pdf

Statistics Canada. (2013). Ethnic origin reference guide, National Household Survey, 2011. Retrieved from *Statistics Canada*: http://www12.statcan.gc.ca/nhs-enm/2011/ref/guides/99-010-x/99-010-x2011006-eng.cfm

Stoddard, A. (2012). The birth of Canada's multicultural policy: Plotting the official acceptance of diversity. Retrieved from *Dalhousie Faculty of Arts and Social Sciences*: http://arts.dal.ca/Files/Froese_Stoddard_Birth_of_Canada's_Multiculturalism_Policy_FA.pdf

Sun News Video Gallery. (2011, July 19). Edmonton worries. Retrieved from http://www.sunnewsnetwork.ca/video/search/somali/edmonton-worries/1066103814001

Todd, D. (2010, June 26). Multicultiphobia in Canada and beyond. Retrieved from *Vancouver Sun*: http://blogs.vancouversun.com/2010/06/26/multicultiphobia-in-canada-and-beyond/

Todd, D. (2011a, August 8). Trans-Atlantic poll shows Canadians have much to learn about immigration. Retrieved from *Vancouver Sun*: http://blogs.vancouversun.com/2011/08/08/trans-atlantic--poll-shows-canadians-have-much-to-learn--about-immigration/

Todd, D. (2011b, December 20). Ethnic mapping conclusion: As enclaves grow, will Metro residents' trust fade? Retrieved from *Vancouver Sun*: http://blogs.vancouversun.com/2011/10/20/ethnic-mapping-conclusion-as-enclaves-grow-will-metro-residents-trust-fade/

Trudeau, P. E. (1971, October 8). On multiculturalism: To the House of Commons. Retrieved from *Canada History*: http://www.canadahistory.com/sections/documents/Primeministers/trudeau/docs-onmulticulturalism.htm

Trudeau, The Right Honourable (2016, January 21). Justin Trudeau at World Economic Forum. Davos, Switzerland. Retrieved from http://qz.com/602525/justin-trudeau-perfectly-articulates-the-value-of-diversity-in-childhood-not-just-in-the-workforce/

Urzúa, F. (2012, March). The Canadian multiculturalism policy within the Colombian community. Retrieved from *Concordia Working Papers in Applied Linguistics*: http://doe.concordia.ca/copal/documents/4_Soler-Urzua.pdf

Wong, L. (2011). Questioning Canadian multiculturalism: Debunking the fragmentation critique of multiculturalism. Retrieved from *Diverse*: http://www.diversemagazine.ca/Culture-Questioning%201.htm

Yalnizyan, A. (2011, February 18). Canada's immigration policy: Who is on the guest list? Retrieved from *The Globe and Mail*: http://www.theglobeandmail.com/report-on-business/economy/economy-lab/the-economists/canadas-immigration-policy-who-is-on-the-guest-list/article1913178/

Yi, Sun-Kyung. (2004). An immigrant's split personality. In *The maple collection.* Toronto: Nelson Education Ltd.

CHAPTER 10

Religion

"Now some point a finger and let ignorance linger
If they'd look in the mirror they'd find
That ever since the beginning to keep the world spinning
It takes all kinds of kinds."

(Miranda Lambert, 2011)

LEARNING OUTCOMES

By mastering this unit, students will gain the skills and ability to:

- identify the diverse nature and the geographic distribution of major world religions

- demonstrate awareness of the changing religious demographics in Canada and the factors contributing to these changes

- demonstrate knowledge of the history of Canada's religious communities and challenges facing religious communities today

- examine the main components of religious accommodation as dictated by provincial laws

At a very basic level, religion divides the world into believers and non-believers, the enlightened and the unenlightened, or—at the very least—those who "know" and those who don't. At the same time, it has historically been one of the strongest sources of social cohesion because of that very fact. Communities were united as much in their beliefs and traditions as they were by their exclusion of the "other." But unlike race and gender, religion is—at least to a degree—a matter of choice.

According to the Pew Research Centre, 84 percent of the world's population identifies with a religious group (2012) (see Figure 10.1).

According to the 2010 demographic study, the approximately 5.8 billion people include 2.2 billion (32 percent) Christians, 1.6 billion (23 percent) Muslims, 1 billion (15 percent) Hindus, 500 million (7 percent) Buddhists, and 14 million (0.2 percent) Jews (2012). Additionally, more than 400 million (6 percent) practise traditional or folk religions such as African traditional religions, Chinese folk religions, and Native American and Aboriginal religions (Pew Research Centre, 2012). Approximately 58 million people (less than 1 percent) belong to other religions, such as the Baha'i faith, Jainism, Sikhism, Shintoism, Taoism, Tenrikyo, Wicca, and Zoroastrianism (Pew Research Centre, 2012). Additionally, the demographic study found that nearly 1.1 billion (16 percent) of the world's population have no religious affiliation, including many who have some religious or spiritual beliefs, but do not identify with a particular faith (Pew Research Centre, 2012). Geographically, the distribution of religions varies greatly, with many religious groups heavily concentrated in the Asia Pacific region (See Figure 10.2).

Many Canadians who embrace multiculturalism in theory draw the line at what they feel is a bending of Canada's laws to accommodate religious minorities. Combined with international and domestic news of religiously fuelled terror plots and child abuse scandals, it is no wonder that religion in the public sphere has become a perennial lightning rod issue. Many Canadians wonder if a united Canada is possible with various groups clinging to their different religious beliefs, values, and practices. This is not a matter of different ethnic food or dress—these are fundamental beliefs that seem to periodically conflict with what some would call "fundamental Canadian values." In fact, when opponents of multiculturalism speak about its failures, the examples they cite tend to revolve around religious disputes—either in the workplace, in domestic situations, or on the legal stage.

Most Canadians have a vague notion that religious freedom is protected by our Constitution, but the

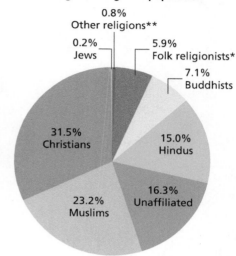

FIGURE 10.1

Major World Religions, 2010

Percentage of the global population

- 0.8% Other religions**
- 0.2% Jews
- 5.9% Folk religionists*
- 7.1% Buddhists
- 31.5% Christians
- 15.0% Hindus
- 16.3% Unaffiliated
- 23.2% Muslims

*Includes followers of African traditional religions, Chinese folk religions, Native American religions and Australian aboriginal religions.

**Includes Bahai's, Jains, Sikhs, Shintoists, Taoists, followers of Tenrikyo, Wiccans, Zoroastrians and many other faiths.

Percentages may not add to 100 due to rounding.

Source: "The Global Religious Landscape" Pew Research Center, Washington, DC (December, 2012), http://www.pewforum.org/2012/12/18/global-religious-landscape-exec/.

details are unclear. You will often hear people cite the concept of separation of church and state—assuming that Canada's stance is the same as that of the United States. But our history of religious freedom in Canada is different from that in the United States, and so are our successes and failures and the ways in which we have sought to protect it.

RELIGIOUS AFFILIATION IN CANADA

Before delving into issues of **religious accommodation**, or the reasonable duty that employers have to meet the needs of their employees' work and faith requirements, it is important to get an idea of where Canadians stand religiously speaking, according to the latest research. Although statistics are changing on an almost daily basis

> **Religious accommodation:** Arrangements made by an employer so that employees can do their jobs and practise their faith at the same time.

FIGURE 10.2

Major World Religions, by Country

Countries are colored according to the majority religion. Darker shading represents a greater prevalence of the majority religion.

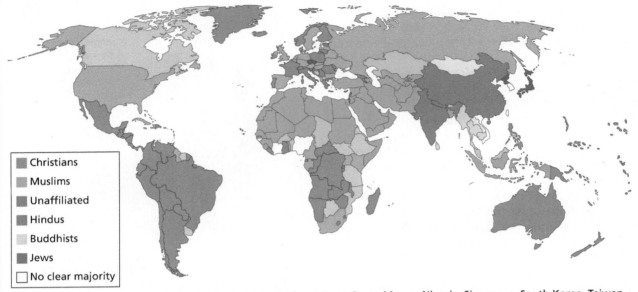

Legend:
- Christians
- Muslims
- Unaffiliated
- Hindus
- Buddhists
- Jews
- No clear majority

Nine countries have no clear religious majority: Guinea Bissau, Ivory Coast, Macau, Nigeria, Singapore, South Korea, Taiwan, Togo and Vietnam. There are no countries in which adherents of folk religions make up a clear majority. There are also no countries in which followers of other religions (such as Bahai's, Jains, Sikhs, Shintoists, Taoists, followers of Tenrikyo, Wiccans or Zoroastrians) make up a clear majority.

Source: "The Global Religious Landscape," Pew Research Center, Washington, DC (December, 2012), http://www.pewforum.org/2012/12/18/global-religious-landscape-exec/.

because of an aging population and immigration, the numbers tell an interesting story of Canadian society. As of the 2006 census, the question on religious affiliation was not specifically addressed, so any numbers beyond the 2001 census are scientific and statistical projections. The latest actual data we have are from the 2001 census; however, the National Household Survey conducted by Statistics Canada in 2011 gives us a more detailed picture of religious affiliation in Canada (see Figure 10.3).

With recent immigrants (47.5 percent) and the overall Canadian population (76.3 percent), Christianity remains the largest religious group in Canada (Statistics Canada, 2011). The Roman Catholic Church maintains primacy as the oldest and most adhered-to church in Canada, claiming 43 percent of the population. But this was not always the case. For more than 100 years, the Protestant churches of Canada far outnumbered the Catholic churches. Studies of immigration patterns made the reasons obvious; until the early 1960s, immigration pools came primarily from Protestant-majority countries.

Protestants remain the second-largest religious group in Canada at 27 percent of the Canadian population (Statistics Canada, 2001). Together, Roman Catholics and Protestants represent roughly 70 percent of Canadians (Statistics Canada, 2001). This percentage may seem high; but it should be noted that in 1851, these two churches claimed 98 percent of the population. Even in 1951, this number was still 96 percent (Statistics Canada, 2001).

However, there is a group whose numbers fall below those self-identifying as Catholic but above those claiming Protestantism. One of the fastest-growing groups in Canada is made up of those marking "no religion" on census forms. The National Household Survey indicates that 23.9 percent of the overall population and 19.5 percent of recent immigrants do not identify with a particular faith (Statistics Canada, 2011). Of the almost 2 million immigrants to arrive in Canada between 1991 and 2001, fully one-fifth claimed no religious affiliation. This was particularly true of immigrants arriving from China, Hong Kong, and Taiwan. Of those reporting no religious

FIGURE 10.3

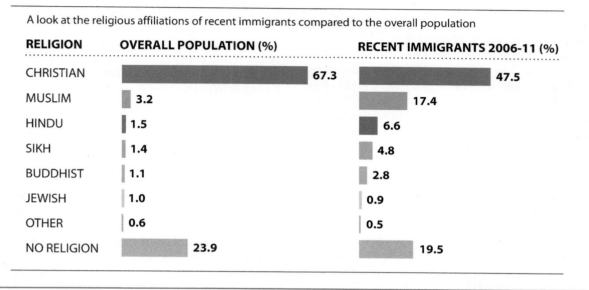

Religious Affiliation in Canada

A look at the religious affiliations of recent immigrants compared to the overall population

RELIGION	OVERALL POPULATION (%)	RECENT IMMIGRANTS 2006-11 (%)
CHRISTIAN	67.3	47.5
MUSLIM	3.2	17.4
HINDU	1.5	6.6
SIKH	1.4	4.8
BUDDHIST	1.1	2.8
JEWISH	1.0	0.9
OTHER	0.6	0.5
NO RELIGION	23.9	19.5

Statistics Canada and Sean Vokey/The Canadian Press

affiliation, younger Canadians are far more represented than older Canadians. For Canadians over the age of 65, 94 percent reported some degree of religious affiliation in 1991 versus 86 percent of those between the ages of 15 and 44. Further, those over 65 were more likely to be affiliated with Catholicism, Protestantism, or Judaism. Those between 15 and 44 were more likely to be affiliated with Eastern religions, such as Islam, Hinduism, Buddhism, and Sikhism (Statistics Canada, 2001).

The fastest-growing religious groups in Canada represent the latest immigration demographics. Between 1991 and 2001, the population of Muslims in Canada doubled from approximately 250 000 to over 500 000. Muslims went from 1 percent to 2 percent of the Canadian population in only 10 years (Statistics Canada, 2003) and the National Housing Survey indicates that as of 2011, 3.2 percent of the overall Canadian population are Muslims (Statistics Canada, 2011). According to Pew Research, the estimated number of Muslims in 2010 was close to 1 million, and projections for 2030 put that number at well over 2 million (Pew Research Center, 2011). Hindu, Sikh, and Buddhist populations also showed significant increases, due mostly to immigration patterns. Each of these groups rose by over 80 percent between 1991 and 2001, and the National Housing Survey illustrates that Hindus comprise 1.5 percent

of the overall Canadian population, with Sikhs at 1.4 percent, and Buddhists at 1.1 percent (Statistics Canada, 2011). Future projections by Statistics Canada predict that in the near future, one of every six residents in the greater Toronto area will either be majority Muslim or Hindu, and that together these two groups in Toronto will make up 1 million people (Jedwap, 2005). While these numbers correspond to the increase of immigration from predominantly Muslim and Hindu countries (as well as Sikh and Buddhist countries to a slightly lesser degree), they also represent the age of the immigrants as well as birth rates among them. While the Anglican population is a predominantly aging one, for example, the median age of Muslims, Sikhs, and Hindus in 2001 was 28, 30, and 32 respectively (Statistics Canada, 2003).

Rounding out the major religious denominations in Canada are those who report adherence to one of the Orthodox Christian denominations. Greek, Ukrainian, Serbian, and Russian Orthodox Church members numbered about half a million in 2001, with the latter two denominations more than doubling between 1991 and 2001. Though historically the largest Orthodox groups, both Greek and Ukrainian churches reported a steady decline (Statistics Canada, 2003).

Finally, the number of adherents of the Jewish faith also increased slightly and now accounts for 1.0 percent of the Canadian population, with half of these adherents living in Ontario (Statistics Canada, 2011).

The Future of Religion in Canada

Churches have long been tied to Canadian heritage, and the loss of the central place of their structures has many people worried about the loss of Canadian heritage as a whole. The panic over the disappearance of a Christian Canada may be legitimate, but religious accommodation is not to blame.

Over the last few decades, mainstream church attendance has declined so steadily that "For Sale" signs on church lawns have become a common sight. A report prepared for the Anglican diocese of British Columbia reiterated the dire findings of a previous study claiming that with current rates of decline, there would only be one Anglican church left in Canada by 2061. While the study does not go into specifics of replacement numbers either by birth or immigration, the results—even if taken with scepticism—show that the once socially, morally, and even politically powerful Anglican church has suffered a substantial blow (Valpy, 2012).

In Montreal, the Très-Saint-Nom-de-Jésus Church—once bursting at the seams with over 1000 parishioners—is now barricaded with signs warning of safety hazards as the building's century-old structure crumbles. Its $2.5 million organ is having a hard time finding a new home, despite the church's willingness to give it away for free.

By 1985, the percentage of Canadians between ages 45 and 63 who attended church on a weekly basis had already dropped to 39 percent. By 2005, this number had plummeted further to only 22 percent of the population (Lindsay, 2008). The largest Protestant denomination in Canada—the United Church—averages one church closing per week (Peritz, 2010).

These realities and projections have created a shift in Canadian society, where some are more likely to call themselves "Christian" rather than ascribing themselves to specific denominations, and many are more likely than ever to check the box marked "no religion" on census forms (Grossman, 2010).

Compare these numbers to those of recent immigrants, and a different trend emerges. Canadians who have immigrated in the last 20 years tend to be at least as regular in religious attendance as they were in 1985. It is no wonder that the narrative spun is often one of encroaching foreign traditions taking over Canadian ones. The irony here is that religious accommodation and pluralism are actually a hallmark of the most Canadian value of all—multiculturalism. And when immigrants come to Canada, they come with this much vaunted value in their minds.

PLURALISM AS A FUNDAMENTAL PRINCIPLE

Among the four fundamental rights guaranteed by the Canadian Charter of Rights and Freedoms is the freedom of conscience and religion. Article 27 states: "This Charter shall be interpreted in a manner consistent with the preservation and enhancement of the multicultural heritage of Canadians." Ideally, this means that everyone living in Canada is free to believe (or not believe) whatever he or she chooses. It also means that the government is responsible for eliminating barriers for those wishing to practise their religion—even if governmental inconvenience or cost is the result. From a diversity perspective, the problems surface when the borders of one Canadian's rights infringe on the borders of another's. The Supreme Court reviews case after case where individuals or groups deem their rights of religion or conscience to have been violated. We will review some of these cases later in the chapter.

Whether claimed by the French or the British, early Canadians were not seeking religious autonomy from their mother countries. For the most part, French colonists carried on the Roman Catholic traditions of France while their English counterparts held fast to the Church of England. As England gained the political upper hand, Anglicanism also became part of the Canadian identity. In fact, part of the process of "Anglicization" of Aboriginal children included indoctrination of religious beliefs and removal of any beliefs or practices not sanctioned by the Church. The Reading selection at the end of this chapter explores Aboriginal spirituality further. Though the Church has since apologized and attempted reparations against the abuses of the residential schools, the message seems to have remained in the cultural ether: to be Canadian meant to belong to the Church.

In his book *Foreigners or Canadians*, C. J. Cameron, the 1913 assistant superintendent of the Baptist Home Mission, draws this point out very clearly:

> We must endeavor to assimilate the foreigner. If the mixing process fails, we must strictly prohibit from entering our country all elements that are non-assimilable. It is contrary to the Creator's law for white, black or yellow races to mix together. If the Canadian civilization fails to assimilate the great mass of foreigners admitted to our country, the result will be destruction to the ideals of a free and nominally Christian nation which will be supplanted by a lower order of habits, customs and institutions ... we shall Canadianize the foreigner by Christianizing him. (Slater, 1977, p. 29)

In an attempt to be more inclusive, some public organizations are not only decorating for Christmas, but also for the Jewish festival of Hanukkah, the Muslim Festival of Eid al-Adha, and the Hindu Festival of Diwali, even though most occur months before Christmas. Is this an example of inclusivity or can it be regarded as tokenism? Should public organizations do away with Christmas decorations all together?

With the Charter of Rights and Freedoms, this message was unambiguously knocked down, but not everyone in Canada would disagree with Cameron's notions. In 1985, the Supreme Court of Canada ruled in *R. v. Big M Drug Mart Ltd.* that a law requiring businesses to shut down on Sundays—the Lord's Day Act—was a violation of Charter Rights. While you can still find businesses that choose to close on Sundays, the court ruled that requiring a business to close on the traditional Christian day of rest was illegal and unconstitutional. Since that case, many of the inherent assumptions about Canada being a "Christian nation" have been challenged. As the faces of immigrants become more and more prominent on the streets of Canada, as city halls across the country add Diwali, Hanukah, and Eid decorations to the traditional nativity scenes, the media reports widespread panic about the loss of a "Canadian identity."

Are fundamental Canadian values being threatened by the demise of Christian Canada? Is our social cohesion and democratic way of life being sacrificed to pander to religious diversity? What about issues of gender equality, where foreign values clash with Canadian ones?

LEGISLATING ACCOMMODATION

While the Supreme Court judges large landmark cases that determine the constitutionality of lesser court decisions, the provincial **human rights commissions**, organizations set out to investigate, protect, and advocate for the rights of individuals, do the everyday work of determining the limits and requirements of religious accommodation. These commissions have jurisdiction over issues around accommodation that arise in the workplace and the marketplace. According to provincial Human Rights Codes (HRCs), discrimination based on creed (religion) is against the law. For example, according to the Ontario Human Rights Commission, it is the responsibility of the employer to accommodate someone's religious requirements, as long as the accommodation does not cause undue hardship, cost, or safety issues (OHRC, 2012). Seems pretty straightforward, right? But everything in between—from defining what constitutes a creed to what constitutes hardship and cost—can get murky and inflame tempers like almost no other issue can.

According to the Policy on Creed and Accommodation of Religious Observances, creed:

> ... is interpreted to mean "religious creed" or "religion." It is defined as a professed system and confession of faith, including both beliefs and observances or worship. A belief in a God or gods, or a single supreme being or deity is not a requisite.
>
> Religion is broadly accepted by the OHRC to include, for example, non-deistic bodies of faith, such as the spiritual faiths/practices of aboriginal cultures, as well as *bona fide* newer religions (assessed on a case by case basis). (OHRC, 1996, p. 4)

The definition of creed itself is a subjective one. Making matters even more complicated is that the practices people sincerely believe to be necessary to their creedal beliefs are protected—whether that system of beliefs is officially considered a religion or not. For example, Falun Gong—defined and outlawed in its native China as a cult—is given creedal status and its members are protected against discrimination. As a result, in May 2011, Ottawa resident Daiming Huang was awarded $15 000 in damages when the Ontario Human Rights Commission ruled that she had been discriminated against by her local senior centre for practising Falun Gong (CBC News, 2011). At the heart of the law is the prohibition against religious discrimination based on the two parties not sharing the same religion. This applies regardless of whether either or both parties are

Human rights commission: A national or international organized body that investigates, protects, and advocates for the rights of human beings.

minority groups or whether either party has no religious beliefs. In other words, religious accommodation laws protect atheists from discrimination, too.

The duty to accommodate is most often seen in the workplace in issues of dress, holidays, and break times. We will review specific cases of accommodation; but in general, employers are duty-bound to accommodate unless there are reasonable occupational reasons for them not to do so. As for what constitutes "hardship," this, too, is considered on a case-by-case basis. There are no universal monetary standards above which an employer is not obligated to expend. Also, what is possible in accommodating one employee with religious requirements may not be possible in future cases with multiple employees. Safety may constitute another form of hardship. If a particular uniform must be worn as part of a job necessity and cannot be modified in such a way as to accommodate religious requirements (e.g., a head covering), then employers may cite this matter as a legitimate lifting of their duty to accommodate.

The duty to accommodate is based in a good faith understanding that the employee's request is a legitimate religious one. The employee is not required to give evidence of the validity or obligatory nature of the accommodation, whether in dress code or time off. It is enough that the employee holds a sincere religious belief and holds it consistently in other areas of his or her life. An obvious violation of this trust would be someone asking for religious time off and using that time off to engage in other employment.

What Is Not Covered by the Law

No matter how strongly someone believes in certain political views, they are not protected by religious accommodation laws. An employee cannot demand time off to attend a political rally in which he or she fervently believes. Also not applicable are practices that incite violence against others or in any way violate international human rights standards or national criminal codes. A religious tradition that involves virginal sacrifices on a yearly basis would, by this principle, be as illegal as any other homicide in Canada.

Finally, religious accommodation is not intended to elevate one set of religious practices over another. It is intended to level the playing field so that all Canadians have equal access to the Charter of Rights and Freedoms, regardless of their religious views.

You Be the Judge

Many discussions of Canada's religious accommodation standards fail to recognize how nuanced and well researched the judicial decisions are. Looking at

IN THEIR SHOES

If a picture can say a thousand words, imagine the stories your shoes could tell! Try this student story on for size – have you walked in this student's shoes?

Although I am not a religious person, my experience was definitely inspiring. I had a bout with depression and decided to move back in with my mother for a short time to recover. Finding it difficult to find work in my hometown, I was offered a job through my sister-in-law at Digital Attractions at the Falls. It was here that I met two very influential people who would change my life forever. Their names were Siddiq and Ahmed. They were from Pakistan and they were Muslim. Now, my first thoughts on them were stereotypical and disrespectful. Only a few years had passed since 9/11, and it was clear that I had a biased opinion that was more media driven than factual.

After several weeks of working side by side with them, I realized that I had a lot in common with them. We shared common interests like music, sports, news, TV shows, and we all became very close friends.

I was asked to come to an all-you-can-eat Chinese buffet one Saturday night with Siddiq, Ahmed, and another friend from Turkey named Tai. I loaded up my plate, unaware that Muslims do not eat pork. I offered some sweet and sour pork to Tai and he refused, saying that he could not eat it. This is where I began to ask many questions about their faith. I have always been the type of person to seek out knowledge of things I do not fully understand, and Ahmed took a more personal interest. He invited me to come with his wife Sarai and their son Hamad to the Islamic Society of Niagara on the following Friday, to learn about their religion. Naturally, I did not want to feel out of place, so I decided to meet with them at their house for dinner on the Thursday to ensure my understanding of their customs.

It was like stepping into a different world being there. Sarai took me under her wing and taught me an English version of prayer and showed me how to perform *wudu*. Wudu is a cleansing ritual that all Muslims perform before they enter into prayer. She also taught me how to wear a *hijab*, which, I admit, I found constricting and degrading. I was told prior to my visit that I would not be required to wear it, as it is not mandated by the *Qur'an*. It was adopted from the teachings of Muhammad and it is more so a personal choice for modest female dress. Given my want for a full cultural experience, I donned the garb and met with them Friday morning at the mosque.

Now, I felt uncomfortable because I was one of two white women in the building, and I had to pray in a room separate from Ahmed with Sarai. There were speakers in the room so we could hear the Imam and a window that showed the prayer floor. I had originally thought that women in all Islamic societies were not allowed to pray with the men. On closer inspection, I realized that the reason I was in the room was because Sarai had her young son with her and I was actually in a room for "little ones" who are learning with their mothers. The prayer floor was a myriad of the sexes, although the women were closer to the back, and there wasn't actually any separation or discrimination. I felt ashamed of my preconception and was quite relieved that I was in the room to practise without making a fool of myself for doing any actions wrong.

After the service, it was time to meet in the cafeteria. I had assumed that I would be stared at because of my obvious minority, and the fact that I was having difficulty with my hijab as it kept sliding down the back of my head. I was surprised as I was welcomed by their congregation, and the attention was more than surprising. I was lavished with attention and praise, and although I made it clear that I was there for educational understanding, the people were happy that I was taking the time to understand their culture and religion.

Islam is not just a religion to Muslims. It is a way of life and a peaceful understanding of their community. It was not what I thought it was going to be at all. I felt spiritual gratification, which I will admit aided in my depression recovery. I did not need Islam for this spiritual gratification; however, it needed me. I say this because most of the people I talked to after that were disgusted that I was practising Islam, including family members. The words *Al-Qaeda*, *Osama Bin Laden*, *terrorist*, and *anti-Christian* were thrown around regularly. I realized my part in this experience was to educate the Islamophobics on what I had learned. I do not make a habit of preaching religion. I consider myself an atheist; however, this experience helped me understand Islam and help others understand that it's not what the media and word-of-mouth makes it out to be.

My experience has aided in my understanding and appreciation for Islam and I wouldn't have changed it for the world.

Shannon Engemann

some landmark cases gives a fuller picture of Canada's means of accommodating its increasing religious diversity. The success of the Canadian multicultural dynamic has encouraged many European countries to re-evaluate their own policies regarding diversity in recent years.

Case i: Chambly v. Bergevin

In Chambly, Quebec, in1994, three Jewish teachers working for the Catholic school board were given an unpaid day off to celebrate Yom Kippur. The teachers' union raised a grievance, claiming that the loss of pay was discriminatory. The courts agreed. Their finding was that although Good Friday and Christmas had essentially become statutory (non-religious) holidays, they are historically and remain Christian-based holidays, and those who did observe them were not penalized for not working on those days. The Jewish teachers, however, were in essence being monetarily penalized for seeking the same accommodation as their Christian co-workers. Supreme Court Judge Justice Peter Cory explained:

> If a condition of work existed which denied all Asian teachers one day's pay, it would amount to direct discrimination. ... The loss of one day's pay resulting from direct discrimination would not be tolerated ... and would fly in the face of human rights legislation. Similarly adverse effect discrimination resulting in the same loss cannot be tolerated unless the employer takes reasonable steps to accommodate the affected employees. (OHRC, 1996, p. 13)

The general principle that emerged from this case has and continues to have far-reaching effects. Employers are required to honour employees' requests for religious days off as long as the request does not cause undue hardship. To demonstrate equal treatment requires that at least two (in some cases three) paid days off be available to those requesting religious leave.

Case ii: Saadi v. Audmax Inc.

In 2009, Seema Saadi brought a religious discrimination suit against Audmax Inc., a corporation that places Canadian newcomers in the workplace. Ms. Saadi's placement with Audmax was terminated because the company deemed that she was not a "good fit." Part of the misfit was caused by Ms. Saadi's wearing of a Muslim hijab or headscarf, which Audmax referred to as a "cap" and inappropriate business attire. Ms. Saadi took her case to the Ontario Human Rights Tribunal, which found that Audmax was guilty of discrimination (CanLII, 2009).

Case iii: Jehovah's Witness v. Hospital

In Manitoba, a young Jehovah's Witness claimed that her rights had been violated when she was hospitalized and given a blood transfusion without her consent. Because she was a minor at the time of her hospitalization, her refusal of a transfusion was overruled because the transfusion was deemed a medical necessity. The doctors contacted Manitoba Children's Services, who intervened and got the courts to order the transfusion. In 2009, as an adult, she took matters to the Supreme Court. Although they upheld the original court's decision and the hospital's actions, the judges concluded that future cases would have to take into account a minor's maturity level in cases of enforced treatment and religious refusal (CBC News, 2009).

Case iv: Hutterian Brethren of Wilson County v. Province of Alberta

In 2003, Alberta modified a driver's licence policy that had previously allowed religious groups opposed to having their photograph taken to obtain a photo-less licence—a special condition "G" licence. The new regulations required that a photo be provided and stored in the province's facial recognition data bank. These new regulations were designed to combat the growing problem of driver's licences used in identity theft.

Members of the Hutterian Brethren of Wilson County objected to the photographs on the grounds of religious principle. In their understanding of the Bible, the second commandment prohibits the taking of images. While the Supreme Court did concede that their beliefs were sincerely held and that the taking of the photographs would violate their religious rights, the court ruled in favour of the universal photo requirement anyway. Their decision stated that the province's necessity of a universal photo bank to combat fraud outweighed the religious rights of the Hutterian Brethren. Chief Justice Beverley McLachlin wrote:

> The law does not compel the taking of a photo. It merely provides that a person who wishes to obtain a driver's license must permit a photo to be taken for the photo identification data bank. Driving automobiles on highways is not a right, but a privilege. While most adult citizens hold driver's licenses, many do not, for a variety of reasons. (Ceballos, 2009)

Case v: Friesen v. Fisher Bay Seafood Ltd.

The courts have also ruled that someone's sincerely held beliefs requiring them to preach in the workplace are not covered by the Charter of Rights and Freedoms. In Sidney, Vancouver Island, in 2008, Seann Friesen's employment with Fisher Bay Seafood was terminated because he refused to stop preaching to fellow employees. A number of employees complained and even threatened to walk off the job if Mr. Friesen was not stopped. By definition, Friesen's termination was religious discrimination because his religious beliefs and practices—not his competence as an employee—ended his employment. However, upon reviewing the case, the British Columbia Human Rights Tribunal found that Fisher Bay Seafood had tried to accommodate Friesen in a reasonable manner: the company allowed him to preach during non-work hours to those employees willing to listen, for example (CanLII, 2009).

"CAN'T WE ALL GET ALONG?"

Hate crimes in Canada, targeted against victims belonging to specific social groups, have shown a steadily increasing trend over the last few decades, with a slight decline between 2009 and 2010. Although race and ethnicity continue to be the highest motivators for police-reported hate crimes, religion is a close second (see Table 10.1).

Religion accounts for approximately 29 percent of all hate crimes, and this percentage did not decrease with the overall decrease in hate crimes between 2009 and 2010. The Jewish community continues to have the highest number of crimes directed toward them, but the percentage of Jewish-targeted crimes has declined, while the number of hate crimes directed toward Muslims (or those "suspected" of being Muslim) has increased significantly between 2012 and 2013 (see Table 10.2).

In Canada, the number of **interreligious**, or **interfaith**, **marriages** continues to climb, with nearly one in five Canadians married to someone from outside their faith tradition. Before we read too much into these numbers, it is important to note that half of these marriages are between Protestants and Catholics. However, even as recently as the middle of the last century, these unions would have been unacceptable to most Canadians. Also increasing in interreligious unions are marriages where one partner—usually the man—is Jewish, while the other partner is not. Sikhs, Hindus, Muslims, and evangelical Protestants were the least likely to be engaged in interreligious

Interreligious, or interfaith, marriages: Occur when two individuals who believe in different faiths marry.

TABLE 10.1

Police-Reported Hate Crimes in Canada, 2013

Type of Motivation	Percentage of Hate Crimes			
	2012		2013	
	Number	%	Number	%
TOTAL	1414	100	1167	100
Race or Ethnicity	704	51	585	51
Religion	419	30	326	28
Sexual Orientation	185	13	186	16
Language	13	1	15	1
Sex	10	1	9	1
Disability	8	1	6	1
Age	4	0	3	0
Other (Occupation or Political Belief)	47	3	27	2

Source: Statistics Canada, Police-reported hate crimes, 2013, http://www.statcan.gc.ca/daily-quotidien/150609/dq150609a-eng.pdf. Reproduced and distributed on an "as is" basis with the permission of Statistics Canada.

TABLE 10.2

Police-Reported Religiously Motivated Hate Crimes in Canada, 2012–2013

Targeted Religion	Percentage of Hate Crimes			
	2012		2013	
	Number	%	Number	%
TOTAL	419	30	326	28
Jewish	242	17	181	16
Muslim	45	3	65	6
Catholic	37	3	29	3
Other	54	4	41	4

Statistics Canada, Police-Reported Hate Crime in Canada, 2010. Reproduced and distributed on an "as is" basis with the permission of Statistics Canada.

marriages, with the latter two groups representing 11 percent of the total numbers of interreligious marriages. One study notes that when a person's parents are in an interfaith relationship, there is a higher likelihood of that person engaging in an interfaith marriage (Clark, 2006). And as with all types of mixed marriages, the broadening effects of the relationship go far beyond the couple themselves and extend to immediate and distant family members.

The Elephant in the Mosque

It would be disingenuous to pretend that all religions are looked at with equal levels of antipathy and suspicion. The concern with religious identities obstructing national social cohesion and threatening democratic values and principles centres on the fear of an Islamic takeover of the Western world.

Though Muslims have been in Canada since shortly after Confederation, the last few decades have seen a growing unease about their ability to live peaceably and integrate seamlessly in Canada. According to a recent survey (Csillag, 2012), 52 percent of Canadians distrust their Muslim compatriots. The vast majority blamed Muslims themselves for this distrust; despite the fact that when prodded, "nearly half of those surveyed, 49 percent, listed the Internet as the number one source of racism and prejudice" (Csillag, 2012).

With the Canadian Muslim population expected to triple by 2031 (Pew Research Center, 2011), distrust of nearly 3 million Canadians will only be exacerbated as communities become more isolated, and as rhetoric—especially on anonymous Internet pages—becomes more radical. The facts are a great deal more reassuring: Canadian Muslims, more than their counterparts in Europe and even in the United States, report great satisfaction with their lives in Canada. Fully 80 percent of Canadian Muslims, polled by the CBC and Environics, reported that they were satisfied. This number is higher than the general Canadian satisfaction level, with only 61 percent of the population reporting satisfaction with their lives. As for how Muslims think they are perceived by their fellow Canadians, only a small minority (17 percent) felt they were perceived with hostility (CBC, 2007). The picture that emerges from the survey is one of a community that experiences Canada and Canadian values positively and has no intention of isolating itself.

When ethnic enclaves are created, it is usually for the same reasons we find in other communities— language, culture, and (most importantly) economics. Though not all Canadian Muslims are immigrants, many are, and a sizable number have immigrated under duress, leaving behind war-torn homelands or social conditions in which they or their families were in danger. Canada witnessed large numbers of Lebanese refugees in the 1980s, and many Somali and Bosnian refugees came to Canada in the 1990s. The circumstances of immigration and landing play a huge role in how long it takes immigrants to establish themselves in their new country. For those who came as refugees, that process naturally takes longer, as basic survival and living needs take precedence over anything else. In Canada, gateway cities like Toronto, Vancouver, and Montreal do have neighbourhoods that are heavily populated by distinct ethnic communities. For many immigrants, these neighbourhoods are natural stepping-stones where they find others with whom they can speak their language, eat similar foods, and share similar values, until they are settled enough to branch out away from these neighbourhoods.

Muslim immigrants in Canada face the same economic struggles we see with other immigrants, if not more of them. The result of a major study conducted by University of Toronto professor Jeffrey Reitz found that "skin colour—not religion, not income—was the biggest barrier to immigrants feeling they belonged here. And the darker the skin, the greater the alienation" (Taylor, 2009). As many Muslim immigrants have darker skin, they have to contend with anti-Muslim sentiment as well as general issues of discrimination faced by darker-skinned non-Muslims.

Finally, one of the most commonly cited concerns with regards to the Muslim community in Canada is the accusation that Muslim immigrants will bring with them the stereotypical misogynistic tendencies with which their community has been associated. Will Canada's liberal attitude toward religious minorities lead to gender discrimination, forced veiling and forced marriages, honour killings, and the like? The well-publicized case of Aqsa Parvez, a 16-year-old girl of Pakistani descent who was murdered by her father in a so-called "honour killing," certainly shone a spotlight on these fears. Dr. Amin Muhammad, a psychiatrist at Memorial University in St. John's, Newfoundland, suggests that while honour killings have no place in Islamic law, defence teams may be attempting to use the term and its cultural connotations to present a case for more lenient sentences (Cohen, 2010).

For opponents of multiculturalism and diversity, these events—however rare—fuel the misconceptions that Muslims struggle to integrate and allowing immigration from Muslim countries to continue would be dangerous. It is important to consider that whether or not certain cases of honour killings lead to guilty verdicts, the courts' message must be clear—that there will be no leniency toward misplaced cultural sensitivity when it comes to violating the laws of the land.

INTERSECTIONALITY

Race and Religion

In *Psychology Today*, Dr. Todd Essig warns of the dangers of confusing race and religion. He speaks of a "race-equals-religion" ideology that has emerged in Western culture, calling it dangerous and outrageous (Essig, 2010). He explains that regarding people of different religions as distinct races, is problematic because it erases racial differences altogether (Essig, 2010). Lumping together Latin American Catholics with Irish Catholics or Indonesian Muslims with Lebanese Muslims ignores aspects of race and ethnicity that may be central to a person's identity formation (Essig, 2010). Simran Jeet Singh agrees, arguing that when the media lumps together people from different parts of the world who practise entirely different religions, it results in an increase in hate crimes (Singh, 2014). Singh argues that a new racial category has emerged, which he calls the "apparent Muslim," a category that combines supposed racial and religious features, associating brown skin, facial hair, and turbans with supposed notions of terrorism (Singh, 2014). He warns that anyone who makes these assumptions about the "apparent Muslim" is not only guilty of overgeneralizing about an entire community based on the actions of a small minority of the Muslim population, but is also including a number of communities, including Arabs and Sikhs who do not identify with Islam in that misconception (Singh, 2014). Singh argues that this lack of understanding of global cultures leads to misidentification and misconceptions that are based on "little more than xenophobia" (Singh, 2014).

Although the intersection of race and religion often compounds oppression, it can also magnify privilege. Elizabeth Durant, a pastor at the United Church of Christ, speaks of her experiences on the intersection of race and religion. She talks about her awareness of her "white privilege" and how she became conscious of the way her spiritual journey is also a form of privilege, because in her cultural context, there is a "subtle, pervasive assumption that if you practise a religion, that religion is Christianity" (Durant, 2013). She argues that this bias is not only apparent in popular culture, but also in work holidays, children's stories, and political jargon (Durant, 2013). She states that while she cannot single-handedly change the cultural context, she can resist it by holding herself and others accountable for "acts of religious racism: tokenizing other faiths or appropriating indigenous religions" (Durant, 2013). You will have a chance to explore the concept of religious privilege in more detail in the KWIP section at the end of this chapter.

Age and Religion

When we look at the intersection of age and religion in Canada, age seems to be influencing the growth of the religiously unaffiliated, where recent generations are significantly less affiliated than earlier generations (Pew Research Centre, 2013). (See Figure 10.4.)

Tendisai Cromwell

Photographer: Reza Dahya

Tendisai Cromwell came to Canada from Zimbabwe when she was six years old and has become passionate about telling the untold stories of the Muslim world. She wasn't born Muslim, but embraced the Islam faith in her first year of university after a high school philosophy class broadened her thinking about faith and spirituality (Mirza, 2015). Cromwell is a Toronto-based filmmaker, writer, and journalist who began her career in Palestine as a reporter in 2009 (Mirza, 2015). Since then, she has written and produced segments for CBC's *The Current*, while completing her Masters of Journalism at Ryerson University (New Narrative Films, 2016). In 2014, she founded New

Narrative Films, a production company that is "committed to creating films that explore faith & spirituality, involve diverse communities,and examine the nuances of identity and belonging" (New Narrative website). New Narrative Films has produced three documentaries to date: *Mosque One: Home of Toronto's First Muslims*, a documentary that tells the story of Toronto's first mosque, profiling some of the early pioneers and supporters; *Educational Attainment West: From Pilot to Legacy*, a documentary that highlights the achievements of Educational Attainment West, an organization committed to promoting growth with youth in Toronto's underserved neighbourhoods; and *The Spirit of Social Change*, a documentary that follows two social activists as they explore their evolving relationship with spirituality and its influence on personal transformation and social change (New Narrative Films, 2016). Her current project is a film illustrating the diversity in roles of Muslim women. Tentatively titled *The Way of Aisha*, this documentary explores the ways in which women contribute to Islam's scholarly tradition and shape religious discourse in North America (Mirza, 2015). Through her work, Tendisai Cromwell is fostering conversation and dialogue around social issues involving faith and spirituality, and promoting pluralism in Canada. To learn more about her work, visit her website: http://www.newnarrativefilms.com/.

FIGURE 10.4

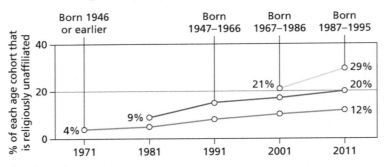

Trends in Canadian Religious Disaffiliation, by Generation

Source: "Canada's Changing Religious Landscape," Pew Research Center, Washington, DC (June, 2013), http://www.pewforum.org/2013/06/27/canadas-changing-religious-landscape/.

According to the Pew Research Centre, 29 percent of Canadians born between 1987 and 1995, and 29 percent of those born between 1967 and 1986 identify as having no religious affiliation as of 2011 (Pew Research Centre, 2013). These numbers are much higher than earlier generations, with only 12 percent of the religiously unaffiliated born in 1946 or earlier, and 20 percent born between 1947 and 1966 (Pew Research Centre, 2013). Additionally, as some generations age, disaffiliation rates also increase. For example, rates of disaffiliation for those born between 1947 and 1966 have doubled from 10 percent in 1981 to 20 percent in 2011, and rates of disaffiliation with Canada's older adults, those born in 1946 or earlier, have also increased from 2 percent in 1970 to 12 percent in 2011 (Pew Research Centre, 2013).

Finally, when we look at age and religious affiliation, according to the 2001 Canadian census, the median age of Canadians who identified as religious was 37.3 years old (Pew Research Centre, 2013). Canadians who identified as Protestant, Jewish, Greek Orthodox, Buddhist, and Roman Catholic had a higher median age than the total Canadian population, while those who identified as Hindu, Sikh, Muslim, and no religion had a lower median age than the total Canadian population (Pew Research Centre, 2013).

ENDING THOUGHTS

What is the future of religion in Canada? Monica Toft, associate professor at Harvard's Kennedy School of Public Policy, argues that religions have outlived any number of social and political ideologies of the past few centuries. The inconvenience and violence caused by religious adherents is less a sign that religion is on its way out than an indication that religion will continue to make itself seen and heard in a modern, **secular** world. Those waiting for religion to go away and take with it all the problems that religious differences cause will have to wait a long time. Those who worry that Canada's religious reality will be unrecognizable compared to the demographic realities at the time of Confederation have reason to worry; however, as Canada marches toward **secularism**, its immigrants come with their religious traditions intact, creating more possible fault lines in the multicultural divide, and also creating the growth that diversity can bring (Friesen & Valpy, 2010).

> **Secular:** Not related to anything religious or spiritual.
>
> **Secularism:** The belief that religion should play no role in public life.

READING

NATIVE SPIRITUALITY AND CHRISTIAN FAITH – BEYOND TWO SOLITUDES

by Donna sinclair

A certain corner of my garden makes me think of Kimberley Robinson, who is one of the pillars holding up the congregation I attend. She gave me the sweetgrass flourishing here, filling the air with its scent. Robinson is Algonquin—her grandmother was from the Golden Lake First Nation—and a member of St. Andrew's United in North Bay, Ontario, where she sings in the choir and volunteers generously. She is also part of a Native women's drumming group that meets by Lake Nipissing in summer and in the church parlour in winter. She honours both traditions, holding them gracefully together in her own person.

I wish this holding-together for our whole United Church. I wish that we could pick up the pieces of Native spirituality shattered by Christianity centuries ago, unhook our faith from our culture, and offer it as a gift and receive gifts in return. I don't think I am alone. It's a yearning named in 1986, when this church apologized to Native congregations for not hearing "when you shared your vision."

How can the spiritualities of Native and non-Native people co-exist in one church, though, when our ways of being are so deeply different, when one cherishes sweetgrass and tobacco and the other bread and wine?

They can, says Robinson. But it requires being very open. When she set out to honour her grandmother's Native heritage and her own, she "made an intention, that I would open myself to learning the teachings as they were offered." Not long afterward, at a Manitou Conference women's event, two Ojibwa women offered a workshop. "They said, 'We're going to invite the grandmothers into the room,'" and soon Robinson had picked up a drum and found her voice. "This feels right," she thought. "This has to be the best thing that's happened to me in a long time."

The teachings cascaded from there. She arrived at church one morning covered with blackfly bites, contented and tired from a moon ceremony the night before, with sacred fire and tobacco, a circle of women in the trees. Last Sunday, she offered the prayers of the people, as she often does. "May we take care of the water," she said in part. "Water cleanses us. It is sacred and nourishing. Humbly we ask for wisdom and understanding. Megwetch. Amen." Robinson always feels connected to the earth when she prays, she says, and "the drumming and the women's circle, it fits with these prayers."

You don't have to be Native to be open to that spirituality. Many of the women in Robinson's drumming circle are non-Native. That's also the experience of Ruth McQuirter Scott, a member of Port Nelson United in Burlington, Ontario. She's been studying with a Métis shaman for many years. When she began, she found it required a "willingness to be challenged, to see spirituality in a much broader sense than I had previously learned though Christianity." At first, she was shocked by "the immediacy, the sense of the cosmos" that was filled with "angels, or spirit guides, surrounding you."

McQuirter Scott had learned about angels in Sunday school, certainly, but they were "far removed beings with wings, that didn't connect with me."

But she soon discovered that the more she learned of Native spirituality, the more meaningful her own Christianity became.

"I learned," she says slowly, "that we have underestimated Jesus."

While his miracles are sometimes labelled "metaphor" by conventional Christianity, she is now more inclined to see Jesus "as one of the ultimate shamans." And when it comes to a miracle, "shamans could do this. It could have really happened."

Like Robinson, McQuirter Scott emphasizes the importance of listening carefully for "each unique opportunity to learn." Her studies have instilled in her a strong sense that we have "a personal responsibility to live a life that is meaningful." That means there is no one moment of "being saved, I've got it." She sees in Native teachings "a school for spirituality. Life is a process, you always look at 'What has this lesson been put in place to teach me?' It is more respectful. You don't look at life as having arrived. You are just at a different point on the path."

We are more able to be open to each other if we know each other well. Métis traditional elder Jules Lavallee from Gateway, Manitoba, is a strong supporter of the Dr. Jesse Saulteaux Resource Centre in Beausejour, Manitoba. "Dr. Jesse Saulteaux wanted to encourage people to honour both the traditional ways of Aboriginal people and Christian teachings," he says. "That was her vision."

Lavallee is clearly sympathetic. "Absolutely, it can be done," he says, but "you need to know the protocol." Just as it takes a while to become a minister, it takes "quite a while for people to acquire the right to do the ceremonies," and it's better if "you feel comfortable with that individual and know you are going to be honoured and appreciated."

But he is also clear that this is never a matter of blending the two traditions. "That wouldn't work." We have to understand "there is a difference. But the more we understand each other's way, the more we realize there are many similarities."

Kimberley Robinson agrees and points to some things that are the same. The sense of belonging and security that clings to both church and the drumming group, each "a safe place to fall." The infusion of energy from drumming that sustains her "until the following circle. It is like Wednesday worship (at church); you get an energy that carries you for the rest of the week." Christians steeped in Celtic worship find especially familiar the Native understanding of prayer as part of daily life.

Knowing each other well means we don't ignore either spirituality. Moderator Rt. Rev. David Giuliano, whose Marathon, Ontario, congregation is near the Pic River Reserve, remembers an Elder at a sweat lodge and healing circle to which Giuliano was invited: "When non-Native people come to learn," the Elder said, "I want to ask 'What are you bringing to the circle?'"

"What can I contribute," Giuliano recalls thinking at the time, "if I don't bring my own tradition as well?" What is required is offering that tradition "with humility. The mystery is much bigger than we can enact symbolically; two conflicting things can be true. That kind of humility makes the conversation possible," Giuliano says. Then, well-acquainted, we are able "to trust that we want good for each other," and we can be "citizens of a place we imagine together."

Getting to that place is a matter of mutual hospitality. "People have to be willing to open their space—not just their building—to allow other people to nurture their spirits, however that will happen," says Robinson.

Last Sunday, at lunch after church, Robinson was holding Vijanti and Dana Murphy's baby, singing a grandmother's lullaby to him that she had learned at her women's drumming circle. "Humma, humma, humma, hya," she sang. "Hi yo way hi yo." The baby was manifestly content.

I listened and thought about the sweetgrass Robinson had given me, spreading freely in my garden. I have not learned the rituals of picking it, the braiding and smudging, so I don't attempt them. It would not be respectful.

But when I inhale its scent, I understand that the land is holy and full of spirits. I cannot pretend another spirituality is my own. But—as a citizen of this church—I can honour and befriend all the spiritualities within it, and let them teach me what I need to know.

Source: Donna Sinclair, "Native spirituality and Christian faith beyond two solitudes," *UC Observer*, November 2006. Found at: http://www .ucobserver.org/faith/2006/11/native_spirituality_and_christian_faith/. Used with permission.

DISCUSSION QUESTIONS

1. Like politics, religion and spiritual beliefs are often regarded as an off-limits topic. Why are some people afraid to discuss their faith? How did it become such a taboo subject? As the author suggests, can we learn about multiple spiritualities, recognize their differences, and still celebrate their similarities? Does the answer lie with "mutual hospitality"—where we open up our space to allow people to nurture their spirits?

2. The author wishes "we could pick up the pieces of Native spirituality shattered by Christianity centuries

ago, unhook our faith from our culture, and offer it as a gift and receive gifts in return." Do you think faith can be "unhooked" from culture? Is it possible for two distinct spiritualities to co-exist in one person? Can we move away from regarding spiritual beliefs with the false dichotomy of "either-or" and instead approach them with a "both-and" way of thinking? Can two conflicting things be true?

KWIP

KNOW IT AND OWN IT: WHAT DO I BRING TO THIS?

The "K" in the KWIP process involves examining aspects of your own identity and social location as the first step in becoming diversity competent.

ACTIVITY: JOURNAL

Religion is one of the most personal choices a person can make.

What are your feelings about religion? Do you identify with a particular faith or hold religious beliefs of any kind? Where and when were they formed? How much choice did you have in the matter? Reflect upon how much of your day and life is determined by your religious background or your rejection of that background. How did these experiences (or lack thereof) impact you as a child or influence who you are today? What do you think of the religious beliefs of others? Are your views on the various issues brought up in this course determined by religion in any way? Are you open to others having views that are determined by religious views at odds with yours? Have you ever engaged in a conversation about conflicting faiths? If so, how did it go? If not, is religion a topic you tend to avoid? If so, why?

WALKING THE TALK: HOW CAN I LEARN FROM THIS?

The "W" in the KWIP process presents a scenario or case study that challenges you to "walk the talk" through problem-based learning.

ACTIVITY: CASE STUDY

You have met the person of your dreams. You have only been dating for a short period of time, but you know this person is the one for you and you are both madly in love. You and your significant other come from drastically different religious backgrounds and both of your families have reservations about the long-term intentions of your relationship. You have discussed this with your partner and you both feel confident that you can pursue a future together, despite your parents' reservations. You agree that the first step is to introduce your parents to each other and get the dialogue started. You and your partner have it all worked out—you are ready to show them that you can make an interfaith relationship work. During dinner, both sets of your parents are listening attentively and being patient as you and your partner discuss strategies for finding common ground, but when you are finished and you ask them for their thoughts, they both ask "what about if you have children? How will you reconcile how you will raise your children?"

1. What was said during the conversation? How would you make an interfaith relationship work?

2. You and your partner have not discussed children. How do you respond to your parents' question? What are your thoughts on children raised in interfaith households?

IT IS WHAT IT IS: IS THIS INSIDE OR OUTSIDE MY COMFORT ZONE?

The "I" in the KWIP process requires you to honestly confront and identify ways in which our complex identities result in experiences of privilege and oppression, and to reflect on how we can learn to honour that privilege.

As with other characteristics of diversity, religion and/or spiritual beliefs can sometimes create privilege for some and barriers for others. While religious oppression is often seen and heard, privilege can many times go unnoticed. What will it take to make Canada a place that is safe and accessible for people of all belief systems?

ASKING: Do I have privilege?

1. You can expect to have time off of work to celebrate your religious holidays.

2. You can worship (or not worship) freely without being questioned, mocked, or inhibited, and without fear of threats or violence.

3. You can reasonably assume that anyone you encounter will have a decent understanding of your faith.

4. You are never asked to speak on behalf of all the members of your faith.

REFLECTING: Honouring Our Privilege

Consider the religious demographics in Canada and how you identify when it comes to religion or spirituality. Describe two circumstances in which you feel disadvantaged because of your beliefs. Then describe two circumstances in which your beliefs give you privilege over someone else. How can you use the advantages you experience because of your beliefs to combat the disadvantages experienced by others within or outside of your religious/spiritual group?

PUT IT IN PLAY: HOW CAN I USE THIS?

The "P" in the KWIP process involves examining how others are practising equity and how you might use this.

ACTIVITY: **CALL TO ACTION**

Founded in 2012, the Inspirit Foundation is a national, grant-making organization that works to encourage pluralism in Canada among young adults of different spiritual, religious, and secular beliefs. Inspirit views pluralism as an approach that strives to recognize, understand, and engage with each other's differences of culture, ethnicity, beliefs, and other elements of identity (Inspirit Foundation, 2013).

The foundation supports projects that bring young people of different beliefs together to achieve a common goal in their communities. They support activities that encourage dialogue about the role of belief in Canadian society and provide space where young people can learn about the variety of beliefs in Canada. The Foundation has partnered with the P.E.I Government to host The New Canada Conference, which brought together 100 delegates from across cultures, regions, and faith groups, to explore Canada's past, current issues, and future possibilities (Inspirit Foundation, 2013). They also partnered with the University of Toronto Department for the Study of Religion for The Elements Experiment—a project that aimed to open conversations about faith (religious and secular) within the academic and public spheres.

To learn more about the Inspirit Foundation, visit https://inspiritfoundation.org/

Located at www.nelson.com/student

Study Tools
CHAPTER 10

- Review Key terms with interactive **flash cards**
- Check your Comprehension by completing **chapter review quizzes**
- Gauge your understanding with **Picture This** and accompanying short answer questions
- Develop your critical thinking/reading skills through compelling **Readings** and accompanying short answer questions
- Apply your understanding to your own experience with **Connect A Concept** activities
- Evaluate Diversity in the Media with engaging **Video Activities**
- Reflect on your Understanding with **KWIP** activities

Boyd, M. (2004, December 1). Dispute resolution in family law: Protecting choice, promoting inclusion. Retrieved from *Ministry of the Attorney General*: http://www.attorneygeneral.jus.gov.on.ca/english/about/pubs/boyd/section1.pdf

CanLII. (2009, January 6). *Friesen v. Fisher Bay Seafood and others*, 2009 BCHRT 1. Retrieved from *Canadian Legal Information Institute*: http://www.canlii.org/en/bc/bchrt/doc/2009/2009bchrt1/2009bchrt1.html

CanLII. (2009, October 7). *Saadi v. Audmax*, 2009 HRTO 1627. Retrieved from *Canadian Legal Information Institute*: http://canlii.ca/t/262jc

CBC News. (2007, February 13). Glad to be Canadian, Muslims say. Retrieved from *CBC News*: http://www.cbc.ca/news/canada/story/2007/02/12/muslim-poll.html

CBC News. (2009, June 29). Girl's forced blood transfusion didn't violate rights: Top court. Retrieved from *CBC News*: http://www.cbc.ca/news/canada/story/2009/06/26/supreme-blood026.html

CBC News. (2011, May 5). Falun Gong senior booted from club gets $15K. Retrieved from *CBC News*: http://www.cbc.ca/news/canada/ottawa/story/2011/05/05/ottawa-falun-gong-seniors-347.html

Ceballos, A. (2009, August 7). Province's goals trump religious rights: SCC. *The Lawyer's Weekly*.

Clark, W. (2006, October 1). Interreligious unions in Canada. Retrieved from *Statistics Canada*: http://www.statcan.gc.ca/pub/11-008-x/2006003/9478-eng.htm

Cohen, T. (2010, June 10). "Honour killings" on the rise in Canada: Expert. Retrieved from *The Vancouver Sun*: http://www.vancouversun.com/life/Honour+killings+rise+Canada+Expert/3165638/story.html

Csillag, R. (2012, March 26). 52% of Canadians distrust Muslims, according to latest poll. Retrieved from *Huffington Post*: http://www.huffingtonpost.com/2012/03/26/canadians-distrust-muslims_n_1381239.html

Dowden, C., & Brennan, S. (2013, June 5). Police-reported hate crime in Canada, 2010. Retrieved from *Statistics Canada*: http://www.statcan.gc.ca/pub/85-002-x/2012001/article/11635-eng.htm

Durant, Elizabeth. (2013). On the intersection of race and religion. Retrieved on July 11, 2016, from *State of Formation*: http://www.stateofformation.org/2013/10/on-the-intersection-of-race-and-religion/

Essig, Todd. (2010). Confusing race and religion is dangerous. Retrieved on July 11, 2016, from *Psychology Today:* https://www.psychologytoday.com/blog/over-simulated/201008/confusing-race-and-religion-is-dangerous

Friesen, J., & Valpy, M. (2010, December 10). Canada marching from religion to secularization. Retrieved from *The Globe and Mail*: http://www.theglobeandmail.com/news/national/canada-marching-from-religion-to-secularization/article1833451/

Grossman, C. (2010, February 11). Christian churches in Canada fading out: USA next? Retrieved from *Faith and Reason*: http://content.usatoday.com/communities/Religion/post/2010/02/christian-churches-in-canada-fading-out-usa-next/1#.UbeyAZz3Nug

Inspirit Foundation. (2013). What is pluralism? Retrieved on July 10, 2016, from *Inspirit Foundation*: https://www.inspiritfoundation.org/application/files/7614/5626/0743/What-is-pluralism-paper-EN-2.pdf

Jedwap, J. (2005, March 30). Canada's demo-religious revolution: 2017 will bring considerable change to the profile of the Mosaic. Retrieved from *Association for Canadian Studies*: http://www.acs-aec.ca/pdf/polls/30-03-2005.pdf

Lindsay, C. (2008, November 21). Canadians attend weekly religious services less than 20 years ago. Retrieved from *Statistics Canada*: http://www.statcan.gc.ca/pub/89-630-x/2008001/article/10650-eng.htm

Mirza, Hina. (2015). Through the Lens of Tendisai Cromwell. Retrieved on July 10, 2016 from *Lanterns*: http://isnalanterns.com/2015/02/through-the-lens-of-tendisai-cromwell/

New Narrative Films. (2016). About the Founder. Retrieved on July 10, 2016, from *New Narrative Films*: http://www.newnarrativefilms.com/#nnf

OHRC. (1996, October 26). Policy on creed and the accommodation of religious observances. Retrieved from *Ontario Human Rights Commission*: http://www.ohrc.on.ca/sites/default/files/attachments/Policy_on_creed_and_the_accommodation_of_religious _observances.pdf

OHRC. (2012, May 01). Creed case law review. Retrieved from *Ontario Human Rights Commission*: http://www.ohrc.on.ca/en/creed-case-law-review

Peritz, I. (2010, December 13). As churches crumble, communities fear loss of heritage. Retrieved from *The Globe and Mail*: http://www.theglobeandmail.com/news/national/as-churches-crumble-communities-fear-loss-of-heritage/article1320111/

Pew Research Center. (2011, January 27). The future of the global Muslim population. Retrieved from *Pew Research Religion & Public Life Project*: http://www.pewforum.org/The-Future-of-the-Global-Muslim-Population.aspx

Pew Research Center. (2012, December 18). The global religious landscape. Retrieved from Pew Research Centre: http://www.pewforum.org/2012/12/18/global-religious-landscape-exec/

Pew Research Centre. (2013, June 27). Canada's changing religious landscape. Retrieved July 11, 2016 from *Pew Research Centre*: http://www.pewforum.org/2013/06/27/canadas-changing-religious-landscape/

Sinclair, Donna. (2006). Native spirituality and Christian faith – Beyond two solitudes. Retrieved July 10, 2016, from *UC Observer*: http://www.ucobserver.org/faith/2006/11/native_spirituality_and_christian_faith/

Singh, Simran Jeet. (2014). 9/11 era ignorance of Islam is infecting the Age of Isis: We should know better. Retrieved July 11, 2016, from *The Guardian*: https://www.theguardian.com/commentisfree/2014/sep/09/ignorance-islam-isis-hate-crimes

Slater, P. (1977). *Religion and culture in Canada.* Waterloo, Ontario: Waterloo University Press.

Statistics Canada. (2001). Median age of Protestants well above national level. Retrieved July 11, 2016, from *Statistics Canada*: http://www12.statcan.ca/english/census01/Products/Analytic/companion/rel/canada.cfm

Statistics Canada. (2001, June). Religious groups in Canada. Retrieved from *Statistics Canada*: http://www.statcan.gc.ca/pub/85f0033m/85f0033m2001007-eng.pdf

Statistics Canada. (2003, May 13). Religions in Canada. Retrieved from *Census 2001*: http://www12.statcan.gc.ca/english/census01/products/analytic/companion/rel/contents.cfm

Statistics Canada. (2011). Religion (108), Immigrant Status and Period of Immigration (11), Age Groups (10) and Sex (3) for the Population in Private Households of Canada, Provinces, Territories, Census Metropolitan Areas and Census Agglomerations. Retrieved July 6, 2016, from Statistics Canada: http://www12.statcan.gc.ca/nhs-enm/2011/dp-pd/dt-td/Rp-eng.cfm?LANG=E&APATH=3&DETAIL=0&DIM=0&FL=A&FREE=0&GC=0&GID=0&GK=0&GRP=0&PID=105399&PRID=0&PTYPE=105277&S=0&SHOWALL=0&SUB=0&Temporal=2013&THEME=95&VID=0&VNAMEE=&VNAMEF=

Taylor, L. (2009, May 10). Darker the skin, less you fit. Retrieved from *The Star*: http://www.thestar.com/news/gta/2009/05/14/darker_the_skin_less_you_fit.html

Tyshynski, M. (2009, January 6). Friesen v. Fisher Bay Seafood. Retrieved from *BC Human Rights Tribunal*: http://www.bchrt.gov.bc.ca/decisions/2009/pdf/jan/1_Friesen_v_Fisher_Bay_Seafood_and_others_2009_BCHRT_1.pdf

Valpy, M. (2010). Anglican Church facing the threat of extinction. Retrieved from *The Globe and Mail*: https://beta.theglobeandmail.com/news/british-columbia/anglican-church-facing-the-threat-of-extinction/article4352186/?ref=http://www.theglobeandmail.com&

Ability

> "Forgiveness for the ones who cut us deeply and wish that we could love us all the same"
>
> *Justin Hines, 2011*

LEARNING OUTCOMES

By mastering this unit, students will gain the skills and ability to:

- reflect upon the universalizing implications of the World Health Organization's definition of disability as something every human being can at some point in their lifetime experience

- reconstruct attitudes and approaches whereby people with disabilities move from being "objects" of charity, medical treatment, and social protection to instead being "subjects" with the ability and rights to make their own decisions and actively contribute to their community

- assess strategies for the civic engagement of all citizens through an understanding that people with disabilities may experience limited opportunities for social inclusion

- appraise universal design practices and principles as an approach to social inclusion and accessibility

The term "dis"ability means "not" able and can lead us to believe that we can categorize people as either able-bodied or disabled. The binarism imbued in the term **disability** leads us to believe that you either are or you are not, when the reality is we all live on a continuum of ability that can fluctuate in form and permanence, whether by birth or acquired through life events such as injury, illness, or aging. So, in reality, most of us will experience "disability" throughout our lifetime.

During the early 20th century, persons with disabilities were often segregated and institutionalized. It wasn't until veterans began returning home following World War I that the social and political context for persons with disabilities began to change in Canada. Over time, improvements in public health and safety and disease prevention also saw certain categories of disability diminishing and disappearing. Today, as more people are affected and diagnosed with traumatic brain injuries, spinal cord injuries, depression, anxiety disorders, eating disorders, cancers, autoimmune disease, attention-deficit disorder, autistic spectrum disabilities, and dementia, we have seen the development and expansion of new disability categories such as neurodiversity, psychiatric disabilities, learning disabilities, and disabilities of aging.

As we look at the complex and dynamic interactions that exist between our bodies and the society we live in, it is clear that our social institutions will need to respond accordingly. We live in a time where Canadian's are calling for a comprehensive national strategy for dealing with dementia and building dementia-friendly communities (Standing Senate Committee on Social Affairs, Science and Technology, 2016). We live in a time when 2 in 5 Canadians will develop cancer, 1 in 4 Canadians will die of cancer, and the total number of new cases of cancer in Canada are expected to significantly increase over the next 15 years (Canadian Cancer Society's Advisory Committee on Cancer Statistics, 2015). We live in a time where Canadians have had to fight for door-to-door mail delivery service, a vital part of Canada's infrastructure for older Canadians and Canadians with disabilities. We live in a time where Canadians are calling for a national strategy on autism. Evidence shows that early diagnosis and treatment are essential, but long wait times for assessment and diagnosis and bottlenecks for funded treatment results in serious consequences

World Down Syndrome Day uses a poster campaign slogan "See the ability" in addressing attitudinal barriers to social inclusion.

Source: Canadian Down Syndrome Society

for children and their families. We live in a time where people with disabilities in Canada are more likely to live in poverty, be unemployed, or be underemployed. We live in a time where Canadian students with disabilities are less likely to graduate from high school or postsecondary institutions. So, clearly, we live in a time where we need to continue to work to remove barriers to social inclusion, equity, and accessibility.

DEFINING DISABILITY

World Health Organization Definition

The International Classification of Functioning, Disability and Health of the **World Health Organization (WHO)** describes disability as a universal human experience, noting that all human beings can at some point in their lifetime experience a problem with mental or physical health and, therefore, some degree of disability (World Health Organization, 2013). This is the broadest description of disability. According to the WHO description, a person might expect his or her abilities to change

Disability: A universal human experience that anyone can experience at any time; disabilities can limit a person's ability to engage in daily activities; disabilities can be visible or invisible, temporary or permanent.

World Health Organization (WHO): United Nations authority on health issues.

over the course of a lifetime according to their stage of life span development, individual experiences, ecological issues, and even the time of day (Hoyle, 2004).

Disability Binarism

Defining disability as a universal human experience can help mitigate against **disability binarism** that reinforces an *us*-versus-*them* approach to disabilities. Victoria Venable (2016) talks about her experience of living in between the dichotomy of being deemed *able-bodied* and *disabled*, what she refers to as living in the grey area of disability. When we use the term disability, what often comes to mind are persons who use assistive devices like wheelchairs, walkers, or canes because of mobility issues or sensory disabilities. For Venable, there are times when she uses a wheelchair for mobility and other times when she is able to walk for short periods of time. This highlights the fact that everyone has varying abilities as well as challenges, and so-called disabilities can be as unique as a fingerprint. Variance in ability can be apparent at times for some and invisible at times for others.

Visible and Invisible Disabilities

The term **invisible disabilities**, sometimes referred to as non-evident disabilities, is used to refer to symptoms or conditions that are not always apparent to others, such as chronic pain, mental health issues, debilitating fatigue, sleep disorders, motor neuron disorders, learning disabilities, brain injuries, dizziness, cognitive disabilities, hearing loss, chronic illness, and many more. All of these conditions can be just as life-affecting and debilitating as **visible disabilities**, but persons with invisible disabilities can be reluctant to self-identify, especially when they have had past negative experiences or have recently acquired the disability (Baldridge & Swift, 2013).

Persons with invisible disabilities must disclose these disabilities in order to receive accommodations at school or work. Even where protection against discrimination exists within the law, persons with invisible disabilities might avoid disclosure and forego accommodations because of the social stigma attached to having a disability, especially for psychological or psychiatric conditions (Santuzzi, 2013). Another reason why persons with invisible disabilities are sometimes reluctant to disclose is the additional stigma of being perceived as someone who is trying to take advantage of a situation (Anzovino, 2012). When someone who "looks normal" discloses a disability and requests an accommodation, others sometimes will question the legitimacy of their disability, even when presented with medical evidence (Santuzzi, 2013). For persons with invisible disabilities, the decision about disclosure is not always an easy one. Individuals often carefully weigh the benefits of disclosure against the costs of stigmatization (Chaudoir & Quinn, 2010).

Persons with visible disabilities may often experience treatment that is characterized as patronizing and pitying as relayed by Jessica Hendriks who writes this chapter's In Their Shoes narrative. Persons with invisible disabilities are often overlooked and misunderstood. A student who uses a wheelchair explained it well when she told another student with a learning disability, "People have high expectations of you when you have invisible disabilities, like, 'you're just not trying hard enough,' but they have such low expectations of me in a wheelchair, like I can't do much" (Anzovino, 2012).

PEOPLE FIRST APPROACH AND INCLUSIVE LANGUAGE

While ability is a part of a person's identity, so called "dis" ability does not define who a person might be. When using inclusive language, it is important to put the person first and avoid terms which equate the person with the disability. People do not want to be defined by a diagnosis. For example, this chapter's Agent of Change is Bill MacPhee. Person-first guidelines to inclusive language would describe Bill as a *person living with* schizophrenia, not a schizophrenic. A person-first approach would avoid unnecessary reference to a person's ability unless relevant to the conversation. This does not suggest that a person's disability is irrelevant and should be ignored. It means that unless the person's ability is relevant to the discussion, their disability should not define them. If, for example, Bill MacPhee is speaking publicly on the topic of schizophrenia, he may be introduced as a person living with schizophrenia. But let's say you and Bill are out socially and you run into some friends that you wish to introduce him to; you would simply say, "Meet my friend Bill." You would not say, "Meet my friend Bill. He's schizophrenic."

A people-first approach also does not utilize terminology that implies a person is a suffering victim of disability, illness, or disease. You would not say, "Meet Bill who suffers from schizophrenia." A people-first approach does not

> **Disability Binarism:** Mode of thought that classifies people as either able-bodied or disabled.
>
> **Invisible Disability:** When a person has a disability that is not apparent to others.
>
> **Visible Disability:** When a person has a disability that is apparent by looking at them.

IN THEIR SHOES

If a picture can say a thousand words, imagine the stories your shoes could tell! Try this student story on for size – have you walked in this student's shoes?

YOU KNOW MY DISABILITY NOT MY STORY.

My disability does not define me and it shouldn't. I've been in a wheelchair since I was two years old, but it doesn't really bother me … until people become ignorant, insensitive jerks, staring at me in public and asking my friend questions about me when I am right there.

"I don't know. Ask her," she says.

Over the Christmas break, I took a few trips to the mall, a few times with my friend and one without. I went to pay for pictures I ordered but the machine kept declining. The salesperson called my friend over, who was a few feet away looking at something.

"Does she know what she is doing?" she asked my friend.

The same day, I paid for my things at the dollar store. The salesperson looked over at my friend who was beside me and asked,

"Where does she want her bags?"

"I don't know, ask her," my friend said with a sarcastic smile and a bit of a glare.

The last time I went, I went by myself, because I simply wanted some alone time. I got food and found a seat in the rather busy food court. Someone who I don't know came up to me and unwrapped my food. What are you doing? Stop touching my stuff.

"Do you need anything else?" she asks.

"Um, no thanks." I said. I hadn't asked for help to begin with.

The whole time I was eating, a few people were on the edge of their seats "keeping an eye on me." Every time I coughed, or seemed like I was having trouble, they flinched.

"Are you OK?"

I think at this point anyone in my position would be uncomfortable. But since I have anxiety, it was the worst and I couldn't finish eating my meal.

If you genuinely want to help, ask first. If the response is no thanks, accept that we don't need your help. If I couldn't do the simplest things by myself, why do they think I am out alone? A woman admired me for being independent.

"Are you here by yourself? Wow, you are really independent. Good for you."

But I'm just a person, eating pizza and cheese bread, or at least trying to.

I always make sure I'm wearing my college ID so that people would automatically know that I'm older than I look and that I'm mentally aware and capable. Once, it actually did show someone that. A guy looked over at the person beside him, who I think asked if they should check on me, and said, "I'm sure she's fine."

Other incidents include people petting me like a dog, talking to me like a baby or toddler, and asking if they can pray for me right there in front of me, which is really awkward.

One time, about a year ago, someone actually sat with me when I ate. Would you sit with an abled person?

The one thing I don't mind is when people ask me questions. In fact, I encourage it. I'm an open book. It's better than people making assumptions. In case you are wondering, I can do everything everyone else can, just sometimes in a different way. I am 100 per cent mentally aware and capable.

I am tired of hearing people compliment me and be proud of me for doing "regular tasks." I'd rather be known as a good writer or advocate, because that's who I am.

Discrimination and narrow-mindedness, or just ignorance, come with having a difference, whether it's disability, race, sexuality, religion, income, health, or mental health. But it shouldn't and doesn't have to be this way. I know many people, with various differences, feel the same way as I do.

Discrimination has been around since the beginning of time. Thankfully, because of advocates, people with differences have the same rights as everyone else, and people have become more and more educated, but we're still not where we should be.

You have a choice. Don't go by society's standards. Don't judge, stare, assume, or ignore. If you treat everyone the same, and with the same respect, you're already a better person, and maybe it will cause a chain reaction.

Get to know the person, not the difference.

Source: Hendriks, Jessica, "Don't Judge Based on Disability," January 21, 2016. First published in the Niagara News.

patronize persons with disabilities as heroic or courageous when engaging in regular daily activities. This approach also does not use the banner of "special," as it implies segregation needed for "special handling."

Jessica Hendriks shares her experience of how others have used her disability to define her, patronize her, and segregate her. Her advice: "*Get to know the person, not the difference.*"

DISABILITY HISTORY IN CANADA

FIGURE 11.1

Disability Rights in Canada

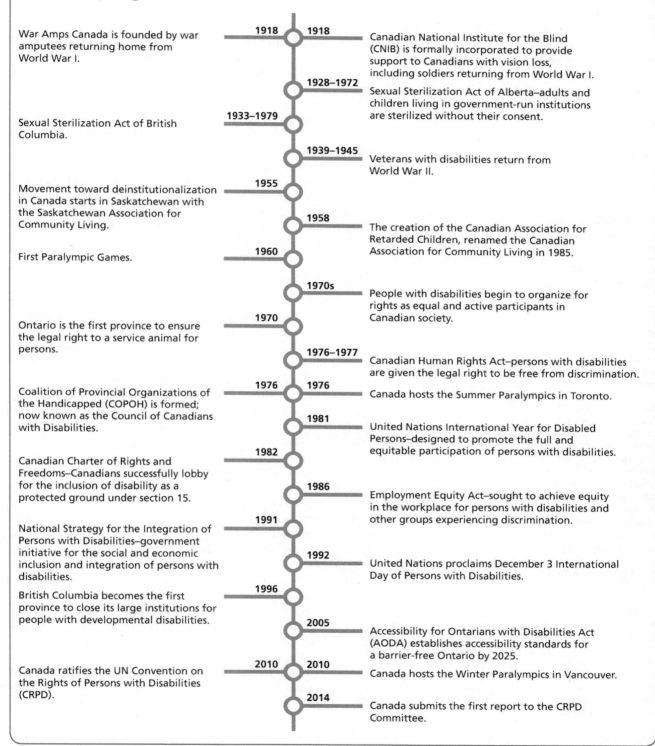

War Amps Canada is founded by war amputees returning home from World War I.	1918 / 1918	Canadian National Institute for the Blind (CNIB) is formally incorporated to provide support to Canadians with vision loss, including soldiers returning from World War I.
	1928–1972	Sexual Sterilization Act of Alberta–adults and children living in government-run institutions are sterilized without their consent.
Sexual Sterilization Act of British Columbia.	1933–1979	
	1939–1945	Veterans with disabilities return from World War II.
Movement toward deinstitutionalization in Canada starts in Saskatchewan with the Saskatchewan Association for Community Living.	1955	
	1958	The creation of the Canadian Association for Retarded Children, renamed the Canadian Association for Community Living in 1985.
First Paralympic Games.	1960	
	1970s	People with disabilities begin to organize for rights as equal and active participants in Canadian society.
Ontario is the first province to ensure the legal right to a service animal for persons.	1970	
	1976–1977	Canadian Human Rights Act–persons with disabilities are given the legal right to be free from discrimination.
Coalition of Provincial Organizations of the Handicapped (COPOH) is formed; now known as the Council of Canadians with Disabilities.	1976 / 1976	Canada hosts the Summer Paralympics in Toronto.
	1981	United Nations International Year for Disabled Persons–designed to promote the full and equitable participation of persons with disabilities.
Canadian Charter of Rights and Freedoms–Canadians successfully lobby for the inclusion of disability as a protected ground under section 15.	1982	
	1986	Employment Equity Act–sought to achieve equity in the workplace for persons with disabilities and other groups experiencing discrimination.
National Strategy for the Integration of Persons with Disabilities–government initiative for the social and economic inclusion and integration of persons with disabilities.	1991	
	1992	United Nations proclaims December 3 International Day of Persons with Disabilities.
British Columbia becomes the first province to close its large institutions for people with developmental disabilities.	1996	
	2005	Accessibility for Ontarians with Disabilities Act (AODA) establishes accessibility standards for a barrier-free Ontario by 2025.
Canada ratifies the UN Convention on the Rights of Persons with Disabilities (CRPD).	2010 / 2010	Canada hosts the Winter Paralympics in Vancouver.
	2014	Canada submits the first report to the CRPD Committee.

Source: Adapted in part from the Canadian Disability Policy Alliance (2012), Timeline of Disability Policy Events.

Institutionalization and Compulsory Sterilization

During the late 19th and early 20th century, persons with mental health issues and physical and intellectual disabilities were often perceived as incapable and were treated as objects of charity. Within this context, many persons with disabilities were segregated from the rest of society and **institutionalized** in large church or state-run facilities.

At this time in Canadian history, the provinces of Alberta and British Columbia engaged in the **compulsory sterilization** of people deemed "unfit" or "mentally defective" (Office for Disability Issues HRDC, 2003). This included people who were institutionalized and deemed "capable of propagating undesirable social characteristics" (Grekul, Krahn, & Odynak, 2004).

The Sexual Sterilization Act of Alberta was in effect in that province from 1929 until 1972 (Grekul, Krahn, & Odynak, 2004); the Sexual Sterilization Act of British Columbia was in effect from 1933 until 1973 (Grekul, Krahn, & Odynak, 2004).

Veterans with Disabilities

As veterans from World War I began to return home with vision and hearing loss, mobility challenges, and mental health issues as a consequence of war, the social and political context in Canada for persons with disabilities began to change (Galer, 2015). Disability emerged as a government policy issue in Canada during World War I as many Canadian soldiers returned home injured, and rapid industrialization produced many work-related injuries (Office for Disability Issues HRDC, 2003). In 1918, War Amps Canada was founded by war amputees returning home to Canada after World War I. (War Amps Canada, 2016). In this same year, the Canadian National Institute for the Blind was incorporated to provide support to Canadians with vision loss, a need underscored by veterans with vision loss returning from World War I (Canadian National Institute for the Blind, 2016).

In the aftermath of World War II, Canadian veterans with disabilities, seeing disparities in services and supports between themselves and civilians with disabilities, began to work as allies to ensure equity for all persons with disabilities (Galer, 2015). New coalitions of activists involving veterans, civilians, professionals, and families of children with disabilities organized to fight for the equal rights of persons with disabilities to participate in society and access the services needed to do so (Galer, 2015). New assistive technologies were developed, including the first electric wheelchair invented by Canadian George Klein to assist veterans who were injured in World War II (Government of Canada, 2005).

Deinstitutionalization and the Anti-Psychiatry Movement

In the 1960s and 1970s, a new social movement toward **deinstitutionalization** in Canada began by advocating for the removal of people from residential institutions and the replacement of these institutions with community-based services (Galer, 2015). Community agencies and group homes (known today as Associations for Community Living) were established to support persons with intellectual disabilities (Canadian Association of Community Living, 2016).

The ideological roots of deinstitutionalization can be traced to **anti-psychiatry** whose origins some argue began with historical examples of psychiatry being used as an oppressive arm of the state in Nazi Germany and Soviet Russia (Furnham, 2015). Anti-psychiatry took hold as a movement in the early 1950s as a result of conflict within the field of psychiatry. Psychiatrists using "talking cures" like psychoanalysis as treatment were challenged by biological physical psychiatrists who saw surgical and pharmacological treatment as more effective and scientific (Furnham, 2015).

By the 1960s, the anti-psychiatry movement focused on the use of psychiatry as a means of controlling and pathologizing social deviance (Furnham, 2015). The writings of sociologist Erving Goffman (notably *Asylums* in 1961) and of Michel Foucault provide a valuable critique of the use of power within psychiatry and its effects on the labelling and stigmatization of persons with mental health issues (Furnham, 2015). Issues of psychiatric processing

---

Institutionalized: Placed in or confined to a residential institution.

Compulsory Sterilization: Government program that forces or coerces the sterilization of persons so that they cannot have children; often part of a eugenics program to prevent the reproduction of members of a population who are deemed undesirable.

Deinstitutionalization: A social movement that continues to reverse the institutionalization of persons with mental illness and intellectual disabilities that began in the late 19th and early 20th centuries in Canada. It involves a process of removing residents from long-term institutions to integrated community-based settings.

Anti-Psychiatry: A term coined by David Cooper in 1967 and reflective of a movement that was critical of and opposed to many forms of psychiatric treatment, institutionalization of persons with mental illness and intellectual disabilities, the unequal power relationship between psychiatrist and patient, and the highly subjective process for diagnostic labelling.

and control, and critiques of diagnostic labelling were also highlighted in the writings of Thomas Szasz's *The Myth of Mental Illness* and David Cooper's *Psychiatry and Anti-Psychiatry*. In 1973, Stanford University professor and psychologist David Rosenhan's famous experiment and published research, *On Being Sane in Insane Places, raised* questions and scepticism about the validity of psychiatric diagnosis and became one of the most famous anti-psychiatry studies of all time. The Academy Award-winning film, *One Flew Over the Cuckoo's Nest*, became a classic that highlighted many of the concerns within the anti-psychiatric movement, including the existence of mental illness, the medicalization of madness, and the abuse of power by health professionals within mental health institutions. This included concerns around the power of health professionals to detain for indeterminate periods of time, the prison-like environments of mental health facilities, and the use of electroconvulsive therapy (ECT), and prefrontal lobotomies as treatment (Furnham, 2015).

Disability Human Rights Paradigm

The disability social movement arose in the 1970s as a human rights-based response to the discrimination that people with disabilities had historically faced in Canada, and was premised on the inherent right for all persons to be treated with human dignity (ARCH, 2013). One organization that has sought to address these issues is the Council of Canadians with Disabilities, which is a coalition that defines itself as "a national human rights organization of people with disabilities working for an accessible and inclusive Canada" (Council of Canadians with Disabilities, 2007).

During the 1970s and 1980s, a number of significant pieces of new legislation were adopted at a national and international level to ensure people with disabilities were able to live and work as equal and active participants in society (ARCH, 2013). In 1976, the Canadian Human Rights Act gave Canadians with disabilities the legal right to be free from discrimination (ARCH, 2013). The United Nations proclaimed 1981 the International Year of Disabled Persons, promoting their full and equal participation in society and increasing public awareness and understanding of issues faced (ARCH, 2013). The Federal Special Committee on the Disabled and the Handicapped published the *Obstacles Report* in 1981, outlining 130 recommendations to work toward the full participation and integration of persons with disabilities in Canadian society (ARCH, 2013). In 1982, Canadians were able to lobby to have disability included as protected ground within the Canadian Charter of Rights and Freedoms, guaranteeing a constitutional right

under Section 15 to equal treatment and equal protection against discrimination:

(1) Every individual is equal before and under the law and has the right to the equal protection and equal benefit of the law without discrimination and, in particular, without discrimination based on race, national or ethnic origin, colour, religion, sex, age, or mental or physical disability.
(2) Subsection (1) does not preclude any law, program, or activity that has as its object the amelioration of conditions of disadvantaged individuals or groups, including those that are disadvantaged because of race, national or ethnic origin, colour, religion, sex, age. or mental or physical disability.

In addition to the guarantees of equal treatment and protection from discrimination, the Charter explicitly recognized mental disabilities. Until 1982, Canadians with mental disabilities had limited protection (Office for Disability Issues HRDC, 2003).

Neoliberalism and Cutbacks

During the late 1980s and early 1990s, the disability community in Canada was deeply affected by a global recession, growing deficits, and concomitant changes in the political climate (Galer, 2015). Deinstitutionalization continued under new ideological and fiscal purposes, including the neo-liberal privatization of care and cutbacks in government spending. This period of fiscal austerity hit hard for Canadians with disabilities as social assistance rates were slashed, government grants to disability organizations dwindled, and many disability rights organizations were disbanded (Galer, 2015).

The privatization of responsibility and cost of care had a significant impact on families as caregivers. From the 1980s until present day, families were cited as the most common source of help for persons with disabilities in Canada (Statistics Canada, 2012). Caring for children, parents, and other family members with disabilities can result in a great deal of pressure for caregivers and often impacts their employment status. In addition to juggling family responsibilities, employment, child care, household chores, medical appointments, and caregiving, families with disabled children struggle with increased financial pressures because of additional costs and lost wages.

Entering the 21st Century

In 1992, the United Nations proclaimed December 3rd as International Day of Persons with Disabilities.

In 2010, Canada ratified the **United Nations Convention on the Rights of Persons with Disabilities (CRPD),** which represented a new opportunity for all levels of government to be proactive in eliminating disadvantage and achieving full inclusion for persons with disabilities (Brodsky, Day, & Peters, 2012). In 2014, Canada submitted its first report to the UN's Convention on the Rights of Persons with Disabilities Committee, outlining federal and provincial measures taken toward ensuring the rights of persons with disabilities under the Convention (Galer, 2015). The federal government was currently considering a Canadians with Disabilities Act that would aid in Canada's implementation of the UN Convention on the Rights of Persons with Disabilities. The volume and influence of critical disabilities studies grew across Canadian postsecondary institutions with new interdisciplinary scholarly discourse emerging during this period as well as the empowerment of persons with disabilities to take control of their own narratives and histories (Galer, 2015). But the issue of poverty for persons with disabilities in Canada remains a serious concern. In a Council of Canadians with Disabilities (2016) research project, *Disabling Poverty, Enabling Citizenship,* that was funded by the Social Sciences and Humanities Research Council, one key finding was that "throughout the working years (15–64 years of age), people with disabilities remain about twice as likely as those without disabilities to live with low income."

United Nations Convention on the Rights of Persons with Disabilities (CRPD): A human rights instrument adopted by the United Nations in 2006 for the purpose of protecting the human rights and dignity of persons with disabilities; signatories are required to ensure that people with disabilities are equal under the law.

Impairment: According to the biomedical perspective, a medical condition that leads to disability; according to functional perspective, it is any loss or abnormality of physiological, psychological, or anatomical structure or function, whether permanent or temporary.

DISABILITY PERSPECTIVES

The Canadian government and its agencies use many different definitions of disability and consequently many different criteria for assessing eligibility for programs and services. The lack of consistency among definitions of disability in Canada can be attributed in part to the different perspectives used in understanding this complex and multidimensional concept. When interpreted using a biomedical perspective, the definition of disability is approached as illness or impairment in function that is located within the individual (Office for Disability Issues HRDC, 2003). Using this perspective, disability is viewed as something objective that is fixed in an individual's body or mind and that can be measured in terms of ability to perform specific tasks (Office for Disability Issues HRDC, 2003). When defining disability using a social perspective, the concept can be interpreted as a socially constructed disadvantage that results in a person's social exclusion and possible violation of their legal rights. There is clear evidence at an international level that there is a movement toward including a social perspective within a definition of disability. The United Nations Convention on the Rights of Persons with Disabilities includes a social perspective when it defines persons with disabilities as "those who have long-term physical, mental, intellectual or sensory impairments which in interaction with various barriers may hinder their full and effective participation in society on an equal basis" (United Nations, 2006). The United Nation's International Day of Persons with Disabilities on December 3rd offers an opportunity to examine the social, economic, physical, and attitudinal barriers affecting one billion people every day worldwide.

How we define disability has evolved over the years through different perspectives that will be referred to as the biomedical perspective, functional limitations perspective, socio-constructionist perspective, socio-economic perspective, legal rights perspective, and the social inclusion perspective (see Table 11.1). These perspectives have influenced the approaches used to accommodate disability within society, including the development of programs and services and their eligibility criteria.

Biomedical Perspective

The biomedical perspective considers disability as an **impairment** to an individual's body or mind that is a result of a health problem, illness, disease, or abnormality (Nixon, 1984). According to this perspective, people with disabilities are sick and therefore require help in the form of medical treatment and specialized care (Bickenbach, 1993). The approach used to accommodate those with disabilities is to fix the individual's impairment through medical interventions focused on the illness or abnormality. An example of this approach is reflected in the branch of psychology referred to as *abnormal psychology*—a term that pathologizes mental health issues as medically or psychologically abnormal. Criticism of this perspective is that it "locates the defect in a person's body or mind, and that person may be defined as defective, abnormal and by extension biologically or mentally inferior"

TABLE 11.1

Disability Perspectives

Perspective	Approach to Disability	Critique of Perspective
Biomedical	Considers disability as an *impairment* to an individual's body or mind that is a result of a health problem, illness, disease, or abnormality.	Ignores social and environmental factors that affect people with disabilities.
Functional limitations	Quantifies the type and degree of disability a person might have to measure his or her functional limitations and ability to perform certain tasks.	Focuses on individual inability; creates inconsistency in the determination of eligibility for federal programs; approaches disability as pathology (abnormalcy).
Socio-constructionist	Disability is a concept we socially construct as different from the norm; locates disablement issues within society.	Its exclusive focus on the social neglects issues of the physical body and mind.
Socio-economic	Focus on inclusion of people with disabilities within the economic framework of a society; goal of programs is to make people with disabilities more employable.	Does not address impairment nor process of stigmatization experienced by people with disabilities.
Legal rights	Focus is on rights to full participation and non-discrimination for people with disabilities using a human rights and social justice paradigm; perspective used by disability social movement.	Is sometimes critiqued for its neglect of individualized experiences of body and mind (medical model) and functional limitation.
Social inclusion	Focus is on contributions all people can make to the community as a whole as valued and respected members; challenges the notion of disabled and non-disabled that reinforce an *us*-and-*them* mentality.	Sometimes fails to acknowledge differences arising from intersectionality of identities (e.g., women with disabilities, Aboriginal people with disabilities, people with disabilities from visible minority communities).

(Office for Disability Issues HRDC, 2003). In viewing disabilities solely as a medical issue, this perspective ignores social and environmental factors that affect people with disabilities. This can result in the social marginalization and isolation of people with disabilities because this model believes in a service-based approach that focuses on caring for deficiencies using specially trained health professionals (Kretzmann & McKnight, 1993).

Functional Limitations Perspective

The functional limitations perspective, like the biomedical perspective, uses objective standards and is rooted in the concept that impairment of the mind or body is the direct cause of disability (Office for Disability Issues HRDC, 2003). This perspective is used to quantify the degree of disability a person might have and to measure his or her functional limitations (Office for Disability Issues HRDC, 2003). What distinguishes this perspective from the biomedical one is that it utilizes social and environmental criteria in the evaluation of the individual's functional limitations: "Disability is seen as influenced not only by the characteristics of impairments, such as type and severity, but also by how the individual defines a given situation and reacts to it, and how others define that situation through their reactions and expectations" (Office for Disability Issues HRDC, 2003, p. 6). An example of a program that utilizes the functional limitations perspective is the Ontario Disability Support Program. If a person with a disability living in Ontario is applying for financial support through this program, the doctor would be required to complete a medical assessment that includes a Health Status Report and Activities of Daily Living Index. But in addition to the medical assessment, the applicant completes a Self-Report Form that allows him or her to define and describe the situation and detail how the disability affects personal care, participation in the community, and ability to work.

Critiques of the functional limitation perspective have suggested that the attempts to quantify the type and severity of disabilities have contributed to the inconsistency in eligibility criteria for government programs for people with disabilities. Dr. Blake Woodside of the Canadian Psychiatric Association notes that the current system discriminates against those with mental illness because "the use of primitive efforts, phrases such as 'almost all of the time,' 'greater than 90 percent' or 'prolonged,' simply does not address the complex issue of characterizing psychiatric disability" (Office for Disability Issues HRDC, 2003, p. 52). The functional limitations perspective fails to account for recurrent, cyclic, or episodic disabilities, or for the variation in lived experiences of people with disabilities. The other criticism of this perspective is its focus on inability and not ability. Using a standard of "normalcy" for functional assessment, disability is considered pathology, and those people who live outside of these standards become disadvantaged (Oliver, 1990).

Socio-Constructionist Perspective

According to this perspective, disability is something that is socially constructed by the ability-orientated and ability-dominant society we live in (Office for Disability Issues HRDC, 2003). The **pathologization** of disability is something that we create socially, and when it is reinforced through ableism, it often results in social isolation and marginalization (Bickenbach, 1993). Stereotypes of people with disabilities focus on what we perceive to be different from the "norm"—with the result that similar needs for dignity, friendship, hope for the future, and valuation in the community can then be overlooked (Kretzmann & McKnight, 1993). Our belief about what is normal becomes embedded in our thoughts and is used to legitimize social policies (Oliver, 1990). Ron Mace, founder of the Centre for Universal Design, discusses how socially constructed concepts of normalcy can disadvantage those who fail to measure up:

We discount people who are less than what we popularly consider to be "normal." To be "normal" is to be

Pathologization:
Characterization as medically or psychologically abnormal.

perfect, capable, competent and independent. Unfortunately, designers in our society also mistakenly assume that everyone fits this definition of "normal." This just is not the case. (Social Planning Council of Kitchener Waterloo, 2001, p. 1)

According to the socio-constructionist perspective, if society constructs disability as inability, then we limit human potential.

One of the contributions of the socio-constructionist perspective is that it locates disablement issues within society and there is the opportunity, therefore, to socially construct meaning and value around abilities (rather than disabilities) that are manifest in ways different from what is socially defined as "normal." A critique of the socio-constructionist perspective of disability is its exclusive focus on the social neglects issues of the physical body and mind (Hoyle, 2004).

Socio-Economic Perspective

The socio-economic perspective focuses on the inclusion of people with disabilities in the economic framework of society through policies and programs (Hoyle, 2004). Poverty is a significant issue for people with disabilities. According to Statistics Canada's Canadian Survey on Disability (2012), less than half (47 percent) of persons with disabilities aged 15 to 64 were employed. Compared with persons without disabilities, those with disabilities were significantly more likely to be unemployed or not in the labour force (Government of Canada, 2015; Statistics Canada, 2013). For those people with disabilities who must then rely on disability income support programs, they often find themselves stigmatized and faced with financial limitations. A forum called "Unlocking the Possible" examined the issue of civic participation for people with disabilities and participants discussed poverty as a barrier:

Disability dollars don't go far.... For people receiving disability assistance from the government, their disability assistance didn't go far. The rent that some people paid was more than what was allotted by disability assistance. This meant there was little or no money left for other activities so people stayed home. (Hoyle, 2004)

In its 2014 report to the United Nations Committee on the Convention on the Rights of Persons with Disabilities, federal, provincial, and territorial governments noted that poverty rates among persons with disabilities in Canada remain a challenge, as does ensuring more disabled Canadians find work (Government of Canada, 2014). The Council of Canadians with Disabilities in its report, *As a Matter of Fact: Poverty and Disability in Canada* (2009),highlights a further relationship between the types of disability a person has and the likelihood of that person living in poverty (see Figure 11.2).

The goal of the socio-economic perspective is to ensure economic well-being of persons with disabilities, and one of the strategies used by governments to do this is to make people with disabilities more employable (Bickenbach, 1993). There are a number of government programs at various jurisdictional levels that are designed to assist with employment barriers. Two examples of federal initiatives include the Opportunities Fund and the CPP Disability Vocational Rehabilitation Program. The Opportunities Fund is a federal government program that uses a socio-economic perspective in helping people with disabilities find employment or self-employment (Government of Canada, 2016). The Canada Pension Plan Disability Vocational Rehabilitation Program offers training support and job search services to recipients of Canada Pension Plan (CPP) Disability Benefits to help them return to work (Government of Canada, 2016).

One criticism of the socio-economic perspective is that it is rooted in supporting individuals who experience activity limitations and participation restrictions in specific environments such as work, but it does not acknowledge how the socio-economic environment can *create* disability (Hoyle, 2004). While this perspective can change the experience of disability in an employment context, it "does not change the impairment or the process of stigmatization experienced by people with limited functional abilities" (Hoyle, 2004, p. 8).

Legal Rights Perspective

The legal rights perspective approaches disability as a human rights and social justice issue. The primary concerns of this perspective are the rights to full participation, equity, and non-discrimination for people with disabilities. Important is the fact that "equality" is not defined legally as treating all people the same way; the law recognizes that special measures or special programs are sometimes necessary to correct disadvantage. Canada has adopted a number of laws to ensure that people with disabilities are able to fully participate in society and are protected against discrimination, including the Canadian Human Rights Act in 1977 and the Canadian Charter of Rights and Freedoms in 1982. In 2010, Canada ratified the United Nations Convention on the Rights of Persons with Disabilities and was required to report internationally on the progressive implementation of rights (ARCH, 2013).

One of the critiques of the legal rights perspective is that laws and human rights machinery cannot

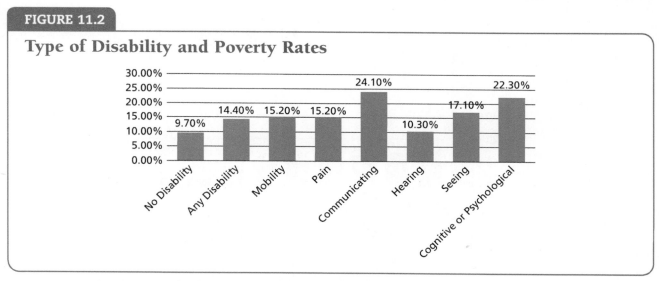

FIGURE 11.2

Type of Disability and Poverty Rates

Note: "Cognitive or Psychological" includes learning, memory, or developmental/intellectual disability, or psychiatric diagnosis.
Source: Adapted from Council of Canadians with Disabilities, "As a Matter of Fact: Poverty and Disability in Canada."

alone eliminate barriers and reduce discrimination for people with disabilities (Brodsky, Day, & Peters, 2012). In a famous Supreme Court decision (*Eldridge v. British Columbia [Attorney General]*, 1997), the court ruled that failure by the province to pay for and provide sign language interpretation during pregnancy and childbirth to the three appellants who were deaf was discriminatory, as the government had an obligation to address their needs as a disadvantaged group, and by failing to do so, denied them equal access to public medical services (Office for Disability Issues HRDC, 2003). In 1999, the Supreme Court of Canada, in two rulings, referred to as the *Meiorin* and *Grismer* decisions, reinforced the duty to accommodate individuals who cannot meet an employment or serviced delivery standard for any reason related to a ground protected by the Canadian Human Rights Act, such as disability.

Despite all the conventions, laws, and legal decisions made by tribunals and courts, the question remains as to why the largest part of the caseload of many human rights commissions continues to be complaints dealing with disability (Brodsky, Day, & Peters, 2012). The complaints to the Canadian Human Rights Commission (CHRC) consistently identify disability as the predominant reason for which a person was discriminated against (see Figure 11.3). In 2015, almost 60 percent of the complaints received by the CHRC were disability related (Canadian Human Rights Commission, 2016). Forty percent of these were related to mental health issues; this means that almost one quarter of all the complaints received in 2015 by the CHRC were related to mental health (Canadian Human Rights Commission, 2016).

Clearly, discrimination faced by people with disabilities is an enduring facet of their lived experience in Canada. While the legal rights perspective on disability is important in terms of protection of individual legal rights, it has also been important at a broader societal level, as it has served as the impetus for disability rights groups. The major criticism of this model is that its focus on the common experience of discrimination fails to consider the individual experiences of one's body as unique and diverse (Wendell, 1989). In its exclusive focus on legal and social aspects of disability, this perspective neglects aspects of the medical model of disability that need to be considered, as well as the lived realities of the physical and mental experiences of disability, such as living with pain and limitations (Hoyle, 2004).

FIGURE 11.3

Grounds of Discrimination Complaints Received by Canadian Human Rights Commission 2015

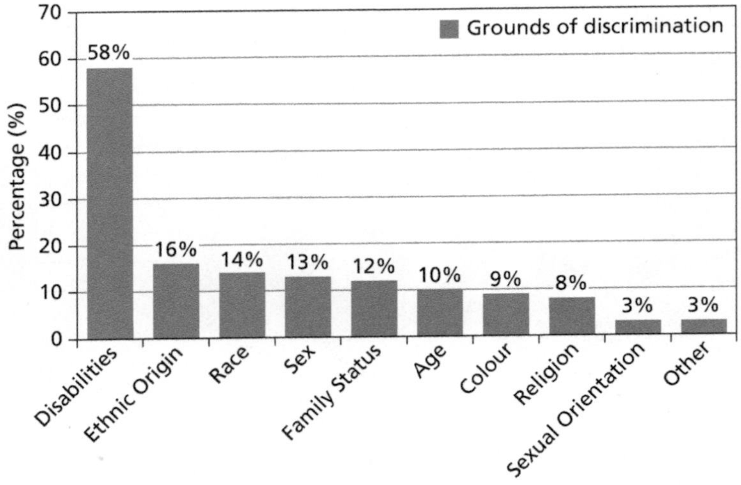

Note: Total number of grounds cited exceeds the total number of received complaints because some complaints dealt with more than one ground.
Source: Canadian Human Rights Commission 2015 Annual Report, Canadian Human Rights Commission, 2016.

Social Inclusion Perspective

Many Canadians with disabilities experience barriers to their full and equal participation in Canadian society (Council of Canadians with Disabilities, 2007). All of the other perspectives differentiate people with disabilities from people without disabilities. Social **accommodation** then focuses on these differences (Hoyle, 2004). The *us-versus-them* dichotomy can lead to ableism, which excludes and discriminates against those who don't measure up to societal norms of ability:

> PWD's [Persons with disabilities] must be seen for who we are: Regular people, neither pathetic poster children nor superheroes "overcoming" the unimaginable. And regular people need regular things: transportation, be it bus or wheelchair; help around the house, be it from their kids or personal assistant; information, be it gleaned from print or sign language or Braille; relief from pain, be it an aspirin or a prescription for morphine; and a decent standard of living, be it from a job or a government check. When all people are provided with such necessities, they will be assured the opportunity for a good quality of life. This is what PWDs deserve and require— tangible assistance that provides freedom, independence, and control over our lives as disabled people, not adulation, pity, or encouragement to focus on a cure that will make us nondisabled. (Wachsler, 2007, p. 14)

In inclusive settings, people with disabilities "have the opportunity to discredit the negative stereotypes and challenge the notions of disabled and non-disabled that reinforce the boundaries between sameness and difference" (Hoyle, 2004). Models for social inclusion recognize that a difference approach separates and disengages persons with disabilities from active participation in their communities. Social inclusion is premised on the belief that all members of a community have a valuable contribution to make (Laidlaw Foundation, 2002–2003): "A true community is only able to grow and strengthen itself by including all of its members and finding room for them to develop their capacities within its own pattern of growth" (Kretzmann & McKnight, 1993). The social inclusion perspective is sometimes criticized for failing to account for the complexity and intersectionality of aspects of identity, such as women with disabilities, Indigenous people with disabilities, and people with disabilities from visible minority communities.

10 FACTS AND FIGURES ABOUT HAVING A DISABILITY IN CANADA

1. One in Seven Canadians Lives with a Disability
According to Statistics Canada (2012), almost 14 percent of Canadians over the age of 15, that's 3.8 million individuals, reported having a disability that limited their daily activities.

2. Half Live with Severe Disabilities
Of the 3.8 million Canadians who identified as having a disability, 32 percent were classified as having a mild disability; 20 percent, a moderate disability; 23 percent, a severe disability; and 26 percent, a very severe disability (Statistics Canada, 2012).

3. Pain Most Common Type of Disability in Canadian Adults
Pain is the most prevalent type of disability reported by Canadian adults from among the 11 types of disability included in the Canadian Survey on Disability (2012). (See Figure 11.4.) Next to pain, flexibility and mobility are also common types of disability among Canadians (Canadian Survey on Disability, 2012).

4. Age Matters When It Comes to Disabilities
The 2012 Canadian Survey on Disability reveals that disability rates increase steadily with age. It also reveals that the most common types of disability also vary by age. Among school-aged boys and girls from 5 to 14 years old, learning disabilities were the most common type of disability (Statistics Canada, 2012). For Canadians between the ages of 15 to 24, mental/psychological disabilities were the most commonly reported type of disability (Statistics Canada, 2012). Among those aged 45 to 64, the most commonly reported disabilities were pain, flexibility, and mobility (Statistics Canada, 2012). Seniors reported these same three types of disabilities but at higher rates (Statistics Canada, 2012). Hearing and memory disabilities were also more prevalent among the senior population (Statistics Canada, 2012).

5. Family Members Are Primary Caregivers for Persons with Disabilities
Most Canadians can expect at some point in their lives to help a family member or a friend with a disability or with problems related to aging (Turcotte, 2013). For Canadians with disabilities who did not live alone, almost 80 percent reported they received help with their everyday activities from family members (Turcotte, 2013). For

> **Accommodation:** The specialized design of products, environments, and individualized strategies, which are uniquely adapted to an identified limitation for the purpose of ensuring access.

FIGURE 11.4

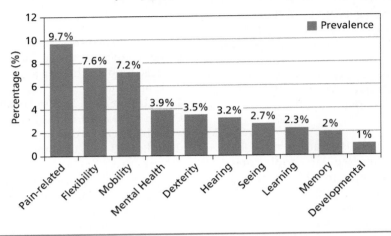

Prevalence of Disabilities by Type, Canadians Aged 15 Years or Older

Source: Statistics Canada, Canadian Survey on Disability 2012, http://www.statcan.gc.ca/pub/89-654-x/89-654-x2013002-eng.htm. Reproduced and distributed on an "as is" basis with the permission of Statistics Canada.

Canadians with disabilities who lived alone, 56 percent reported that they received help with everyday activities from family members (Turcotte, 2013). Significant help was also received from friends, neighbours, and volunteers. While there are many benefits to having family members as caregivers, there can also be negative impacts on these caregivers in terms of their physical and mental health, reduced ability to participate in the workforce, pressures on personal finances, and limited time for other activities (Turcotte, 2013).

6. More Help Needed with Daily Living Activities

The most commonly reported help received by Canadians with disabilities includes assistance getting to appointments, running errands, and doing everyday housework, particularly heavy household chores (Statistics Canada, 2012). The Canadian Survey on Disability (2012) found that unmet needs for daily living rose with the severity of a person's disability. "The prevalence of receiving help increased with disability severity, but so did the prevalence of needing but not receiving help" (Statistics Canada, 2012).

7. Less Likely to Graduate

Canadians with disabilities were less likely than Canadians without disabilities to be high school or university graduates (Statistics Canada, 2012). According to the Canadian Survey on Disability (2012), the percentage of university graduates declines as the severity of disability increases. Some key barriers cited by postsecondary students with disabilities include

difficulties with learning accommodations, transportation, housing, finances, and unsupportive and inaccessible learning environments (Anzovino, 2012). The Canadian Survey on Disability (2012) found that for students with disabilities,

> one-third (34 percent) reported that they took fewer courses/subjects; 30 percent reported that it took them longer to achieve their present level of education; 30 percent discontinued their studies; and 23 percent reported that their education was interrupted for long periods. About 40 percent indicated that people avoided or excluded them at school, and 27 percent experienced bullying.

8. More People with Disabilities Are Unemployed and Underemployed

In addition to being more likely to live in poverty, people with disabilities have a long history of high unemployment and underemployment in Canada (Government of Canada, 2014). Half of working-age adults with disabilities were employed (47 percent), compared to 74 percent of their contemporaries without disabilities (Statistics Canada, 2012).

9. Many Companies Are Not Hiring People with Disabilities

In a survey of Canadian small businesses conducted in 2013 by the BMO Financial Group, only 1 in 3 small business owners hired people with disabilities, and the majority of small businesses surveyed had never hired a person with a visible or invisible disability (BMO Financial Group, 2013). The survey found that one of the major reasons why small businesses did not hire persons with disabilities was because of unfounded myths and misunderstandings about their capabilities (BMO Financial Group, 2013).

10. Workers with Disabilities Meet Employer Expectations

In a survey conducted by BMO Financial Group in 2012, more than 75 percent of employers who recruited persons with disabilities to work for them were happy with their performance (BMO Financial Group, 2013). There are also examples of Canadian employers who are creating unique employment opportunities for persons with disabilities. Take, for example, the restaurants O.Noir in Montreal and Toronto that provide a "dining in the dark" experience for customers. All servers are legally blind. As the Canadian National Institute for the Blind (2016) notes, only one-third of Canadians with vision loss are employed and almost half of Canadians with vision loss are living in poverty. So companies like O.Noir are meeting the employment needs of persons with vision loss by hiring those most capable of working in the dark. "It is a true 'transfer of trust' and an amazing approach to raise awareness about blindness and disabilities in general" (O.Noir, 2016).

THE CONTEXT OF DISABILITY WITHIN OUR SOCIAL ENVIRONMENT

When we conceptualize disability as something different from what is "normal," then the approach we use in dealing with disability issues is one that employs individual accommodations. These accommodations are designed to meet what is conceived of as a "normal" standard. When we conceptualize disability as a universal experience, then the approach we use employs the goals of accessibility and social inclusion for everyone. For any approach to be effective, it is important to begin with the lived experiences of people with disabilities.

The Difference Approach and Accommodation

When disability is constructed as "difference from the norm"—that is, different from the "normal" body or "normal" mind—then the approach used is one that focuses on how to accommodate difference. This approach maintains an *us-versus-them* mentality and differentiates people with disabilities from people without disabilities. An accommodation approach often requires people with disabilities to be their own advocates in initiating adaptations for their unique needs; it often results in reactive and retrofit solutions.

Using the familiar context of the postsecondary classroom, what does equal access to learning through individualized accommodations mean for the college or university student with disabilities? Using the approach of recognizing disability as difference from the norm, individualized accommodations rely on the student to initiate the process of having their unique needs met (Dosis, Coffey, Gravel, Ali, & Condra, 2012). Students have to identify themselves as having a disability and provide documentation to their professors and instructors as well as to the institutional service supporting students with disabilities. One of the barriers that can be created for students through individualized accommodations is that this approach can leave them feeling as though they are asking for a special favour (Dosis, Coffey, Gravel, Ali, & Condra, 2012). Students may be reluctant to request accommodations as they can feel labelled in a negative way by this process. As one college student states:

> I was bullied in high school because of my learning disability. People called me names like "tard and "slow" and I was pushed and punched by a group of girls on a regular basis. They even bullied other people who tried to be friends with me, so I basically spent four years in high school alone. I went away to college so I could have a fresh start. I just want to fit in, so there is no way I am going to label myself as a student with a disability. I would rather fail than go through that again. (Anzovino, 2012)

Bullying behaviour is rooted in power and aggression and is often based on perceived differences in appearance, sexual orientation, ethnicity, or ability. If disabilities are perceived as different from the norm of ability, then they can reinforce an *us-versus-them* mentality that can create a fertile environment for bullying behaviour. Violence in Canadian schools against children with disabilities is a prevalent reality. In *Abilities Magazine*, Melissa Martz details her childhood experiences with bullying:

> Bullies can find any reason to target someone— maybe because he or she wears glasses, is overweight or wears hand-me-downs. Children with disabilities are frequently singled out for

being different. They may speak differently, walk differently, use assistive devices or have difficulty with social interaction—all of which bullies can turn into jokes.... It happened to me. While growing up in Kitchener, Ontario, I was tormented because I stuttered (something that I still struggle with) and required special education for math and language arts. (Martz, 2008–2009)

The case of Mitchell Wilson has garnered media attention around the issue of bullying children with disabilities. Mitchell Wilson was an 11-year-old boy from Pickering, Ontario, who had muscular dystrophy. Children at his school bullied him socially and physically. Prior to testifying in court, Mitchell committed suicide (Kennedy, 2011). This tragedy has highlighted the issue of bullying of children with disabilities. Approximately 30 percent of parents with school-aged children with disabilities report that their children have been physically assaulted by other children at school, and this number increases to 38 percent for children with severe or very severe disabilities (Statistics Canada, 2007).

In the context of the postsecondary classroom, individualized accommodations in this academic setting can result in retrofit solutions after the program, curriculum, course, and instruction have been designed (Dosis, Coffey, Gravel, Ali, & Condra, 2012). Accommodations such as providing a student with more time to write a test or arranging for a note-taker can become a "one-size fits all" adaptive measure that can ensure equality but not equitable access to education.

When access in society is based on norms of ability, the danger is "it can lead to ableism in policy and practice as it negates the commonalities of our shared lived experiences" (Hoyle, 2004). What if, instead of measuring people against norms of ability, a society reframed itself on a universal acknowledgment that all members can, over the course of their lifetime, expect to experience disability? What if, based on this viewpoint, we decided to create a society where everyone would feel included, would be able to develop their own capabilities, and would be valued and contributing members of society based on who they really are and not on some normative ideal physicality? Picture

Accessibility: The degree to which a product, device, service, or environment is available to as many people as possible.

Barriers: Policies or practices that prevent full and equal participation in society; barriers can be physical, social, attitudinal, organizational, technological, or informational.

This... looks at the work of fashion photographer Rick Guidotti as he challenges the industry's notion of ideal physicality and beauty.

In creating inclusive environments, people can challenge negative stereotypes and the assumption that people with disabilities are a separate social group because of differences in ability (Hoyle, 2004). Social inclusion challenges the "the notions of disabled and non-disabled that reinforce the boundaries between sameness and difference" (Hoyle, 2004). Social inclusion as a perspective helps us to move from accommodation as special treatment for people with disabilities to accessibility as equitable treatment for all.

Moving from Accommodations to Accessibility

When developing inclusive policy and practice, it is important that policy and practice be premised on the understanding that **accessibility** is multifaceted. Accessibility can mean different things to different people based on their experiences of the body, their stage of life, their physical environment, and so on. Creating a society where all members can participate fully in their community is one of the goals of accessibility.

In some universities and colleges that use an accommodation model, you can see students with disabilities having to self-identify with teachers to request accommodations that focus on their differences. Common examples of accommodation include extending time allotted for a test or arranging for a note-taker during classes. Students with disabilities can be reluctant to self-identify, especially when disabilities are invisible or temporary, because of the stigmatization and disadvantages that result from being labelled as different.

A study entitled "Unlocking the Possible" (Hoyle, 2004) is an interesting model of how a municipality might work at becoming more accessible and inclusive to people with disabilities. At a citizen's forum, **barriers** to social inclusion and full participation in civic life were identified:

- *Experiences of the body*, including fear of ridicule over unpredictable body movements and behaviours, variation in physical strength and energy, and personal care needs.
- *Access to physical space*, including both the accessibility of the space itself and how people move within that space in terms of the amount of space, placement of items, transportation, and types of devices available for mobility.

- *Access to time* needed for daily routine, equipment use, and full participation when physical or cognitive issues require this.
- *Access to information*, such as the technological adaptations people need to listen and communicate.
- *Access to services*, including equitable access to employment, recreation, and healthcare services.
- *Access to social networking opportunities*, which include being able to make social connections, having a sense of belonging to the community, feeling valued for one's contributions, feeling a sense of interdependence and reciprocity with other citizens, and experiencing a people-first approach (Hoyle, 2004).

This study highlights that accessibility is a multifaceted concept whose practice would eliminate those barriers and would hold the promise of community where all members can fully participate in civic life. Accessibility is a model for practising diversity as a framework for equity and social inclusion at a municipal level.

Universal Design

Accessibility and inclusion can begin with **universal design**, using design principles which can help to eliminate the need for adaptation or specialized accommodation. You may have heard these ideas referred to as barrier-free design, accessible design, adaptable design, trans-generational design, and inclusive design. Universal design considers all members of the community and plans for the full participation of all members. Universal design has been utilized in architecture, urban planning, ergonomics, and education. There are seven basic principles of universal design that originated as architectural standards in the design of physical spaces:

1. *Equitable use.* The design is useful and marketable to any group of users.
2. *Flexibility in use.* The design accommodates a wide range of individual preferences and abilities.
3. *Simple and intuitive use.* Use of the design is easy to understand, regardless of the user's experience, knowledge, language skill, or current concentration levels.
4. *Perceptible information.* The design communicates necessary information effectively to the user regardless of ambient conditions or the user's sensory abilities.
5. *Tolerance of error.* The design minimizes hazards and adverse consequences of accidental or unintended actions.
6. *Low physical effort.* The design can be used effectively and comfortably and with a minimum of fatigue.

7. *Size and space for approach and use.* Appropriate size and space is provided for approach, reach, manipulation, and use, regardless of user's body size, posture, or mobility (Connell et al., 1997)*.

Universal Design and Postsecondary Education

The focus of accessible education is to create sustainable access based on the following principles (Dosis, Coffey, Gravel, Ali, & Condra, 2012):

- dignity for students, by maintaining privacy and not singling them out;
- equitable opportunities and advantages for all students for learning, not just for information;
- independence for students, so they can complete learning tasks without specialized help; and
- integration, so that each student is able to benefit from the same learning experiences.

Universal Design for Learning (UDL) and Universal Instructional Design (UID) are educational frameworks used for program development, curriculum design, and delivery for the purpose of creating inclusive and accessible learning opportunities for all learners. UDL and UID are learner-centred approaches, which recognize "one size does not fit all." Being purposeful in design and intent, UDL and UID strategically address the diverse learning needs of all students by minimizing barriers and maximizing learning opportunities for all students. Universal design emphasizes flexibility in curriculum design; it uses multiple modes of engaging students, presenting content, and assessing comprehension; and it utilizes technology to maximize equitable learning opportunities for all students (Dosis, Coffey, Gravel, Ali, & Condra, 2012).

Universal Design for Learning

Universal Design for Learning (UDL) assists in creating learning objectives, methods of instruction, course materials, learning activities, and methods of assessment that work for everyone. UDL is not "a single, one-size-fits-all solution but rather flexible

> **Universal design:**
> "The design of products, environments, programs and services to be usable by all people, to the greatest extent possible, without the need for adaptation or specialized design (but does not exclude assistive devices for particular groups of persons with disabilities where this is needed)" (United Nations, 2006).

*Copyright © 1997 NC State University, The Center for Universal Design.The Center for Universal Design (1997). The Principles of Universal Design, Version 2.0. Raleigh, NC: North Carolina State University."

approaches that can be customized and adjusted for individual needs" (National Centre on Universal Design for Learning, 2014). UDL engages students through multiple and flexible ways of learning and expressing what they know. The three main principles of UDL developed by the Center for Applied Special Technology (CAST, 2011) are an evidence-based approach to improving the presentation of information, the engagement of learners, and the creation of flexible assessment and evaluation. These principles include providing multiple means of representation, providing multiple means of action and expression, and providing multiple means of engagement. (See Figure 11.5.)

Universal Instructional Design (UID)

Universal Instructional Design is an educational framework that utilizes the seven principles of Universal Design tailored to creating more inclusive, equitable, and accessible learning environments. In the familiar context of the postsecondary classroom, Universal Instructional Design is also a proactive

---

FIGURE 11.5

Universal Design for Learning Guidelines

I. Provide Multiple Means of Representation	II. Provide Multiple Means of Action and Expression	III. Provide Multiple Means of Engagement
1: Provide options for perception 1.1 Offer ways of customizing the display of information 1.2 Offer alternatives for auditory information 1.3 Offer alternatives for visual information	**4: Provide options for physical action** 4.1 Vary the methods for response and navigation 4.2 Optimize access to tools and assistive technologies	**7: Provide options for recruiting interest** 7.1 Optimize individual choice and autonomy 7.2 Optimize relevance, value, and authenticity 7.3 Minimize threats and distractions
2: Provide options for language, mathematical expressions, and symbols 2.1 Clarify vocabulary and symbols 2.2 Clarify syntax and structure 2.3 Support decoding of text, mathematical notation, and symbols 2.4 Promote understanding across languages 2.5 Illustrate through multiple media	**5: Provide options for expression and communication** 5.1 Use multiple media for communication 5.2 Use multiple tools for construction and composition 5.3 Build fluencies with graduated levels of support for practice and performance	**8: Provide options for sustaining effort and persistence** 8.1 Heighten salience of goals and objectives 8.2 Vary demands and resources to optimize challenge 8.3 Foster collaboration and community 8.4 Increase mastery-oriented feedback
3: Provide options for comprehension 3.1 Activate or supply background knowledge 3.2. Highlight patterns, critical features, big ideas, and relationships 3.3 Guide information processing, visualization, and manipulation 3.4 Maximize transfer and generalization	**6: Provide options for executive functions** 6.1 Guide appropriate goal-setting 6.2 Support planning and strategy development 6.3 Facilitate managing information and resources 6.4 Enhance capacity for monitoring progress	**9: Provide options for self-regulation** 9.1 Promote expectations and beliefs that optimize motivation 9.2 Facilitate personal coping skills and strategies 9.3 Develop self-assessment and reflection
Resourceful, knowledgeable learners	**Strategic, goal-directed learners**	**Purposeful, motivated learners**

Source: © 2011 by CAST. All rights reserved. www.cast.org, www.udlcenter.org.

I AM NOT OK

An in-depth investigation into student mental health issues

'I feel like I'm bothering them when I tell them.'

Courtesy of Niagara News

Second-year Journalism students at Niagara College won the Emerge Media Award in the category of Multimedia Production for their work on Crisis on Campus, a website, iamnotok.ca, that addresses the topic of depression and anxiety in college and university students. Results of a recent survey of more than 25 000 Ontario college and university students conducted by the Ontario University and College Health Association (OUCHA) reported 65 percent of students experienced overwhelming anxiety and 46 percent reported feeling so depressed they had difficulty functioning. Some mental health professionals are suggesting we have a crisis on college and university campuses. Do you agree? What is your college or university doing to address and support those with mental health issues?

way to remove barriers to learning at the outset, as opposed to a retrofit solution after the fact, thereby reducing the need for individual accommodations (Dosis, Coffey, Gravel, Ali, & Condra, 2012). This proactive approach designs curriculum and instruction to meet the diverse learning styles and needs of students (Dosis, Coffey, Gravel, Ali, & Condra, 2012). The Seven Principles of UID describe how instructional materials and activities should be accessible and fair, flexible, explicit, straightforward, and consistent (Durham College, 2016). The Seven Principles of UID also describe how the learning environment should be supportive, minimize unnecessary physical effort, and accommodate students and multiple teaching methods (Durham College, 2016).

MENTAL HEALTH AND STIGMA

One in five Canadians experiences a mental health problem (Mental Health Commission of Canada, 2012). Most Canadians can expect to be affected by mental health issues directly or indirectly through family, friends, or co-workers (Mental Health Commission of Canada, 2012). And yet despite how common mental illness is, stigmatization of and discrimination against persons with mental health issues is widespread.

One quarter of all human rights complaints received in 2015 by the Canadian Human Rights Commission were related to discrimination based on mental health (CHRC, 2016). Some of the most serious consequences can be evidenced in the fact that some people with serious mental health issues are represented in the statistic that every day in Canada, almost 11 people end their lives through suicide (Canadian Association for Suicide Prevention, 2016).

For persons with mental health issues, stigma comes from the negative stereotypes and labels associated with having a mental illness, and discrimination is the behaviour that results from this (Canadian Mental Health Association, 2016). Stigma is one of the major reasons why more than 60 percent of Canadians with mental health issues won't seek the help they need (Mental Health Commission of Canada, 2012). The stigma attached to mental health conditions is so pervasive that it can affect all aspects of a person's life, from employment, housing, and education to feelings of self-worth and interpersonal relationships with family and friends (Canadian Mental Health Association, 2016). In a study by the Canadian Medical Association, only 49 percent of Canadians indicated they would socialize with a friend who had a serious mental health issue (Bell Let's Talk, 2016).

Many myths contribute to the stigmatization of persons with mental health issues. One very common misperception that is perpetuated by the media and within popular culture is that people with mental illness are typically violent. According to evidence, people with mental health issues are more likely to be the *victims* of crime, bullying, hate, and discrimination (Mental Health Commission of Canada, 2012).

Stigma associated with mental health issues can be reduced. This is key to improving individual health outcomes and making improvements in mental health systems. "Reducing stigma requires a change in behaviours and attitudes towards acceptance, respect and equitable treatment of people with mental health problems and mental illnesses" (Mental Health Commission of Canada, 2012).

Courtesy of Bell Canada

There are many strategies being utilized to reduce stigma. New systems and service strategies are emerging from coalitions of stakeholders. Public education and awareness campaigns are being used to target the fear of mental health as something "abnormal." We see examples of celebrities such as Demi Lovato, Dwayne Johnson, Howie Mandel, Collin Farrell, Catherine Zeta Jones, Brooke Shields, JK Rowling, Jake Padalecki, and Larry Sanders using their social influence to talk about their own mental health issues, thus creating a form of a new "normal." Bell Let's Talk is one example of a Canadian mental health initiative that provides support for care and access at a community level, research, and the encouragement of greater corporate responsibility for workplace health across Canada, but it is probably best known for its fourth pillar of support whose anti-stigma awareness campaign hosts a national conversation about mental illness (Bell Let's Talk, 2016). As spokesperson Howie Mandel argues, if physical health, dental health, and mental health were all viewed as simply health issues, perhaps the stigmatization of persons with mental illness would be diminished. Agent of Change Bill MacPhee is a further example of a person using their own story to reduce stigma by demonstrating that recovery is possible with treatment and supports. What appears common among most strategies designed to reduce the stigma associated with mental health is increased public understanding that mental illness is not a personal choice and that recovery is possible (Mental Health Commission of Canada, 2012).

ENDING THOUGHTS

One of the ways of understanding the context of disability within our social environment is through the lived experiences of people with disabilities. This is the foundation upon which *Walk a Mile* was first conceived. Any strategies used to promote social inclusion of people with disabilities means respecting their varied lived experiences.

Human rights commissions and tribunals in Canada have reported discrimination based on disability as the predominant type of discrimination for many years now (Canadian Human Rights Commission, 2012). To change this fact, it is necessary to demonstrate to society the many ways that disability touches people's lives and to promote improved reflective practices within all professions around issues related to ableism (Shier, Sinclair, & Gault, 2011).

AGENT OF CHANGE

Bill MacPhee

Bill MacPhee

Bill MacPhee describes himself as a mental health recovery expert and defines recovery as "when you wouldn't want to be anyone other than who you are today." Why an expert? Many reasons, in fact. Bill is the founder of Magpie Media Inc. that publishes *SZ Magazine*, a publication containing the latest research and information on schizophrenia, schizoaffective, and psychosis. In 2008, he published *Anchor* magazine to aid in recovery from bipolar disorders, depression, and anxiety. These magazines have recently been replaced with a daily video vlog that

Bill provides called Life after Mental Illness. Bill is the recipient of numerous awards recognizing his dedication to mental health recovery, including the 2012 National Council Reintegration Award, 2011 Kaiser Foundation National Awards for Excellence—Media Reporting, 2009 Transforming Lives Award (Center for Addiction and Mental Health), 2007 Ontario Psychiatric Association Theodore Allen Sweet Award, 2005 National Alliance on Mental Illness New York State Distinguished Media Award, 2005 FAME (Family Association for Mental Health Everywhere) Award, and many more. Bill has spent the last 20 years of his life speaking to audiences across North America, including college and universities, in an effort to educate others on mental health recovery and to help eliminate the stigma associated with mental health. It is not uncommon to find Bill at Tim Hortons in his community lending a supportive ear to someone in need, and the next day see him as keynote speaker at a national conference. So by now, most are nodding their heads that Bill is indeed a mental health recovery expert and an exemplary agent of change. But perhaps most important of all is that Bill has experienced recovery himself. At the age of 24, he was diagnosed with schizophrenia and was pacing in a psychiatric ward, trapped in psychosis. His inspirational personal journey to recovery is chronicled in his personal biography titled, *To Cry A Dry Tear.*

Bill's life with schizophrenia and the stigma attached to the illness is also the focus of Canadian filmmaker Mark Ashdown's 2009 film, *Life After Mental Illness: The Story of Bill MacPhee.* Bill has campaigned to change the name of schizophrenia in an attempt to remove the myths and stereotypes associated with the diagnostic label of schizophrenia. An agent of change in Canada who has "walked a mile" and then some.

Diversity-competent practitioners must honour the voices and lived experiences of all people. When we define *diversity* in a way that privileges the concepts of culture, ethnicity, and race, we further eclipse the lived experiences of people with disabilities in a manner that reinforces their social exclusion as members of our communities. When the concept of *ability* is included in the framework for achieving social equity and justice, we move closer to the goal of social inclusion. As part

of this discussion, it becomes important to acknowledge the fact that "all people live on a continuum of ability and disability with our position on this continuum dependent on time of life, the time of day, the situation and the environment … ability is not an either/or situation where one is either able or not able" (Hoyle, 2004). When we consider these universalizing lived experiences, we can begin to look at our shared humanity with a people-first approach that sees ability in everyone.

LIVING WITH POST-TRAUMATIC STRESS DISORDER

By Vesna Plazacic

There has been a lot of media coverage about post-traumatic stress disorder (PTSD), with first-responders, war correspondents and former soldiers sharing personal stories publicly and often for the first time. Experts agree that this is not only lifting the veil of silence surrounding the condition, but also helps to promote healing.

Most of us will experience some form of stress as a child or adult that affects us physically, but our reactions and the healing process are usually brief. For people who directly experience or witness events such as crimes, natural disasters, car accidents, war or conflict, sexual violence or other threats to life or safety, that stress can become life-altering. In fact, eight per cent of us will experience PTSD after a traumatic event. Those affected tend to re-experience the traumatic event through vivid flashbacks and nightmares, which can lead to anxiety, depression and avoidance of social situations.

According to the Canadian Mental Health Association (CMHA), people affected with PTSD often feel guilt and shame and, because mental health illnesses are often not physically visible, are told to try to "get over" the difficult experience.

Complex PTSD

While PTSD forms as a mental disorder occurring after a single event, some people will develop what is referred to as complex (C)-PTSD. This form of PTSD is not well known, and is a much more complicated version of the illness. It is characterized by prolonged exposure to trauma, usually in childhood, and the experience of events in which the victim feels he or she has no control or opportunity for escape. C-PTSD is often seen in victims of domestic, emotional, physical or sexual abuse, and in those who have experienced entrapment, kidnapping, slavery, long-term exposure to a crisis or neglect. For individuals with physical disabilities, C-PTSD can develop from traumatic events such as major surgery, a life-threatening illness, time spent in intensive care units (ICUs) and prolonged abuse by caregivers.

On her blog "Psychology in Everyday Life," Dr. Deborah Khoshaba explains that a person who experiences a trauma in childhood is also more susceptible to developing C-PTSD later in life. "The ongoing trauma takes up so much of the growing child's heart, mind and spirit," she says, "that the brain cannot attend to needs outside of the trauma, to grow and strengthen. This is especially true of the brain's frontal lobe region that helps children to learn, control impulses, regulate our emotions, reason, concentrate and problem-solve, or use their imagination toward goal-achievement. Simply put, C-PTSD has a devastating impact on people's lives, as their intelligence, will and talent are seriously undermined by the disorder."

Natacha's experience

Retired Cpl Natacha Dupuis is still dealing with the aftermath of the death of two friends and fellow solders who were killed by an improvised explosive device in Afghanistan in 2009. Three others were injured. Natacha was the first responder on the scene.

"The whole thing lasted for about 20 minutes, from the explosion to the evacuation of the injured and the deceased," she says. "But that 20 minutes left me with mental scars." Following the event, Natacha says she started experiencing nightmares, flashbacks and panic attacks. When she returned home to Quebec, she was diagnosed with PTSD.

"When I came home from Afghanistan I really fell apart," she says. "I think the stress and the adrenaline kept me going through the mission, but when back in Canada, it got worse. I suffered from flashbacks that were so real I would faint." Natacha even ended up in the hospital after a particularly vivid, violent flashback.

She says it took her five years to learn how to cope with her symptoms, find help through therapy and begin to put the pieces of her life back together. Through the healing process, in 2011 she came across Soldier On, a Canadian Armed Forces program that supports currently serving members and veterans. Soldier On helps people to overcome physical or mental illnesses and to cope with symptoms. Natacha talks about attending events and beginning to see a positive change in herself. Having gained more strength and developed new goals for success, she then found her way into the Invictus Games, the international competition for military personnel injured during service that was founded by Prince Harry in 2014.

Natacha took home two gold medals in the IT7 100m and 200m and a bronze in weight-lifting in last year's competition in Florida.

"I gave it all I had for my country and for the fallen, those two friends I lost," she says of the emotional moments crossing the finish line and winning her medals. She is currently training for the 2017 games, which will be held in Toronto this September, and Natacha will be at the helm as one of Team Canada's two captains.

With respect to PTSD, Natacha believes that "It's really important to reach out and get the support you need. I don't think this will ever go away completely, but with time, hard work and therapy, I will learn how to cope."

Source: Plazacic, Vesna. "Post-Traumatic Stress Disorder." *Abilities Magazine,* found at: http://abilities.ca/post-traumatic-stress-disorder. Reprinted with permission from *Abilities Magazine.*

DISCUSSION QUESTIONS

1. Who gets PTSD? What are the symptoms of PTSD?
2. What is complex C-PTSD and what types of events may cause this specific form of PTSD to develop?
3. What strategies has Natacha used to cope with the symptoms of PTSD in her life? Do you think as people like Natacha share their experiences, the stigma and silence surrounding PTSD will diminish?

KWIP

KNOW IT AND OWN IT: WHAT DO I BRING TO THIS?

The "K" in the KWIP process involves examining aspects of your own identity and social location as the first step in becoming diversity competent.

ACTIVITY: JOURNAL

This journal activity is based on a true story. Two sisters decide to go to the local shopping mall to pick something up. One sister was diagnosed with multiple sclerosis at a young age. This chronic progressive disease makes many aspects of daily living a struggle, including walking. They park in an accessible parking space close to the mall entrance and display their accessibility permit as required by law. When they return to their vehicle, they find this handwritten note on the windshield.

1. In your journal, write a letter responding to the author of this handwritten note.

2. We know that some disabilities are visible, some are invisible, some are permanent, and others are temporary. Have you had personal experience with disability, either yourself or through someone you know? What effect has this had on your understanding of what the daily lived experiences are for people with disabilities?

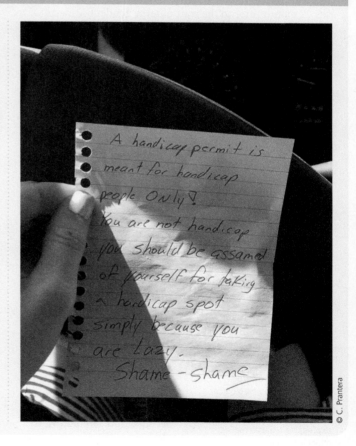

© C. Prantera

WALKING THE TALK: HOW CAN I LEARN FROM THIS?

The "W" in the KWIP process presents a scenario or case study that challenges you to "walk the talk" through problem-based learning.

ACTIVITY: CASE STUDY

The issue of academic accommodations in the postsecondary learning environment is an important one for students with disabilities, particularly in the face of statistical data showing they are less likely to graduate than students without disabilities (Statistics Canada, 2012). In addition to dealing with issues relating to their disability, students with disabilities are also faced with dealing with barriers to accessible education. This case study is based on a true story of a student studying in a Canadian university.

After completing a college diploma, Student X goes on a vacation they had received as a graduation gift. While on vacation, Student X is injured in a skiing accident and acquires a brain injury. After a year of physical and neuropsychological rehabilitation, Student X wishes to continue their education and enrols in a degree program at university. Student X registers with the centre for students with disabilities at the university. A month prior to the start of term, Student X meets with the centre's staff and submits all required documentation, including medical reports and an educational assessment completed by their neuropsychologist. Based on extensive diagnostic assessment, the neuropsychologist recommends the following academic accommodations for Student X:

- Note-taker

- Permission to record lectures (student signs agreement that audio- recordings are for personal and educational purposes only and cannot be distributed in any way to another party).

- Because Student X has **expressive aphasia**, it is recommended that alternative assignments be given to take the place of oral presentations.

- Because of cognitive issues relating to attention and concentration, a modified exam schedule of no more than one exam per day is recommended, with at least one day between exams, and exams taking longer than one hour to be completed at separate exam times of not more than one hour.

- Because of deficits with working memory, evaluations need to be multiple-choice methods. Where this is not possible, a take-home examination would be a second alternative.

While most of the academic accommodations are facilitated for Student X, they do have difficulty with some faculty refusing to follow the letter of accommodation with respect to evaluation

Expressive aphasia: A communication disorder resulting from a neurological issue affecting a part of the brain that deals with language; expressive aphasia makes it difficult for the person to find the words to say or write what they mean.

methods. This has an effect upon Student X's ability to complete the program. Here are a few of the situations that Student X faces trying to complete their courses:

- Student X is forced in one class to do an oral presentation and as they struggle to find the words to speak because of expressive aphasia, classmates begin to laugh. Student X leaves the classroom and drops the course.

- In another course, the professor tells Student X that they are not going to make up a special test just for them, so the student is forced to do a test made up of short answer, fill in the blank, and essay questions. Student X receives a grade of 58 in the course. Student X retakes the course a year later with a professor who evaluates their performance with multiple-choice testing and receives a grade of 92 in the course.

- Student X takes a course that requires the application of mathematical equations. When Student X asks the course professor if they can use a cue sheet for equations for the exam, they are refused. The professor tells Student X that it would be unfair for them to have a cue sheet when all other students had to memorize equations—a kind of "reverse discrimination" for students without disabilities in the course.

- When Student X approaches the disability counsellor assigned to them through the university's centre for students with disabilities, Student X is told by their staff that professors have the academic freedom to evaluate a student's performance in whatever manner they deem necessary and are not required to follow academic accommodation relating to evaluation.

1. Using human rights legislation or other relevant legislation in your province or territory, discuss how "duty to accommodate" would be relevant in this particular case.

2. Using human rights legislation or other relevant legislation in your province or territory, discuss how the concept of "academic freedom" and "essential requirements" would be relevant in this particular case.

3. Imagine now that you are a professor in a postsecondary institution. How would you specifically deal with Student X's need for accommodation with course evaluation methods? What would you do to ensure that *all* students in your classroom are included? What would you do to account for differences in learning styles and learning needs? How would you evaluate students' understanding in ways that allow all students to demonstrate what they know?

IT IS WHAT IT IS: IS THIS INSIDE OR OUTSIDE MY COMFORT ZONE?

The "I" in the KWIP process requires you to honestly confront and identify ways in which our complex identities result in experiences of privilege and oppression, and to reflect on how we can learn to honour that privilege.

ACTIVITY: GOT PRIVILEGE?

As self-reflective practice helps us to examine our biases, stereotypes, and prejudices, we often realize that many of our feelings about disability may be unconscious and unintentional. But at some point in your life, you are likely to experience a disability, based on age, accident, or circumstance. It can be visible or invisible, temporary or permanent. You will hope you will be able to get to where you need to go. You will not want to be praised for your courage, nor will you want to be pitied. You will hope that your friends and loved ones won't treat you differently and begin to withdraw. You will hope that other people you meet won't define you by your disability, but will rather see you as a whole person. In this exercise, we ask you to examine your privilege and power in relation to your abilities.

ASKING: Do I have privilege?

1. I am reasonably sure that I will have physical access to all areas of my classrooms and to all parts of the campus.

2. I can do well in situations without being told I am an inspiration to other people of my ability status.

3. If I am not hired for a job, I do not have to question whether this had anything to do with my physical or mental ability.

4. I can easily access and implement the appropriate accommodations to help me learn and succeed in the postsecondary institution I attend.

REFLECTING: Honouring Our Privilege

In relation to your own physical, mental, sensory, developmental, or cognitive abilities, describe two traits that you possess that disadvantage you. Then describe two abilities that privilege you over someone else. How can you use the advantages you experience because of your abilities to combat the disadvantages experienced by others with different abilities?

PUT IT IN PLAY: HOW CAN I USE THIS?

The "P" in the KWIP process involves examining how others are practising equity and how you might use this.

ACTIVITY: CALL TO ACTION

When Canada Post announced that it would be ending door-to-door delivery in Canada, there was concern expressed by seniors, persons with disabilities, and persons living in remote communities about the impact of this business decision on what had become an access issue for many Canadians with issues relating to mobility. Susan Dixon, a mother with a young son with cerebral palsy, launched a change.org petition against the move. In her petition, Dixon issued a challenge to Canada Post bosses to try and spend a week retrieving their mail from a community box using a wheelchair. While the petition claimed victory and did indeed make door-to-door mail delivery an issue in the last federal election, the victory chant may be a little premature. You can view the petition on change.org at https://www.change.org/p/don-t-let-canada-post-end-door-to-door-delivery/responses/30760.

Study Tools
CHAPTER 11

Located at www.nelson.com/student

- Review Key terms with interactive **flash cards**
- Check your Comprehension by completing **chapter review quizzes**
- Gauge your understanding with *Picture This* and accompanying short answer questions
- Develop your critical thinking/reading skills through compelling **Readings** and accompanying short answer questions
- Apply your understanding to your own experience with **Connect A Concept** activities
- Evaluate Diversity in the Media with engaging *Video Activities*
- Reflect on your Understanding with *KWIP* activities

Anzovino, T. (2012). Universal design in the post-secondary learning environment. Unpublished raw data.

ARCH Disability Law Centre. (2013). *A brief history of disability rights in Canada.* Toronto: ARCH.

Baldridge, D. C., & Swift, M. L. (2013). Withholding requests for disability accommodation: The role of individual differences and disability attributes. *Journal of Management*, 39, 375–385.

Bell Let's Talk. (2016). 5 simple ways to help end the stigma around mental illness. Retrieved from *Bell Let's Talk: End The Stigma*: http://letstalk.bell.ca/en/end-the-stigma/

Bickenbach, J. E. (1993). *Physical disability and social policy.* Toronto: University of Toronto Press.

BMO Financial Group. (2013, October 24). About BMO:Canadian businesses slow to hire people with disabilities. Retrieved from *BMO*: http://newsroom.bmo.com/press-releases/canadian-businesses-slow-to-hire-people-with-disab-tsx-bmo-201310240906438001

Brodsky, G., Day, S., & Peters, Y. (2012, March). Accommodation in the 21st century. Retrieved from *Canadian Human Rights Commission*: http://www.chrc-ccdp.gc.ca/sites/default/files/accommodation_eng.pdf

Canadian Association of Community Living. (2016). History. Retrieved from *CACL*: http://www.cacl.ca/about-us/history

Canadian Association for Suicide Prevention. (2016). Understanding: The facts about suicide. Retrieved from *CASP: Sharing Hope and Resiliency*: http://suicideprevention.ca/understanding/

Canadian Cancer Society's Advisory Committee on Cancer Statistics. (2015). *Canadian cancer statistics 2015.* Toronto: Canadian Cancer Society.

Canadian Charter of Rights and Freedoms. (1982). Part I of the Constitution Act, 1982, RSC 1985, app. II, no. 44.

Canadian Disability Policy Alliance. (2012). Timeline of disability policy events. Retrieved from *CDPA Queen's University*: http://www.disabilitypolicyalliance.ca/latest-news/timeline-of-disability-policy-events.html

Canadian Human Rights Commission. (2016). *Annual report.* Ottawa: Minister of Public Works and Government Services, 2015.

Canadian Mental Health Association. (2013). Homelessness. Retrieved from *CMHA*: http://www.cmha.ca/public-policy/subject/homelessness/

Canadian Mental Health Association. (2016). Stigma and discrimination. Retrieved from *Canadian Mental Health Association Ontario*: http://ontario.cmha.ca/mental-health/mental-health-conditions/stigma-and-discrimination/

Canadian National Institute for the Blind. (2016). Who we are: Our history. Retrieved from *CNIB: Seeing Beyond Vision Loss*: http://www.cnib.ca/en/about/who/history/Pages/default.aspx

CAST (2011). *Universal Design for Learning guidelines version 2.0.* Wakefield, MA.

Chaudoir, S. R., & Quinn, D. M. (2010). Revealing concealable stigmatized identities: The impact of disclosure motivations and positive first disclosure experiences on fear of disclosure and well being. *Journal of Social Issues*, 66, 570–584.

Christy, R. (2001, February 5). Life with cerebral palsy. *Maclean's*.

Connell, B., Jones, M., Mace, R., Mueller, J., Mullick, A., Ostroff, E., ... Vanderheiden, G. (1997, April 1). The principles of universal design. Retrieved from *The Centre for Universal Design*: http://www.ncsu.edu/ncsu/design/cud/about_ud/udprinciplestext.htm

Council of Canadians with Disabilities. (2007). From vision to action: Building an inclusive and accessible Canada: A national action plan. Retrieved from *Council of Canadians with Disabilities*: http://www.ccdonline.ca/en/socialpolicy/actionplan/inclusive-accessible-canada

Council of Canadians with Disabilties. (2009). As a matter of fact: Poverty and disability in Canada. Retrieved from *CCD Online: Social Policy*: http://www.ccdonline.ca/en/socialpolicy/poverty-citizenship/demographic-profile/poverty-disability-canada

Dosis, O., Coffey, K., Gravel, D., Ali, I., & Condra, E. (2012). *Accessibility awareness training for educators. Fulfilling our commitment to accessible education. Colleges Ontario.*

Durham College. (2016). Universal Instructional Design versus Universal Design for Learning. Retrieved from *Durham College Curriculum Development: Universal Design*: http://cafe.durhamcollege.ca/index.php/curriculum-development/universal-design-for-learning/udl-vs-uid

Eldridge v. British Columbia (Attorney General), S.C.R. 624; [1997] S.C.J. No. 86 (Supreme Court of Canada 1997).

Furnham, A. (7, May 2015). *The anti-psychiatry movement.* Retrieved from *Psychology Today*: https://www.psychologytoday.com/blog/sideways-view/201505/the-anti-psychiatry-movement

Galer, D. (2015, April 23). Disability rights movement. Retrieved from *Historica Canada*: http://www.thecanadianencyclopedia.ca/en/article/disability-rights-movement/

Government of Alberta. (2010). Training for work: Disability related employment supports. Retrieved from *Government of Alberta: Human Services*: http://www.humanservices.alberta.ca/working-in-alberta/3159.html

Government of Canada. (2005, October 5). NRC helps welcome home a great Canadian innovation: Original electric wheelchair returns to Ottawa. Retrieved from *National Research Council*: http://www.nrc-cnrc.gc.ca/eng/achievements/highlights/2005/electric_wheelchair.html

Government of Canada. (2014). *Convention on the rights of persons with disabilities: First report of Canada.* Ottawa:

Her Majesty the Queen in Right of Canada, represented by the Minister of Canadian Heritage and Official Languages.

Government of Canada. (2015, November 30). *Canadian Survey on Disability*. Retrieved from *Statistics Canada*: http://www.statcan.gc.ca/pub/89-654-x/89-654-x2014001-eng.htm

Government of Canada. (2016). *Employment for people with disabilities*. Retrieved from *Service Canada*: http://www.servicecanada.gc.ca/eng/audiences/disabilities/employment.shtml

Grekul, J., Krahn, A., & Odynak, D. (2004, December). Sterilizing the "feeble-minded": Eugenics in Alberta, Canada, 1929–1972. *The Journal of Historical Sociology, 17* (4), 358–384.

Guidotti, R. (2015). *About the program*. Retrieved from *Positive Exposure: The Spirit of Difference*: http://positiveexposure.org

Hoyle, J. (2004). *Unlocking the possible: A case for inclusion.* Fort Erie: Town of Fort Erie.

Hoyle, J. (1999). Physical activities in the lives of women with disabilities. *Sport and Gender in Canada,* 254–268.

Human Resources and Skills Development Canada. (2011). *Disability in Canada: A 2006 profile.* Gatineau: Human Resources and Skill Development Canada. Retrieved from http://publications.gc.ca/collections/collection_2011/rhdcc-hrsdc/HS64-11-2010-eng.pdf

Information bulletin - What is albinism? (2015). Retrieved from *National Organization for Albinism and Hypopigmentation*: http://www.albinism.org/site/c.flKY-IdOUIhJ4H/b.9253761/k.24EE/Information_Bulletin__What_is_Albinism.htm

Kennedy, B. (2011, September 26). *Disabled Pickering boy took his own life after he was mugged and bullied.* Retrieved from *Toronto Star*: http://www.thestar.com/news/gta/2011/09/26/disabled_pickering_boy_took_his_own_life_after_he_was_mugged_and_bullied.html

Kretzmann, J. P., & McKnight, J. (1993). *Releasing individual capacities. Building communities from the inside out.* ACTA Publications: Chicago.

Laidlaw Foundation. (2002–2003). *Perspective on social inclusion: The working paper series.* Toronto: Laidlaw Foundation.

Let's talk about invisible disabilities. (2015, July 15). Retrieved from *Rick Hanson Foundation*: http://www.rickhansen.com/Blog/ArtMID/13094/ArticleID/81/Lets-Talk-About-Invisible-Disabilities

Magpie Media Inc. (2015). *About Bill*. Retrieved from *Bill MacPhee: Your Mental Health Recovery Expert*: http://billmacphee.ca/

Martz, M. (2008–2009, Winter). Standing up to bullying. Retrieved from *Abilities Magazine*: http://www.abilities.ca/learning/2009/02/11/issue77_article_bullying/

Mental Health Commission of Canada. (2012). *Changing directions, changing lives: The mental health strategy for Canada.* Calgary Alberta: Mental Health Commission of Canada.

The National Centre on Universal Design for Learning. (2014, July 31). *UDL guidelines 2.0.* Retrieved from *The National Centre on Universal Design in Learning*: http://www.udlcenter.org/aboutudl/udlguidelines

Nixon, H. L. (1984). Handicapism and sport: New directions for sport sociology research. *Sport and the sociological imagination,* Nancy Theberge and Peter Donnelly (Eds.), pp. 162–176.

Office for Disability Issues HRDC. (2003). *Defining disability: A complex issue.* Gatineau: Human Resources Development Canada.

Oliver, M. (1990). *The politics of disablement.* London: The MacMillan Press Ltd.

O.Noir. (2016). *What is O.Noir?* Retrieved from *O.Noir*: http://www.onoirtoronto.com/what-is-o-noir/

Plazacic, Vesna. (2017). "Post-traumatic stress disorder." Retrieved from *Abilities Magazine*: http://abilities.ca/post-traumatic-stress-disorder.

Reaume, G. (2012). Disability history in Canada: Present work in the field and future prospects. *Canadian Journal of Disability Studies, 1*(1), 35–81. doi: http://dx.doi.org/10.15353/cjds.v1i1.20

Rosenhan, D. (1973). On being sane in insane places. *Science,* 179, 250–258.

Santuzzi, A. M. (2013, June 26). *Invisible disabilities: The challenges of identifying and disclosing disabilities that others can't see.* Retrieved from *Psychology Today*: https://www.psychologytoday.com/blog/the-wide-wide-world-psychology/201306/invisible-disabilities

Shapiro, J. (1993). *No pity: People with disabilities forging a civil rights movement.* New York: Three Rivers.

Shier, M. L., Sinclair, C., & Gault, L. (2011). Challenging "ableism" and teaching about disability in a social work classroom: A training module for generalist social workers working with people disabled by the social environment. *Critical Social Work 12*(1), 47–64.

Social Planning Council of Kitchener Waterloo. (2001, April). *Disabilities: Universal design: Waterloo Region trends research project.* Retrieved from *Waterloo Region*: http://temp.waterlooregion.org/spc/trends/disabilities/design.html

Standing Senate Committee on Social Affairs, Science and Technology. (2016, November). *Dementia in Canada: A national strategy for dementia-friendly communities.* Ottawa, ON.

Statistics Canada. (2007). *Participation and activity limitation survey 2006: Tables.* Ottawa: Statistics Canada.

Statistics Canada. (2012). *Canadian survey on disability 2012: Tables.* Ottawa: Statistics Canada.

Turcotte, M. (2013, September). *Family caregiving: What are the consequences?* Retrieved from *Statistics Canada*:

Insights on Canadian Society: http://www.statcan.gc.ca/pub/75-006-x/2013001/article/11858-eng.htm

United Nations. (2006). *United Nations convention on the rights of persons with disabilities*. Geneva: United Nations.

United Nations. (2012, September 12). Seeking to advance rights of people with disabilities, UN treaty review starts in New York. Retrieved from *UN News Centre*: http://www.un.org/apps/news/story.asp?NewsID =42860

Venable, V. (2016, March 17). Living my life in the grey area of disability. Retrieved from *The Mighty*: http://themighty.com/2016/03/living-my-life-in-the-grey-area-of-disability/

Wachsler, S. (2007). The real quality of life issue for people with disabilities. *Journal of Progressive Human Services*, *18*(2), 7–14.

War Amps Canada. (2016). About us: History. Retrieved from *The War Amps*: http://www.waramps.ca/about-us/history/

Wendell, S. (1989). Towards a feminist theory of disability. *Hypatia*, Vol. 4, No. 2, 104–124.

World Health Organization. (2013). Health topics: Disabilities. Retrieved from *World Health Organization*: http://www.who.int/topics/disabilities/en

Age

> "So scared of getting older
> I'm only good at being young
> So I play the numbers game
> To find a way to say that life has just begun"
>
> *(John Mayer, 2006)*

LEARNING OUTCOMES

By mastering this unit, students will gain the skills and ability to:

- distinguish chronological age from psychological age and social age

- analyze the ways in which age is socially constructed

- describe age stratification as a hierarchal system of power and privilege

- summarize the differing characteristics that identify each generation

- evaluate the effects of ageism on youth and the aging population.

- understand and identify the intersectionality of age and socio-economic status, ability, and gender and sexuality

Muellek Josef/Shutterstock

Have you ever been told to "act your age"? If so, did you understand what was being asked of you? The frequency with which the phrase "act your age" is used and commonly understood suggests that age is more than just a number, but for most of us, age is simple math. We think of our **chronological age**, or the number of years since birth, as an objective reality and understand that as we get older, we will change physically and biologically. In addition to chronological age, we also commonly refer to our **psychological age**, or how old we feel with respect to maturity and emotional growth. For some, there is a discrepancy between our chronological age and our psychological age; we may not feel as old as we are or we may act more mature than others of similar chronological age. But what influences how we should feel, act, or behave at a certain age? What is someone referring to when they say "act your age"? How are the norms, roles, and expectations of certain age groups created? As you read through this chapter, consider your **social age**—what generation do you belong to and what set of expectations exist for people of your age in your culture? How does one "act" their age?

AGE AS A SOCIAL CONSTRUCTION

If our understanding of age was objective and only based on the biological process of aging, than it would be logical to think that there would be universal agreement on what constitutes young or old, dependent or independent, mature or immature, since we all *biologically* age in similar ways. But we see significant variations in the social construction of age categories and perceptions of age both historically and cross-culturally.

In her essay, "Act Your Age," Professor Cheryl Laz argues that our perception of age and what we accept as the norms, roles, and expectations of various age categories are influenced by more than just biology (2009). Laz suggests

that, similarly to race and gender, age is socially constructed by our interactions, culture, social institutions, and social trends (2009). More specifically, our perception of age has very little to do with the physical changes we experience as we age and more to do with the socio-cultural meaning that we attach to the aging process (Laz, 2009). For example, the presence of grey hair or wrinkles for someone in their twenties means very little, but in a cultural context can indicate that one is aging prematurely, leaving them feeling ashamed. In contrast, the absence of grey hair or wrinkles for someone in their sixties can suggest one is aging gracefully and leave them feeling proud. Laz argues that institutions, such as the media, present images of people in various age categories, and we use these images to compare ourselves and evaluate how we measure up to the ideal (2009). Our legal system places age restrictions on education, marriage, driving, voting, drinking, and sexual intercourse, and we use these restrictions to determine age-appropriate behaviours. Furthermore, trends in employment and labour markets dictate social policy for people in certain age groups. For example, prior to December 12, 2006, the Ontario Human Rights Code did not protect persons 65 years or older against age discrimination in employment, resulting in mandatory retirement policies. Laz suggests that we create our own ideologies about age within the context of a society that heavily influences the construction of our beliefs about age and aging (2009). Consequently, it is not the physical or biological nature of aging that determines the expectations of these social age groups. In the box titled "Picture This," we see a woman who challenges traditional age expectations—what do you think when you see this photo? To what extent should we all "act our age"?

THE GENERATIONAL DIVIDE: AGE STRATIFICATION IN CANADA

A **generation** is a **cohort** or group of people who share similar chronological ages and are shaped by the events, trends, and developments of a particular span of time (McCrindle & Wolfinger, 2010). The world events, popular culture, and media that help to shape the individuals of a generation create the bonds that unite those individuals, while simultaneously constructing barriers that can separate one generation from another. Although generations tend to be associated with a specific chronological age, they are in fact social constructs that are open to any number of definitions and interpretations. It is important to understand that it is the lived experiences that people

Chronological age: The number of years a person has lived.

Psychological age: How a person feels, acts, or behaves with respect to maturity and emotional growth.

Social age: Based on how a person meets the set of expectations that people in a given culture have about when life's major events "should" occur.

Generation: A cohort or group of people who share similar chronological ages and are shaped by the events, trends, and developments of a particular span of time.

Cohort: Any group of people who share a time-specific period, such as graduating in the same year.

Who says you should act your age?

have in common—more so than their age—that make them part of a generation. For that reason, we see different time frames associated with the same cohort or overlap between generational categories throughout the literature.

Like most hierarchal systems, **age stratification** is based on the unequal distribution of power, where some generations are valued more than others, creating advantages for some and disadvantages for others. These divisions often result in ageism, or age-based stereotypes and prejudice that can lead to age discrimination. Throughout this chapter, we will look at how generations are divided and shaped by the social events and trends of each time period, followed by a look at ageism and how it affects people of all ages.

Traditionalist

Born in 1945 or earlier, **Traditionalists** are the parents of the Baby Boomers. Those born specifically between 1939 and 1945 are also known as the World War II generation, and they are the fastest growing cohort

in Canada today. This generation makes up 13.8 percent of Canada's population (Statistics Canada, 2012a). Within Canada's senior population, 73 percent are between the ages of 65 and 79 years, and more than one-quarter (27 percent) are 80 years and older. By 2036, according to a 2010 Statistics Canada projection, between 9.9 million and 10.9 million, roughly 23 percent of the estimated 40 million Canadians, will be seniors (Statistics Canada, 2010a).

Individuals growing up during these times experienced such events as the Great Depression, World War II, Pearl Harbor, the Korean War, and the rise of labour unions (Buahene & Kovary, 2007). During the Great Depression, men, women, and children had to learn to survive without food, clothing, or, in some cases, a roof over their heads until they found work. Some Traditionalists placed a great deal of value on hard

Age stratification: The hierarchal ranking of people into age groups within a society.

Traditionalist: Refers to individuals born in 1945 or earlier.

FIGURE 12.1

Generations at a Glance

Traditionalist	Baby Boomer	Generation X	Generation Y	Generation Z	Generation Alpha
1945 & earlier	1946–1964	1965–1979	1980–1995	1996–2009	2010–
AKA: World War II Generation Parents of Baby Boomers	**AKA:** Generation Jones	**AKA:** Lost Generation, Latchkey Kids	**AKA:** Millennials, Net Generation	**AKA:** Centennials iGeneration	**AKA:** Children of Gen Y & Gen Z, Grandchildren of Baby Boomers & Gen X
13.8% of Canadian Population	**28.6%** of Canadian Population	**8.4%** of Canadian Population	**27.3%** of Canadian Population	**16.9%** of Canadian Population	**5%** of Canadian Population
EVENTS: Great Depression World War II Pearl Harbor Korean War Rise of Labour Unions	**EVENTS:** Civil Rights Movement Women's Liberation Quebec Crisis & Bill 101 Vietnam War Woodstock Neil Armstrong Landing on the Moon	**EVENTS:** MTV/Much Music Personal Computer Canadian Charter of Rights and Freedoms Quebec Separation Referendum Fall of the Berlin Wall AIDS	**EVENTS:** Death of Princess Diana Oklahoma Bombing Digital Age Y2K 9/11 US War on Terror	**EVENTS:** Google/Wikipedia/ Skype Apple TV/Netflix Reality TV Social Media Cyberbullying Laws Occupy Wall Street Boston Marathon Bombing	**EVENTS:** Facetime Siri 3D Printers Google Glass And many more to come...

Source: Statistics Canada (2015), Portrait of Generations, using the age pyramid, Canada, 2011. Retrieved from Statistics Canada: https://www12.statcan.gc.ca/census-recensement/2011/as-sa/98-311-x/2011003/fig/fig3_2-2-eng.cfm. Reproduced and distributed on an ""as is"" basis with the permission of Statistics Canada."

FIGURE 12.2

Evolution of Music Technology

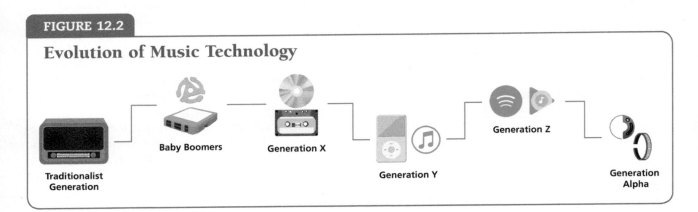

Traditionalist Generation — Baby Boomers — Generation X — Generation Y — Generation Z — Generation Alpha

work and saving money, as it was commonly viewed as the only way to get ahead in life. Despite difficult times, many families remained strong, resulting in a fierce sense of loyalty, pride, and dedication to one another.

If you were fortunate and found a good job, then you stayed loyal to the company; you believed that the company would provide for you, so that you could in turn provide for your family (Buahene & Kovary, 2007).

In terms of technology, this was the golden age of radio. Children's stories, comedy skits, and the names of loved ones lost were broadcast through the airwaves (Canadian Communications Foundation, 2001; BBC, 2012). Women's programming and weather reports, along with regular updates about the war, kept families huddled around the "wireless" radio to hear the latest news. In Hollywood, movies began to emerge on the silver screen. Silent movies became popular in the 1910s, and when members of this generation were in their teens, movies like *The Wizard of Oz, Gone with the Wind*, and *King Kong* would go on to become box office hits (IMDb, 2012). Though telephones were available, not every home had one, and many people read the daily newspaper. The challenging times that Traditionalists lived through defined the individuals they are today.

Baby Boomer

Beginning in 1946, the Baby Boom lasted almost 20 years in Canada and saw women giving birth, on average, to 3.7 children, compared to 1.7 in more recent years (Statistics Canada, 2012b). The children born at the beginning of this generation have reached the age of 65; and one study projects that by 2036, the number of seniors will more than double, ranging between 9.9 and 10.9 million, compared to 4.7 million in 2009 (Statistics Canada, 2010b).

Born between 1946 and 1964 and raised by Traditionalists, **Baby Boomer** children had a very child-focused upbringing (Buahene & Kovary, 2007). Unlike their parents who lived through the Great Depression, Baby Boomers grew up in an economy that was booming after World War II. They lived in optimistic, idealized times and quickly realized that, just by their sheer numbers alone, they could push for social change—and they did. As these individuals grew into adulthood, some began to challenge the norms of society and question the political institutions in power. The events that helped to define the lives of these individuals included the Civil Rights and Women's Liberation Movements, the Quebec Crisis and Bill 101, the Vietnam War, Trudeau's multiculturalism, Woodstock, and Neil Armstrong landing on the Moon. The size of this cohort also affected this generation in terms of the workplace. With limited jobs available, competition became fierce and forced many Baby Boomers to question their commitment to social change. Raised with a strong work ethic, many Baby Boomers soon poured their waning social commitment into their careers. This is the generation that coined the phrase "Thank God it's Monday" (Buahene & Kovary, 2007), and their careers often became a

symbol of their value and self-worth—until the massive downsizing of the 1980s. A huge wake-up call for many of the Baby Boomers, massive layoffs gave them cause to rethink their lives and redefine who they were. Baby Boomers are described as optimistic and career-oriented individuals, whose competitive edge drives them to succeed in most of their endeavours. They often work hard and play hard, and enjoy the benefits of their labour by indulging in consumer goods and vacations (Sheppard, 2006).

In terms of technology, this generation saw television ushered in as the newest form of social media (BBC, 2012; Giedd, 2012). Black-and-white television came into its own in the late 1940s, and Baby Boomers grew up with shows like *Milton Berle, Leave It to Beaver, All in the Family*, and *Hockey Night in Canada* (NBC, 2012; CCF, 2001; BBC, 2012). Movies like *To Kill a Mockingbird, A Clockwork Orange*, and *Apocalypse Now* were entertaining audiences everywhere at this time (IMDb, 2012). Baby Boomers were in their early teens when the first IBM PCs and Apple computers became popular (Rogers, 2009). It was a bit of a paradox because although Baby Boomers seemed very "high-tech," they still had to rent their phones and send letters through the post to keep in touch.

If they woke up in the middle of the night, they would see a test pattern on the television; and if their friends moved away, they would most likely lose touch. This generation had mastery over the technology at the time—they used it for data analysis, word processing, and to make their lives easier, but it did not define who they were (Rogers, 2009). They were like the pioneers or settlers of technology, setting down the foundation for what was to come, by building the networks and software programs that would fuel the digital revolution that would change the world and come to define their children.

Generation X

Many authors consider people born between 1965 and 1979 as **Generation X**, but Statistics Canada has designated it only as 1966–1971, roughly 2.8 million or 8 percent of the total population in 2011 (Statistics Canada, 2012). These individuals were born at a time when the fertility rates were rapidly declining, resulting in a smaller cohort of people. Growing up in the shadow of the Baby Boomers, Generation X was not afforded the same attention as its predecessor. Media, marketers, and government

Baby Boomer: Refers to individuals born between 1946 and 1964.

Generation X: Refers to individuals born between 1964 and 1979.

were not as concerned with this group of individuals and, consequently, some have labelled Generation X as the "lost" generation. (Buahene & Kovary, 2007).

MTV/Much Music and personal computers, the Canadian Charter of Rights and Freedoms, the Los Angeles Riots, the Quebec Separation Referendum, the fall of the Berlin Wall, and AIDS—all influenced the Generation X cohort. Since their Baby Boomer parents were often dual-income earners, divorced, or single, this generation became the first "latch-key kids," forcing them to become independent at a very young age. This independence would become a prominent feature of Generation X and form the foundation for other characteristics, including self-reliance, skepticism, and pragmatism (Buahene & Kovary, 2007).

Rather than forming tight-knit family ties, many members of this generation developed close friendship networks and non-traditional families by bonding with peers. In the workforce, many Generation X individuals found themselves competing for entry-level positions with downsized Baby Boomers. Realizing that competition for employment was fierce, they focused their energies on job employability rather than on job security and did not develop the allegiance to companies that the previous generations had. Often, they chose contract work or started their own businesses and preferred rewards for their efforts rather than their years of service. If they found that the rewards did not equal their efforts, this generation had no problem leaving a job to find another. Unlike their parents, the motto of Generation X is "work to live, rather than live to work" (Buahene & Kovary, 2007).

Essentially, Generation X is the first "techno" generation. Moving from a manufacturing-based economy to a service- and knowledge-based economy required more sophisticated technology, and this generation embraced virtually every form of gadgetry available. PDAs, cellphones, video games, email, Blackberrys, and laptops intertwined their lives at work and at home (Kane, 2012). *Super Mario*, *Pong*, and the dawn of the computer age are some of the events that helped to shape their lives. Major social issues like the Los Angeles riots took on new meaning, as the availability of information about these events was far more extensive than ever before. Individuals became inundated with images and news stories that were broadcast 24 hours a day on bulletin boards, via the Internet, and in the daily news. Generation X experienced the electronic media just as it was emerging as a social institution; print, radio, film, and television each played equally pivotal roles in shaping the lives of these individuals.

Generation Y: Refers to individuals born between 1980 and 1995.

As children, they watched *Sesame Street*, *The Electric Company*, *ABC Afterschool Specials*, and *Schoolhouse Rock* (Garrard, 2008), and at the movies, teens watched films like *Star Wars* and *Pulp Fiction* (IMDb, 2012). Computer-literate, with the ability to multi-task and adapt to new environments, Generation X individuals are described as results-driven and creative people who focus on creating harmony between their work and home lives.

Generation Y

Born between 1980 and 1995, **Generation Y**, also referred to as the Millennials or Net Generation, were very different from the "latch-key kids" of Generation X. Generation Y grew up in a time when violence in society was rising, so parents protected and supervised their Generation Y children, often treating them like peers, as opposed to offspring (Buahene & Kovary, 2007). Many Generation Y children treat this "lateral" relationship with their parents much the same as they would a relationship with a friend their own age—even referring to their parents by their given names, instead of calling them "Mom" and "Dad."

Life-defining events during this time included the legalization of same-sex marriage in Canada, the creation of Nunavut as a new arctic territory, the digital age, Y2K, the invention of the toonie, increased school violence, 9/11, and the U.S. War on Terror. Generation Y was raised in a highly nurturing atmosphere that inspired confidence and hopefulness. Their parents taught them that not only life was going to be challenging and negotiable, but also collaborative, and they entered the workforce with a positive attitude and high expectations. Unfortunately, their enthusiasm was not always reciprocated once employed, and at times, was met with hostility by older, more experienced workers (Pooley, 2006).

In terms of technology, teenagers and twentysomethings today are growing up in an unparalleled era of interconnectivity. According to recent data collected by the Pew Internet Project, 95 percent of young people 12 to 17 are online, 76 percent use social networking sites, and 77 percent have cellphones (Netburn, 2012). Author Donald Tapscott describes Generation Y as the Net Generation "because they were the first to grow up surrounded by digital technology" (Tapscott, 2009, p. 9). They are the first generation in human history to see behaviours such as tweeting and texting, and websites like Facebook, YouTube, Google, and Wikipedia as everyday facets of their lives that play important roles in their search for meaning and understanding in the world.

As a generation, they are racially diverse and more accepting of diversity, confident, and less religious than any of the preceding generations (Keeter & Taylor, 2009). Though they are likely to be the most educated of all the generations, they are also the most likely to spend their money on luxury goods. At a time when youth unemployment has been consistently high in Canada, from 2009 to 2011, Generation Y spenders have been the driving force behind national spending on luxury goods, with increased spending on luxury fashion by 33 percent, travel by 74 percent, and fine dining by 102 percent (Infantry, 2012).

Generation Z

There is some debate over where Generation Y ends and **Generation Z** begins, but the one thing that is consistent is that while they are similar, there are some distinct differences between the older and younger millennials. Most agree that Generation Z begins in 1996, while some argue it starts as early as 1993 or as late as 2000. For our discussion, Generation Z is comprised of individuals born between 1996 and 2009. This generation does not remember 9/11, but grew up in the aftermath, where global conflict and the war on terror are still realities (Center for Generational Kinetics, 2016). These individuals continue to be saturated in technology, and are still heavily influenced by Google, Wikipedia, YouTube, Skype, and social media platforms. Video sharing and streaming with platforms like Vimeo, Ted Talks, Netflix, and Apple TV have caused movie rental businesses like Blockbuster Video to be a thing of the past. Events that influenced members of Generation Z include the SARS outbreak, Canada winning a record number of gold medals at the Vancouver Winter Olympics, and the Human Rights Act amendment prohibiting discrimination based on sexual orientation.

Children of this generation were entertained by Barney, SpongeBob Squarepants, Dora the Explorer, Blue's Clues, and the Backyardigans (TV.com, 2016), while teens were enthralled with series like Twilight, The Hunger Games, and Divergent (Perez, 2016).

When it comes to technology, these "digital explorers" do not remember a time before social media, as technology has always been integrated into every aspect of their daily lives. Similar to Generation Y, Generation Zs do everything from their mobile devices, but unlike their older Millennials, their phones have always been "smart" (Elmore, 2014). Almost all mobile devices used by this generation are touchscreen and capable of performing just like a computer. Generation Z is also distinct from Generation Y in that many value online privacy and prefer more instantaneous or anonymous social media, like Snapchat, Vine, Instagram, Secret, or Whisper (Center for Generational Kinetics, 2016; Elmore, 2014).

Trends in employment and work ethic are still yet to be determined as the older portion of this generation is just starting to enter the workforce, but it is safe to say that their smartphones will play a role in their work environments. Although largely comprised of kids and adolescents, this generation is described as self-aware, self-reliant, innovative, and goal-oriented (Center for Generational Kinetics, 2016). They are more pragmatic, more cautious, and more money conscious than the previous generation (Jenkins, 2015). Some researchers predict that Generation Z will be highly educated, with a larger percentage of those attending and graduating from postsecondary institutions than any previous generation (Center for Generational Kinetics, 2016). Not surprising, Generation Zs prefer a digital approach to learning through self-education and online learning (Jenkins, 2015). Educational institutions are beginning to move in this direction as well, offering more online courses than ever before. But how does this trend affect students of other generations—will we see a generational divide as educational institutions move more and more courses online? See the box titled "In Their Shoes" for one student's account of returning to school at a later age and how that impacted her experience.

Generation Z are described as a generation that values innovation, financial responsibility, racial diversity, and environmental sustainability (Jenkins, 2015; Elmore, 2014), and a large portion of this generation volunteer and have a desire to make a positive impact on the world (Center for Generational Kinetics, 2016).

Generation Alpha

Although it might be difficult to predict characteristics of the generation following Generation Z, some researchers have already named this cohort **Generation Alpha**. In an interview for the *New York Times*, Mark McCrindle argues that it is never too early to start to observe the generation that will be "the most formally educated generation ever, the most technology supplied generation ever, and globally the wealthiest generation ever" (Williams, 2015). Generation Alpha are born in 2010 or later, and based on birth estimates from 2010–2015, make up about 5 percent of the Canadian population (Statistics Canada, 2015).

Generation Z: Refers to individuals born between 1996 and 2009.

Generation Alpha: Refers to individuals born in 2010 and later.

If a picture can say a thousand words, imagine the stories your shoes could tell! Try this student story on for size – have you walked in this student's shoes?

I was content with my life. The kids had grown and moved out of the house, the grandchildren were a joy to babysit, and we were looking forward to my husband's retirement in two short years. My life was turned upside down when he collapsed at work and died of heart failure on the way to the hospital. The months after his funeral were a blur—flowers, casseroles, and sympathy cards in the mail. It wasn't long before I started to notice the bills arriving in the mail as well. My husband had always handled our finances so it was inevitable that I would start falling behind in our payments.

Alone now, I realized that I needed to support myself financially. I realized that I needed to get a job. Prior to having my children, I had worked in an office and had excellent keyboarding skills, but I was using a typewriter! I couldn't imagine using a computer and doubted that I would even know how to turn one on. So back to school I went.

The fear of being alone was nothing compared to my fear of failing. Coupled with my age and computer illiteracy, I was a basket case. When the teacher started talking, my eyes glazed over. The technical terms and jargon did nothing but confuse me, and as I looked around the classroom watching all those kids on their cellphones and laptops, I was ready to give up, but I knew that wasn't an option—I needed to learn this new skill.

I didn't buy a computer right away because they had them at school and at the local employment office. I focused on using Microsoft Office and spent hours creating a recipe index, address book, cover letter, and resumé. I even used Excel to create my monthly budget. Eventually, I bought my own computer and began to send emails. I set up a Facebook account. I renewed old friendships and even learned to Skype with my grandkids.

My computer changed my world. I no longer felt isolated and alone. Eventually, I got Linked In and heard of a part-time receptionist position at a local not-for-profit agency. I applied online and was called for an interview. The interviewer was impressed with my computer skills. We laughed about teaching an old dog new tricks. This "old dog" got the job and is now financially secure. I'm so glad that I overcame my insecurities and went back to school.

My future is definitely brighter than it was three years ago and I'm ready for my next challenge—who knows, I might try online dating next!

Born in the same year that the iPad was introduced, Generation Alpha are part of what McCrindle calls "an unintentional global experiment where screens are placed in front of them from the youngest age as pacifiers, entertainers and educational aids (Williams, 2015).

A YouTube search of "babies using iPads" brings up numerous video clips that show children as young as eight months interacting with computer tablets in various ways. In one clip, a one-year-old cannot figure out why her fingers will not make the images in a magazine move. She even presses her finger on her leg to make sure that her finger is working; then she goes back to trying to make the magazine as interactive as the tablet she was previously holding. "Technology codes our minds, changes our OS (operating system). The video shows how magazines are now useless and impossible to understand, for digital natives" (UserExperiencesWork, 2011). The comments accompanying this video make for interesting reading, as users discuss the controversy surrounding young children and technological devices. Is a child ever too young to start learning new technology?

As with most generations, the environment in which they grow up in shapes who they will become, and generational labels do not really become useful until a portion of time has passed and comparisons can be made within and between generations. However, HR Executive Dan Schawbel believes it is not too early to make some predictions about Generation Alpha. He predicts that Generation Alpha will be the most entrepreneurial generation yet, and every generation moving forward will be more and more entrepreneurial because they will have earlier access to information, people, and resources, they will take more risks earlier in life, and have more time to build reputations and relationships than previous generations (Schawbel, 2014). Schawbel believes that Alpha will be the most tech savvy generation to date with less human interaction than previous generations; consequently, he predicts Alphas will "feel more alone, despite being so connected" (Schawbel, 2014).

Founder and Executive Chairman of the World Economic Forum, Professor Klaus Schwab, predicts this generation will grow up during the **Fourth Industrial Revolution or Industry 4.0** over the next decade, where new technologies that have the ability to combine the physical, digital, and biological worlds will challenge our perceptions of what it means to be human (Marr, 2016; World Economic Forum, 2016). Researchers are predicting that Generation Alpha will see the **Internet of Things (IoT)**, where everyday objects have network connectivity, allowing them to send and receive data (Morgan, 2014). The creation of this technology will drastically change our lives, with "smart factories," leading to "smart communities," and eventually "smart cities" (Morgan, 2014). On an individual level, the ability for devices to communicate with each other could allow your alarm clock to start your coffee maker when it wakes you up in the morning; your car to send a text message to your date informing them you will be late because you hit traffic; and your printer to order ink or paper supplies when it detects they are getting low (Morgan, 2014). Generation Alpha will see improvements in artificial intelligence, with experts in the field predicting that computers will become more intelligent than human beings by 2030 (Connor, 2008). A *Business Insider Intelligence Report* predicts that a fully autonomous car will be released in 2019, making roads safer and people's lives easier (Greenough, 2015). The consequences of these new technologies remains to be seen, but Schwab fears that "organizations could be unable or unwilling to adapt to these new technologies and that governments could fail to employ or regulate these technologies properly" (Marr, 2016). He worries that "shifting power will create important new security concerns, and that inequalities could grow rather than shrink if things are not managed properly" (Marr, 2016).

AGEISM

Age stratification and ageism are not mutually exclusive. According to J.A. Garretson, one cannot exist without the other—age is simply a way of keeping track of how long something has been around, while age stratification separates people into three categories: the young, the old, and the rest (2011). The term **ageism** refers to the stereotyping of and prejudice against individuals or groups because of their age. These ageist attitudes are the basis for age discrimination. Ageist assumptions and beliefs have less to do with the biological process of aging and more to do with *how society responds* to the process of aging (OHRC, 2001). Unlike racism and sexism, ageism is

likely something we will all experience at some point in our lives, and in order to combat ageism, we must view it as a societal issue, not as just a problem for the aging population.

Ageism and Employment

In a news release on September 23, 2014, the Conference Board of Canada reported that age, rather than gender, is becoming the new income divide in Canada (2014). The **income gap** between older and younger workers has never been wider, with the average disposable income of Canadians between the ages of 50 and 54 being 64 percent higher than that of 25- to 29-year-olds (Conference Board of Canada, 2014).

Although it is common for older workers to make more money than those with less experience, that gap is widening at an alarming rate, especially for women. The news release illustrates that between 1984 and 2010, the gap in employment income for men grew from 53 percent to 71 percent, while for women it went from only 9 percent to 43 percent (Conference Board of Canada, 2014). The concern is that with the Baby Boomer generation approaching retirement, Canadians are relying on a smaller portion of the population to drive economic growth and sustain the publically funded services we rely on through taxes (Conference Board of Canada, 2014). This income gap is reflected in the case of *Mackinnon v. Celtech Plastics Ltd.* In 2012, Mackinnon, 67 was recalled from being laid off, and when he returned to work, he was subjected to an unusually heavy workload at unreasonably high standards. He alleged that his employers were trying to make the job so difficult for him that he would quit and they could save money by replacing him with a less experienced worker who could be paid less. The **Human Rights Tribunal of Ontario** awarded $27 000 to Mackinnon for the violation of his inherent right to be free from

> **Fourth Industrial Revolution or Industry 4.0:** A term that describes the creation of new technologies that have the ability to combine the physical, digital, and biological worlds.
>
> **Internet of Things (IoT):** Technology that allows for everyday objects to have network connectivity, allowing them to send and receive data.
>
> **Ageism:** the stereotyping of and prejudice against individuals or groups because of their age.
>
> **Income gap:** A gap in income between one group and another, such as discrepancies in income between older and younger employees performing that same job.
>
> **Human Rights Tribunal of Ontario:** An administrative tribunal in Ontario, Canada, that hears and determines claims based on the violation of rights protected under the Ontario Human Rights Code.

discrimination as per section 5(1) of the Ontario Human Rights Code:

"every person has a right to equal treatment with respect to employment without discrimination based on race, ancestry, place of origin, colour, ethnic origin, citizenship, creed, sex, sexual orientation, gender identity, gender expression, age, record of offences, marital status, family status, or disability" (OHRC, 2015).

David Stewart-Patterson, Vice President of the Conference Board of Canada, worries that if the trend to pay younger workers less for the same work continues, we will see increased conflict between "older haves and younger have nots" (Conference Board of Canada, 2014). For example, paying younger workers less can reinforce the negative misconception that younger workers are less competent and have less value. In order to combat those negative perceptions, some companies are experimenting with **reverse mentoring** programs that allow "younger employees to train their older coworkers in things like social media and technology" (Nelson, 2012). You will get a chance to explore the concept of reverse mentoring at the end of this chapter in the case study presented in the KWIP section.

Ageism and Youth

There is a general misconception that ageism is something that only impacts older people, but ageism does not

discriminate—stereotypes, prejudice, and discrimination based on age can target people of all ages. In this section, we will investigate how youth, in particular, often fall victim to age discrimination. Youth rights activist Kathleen Nicole O'Neal distinguishes between various forms of ageism that are oppressive toward youth. First, she defines **normative ageism** as the assumptions made about a person's capabilities, interests, strengths, or weaknesses based on age; for example, generalizations about "all teenagers" or "typical children" that describe normative behaviour for a specific age group, such as "preventing teen pregnancy" or "stopping underage drinking" (O'Neal, 2012). This type of ageism dictates acceptable behaviours for certain age groups and often labels those who do not conform. **Cultural ageism** occurs when one generation devalues the actions of another generation because it is different from what they are used to (O'Neal, 2012). For example, when older generations criticize younger generations for their reliance on technology and make assumptions about their inability to communicate effectively, they are in fact judging the younger generation based on standards that only apply to people of their own generation. O'Neal describes **paternalistic ageism** as the idea that children and young adults lack maturity and therefore require older adults to make decisions for them (O'Neal, 2012). "Free the Children," founded by Craig and Marc Kielburger, is an example of an organization that challenges paternalistic ageism. Craig and Marc Kielburger created the charity as children themselves—Craig was 12 years old and Marc was 17—and today, "Free the Children" is an internationally recognized charity that works to empower young people to become agents of social change (Free the Children, 2016). **Pedophobic** or **ephebophobic ageism** denotes a fear or hatred of children or adolescents, and can be seen in the common misconceptions that young people are troublemakers, hormonally out of control, and a general burden to society (O'Neal, 2012). **Economic ageism** refers to the ways in which people experience prejudice based on age in employment (O'Neal, 2012). As previously discussed in the section on "Ageism and Employment," we can see economic age discrimination when young workers are often paid less than older workers to do the same job. **Scientific ageism** refers to the ways that scientific fields, like biology, psychology, and psychiatry, are used to reinforce assumptions based on age (O'Neal, 2012). For example, studies on the brains of adolescents often perpetuate the stereotype that teenagers are "impulsive, foolish, and immature" (O'Neal, 2012). Lastly, O'Neal cites examples of **institutionalized ageism** as laws prohibiting people under the age of 18 to vote,

---

Reverse mentoring: Programs that provide younger workers an ability to mentor and teach older colleagues.

Normative ageism: The assumptions made about a person's capabilities, interests, strengths, or weaknesses based on age.

Cultural ageism: Occurs when one generation devalues the actions of another generation because it is different from what they are used to.

Paternalistic ageism: The idea that children and young adults lack maturity and therefore require older adults to make decisions for them.

Pedophobic or **ephebophobic ageism:** Denotes a fear or hatred of children or adolescents, and can be seen in the common misconceptions that young people are troublemakers, hormonally out of control, and a general burden to society.

Economic ageism: Refers to the ways in which people are discriminated against based on age in employment.

Scientific ageism: Refers to the ways that scientific fields, like biology, psychology, and psychiatry, are used to reinforce assumptions based on age.

Institutionalized Ageism: Policies or practices within institutions that discriminate based on age.

retail business practices that limit the number of students permitted entry at one time, and media bias that portrays youth as delinquent (O'Neal, 2012). We can see that ageism comes in various forms and is directed at youth as well as the aging population.

The Freechild Project, co-founded in 2001 by Adam Fletcher with a group of youth advocates from around the world, works to support and encourage young people to change the world. One of the key issues that has emerged from the project is something Fletcher calls **adultism**, or the bias toward adults that often results in age discrimination against youth (Fletcher, 2015). Fletcher suggests that adultism takes place when decisions are made based on age, rather than ability (2015). It includes language that "excludes, belittles, or talks down to youth" and "any behaviour or attitude that is routinely biased against young people simply because they are young" (Fletcher, 2015). (See Figure 12.3.)

Fletcher acknowledges that discriminating based on age is sometimes beneficial, like in emergency or healthcare services, but the problem arises when society assumes this type of discrimination is acceptable *all of the time* (2015). It is often the unconscious sharing of experiences that causes adults to reinforce adultism, and at times, youth even perpetuate it by accepting bias toward adults (Fletcher, 2015). The consequences of adultism are severe—treating young people as inferior, subordinate, and incapable can cause them to internalize these negative perceptions, becoming dependent on adults and limiting their ability to trust their own judgment (Fletcher, 2015). In a sense, it becomes a form of "learned helplessness" in that young people accept that adults simply know best all of the time, creating a revolving cycle of powerlessness (Fletcher, 2015). Fletcher argues that it is this internalized powerlessness that makes children and youth vulnerable to accepting other forms of oppression, including verbal, emotional, and physical abuse (Fletcher, 2015).

Adultcentrism is the view that only adults can make substantial contributions to society, and while adultism is sometimes appropriate, adultcentrism is almost always unacceptable (Fletcher, 2015). Adultcentrism systematizes adultism when social institutions, like schools and government agencies, operate under the assumption that young people do not add value to society until they become adults (Fletcher, 2015). The assumption that youth lack the knowledge and ability to make a meaningful contribution is being felt by youth who are volunteering in Canada. Volunteer Canada reports that there is "a feeling among youth that their opinions and insights aren't valued, respected or taken into account," and although

FIGURE 12.3

10 Signs You Are Experiencing Adultism

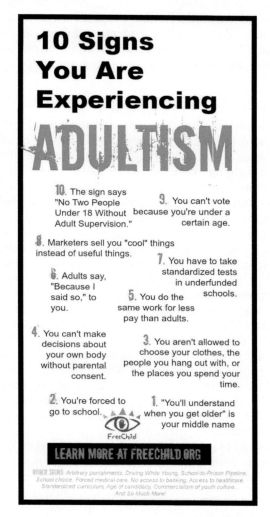

Can you see examples of adultism in your own life?

Fletcher, Adam. (2013). "10 Signs You Are Experiencing Adultism." Found at: https://adamfletcher.net/10-signs-youre-experiencing-adultism. Date accessed: July 12, 2016.

Canadians aged 15 to 24 volunteer more than any other age group, there is an assumption that youth are viewed as needing services, rather than recognized for their ability to contribute to volunteer initiatives (Mehta, 2012).

Adultism: Refers to bias toward adults that often results in discrimination against youth.

Adultcentrism: The assumption that only adults can make substantial contributions to society.

Fletcher suggests that the only way to combat adultism is to explore our personal attitudes and address how we perpetuate adultism (Fletcher, 2015). Once we see adultism in our everyday lives, we can begin to build equitable relationships with young people throughout society. The reading for this chapter further investigates how adults perceive students and discusses how we can all become more critically aware of our own attitudes, behaviours, and perceptions about age.

Ageism and the Aging Population

Ageism and the disdain people feel toward aging can be viewed in a way as a form of self-loathing or self-hate, since we will all eventually become that which we are judging (CARP Canada, 2010). The Revera Report on Ageism questioned Canadians aged 18–32 (Generation Y), 33–45 (Generation X), 46–65 (Baby Boomers), 66–74 (Seniors) and 75+ (Older Seniors) to explore attitudes about aging and experiences with ageism. The report found that more than half of participants agreed that ageism is the most tolerated social prejudice when compared to gender- or race-based prejudice, and one-in-three Canadians admitted to treating someone differently because of their age (Revera, 2012). According to the report, while the majority of seniors 66 and older are optimistic about aging, the opposite is true of younger generations, with Generation Y and Generation X being the most likely to hold negative perceptions of the aging population (Revera, 2012). Sixty-three percent of seniors aged 66 and older say that they have experienced some form of age discrimination, including being ignored or treated as invisible (41 percent); being treated like they have nothing to contribute (38 percent); and assumed to be incompetent (27 percent) (Revera, 2012). Senior participants identified the most common source of the discrimination as younger people (56 percent), followed by healthcare professionals (34 percent), government (27 percent), and employers (20 percent) (Revera, 2012).

Despite these negative experiences, most older Canadians associate aging with positive things, like spending more time with loved ones and feeling wiser and more self-assured (Revera, 2012). Forty-one percent of participants aged 66 and over said, "age is just a number," 36 percent said, "you never stop living life to the fullest," and 40 percent reported, "the best is yet to come" (Revera, 2012). We also see negative associations with aging challenged when 79.2 percent of women and 77.4 percent of men aged 65 and over report feeling happy and interested in life as a whole (Statistics Canada, 2008).

So where do negative perceptions of age come from? Part of the answer to that question can be attributed to the ways in which our society values youthfulness. More than 70 percent of Canadians agree that Canadian society values younger generations over older generations (Revera, 2012). With a billion dollar market of anti-aging campaigns promising to re-establish youthful appearance, we are essentially trying to defy the inevitable and deny the existence of a significant portion of our population (Kingston, 2014). We need to consider the social and economic costs of our efforts to deny or defy age. Anne Kingston refers to this as **cultural greywashing**, "where old age has been erased, sanitized or reduced to extreme stereotypes" (2014). How our society perceives age shapes our experiences of aging and the social systems we put in place to meet the needs of the aging population. If we have anxiety about growing older and negative perceptions of aging, our social systems will reflect those ideologies. By valuing youthfulness to such an extent, we have created a society that is organized based on the assumption that everyone is young. This **systemic ageism** is reflected in mainstream media that bombard us with anti-aging ads and idealized youthful images; birthday cards that poke fun at getting older; healthcare professionals that dismiss ailments as a normal process of aging (Stall, 2012); and the ways in which we use the phrases "grow up" or "act your age." Our resistance of aging has not only left us unprepared for the aging process, but it also has left our healthcare system ill-equipped to meet the needs of a senior population that will soon be larger than ever before (Kingston, 2014). Consequently, there are a number of organizations that believe we are headed for a national health crisis and we need to create a plan to better prepare us for what lies ahead.

The **National Seniors Strategy for Canada** Report suggests that the aging process should be celebrated as a triumph rather than viewed as a disease, because over the next 20 years, one in four Canadians will be over the age of 65 (National Seniors Strategy, 2015).

The report argues that it will take coordinated efforts at all levels of government—municipally, provincially, and federally—as well as between private and public sectors to meet the growing needs of this aging population (National Seniors Strategy, 2015).

Cultural greywashing: Refers to the marginalization and overgeneralizations made about older adults that result in extreme stereotyping.

Systemic ageism: Behaviour, policies, or practices that discriminate based on age; can be intentional or unintentional.

National Senior Strategy for Canada: An organization working to promote and explore policy options that can support the development of a National Seniors Strategy.

The report calls for all proposed policies, programs, and services to adhere to five principles: access, equity, choice, value, and quality. The **Principle of Access** requires that service providers ensure older Canadians, their families, and their caregivers can easily access supports in a timely and efficient way (National Seniors Strategy, 2015). The **Principle of Equity** ensures that older Canadians from different ethno-cultural groups, those from LGBTQ communities, and those with varying abilities are equally supported (National Seniors Strategy, 2015). The **Principle of Choice** ensures that older Canadians, their families, and their caregivers have as many choices as is possible and are empowered with the necessary information to make informed choices (National Seniors Strategy, 2015). The **Principle of Value** ensures that we are spending tax dollars in the most effective and efficient ways to protect the sustainability of programs and services (National Seniors Strategy, 2015). The **Principle of Quality** ensures that the quality of the programs and services is not lost in an effort to keep costs down (National Seniors Strategy, 2015). This framework was created to guide future policy makers in the development of programs and services designed to support the well-being of Canadian seniors.

The **Alzheimer Society of Canada** is also suggesting that dementia is an issue worthy of national concern. Mimi Lowi-Young, CEO of the Alzheimer Society of Canada, states that "dementia is a huge threat to our public health system and to our nation's productivity," predicting that by 2040, Canada will spend $293 billion a year on this disease alone (Alzheimer Society of Canada, 2016). The society is asking the government to partner with experts in the field, healthcare providers, and those living with and caring for someone with dementia to develop a national strategic plan with a mandate to:

- "Increase investment in research, foster collaboration and improve knowledge exchange and translation
- Provide a surveillance system and evidence-based information on all aspects of Alzheimer's disease and other dementias to inform best practices
- Enhance the competency and capacity of those delivering dementia care
- Increase awareness about dementia risk factors, early diagnosis and timely interventions
- Strengthen the integration and coordination of care and service delivery across the health-care continuum
- Recognize family caregiver needs and develop supports that provide options and flexibility" (Alzheimer Society of Canada, 2016)

The Alzheimer Society of Canada believes that Canada has a unique opportunity to become a world leader in dementia, and Canadians agree, with 83 percent recognizing the need for a national dementia plan (Alzheimer Society of Canada, 2016). Lowi-Young, along with Federal Health Minister Rona Ambrose and leading Canadian researchers, were invited to attend the G8 Dementia Summit in December of 2013—the Summit discussed dementia as a global epidemic, citing that between 2013 and 2050, the global number of people with dementia will triple, with those in the Americas increasing by 248 percent (Alzheimer Disease International, 2016). Out of the 193 countries in the World Health Organization, only 13 have national dementia plans; sadly, Canada is not one of them (Alzheimer Disease International, 2016).

Elder Abuse

Although a universal definition does not exist, the Public Health Agency of Canada (PHAC) and the World Health Organization (WHO) define **elder abuse** as "any action by someone in a relationship of trust that results in harm or distress to an older person" (PHAC, 2012; WHO, 2016). In December 2013, CARP reported that approximately 520 000 people 65 years and older were confronting elder abuse in Canada, and they predict that number will increase to 800 000 before 2023 if

Principle of Access: One of the Five Principles of the National Senior Strategy for Canada recommending that service providers ensure that older Canadians, their families, and their caregivers can easily access supports in a timely and efficient way.

Principle of Equity: One of the Five Principles of the National Senior Strategy for Canada recommending that older Canadians from different ethnocultural groups, those from LGBTQ communities, and those with varying abilities are equally supported.

Principle of Choice: One of the Five Principles of the National Senior Strategy for Canada recommending that older Canadians, their families, and their caregivers have as many choices as is possible and are empowered with the necessary information to make informed choices.

Principle of Value: One of the Five Principles of the National Senior Strategy for Canada recommending that tax dollars are spent in the most effective and efficient ways to protect the sustainability of programs and services.

Principle of Quality: One of the Five Principles of the National Senior Strategy for Canada recommending that the quality of the programs and services is not sacrificed in an effort to save money.

Alzheimer Society of Canada: A national charity for people living with and affected by Alzheimer's disease and other dementias.

Elder abuse: A single or repeated act against an older person in which harm or distress is caused.

nothing is done (CARP Canada, 2013). Elder abuse can take on many forms but commonly comes in the form of physical, psychological, and financial mistreatment, with financial abuse being the most commonly reported at 62.5 percent, followed by verbal abuse at 35 percent, physical abuse at 12.5 percent, and neglect at 10 percent (OHRC, 2001; PHAC, 2012). When considering these statistics, it is important to note that elder abuse is often under-reported because of the nature of such abuse. Older adults may feel embarrassed or ashamed when the abuse is coming from someone they know and trust, or there may be hesitation to report abuse from a caregiver for fear of losing that care (OHRC, 2001; PHAC, 2012; CARP, 2013).

Physical elder abuse occurs when someone hits or handles a person roughly, even if there is no injury (CNPEA, 2016). Giving someone too much or too little medication or physically restraining them are also forms of elder abuse (PHAC, 2012; CNPEA, 2016). **Sexual elder abuse** takes place when an older adult is forced to engage in sexual activity; this may include verbal suggestive behaviour, sexual touching, sex without consent, or not respecting a person's privacy (CNPEA, 2016). **Psychological and emotional elder abuse** are actions that limit a person's sense of identity, dignity, or self-worth, including insults, threats, intimidation, humiliation, treating an individual like a child, or isolating them from family, friends, or regular activities (PHAC, 2012; CNPEA, 2016). **Financial elder abuse** occurs when someone tricks, threatens, or persuades an older adult out of their money, property, or possessions (CNPEA, 2016). It can include cashing a person's cheques without permission, forging an older person's signature, or sharing an older person's

home without covering a fair share of expenses when requested (PHAC, 2012). **Elder neglect** takes place when a person fails to provide an older person with the necessities of life, such as food, clothing, shelter, medical attention, personal care, and required supervision (CNPEA, 2016; PHAC, 2016). This type of abuse can be intentional or unintentional in that sometimes caregivers lack the necessary knowledge, experience, or ability to properly care for older adults (CNPEA, 2016). Elder abuse can also take place when there is interference with an older person's ability to make choices. The removal of such autonomy is especially harmful when it violates choices involving protected rights and freedoms (CNPEA, 2016). Some examples include interfering with a person's spiritual practices, withholding mail or information, denying privacy, preventing visitation, or forcing someone into an institution without a legitimate reason (CNPEA, 2016). Sometimes elder abuse can be systemic or institutionalized discrimination when policies or social practices harm or discriminate against older persons (CNPEA, 2016). For example, to save time and effort, a caregiver may diaper a person rather than helping them to the washroom, or an institution may use restraints to prevent an older individual from falling (CNPEA, 2016). These forms of systemic abuse often result from staff shortages and a lack of resources (CNPEA, 2016).

Factors contributing to elder abuse are varied and extensive, but research suggests that elder abuse stems from ageism and a negative attitude toward seniors (OHRC, 2001; PHAC, 2012). Other contributing factors include lack of community services, lack of affordable and accessible housing, and the economic and social vulnerability of the aging population (OHRC, 2001). In January 2013, Bill C-36, the *Protecting Canada's Seniors Act* came into effect (Alberta Law Library, 2013). Under the amendments to the Canadian Criminal Code, "evidence that an offence had a significant impact on the victims due to their age—and other personal circumstances such as their health or financial situation—will now be considered an aggravating factor for sentencing purposes" (Alberta Law Library, 2013). While the *Protecting Canada's Seniors Act* is a step in the right direction, CARP is calling on the Canadian government for a comprehensive strategy to "eradicate elder abuse with adequate financial and legal resources, training and support for law enforcement, caregivers support, and policies and laws that protect older Canadians from abuse" (CARP, 2013). Their recommendations include an Elder Abuse Hot Line—a 1-800 number with the ability to redirect calls to local service agencies equipped and trained in the area of elder abuse and sensitive to cultural and linguistic needs (CARP,

Physical elder abuse: Occurs when someone hits or handles a person roughly, even if there is no injury.

Sexual elder abuse: Occurs when an older adult is forced to engage in sexual activity; this may include verbal suggestive behaviour, sexual touching, sex without consent, or not respecting a person's privacy.

Psychological/emotional elder abuse: Refers to actions that limit a person's sense of identity, dignity, or self-worth, including insults, threats, intimidation, humiliation, treating an individual like a child, or isolating them from family, friends, or regular activities.

Financial elder abuse: Occurs when someone tricks, threatens, or persuades an older adult out of their money, property, or possessions.

Elder neglect: Occurs when a person fails to provide an older person with the necessities of life, such as food, clothing, shelter, medical attention, personal care, and required supervision.

2013). The research indicates that elder abuse is an unfortunate reality for many older Canadians, and it is going to take a coordinated effort between government and non-government agencies and public and private sectors to see impactful social change.

Combating Ageism

The first step in challenging ageism in society is to recognize that it is not the responsibility of any one group. Ageism is a societal issue that requires a collaborative effort among individuals, organizations, and policy makers to see any real change (Revera, 2012). As individuals, we must increase awareness around issues of ageism, challenging negative stereotypes, rather than making assumptions about a person's ability or quality of life based on age. The Ontario Human Rights Commission recommends that in order to increase awareness, those who work with the public "should receive training that dispels negative assumptions and attitudes and serves to increase awareness of how to appropriately respond to the aging process" (OHRC, 2001). It is only with optimism and open-mindedness that we can shift negative perceptions. Organizations and employers need to value the work of older and younger workers, recognizing their abilities and contributions regardless of age (Revera, 2012). Governmental and non-governmental policy makers need to focus on age-inclusive policies that afford people of all ages the autonomy to make the choices they need to live their lives to the fullest (Revera, 2012). The Ontario Human Rights Commission suggests that **intergenerational programs**, where interaction of different age groups is planned and intentional, can effectively break down barriers between generations by encouraging communication, cooperation, and promoting health and well-being (OHRC, 2001). Filmmaker Evan Briggs is creating a documentary that illustrates the power of intergenerational programs. The film, titled "Present Perfect," explores the experiences of older West Seattle residents with children from a daycare in the same building (Lau, 2015). The older residents meet with young children at the Intergenerational Learning Centre, where they interact through music, dance, art, and storytelling (Lau, 2015). A Kickstarter campaign raised $50 000 for the rough edit of the film, and Briggs is hoping to raise another $100 000 to complete the production (Lau, 2015). Briggs filmed at the Intergenerational Learning Centre three days a week over the course of a year and says, "it's nice when you present an issue to also present a possible solution ... one of many solutions that we can offer to close that circle of life loop a little better" (Lau, 2015).

The Ontario Human Rights Commission also suggests that in order to dispel ageist attitudes in the public and healthcare sectors, improved training for doctors, nurses, police, lawyers, journalists, social workers, policy makers, and others who work with the aging population is a good place to start (OHRC, 2001; Stall, 2012). Fourth year medical student Nathan Stall is "troubled by the care being provided to older Canadians and concerned that trainees across the country are being inadequately prepared to meet the unique needs of our aging population (2012). He suggests that medical teaching practices are biased against the management of older patients, in that older patients are not receiving adequate assessment and their concerns are often dismissed as "non-medical" (Stall, 2012). Stall says that medical students tend to "see older patients as sources of frustration and ... they perceive these individuals as impediments to clinical efficiency and medical education" (Stall, 2012). Consequently, ageist attitudes like these make it difficult for Canadian geriatric programs to recruit trainees, resulting in a shortage of geriatricians practising in Canada (Stall, 2012). In the box titled "Agent of Social Change," you will meet an inspiring Canadian who has devoted his career to challenging such age discrimination and promoting social change in this area.

INTERSECTIONALITY

Socio-Economic Status and Age

The aging population in Canada experiences unique barriers arising from the intersection of age and other aspects of social identity, such as socio-economic status, gender, sexuality, and ability. We see the intersection of age and socio-economic status when Statistics Canada reported that as of December 2014, 12 percent of seniors in Canada were living in poverty, with those living alone being particularly vulnerable (Statistics Canada, 2014; CARP Canada, 2014). The Canadian aging population is recognized by Canada's Federal Poverty Reduction Plan as one of the nine demographic groups most vulnerable to low income rates (Hoeppner, 2010). Once we factor in gender, we see that senior women are more likely to experience poverty, with incidences of low income at 15.6 percent, compared to their male counterparts at 11.5 percent (Statistics Canada, 2010).

> **Intergenerational programs:** Programs that have intentional and planned interactions of different age groups, which can be effective in breaking down barriers between generations by encouraging communication, cooperation, and promoting health and well-being.

AGENT OF CHANGE

Moses Znaimer

Reprinted with permission of ZoomerMedia Limited

Moses Znaimer

Moses Znaimer is best known for his achievements in television and groundbreaking concepts like Videography, Speaker's Corner, and the Studioless Operating Environment that were at the core of over 20 independent stations and channels he co-founded, including CityTV, CablePulse24, SPACE, Bravo!, MuchMusic, and MusiquePlus.

Now, Znaimer is a champion for Canada's 15.8 million people aged 45-plus (aka "Zoomers" or "Boomers With Zip!") and is President of CARP, "a national, non-partisan, non-profit organization committed to advocating for *A New Vision of Aging for Canada*, promoting social change that will bring financial security, equitable access to health care and freedom from age discrimination."

In addition to his role as CARP President, Znaimer is the Founder of ZoomerMedia Limited (TSXV: ZUM), a multi-media company uniquely devoted to serving the needs and interests of the 45-plus on all platforms: national and local television channels and stations, regional radio stations, a national magazine, websites, conferences and tradeshows. ZoomerMedia provides CARP with advertising and editorial support to advance its advocacy messages and develop CARP membership.

"My insight was that even that massive, market moving and ultra youthful generation known as the Boomers had inevitably to age; and that if they were considered dominant while they were young (because of their size) then surely they must still be dominant even if they were getting older. And what is old anyway, these days? When does it begin? How should society understand and respond to the deep changes at hand given that we are all, generally speaking, living longer and better? In Canada, that discussion is led, with increasing vigor, by the Advocacy Association known as CARP." – Moses Znaimer

Source: Zoomer Magazine. Used with permission.

In their report titled *Time For Action: Advancing the Rights of Older Persons in Ontario*, the Ontario Human Rights Commission (OHRC) suggests that older women experience such disadvantage because they live longer, continue to experience wage inequalities, and are part of a social structure that is still primarily male-centered (OHRC, 2001). The National Senior Strategy for Canada Report proposes that in order to ensure no older Canadians live in poverty, all levels of government need to coordinate to consider how best to enhance public pension vehicles, such as Canada Pension Plan (CPP)/Quebec Pension Plan (QPP), Old Age Security (OAS), and Guaranteed Income Supplement (GIS), rather then encouraging private savings vehicles, like Registered Retirement Savings Plans (RRSP) and Tax Free Savings Accounts (TFSA), as they tend to only benefit those who have higher than average incomes (National Seniors Strategy, 2015). As Baby Boomers start to retire, there is a "growing divide between the haves, whose retirement is secured by guaranteed workplace pensions, and the have-nots, left to scrape by on their own meagre savings" (McMahon & MacQueen, 2014). The fear is that this "pension envy today will eventually explode into a full-blown crisis as younger workers, saddled with student debt, mortgages, and stagnant incomes, age into retirement" (McMahon & MacQueen, 2014).

Ability and Age

When we investigate the intersectionality of age and ability, we see that the likelihood of disability increases steadily with age, with 33.2 percent of Canadian seniors reporting a disability (Statistics Canada, 2012).

Women have a higher risk of disability in almost all age groups, with 35.2 percent of women 65 years and over reporting a disability (Statistics Canada, 2012). The OHRC suggests that for persons with disabilities, aging can compound disadvantage, "creating additional barriers and limitations on the ability to fully participate in society" (2001). For example, age and disability often create barriers to obtaining employment, so an older person with a disability may experience discrimination in employment as a "double-edged sword" (OHRC, 2001). Furthermore, when it comes to mobility, public transportation is essential for older Canadians with disabilities to remain independent, so the OHRC emphasizes that "conventional transit systems must ensure maximum accessibility and that para-transit services be made available for those who cannot access conventional public transportation (2001).

Sexuality and Age

For the LGBT aging population, there is a real concern that as they age, they will face continued homophobia and discrimination in employment, housing, and community-based or long-term care. CBC's "On the Coast" produced a radio series titled "Gay and Grey," focusing on issues of aging in the lesbian, gay, bisexual, and transgender community. In the series, Alan Herbert, a prominent gay activist in Vancouver living with HIV, is interviewed and describes how he started a support group for aging gay men after being asked to leave his current support group because "people didn't care about HIV, they weren't interested in HIV, and they didn't want to hear about HIV" (CBC News, 2014). He explains how this cohort of gay men in their sixties felt unique "isolation and loneliness" and knew first-hand "the terror of depression" (CBC News, 2014).

Also in the series, Brian deVries, a Professor of Gerontology from San Francisco State University, explains that older gay men are more likely to age alone, age without children, and to live alone with increasing numbers of disabilities and limited support systems (CBC News, 2014). Professor deVries suggests that unlike older gay men, aging lesbian women have slightly greater success at creating communities of care that support independent living (CBS News, 2014). Moreover, deVries proposes that transgender or transsexual seniors are particularly disadvantaged because the extent to which they are not recognized and not supported in society presents a unique struggle to "find a place in a world that does not recognize how they see themselves" (CBS News, 2014).

ENDING THOUGHTS

As an inevitable reality for all of us, is age just a number? Or is that number saturated with socially constructed attitudes, behaviours, and expectations? Do generational differences divide us or can our differing strengths and abilities unite us in a world of work where five generations are working side by side for the first time? Will predictions of the Fourth Industrial Revolution flip our world upside down and change it forever, the way the Digital Revolution did? Will we be able to keep up with the technological advancements and adapt to a new social organization based on the Internet of Things?

When we deconstruct the layers of meaning attached to age and address the consequences of age stratification in Canadian society, we can see that age is much more than just a number or an inevitable biological process. It is a social construct that has very real consequences for people of all ages. So how can we address ageism and dispel negative misconceptions of youth and the aging population? As we move toward a society free from ageist attitudes, we must accept ageism as an issue for people of *all* ages, and work together to challenge negative stereotypes and assumptions based on age. Only in recognizing the role that we all play in ageism, can we break down the barriers of age discrimination.

READING

KEEPING AN EYE OUT: HOW ADULTS PERCEIVE STUDENTS

By Adam Fletcher

Adults who work in schools have many different motivations to become teachers, counselors, administrators, and school support workers. However, few motivations are as strong as

our perceptions: if we actually perceive of students as needing us, we can pretty much justify doing anything in schools. The way adults **see** young people in schools determines how adults **treat** young people in schools; and the way we treat them determines the **outcomes** of our activities with them. It can be hard for adults who are in the middle of busy lives to stop and reflect on the ways we perceive students. This article is a tool that can help make that easier. Over the years, I have reflected a lot on my perceptions of young people. After working in schools, nonprofits, government agencies, and throughout communities for more than 20 years, I have critically deconstructed my actions and assumptions, and worked with others to reflect on their perceptions of young people. Through these reflections, I have seen *five basic ways* that adults in schools perceive students. These ways determine how teachers, administrators, counselors, and other school support staff treat students every single day. That determines how schools feel to students, and this, in turn, fosters student engagement.

Perceptions of Student Voice
Apathy:
The first way in which adults can perceive student voice is with *apathy*. This occurs when adults deliberately choose to be indifferent toward students. This is different from *antipathy*, where one person does not know the other person exists. However, in schools, adults implicitly know student voice exists. Our conscious choice not to perceive it is what determines our apathetic perspective. This can happen throughout school decision-making, affecting both individual students and entire schools. Both students and adults can (and do) express apathy toward student voice.

Pity:
Pity happens when adults perceive student voice from the top down, seeing it as a 'nice thing' to do. Pity makes adults completely superior to students in all ways, including intellectually, morally, and culturally. Adults view students as completely incapable of providing anything for themselves, and see students as fully dependent on adults. By positioning adults in positions of absolute authority, pity dehumanizes students by suppressing their self-esteem and incapacitating their self-conceptions of ability and purpose.

Sympathy:
Perceiving students with *sympathy* is alluring to many adults. Sympathy disengages students from actively creating knowledge or resources by singularly positioning adults to give without acknowledging that they are receiving anything in return. Sympathy is another top-down perception. It allows adults to give to students what they apparently cannot acquire for themselves, whether material, time, money, or otherwise, and to do that from a position of compassion.

Empathy:
Reciprocity is at the core of an *empathetic* perception of student voice. This viewpoint allows adults to see students are giving something as well as receiving it. Each person acknowledges the other as a partner, and each person becomes invested in the outcomes of the others' perception. Empathy is rooted in equity and reciprocity.

Solidarity:
Complete *solidarity* comes from the perception that students are not different from adults simply because of their age. Instead, it allows for complete equity by fully recognizing the benefits and challenges of student voice, and engaging students and adults in complete partnerships. These relationships between adults and students operate from a place of possibilities rather than deficits.

Considerations:
There are many important considerations to recognize about our perceptions of student voice. Following are two of the most important:

- Adults do not maintain one perception of all students all the time. While there are predominant perceptions, there are also exceptions to the rule. When confronting challenging perceptions, it can be important to acknowledge the exception, if it is positive.
- These perceptions are not about "good" and "bad"—they just are. Adults simply cannot operate in complete empathy toward students all the time; likewise, students should not be expected to care for every single adult they ever meet.

Reflections:
Using these perceptions of student voice as a starting point, the challenge for adults becomes whether we can consciously, critically, and creatively reflect on our attitudes, behaviors, and, ultimately, our perceptions. While we do this, it is our obligation to keep an eye toward further developing our practice in order to be more effective in the work we do. Acknowledging how we see students can be the road to changing how we treat students. Isn't that the goal of all student participation work?

Source: Fletcher, A. (2011). Keeping An Eye Out: How Adults Perceive Students. *Connect*, 190 (August 2011), 27–28. Retrieved 24 May 2016, from http://research.acer.edu.au/cgi/viewcontent.cgi?article=1198&context=connect.

DISCUSSION QUESTIONS

1. From your experience, which of the perceptions of student voice do you see in practice most often? Why do you think this perception is most common?
2. Fletcher states that "perceptions are not 'good' and 'bad'—they just are." What are the ramifications for students who are learning in an environment where they are perceived mostly with empathy and solidarity?
3. Fletcher discusses *adult* perceptions of *students* in an educational environment; how do you suppose he defines 'adult' and 'student'? Are these perceptions of student voice relevant to students of all ages?

KNOW IT AND OWN IT: WHAT DO I BRING TO THIS?

The "K" in the KWIP process involves examining aspects of your own identity and social location as the first step in becoming diversity competent.

ACTIVITY: JOURNAL

How old are you? How old do you feel? Is there a gap between your chronological age and your psychological age? If so, in your opinion, what causes that gap?

If you had to use a number to describe your social age, what would it be? How does it compare to your chronological age, or the number you would assign for your psychological age? What factors would you consider in determining your social age?

Consider whether you "act your age" and what that means to you. Are there specific norms, roles, or behaviours that are expected of people your age in your culture?

Think about a time when *you heard or when you used* one of the following statements; describe the scenario and identify how you felt at that time:

- "Do as I say, not as I do."

- "Because I said so."

- "You'll understand when you get older."

- "Grow up."

WALKING THE TALK: HOW CAN I LEARN FROM THIS?

The "W" in the KWIP process presents a scenario or case study that challenges you to "walk the talk" through problem-based learning.

ACTIVITY: CASE STUDY

You landed your dream job—Congratulations! You have met everyone in your department and realize that you are the youngest of the group. You are excited about being the youngest because you are full of great ideas and are confident that you can bring something new to the table. A few of your colleagues have already come to you for help with some of the technological tools you use and are so grateful that you were able to make things run more efficiently for them. With the intention of helping more people in your area, you approach your manager and offer to host a "how-to" seminar on using technology in the workplace. Your manager thinks it is a great idea, as it will allow for more productivity and efficiency if everyone learns how to use these tools. Unfortunately, a few of your colleagues are not happy about this. They say they are comfortable with the way they are currently doing their jobs and are not interested in learning about any new tools. You can feel their hostility toward you and you have even overheard people asking, "Who does this new kid think he/she is, trying to make all these changes? It's only going to cause more work for us!" Surprisingly, everyone in your department shows up to your seminar and you have an audience ranging in age, including Traditionalists, Baby Boomers, Generation Xs, and Generation Ys. Some are eager to learn, some are nervous but willing to learn, and others are outright resistant.

1. What do you do? How do you tailor your seminar to meet the needs of each of these generations?

2. Do you think generational conflict in the workplace is inevitable, or is there a way for multiple generations to work well together?

IT IS WHAT IT IS: IS THIS INSIDE OR OUTSIDE MY COMFORT ZONE?

The "I" in the KWIP process requires you to honestly confront and identify ways in which our complex identities result in experiences of privilege and oppression, and to reflect on how we can learn to honour that privilege.

Age stratification creates a social hierarchy based on age that can be a major source of inequality. The division of social age groups creates a wide range of privileges and entitlements for some, and significant barriers and obstacles for others. Since every age group can experience advantages and disadvantages based on age, in this exercise, we ask you to examine yourself in terms of the unequal distribution of power and privilege based on age.

ASKING: Do I have privilege?

1. I can freely participate in social activities without being restricted based on my age.

2. I am capable of making decisions that affect my body or self without parental consent.

3. I am reasonably sure that when I interview for a job, I will not experience discrimination based on my age.

4. People take my opinion seriously and believe that I have something to contribute to a conversation.

REFLECTING: Honouring Our Privilege

Consider the social age group that you fall into, and describe two circumstances in which you feel disadvantaged because of your age. Then describe two circumstances in which your age gives you privilege over someone else. How can you use the advantages you experience because of your age to combat the disadvantages experienced by others within or outside of your social age group?

PUT IT IN PLAY: HOW CAN I USE THIS?

The "P" in the KWIP process involves examining how others are practising equity and how you might use this.

ACTIVITY: **CALL TO ACTION**

The Youth Restorative Action Project (YRAP), formed in 2001, is an example of a youth initiative that attempts to combat adultism by increasing the youth voice within social institutions. YRAP is a Youth Justice Committee founded and run by a group of Edmonton youth concerned that young people did not have the opportunity to act as advocates for their rights in the larger systems in which they find themselves, such as the child welfare system, the education system, and the criminal justice system.

Their mandate is to work with young people who have caused harm while being affected by a variety of significant social issues, such as intolerance, racism, substance abuse, homelessness, family violence, and prostitution, by providing legal functions, including sentencing and extrajudicial sanctions, judicial interim releases, treatment plans, the supervision of community service and probation orders, and mediation. YRAP is heavily influenced by a community-based, restorative justice approach offering those who have caused harm an opportunity to take responsibility for their actions and to grow positively, in addition to offering victims a meaningful role in the process.

YRAP is the only justice committee worldwide that is run entirely for youth by youth, with members ranging from the age of 15 to 24 years old. YRAP was officially sanctioned as a justice committee in 2003 and has dealt with over 500 cases of all levels of severity.

To learn more about the Youth Restorative Action Project, visit www.yrap.org.

Located at www.nelson.com/student

- Review Key terms with interactive **flash cards**
- Check your Comprehension by completing **chapter review quizzes**
- Gauge your understanding with *Picture This* and accompanying short answer questions
- Develop your critical thinking/reading skills through compelling **Readings** and accompanying short answer questions
- Apply your understanding to your own experience with **Connect A Concept** activities
- Evaluate Diversity in the Media with engaging *Video Activities*
- Reflect on your Understanding with *KWIP* activities

Study Tools
CHAPTER 12

Alberta Law Library. (2013). Bill C-36, Protecting Canada's Seniors Act – In Force January 13, 2013. Retrieved on May 29, 2016, from *Alberta Law Library*: https://www.lawlibrary.ab.ca/staycurrent/2013/01/15/bill-c-36-protecting-canadas-seniors-act-in-force-january-13-2013/

Alzheimer Disease International. (2016). Policy brief: The global impact of dementia 2013–2050. Retrieved May 26, 2016, from *Alzheimer's Disease International:* http://www.alz.co.uk/research/G8-policy-brief

Alzheimer Society of Canada. (2016). Alzheimer Society of Canada calls for Alzheimer's disease and dementia partnership. Retrieved May 26, 2016, from *Alzheimer Society of Canada:* http://www.alzheimer.ca/en/kfla/Get-involved/Raise-your-voice/Where-we-stand/A-new-way-of-looking-at-dementia/alzheimers-disease-dementia-partnership

BBC. (2012). The BBC story. Retrieved from *BBC:* http://www.bbc.co.uk/historyofthebbc/

Buahene, A., & Kovary, G. (2007). *Loyalty unplugged.* Philadelphia: Xlibris.

Canadian Communications Foundation (CCF). (2001). The history of Canadian broadcasting. Retrieved from *Canadian Communications Foundation:* http://www.broadcasting-history.ca/timeline/CCFTimeline.swf

Canadian Network for the Prevention of Elder Abuse (CNPEA). (2016). What is elder abuse? Forms of abuse. Retrieved on May 29, 2016, from *CNPEA:* http://cnpea.ca/en/what-is-elder-abuse/forms-of-abuse

CARP Canada. (2010). They are we: Understanding ageism in Canada. Retrieved May 26, 2016, from *CARP Canada:* http://www.carp.ca/2010/06/23/they-are-we-understanding-ageism-in-canada/2/

CARP Canada. (2013). CARP action against elder abuse. Retrieved on May 29, 2016, from *CARP Canada:* http://www.carp.ca/wp-content/uploads/2013/12/Elder-Abuse-Paper-Dec-2013.pdf?e4b50d

CARP Canada. (2014). 600,000 Seniors in Canada live in poverty, including more than 1 in 4 single seniors according to new Statistics Canada report. Retrieved May 5, 2016, from *CARP Canada:* http://www.carp.ca/2014/12/11/600000-seniors-canada-live-poverty/

CARP Canada. (2016). Moses Znaimer. Retrieved May 10, 2016, from *CARP Canada:* http://www.carp.ca/about-carp/bios/moses-znaimer/

CBC News. (March 14, 2014). On the Coast: Gay and grey. Retrieved May 5, 2016, from *CBC News:* http://www.cbc.ca/news/canada/british-columbia/on-the-coast-presents-its-gay-and-grey-series-1.2573455

Center for Generational Kinetics. (2016). Top 10 Gen Z and iGen questions answered. Retrieved May 27, 2016, from *The Center for Generational Kinetics:* http://genhq.com/igen-gen-z-generation-z-centennials-info/

Conference Board of Canada. (2014). Age rather than gender becoming the new income divide. Retrieved May 26, 2016, from *Conference Board of Canada:* http://www.conferenceboard.ca/press/newsrelease/14-09-23/age_rather_than_gender_becoming_the_new_income_divide.aspx

Connor, S. (2008). Computers 'to match human brains by 2030.' Retrieved June 2, 2016, from *Independent:* http://www.independent.co.uk/life-style/gadgets-and-tech/news/computers-to-match-human-brains-by-2030-782978.html

Elmore, T. (2014). Contrasting Generation Y and Z. Retrieved May 27, 2016, from *Huffinton Post:* http://www.huffingtonpost.com/tim-elmore/contrasting-generation-y-_b_5679434.html

Fletcher, A. (2011). Keeping an eye out: How adults perceive students. *Connect*, 190(August 2011), 27–28. Retrieved May 29, 2016, from http://research.acer.edu.au/cgi/viewcontent.cgi?article=1198&context=connect

Fletcher, A. (2015). *Facing adultism.* Washington: CommonAction Books.

Free the Children. (2016). About us. Retrieved June 2, 2016, from *Free the Children:* http://www.freethechildren.com/who-we-are/about-us/

Garrard, T. A. (2008). Remember, The force will be with you … always: The electronic media and *Star Wars* and the socialization of Generation X. Dissertation Thesis. Michigan, USA.

Garretson, J.A. (2011, February 19). Age stratification and ageism. Retrieved May 11, 2016, from *Uncommon Thought:* http://www.uncommonthought.com/mtblog/archives/2011/02/19/age-stratificat.php

Giedd, J. (2012). Inside the teenage brain. Retrieved from *Frontline:* http://www.pbs.org/wgbh/pages/frontline/shows/teenbrain/interviews/giedd.html.

Greenough, J. (2015). 10 million self-driving cars will be on the road by 2020. Retrieved June 2, 2016, from *Business Insider:* http://www.businessinsider.com/report-10-million-self-driving-cars-will-be-on-the-road-by-2020-2015-5-6

Hoeppner, C. (2010). Federal poverty reduction plan: Working in partnership towards reducing poverty in Canada – Report of the Standing Committee on Human Resources, Skills, and Social Development and the Status of Persons with Disabilities. Retrieved May 26, 2016, from *Government of Canada:* http://www.parl.gc.ca/content/hoc/Committee/403/HUMA/Reports/RP4770921/humarp07/humarp07-e.pdf

IMDb. (2012). IMDb charts: Top movies. Retrieved August 2, 2012, from *IMDb.com:* http://www.imdb.com/chart/

Infantry, A. (2012, June 19). Gen Y Canadians splurging on luxury items, despite high unemployment. Retrieved May 29, 2016, from *The Toronto Star:* http://www

.thestar.com/business/article/1213344--gen-y-guess-who-s-driving-the-luxury-market

Jenkins, R. (2015). Who is Generation Z: Understanding what matters most to the post-millennial generation. Retrieved May 27, 2016, from: http://ryan-jenkins.com/2015/06/04/who-is-generation-z-understanding-what-matters-most-to-the-post-millennial-generation/

Kane, S. (2012). Generation X. Retrieved from *About.com*: http://legalcareers.about.com/od/practicetips/a/GenerationX.htm

Keeter, S., & Taylor, P. (2009, December 11). The millennials. Retrieved May 29, 2016, from *Pew Research Center Publications*: http://pewresearch.org/pubs/1437/millennials-profile

Kingston, A. (2014, October 13). Why it's time to face up to old age. Retrieved May 26, 2016, from *Macleans*: http://www.macleans.ca/society/health/an-age-old-problem/

Lau, A. (2015). When a preschool is located in a nursing home, magic happens. Retrieved June 3, 2016, from *Huffington Post*: http://www.huffingtonpost.ca/2015/06/21/preschool-inside-nursing-_n_7630064.html

Laz, C. (2009). Act your age. In L. J. McIntyre (Ed.), *The practical skeptic: Readings in sociology 2nd ed.* (pp. 75–84). New York: McGraw Hill Higher Education. (Original work published 1998).

Marr, B. (2016). Why everyone must get ready for the 4th Industrial Revolution. Retrieved June 2, 2016, from *Forbes*: http://www.forbes.com/sites/bernardmarr/2016/04/05/why-everyone-must-get-ready-for-4th-industrial-revolution/#5615fd1e79c9McCrindle, M., & Wolfinger, E. (2010). Generations defined. *Ethos*, 8–13.

McMahon, T., & MacQueen, K. (2014). Canada's looming pension wars. Retrieved June 3, 2016, from *McLean's*: http://www.macleans.ca/politics/canadas-looming-pension-wars/
Mehta, D. (2012). Take them seriously: Youth volunteers work to overcome age discrimination. Retrieved May 29, 2016, from *The Canadian Press*: http://www.huffingtonpost.ca/2012/03/22/take-them-seriously-yout_n_1372869.html

Morgan, J. (2014). A simple explanation of the 'Internet of Things.' Retrieved June 2, 2016, from *Forbes*: http://www.forbes.com/sites/jacobmorgan/2014/05/13/simple-explanation-internet-things-that-anyone-can-understand/#37eb7dfa6828

National Seniors Strategy. (2015). Why Canada needs a National Seniors Strategy. Retrieved May 26, 2016, from *National Seniors Strategy*: http://www.nationalseniorsstrategy.ca

NBC. (2012). NBC timeline: Key dates in its history. Retrieved from *Media Biz on NBC News*: http://www.msnbc.msn.com/id/33994343/ns/business-us_business/t/nbc-timeline-key-dates-its-history/#.UBgpQqPCRe4

Nelson, J. (2012). Millennials and age discrimination: Tips. Retrieved on May 29, 2016, from *Canadian Business*: http://www.canadianbusiness.com/business-strategy/millennials-and-age-discrimination-tips/

Netburn, D. (2012, February 29). Pew study: Is the Internet ruining or improving today's youth? Retrieved May 29, 2016, from *Pew Internet*: http://www.pewinternet.org/Media-Mentions/2012/Is-the-Internet-ruining-or-improving-todays-youth.aspx

O'Neal, K. (2012). Youth rights 101: What is ageism? Retrieved on May 29, 2016, from *The Youth Rights Blog*: http://theyouthrightsblog.blogspot.ca/2012/07/youth-rights-101-what-is-ageism.html

Ontario Human Rights Commission (OHRC). (2001). Time for action: Advancing human rights for older Ontarians. Retrieved May 2, 2016, from *OHRC*: http://www.ohrc.on.ca/sites/default/files/attachments/Time_for_action%3A_Advancing_human_rights_for_older_Ontarians.pdf

Ontario Human Rights Commission (OHRC). (2015). Ontario Human Rights Code. Retrieved May 26, 2016, from *OHRC*: https://www.ontario.ca/laws/statute/90h19/v15#BK6

Perez, R. (2016). As 'divergent' trends downward, hopes for post-'Hunger Games' dystopian teen YA franchises look dour. Retrieved May 27, 2016, from *The Playlist*: http://blogs.indiewire.com/theplaylist/as-divergent-trends-downward-hopes-for-post-hunger-games-dystopian-teen-ya-franchises-look-dour-20160321

Public Health Agency of Canada (PHAC). (2012). Elder abuse: It's time to face the reality. Retrieved on May 29, 2016, from *Public Health Agency of Canada*: http://www.phac-aspc.gc.ca/sfv-avf/sources/age/age-abuse-broch/index-eng.php

Revera. (2012). *Revera report on ageism*. Retrieved May 6, 2016, from *Revera*: http://www.reveraliving.com/revera/files/b2/b20be7d4-4d3b-4442-9597-28473f13b061.pdf

Rogers, M. (2009, October 28). Boomers and technology: An extended conversation. Retrieved from *AARP*: http://assets.aarp.org/www.aarp.org_/articles/computers/2009_boomers_and_technology_final_report.pdf
Schawbel, D. (2014). 5 predictions for Generation Alpha. Retrieved June 2, 2016, from: http://danschawbel.com/blog/5-predictions-for-generation-alpha/

Sheppard, G. (2006). *How to be the employee your company can't live without*. New Jersey: J. Wiley and Sons.

Stall, N. (2012). Time to end ageism in medical education. *Canadian Medical Association Journal*, 184(6), 728.

Statistics Canada. (2008). Women in Canada: A gender based statistical report: Senior women. Retrieved May 26, 2016, from *Statistics Canada*: http://www.statcan.gc.ca/pub/89-503-x/2010001/article/11441/tbl/tbl005-eng.htm

Statistics Canada. (2010a). Incidences of low income using the after-tax Low Income Cut Offs. Retrieved May 5, 2016, from *Statistics Canada*: http://www.statcan.gc.ca/pub/75-202-x/2010000/c-g/ct006-eng.htm

Statistics Canada. (2010b). Population projections: Canada, the provinces and territories. Retrieved May 25, 2016, from http://www.statcan.gc.ca/pub/91-520-x/2010001/part-partie3-eng.htm

Statistics Canada. (2012a). Disability in Canada: Initial findings from the Canadian Survey on Disability. Retrieved May 5, 2016, from *Statistics Canada*: http://www.statcan.gc.ca/pub/89-654-x/89-654-x2013002-eng.pdf

Statistics Canada. (2012b). Generations in Canada. Retrieved July 18, 2012, from *Statistics Canada*: http://www12.statcan.gc.ca/census-recensement/2011/as-sa/98-311-x/98-311-x2011003_2-eng.cfm

Statistics Canada. (2014). Canadian income survey, 2012. Retrieved May 5, 2016, from *Statistics Canada*: http://www.statcan.gc.ca/daily-quotidien/141210/dq141210a-eng.htm

Statistics Canada. (2015). Birth, estimates, by province and territory. Retrieved May 27, 2016, from *Statistics Canada*: http://www.statcan.gc.ca/tables-tableaux/sum-som/l01/cst01/demo04a-eng.htm

Tapscott, D. (2009). *Grown up digital: How the net generation is changing your world.* New York: McGraw-Hill.

Tv.com. (2016). Retrieved May 27, 2016, from *Tv.com*: http://www.tv.com/shows/category/kids/decade/2000s/UserExperiencesWork. (2011, October 6). A magazine is an iPad that does not work.m4v. Retrieved from *YouTube*: http://www.youtube.com/watch?v=aXV-yaFmQNk&feature=plcp

Williams, A. (2015). Meet Alpha: The next "next generation." Retrieved May 27, 2016, from *The New York Times*: http://www.nytimes.com/2015/09/19/fashion/meet-alpha-the-next-next-generation.html?_r=1

World Economic Forum. (2016). The Fourth Industrial Revolution. Retrieved June 2, 2016, from *World Economic Forum*: https://www.weforum.org/pages/the-fourth-industrial-revolution-by-klaus-schwab

World Health Organization (WHO). (2016). Elder abuse. Retrieved on May 29, 2016, from *WHO*: http://www.who.int/ageing/projects/elder_abuse/en/

Families

*"Family is family, in church or in prison
You get what you get, and you don't get to pick 'em."*

(Kacey Musgraves, 2015)

LEARNING OUTCOMES

By mastering this unit, students will gain the skills and ability to:

- compare varying definitions of family and explain why family is difficult to define

- examine family diversity in Canada

- identify barriers that different families face within the larger society

- discuss the various forms of family violence in Canada

- identify and discuss assistive reproductive technologies in Canada

Natalia Lukiyanova/Thinkstock.com

NEL

Families exist in all forms, regardless of how they are defined, and we find them everywhere we look. Restaurants and businesses cater to families, television families like those from *The Fosters, Girl Meets World, Modern Family,* and *Switched at Birth* fill our television screens, and our neighbourhoods are inhabited with children from every kind of family. The face of the modern family has changed over the years, along with how we have come to define it. In this chapter, we look at some of the trends in Canadian families in the 21st century. Regardless of how you choose to describe it, your family is uniquely yours; and in the quest for the perfect definition of a family, we quickly realize that maybe no such definition exists.

DEFINING THE FAMILY

Our family is usually our primary **agent of socialization**. We spend the most time with the people in our family, and consequently, they help to shape our values, ideals, and morals. Our **family of orientation** (the family you were born or raised in) gives us our heritage, social status, and history, while our **family of procreation** (the family you create when you are an adult) allows us to pass those traits—and possibly new ones—on to future generations. How much difference will there be between the family you were born into and the one that you will create when you are older? If we look at how the composition of the family has changed over the past few decades in Canada, we see that there is a good chance that your birth family may not look like the family you will end up with.

How do you define family? Does your definition include your pet or your best friend? Unfortunately, neither of those two are included in the definition that Statistics Canada uses when they gather information about families in Canada. For years, it was easy to define what a family was because marriage conferred legal status on the parents' children (Bird, 2010). Today however, the task of defining the family is much more difficult. Consider this definition from the 1971 census definition for family:

> A census family consists of a husband and wife (with or without children who have never been married, regardless of age) or a parent with one or more children never married, living in the same dwelling. A family may consist, also, of a man or woman living with a guardianship child or ward under 21 years for whom no pay was received. (Statistics Canada, 2012a)

How many families do you know of today that would not fit into the 1971 definition? Compare the 1971 definition to the one used in the 2011 census:

> A census family is composed of a married or common-law couple, with or without children, or of a lone parent living with at least one child in the same dwelling. Couples can be of the opposite sex or of the same sex. (Statistics Canada, 2012b)

Clearly, the definition has changed over four decades. What are some of the reasons that Statistics Canada has had to update how it defines the family? Do you agree with their definition? At its most basic level, Statistics Canada defines a family as a couple—married or common-law, with or without children—or a lone parent with at least one child in the same house. According to the 2011 census data, "it takes two people to make a family. Beyond that, anything goes" (Scoffield, 2012).

Statistics Canada uses that definition so that it can count the families in Canada—in all their various forms. If the definition seems a little stark to you, compare it to the definition that is used by the Vanier Institute of the Family:

> Any combination of two or more persons who are bound together over time by ties of mutual consent, birth and/or adoption or placement and who, together, assume responsibilities for variant combinations of some of the following:

- Physical maintenance and care of group members
- Addition of new members through procreation or adoption
- Socialization of children
- Social control of members
- Production, consumption, distribution of goods and services, and
- Affective nurturance—love. (Vanier Institute of the Family, 2012c) (See Figure 13.1.)

This definition clearly differs from the previous two. As you have probably noticed, Statistics Canada just counts the families and their composition, so as Canadians, we can see how the demographics or composition of the family has changed over time.

> **Agents of socialization:** The people and institutions in society that create the social contexts where socialization takes place.
>
> **Family of orientation:** The family someone was born or raised in; gives us our heritage, social status, and history.
>
> **Family of procreation:** The family we create when we become adults.

FIGURE 13.1

Family Diversity in Canada

Family Diversity in Canada 2016

VANIER
L'Institut Vanier de la famille
The Vanier Institute of the Family

9.9 M
Number of families in Canada in 2015

464,000
Number of stepfamilies in Canada in 2011
(12.6% of all couples with children)

"Families, no matter their background or their makeup, bring new and special patterns to our diverse Canadian tapestry."

– His Excellency The Right Honourable David Johnston, Governor General of Canada, at the *Families in Canada Conference 2015*

Proportion of families that were married couples and common-law couples, respectively, in 2011	67%	17%

65,000
Number of same-sex couple families counted in 2011, 9.4% of whom have children at home

Four in five same-sex couples with children were female couples.

1.5 M
Number of lone-parent families in Canada in 2015
(16.3% of all census families)

Proportion of all couples in Canada that were mixed unions[1]	2.6%	4.6%
	1991	2014

Number of foster children aged 14 and under in private households in 2011 **30,000**

6.3 M
Number of people in Canada who reported belonging to a visible minority group[6] in 2011
(approximately 19% of the total population)

102,000
Number of farm families in Canada in 2013

Number of people[4] in Canada who provided care to a family member or friend with a long-term health condition, disability or aging need in 2012 **8.1M**

Number of people in Canada who reported an Aboriginal identity in 2011 **1.4 M**
(60.8% First Nations only, 32.3% Métis only, 4.2% Inuit only, 1.9% other Aboriginal identity, 0.8% more than one Aboriginal identity)

363,000
Number of multi-generational households[2] in Canada in 2011
(2.7% of all households)

3.8 M
Number of people in Canada who reported in 2012 having some type of disability[5]
(approximately 13.7% of the total population)

Number of "skip-generation"[3] households in Canada in 2011 **53,000**
(0.4% of all households)

Proportion of people in Canada who reported in 2011 being born outside the country **1 in 5**

Number of CAF Regular Force members in Canada, 60% of whom have partners and half of whom have children under 18 **68,000**

[1] Statistics Canada defines a mixed union as "a couple in which one spouse or partner belongs to a visible minority group and the other does not, as well as a couple in which the two spouses or partners belong to different visible minority groups."

[2] Statistics Canada defines multi-generational households as those "containing three or more generations of grandparents, parents and children. The middle generation may be comprised of two parents who are part of a couple, a lone parent, or a more complex situation such as both a couple and a lone parent."

[3] Statistics Canada defines skip-generation households as those that consist of "grandparents and grandchildren without the presence of parents in the home."

[4] Aged 15 and older.

[5] ibid.

[6] Statistics Canada defines visible minorities as "persons, other than Aboriginal peoples, who are non-Caucasian in race or non-white in colour."

© 2016 Vanier Institute of the Family

@vanierinstitute

www.vanierinstitute.ca

MARRIAGE AND FAMILY

According to the 2011 census, there were 9 389 700 census families in Canada, up 5.5 percent from 8 896 840 families in 2006 (Milan & Bohnert, 2012; Statistics Canada, 2012c.) Married couples remained the predominant family structure (67.0 percent) in 2011, but that share has decreased over time (Milan & Bohnert, 2012; Statistics Canada, 2012c). This means that even though people are still getting married and having babies, couples are choosing other forms of cohabitation and child-rearing as well.

The majority of Canadians will marry at some point in their lives for many reasons. Most Canadians value love as a precursor for marriage and choose someone they have a great deal of affection for when considering a life partner. The coupling of love and marriage, however, is not necessarily culturally universal. Many cultures look for factors like economic security and religious characteristics when considering marriage, rather than romantic feelings. We can see these factors when looking at the practice of **arranged marriages**, where it is the families, typically the parents, who choose a partner for their child. Many people, consciously or unconsciously, choose to marry someone with similar characteristics when it comes to age, personality, and cultural interests, a pattern referred to as **homogamy**. Cultural norms, and sometimes laws, also influence the selection of a marriage partner. **Endogamy** encourages marriage between people *within* the same or similar social groups, limiting marriage to those of similar age, social class, religion, race or ethnicity. In contrast, **exogamy** encourages marriage between people *outside* of particular social groups. For example, incest is prohibited by section 155 of the Criminal Code of Canada, so we cannot marry someone within our own family.

Legally, marriage in Canada is restricted to the practice of **monogamy**, uniting only two people and **polygamy**, marriage that unites a person with two or more spouses, is prohibited. However, the isolated community of Bountiful, British Columbia, has been challenging anti-polygamy laws in Canada for over a decade (Omand, 2015). Winston Blackmore, the community leader, was formally accused of having 25 marriages between 1975 and 2001 (Omand, 2015). Blackmore's lawyers are arguing that laws prohibiting polygamy infringe on constitutional rights and freedoms, but the Supreme Court of British Columbia upheld the polygamy ban, stating that the practice of polygamy is inherently harmful (Omand, 2015).

COMMON-LAW UNIONS

In Canada, **common-law unions** fall under provincial jurisdiction, so what is considered common law differs from province to province. On March 18th, 2013, the British Columbia provincial government amended their definition of "spouse" to include those living together for two years or more, affording common-law couples the same rights as married couples (Kazia, 2013). Similarly, in Alberta and Newfoundland, a couple is considered common-law after having lived together for two years; however, those living in Ontario and Manitoba have to live together for three years, or one year with a child, before they receive common-law standing (Kazia, 2013). In Alberta, common-law relationships are called "adult interdependent partners" and include those who have lived together for three years or more or those who live together with a child or children (Kazia, 2013). Quebec is the only province in Canada that does not acknowledge common-law relationships (Kazia, 2013). If a common-law union separates, the only province that would enforce a 50/50 split of shared debts and assets is British Columbia, now that they have amended their definition of "spouse"—other provinces that recognize common-law unions would support property claims if the common-law spouse contributed to the cost of a home owned by their partner, but otherwise, assets would not be divided evenly (Kazai, 2013).

Between 1981 and 2011, the number of common-law couples more than quadrupled (+345.2 percent) and accounted for 16.7 percent of all census families in 2011 (Milan & Bohnert, 2012; Statistics Canada, 2012d). In fact, for the first time, the number of common-law families surpassed the number of lone parent families in Canada (Statistics Canada, 2012d). This growth could be attributed to the decline in divorce rates; but increasingly, couples are choosing to live together

Arranged marriages: a type of marital union where partners are selected by family members.

Homogamy: a marital union where partners are similar to each other when it comes to age, personality, and cultural interests.

Endogamy: a marital union encouraged between people within the same or similar social groups.

Exogamy: a marital union encouraged between people of different social groupings.

Monogamy: the practice of being married to one person at a time.

Polygamy: the practice of being married to 2 or more people at one time.

Common-law unions: two individuals living together that are not considered married.

as an alternative to marriage for a variety of reasons. Those most likely to cohabitate are young adults between the ages of 25 and 29, with approximately 1 in 4 young Canadians in this age group living in a common-law union (Statistics Canada, 2012d). For some, marital status is not as important as achieving respect, happiness, career goals, financial security, or a fulfilling sex life (Pew Research Center, 2010).

Attitudes about cohabitation are changing, too. Those in their 60s have shown the largest amount of growth in common-law unions, illustrating that cohabitation is not just a stepping stone for marriage but is becoming a popular alternative to marriage or remarriage (Statistics Canada, 2012d). What do you think? Is living with someone the same thing as being married? Does it represent the same level of commitment?

SAME-SEX UNIONS

Changes in social ideologies, laws, and medical technologies have allowed same-sex couples more opportunities than ever before to create families of their own. In 2005, Canada became the third country in the world to legalize same-sex marriage, following the Netherlands (2000) and Belgium (2003). Since then, same-sex marriage became legal in Spain in 2005; South Africa in 2006; Norway and Sweden in 2009; Portugal, Iceland, and Argentina in 2010; Denmark in 2012; Uruguay, New Zealand, England/Wales, France, and Brazil in 2013; Luxembourg and Scotland in 2014; United States, Ireland, Finland, and Greenland in 2015; and Columbia in 2016 (Pew Research Center, 2016). Additionally, in Mexico, Mexico City legalized same-sex marriage in 2009, the southern state of Quintana Roo in 2011, the northern state of Coahuila in 2014, and neighbouring state of Chihuahua in 2015 (Pew Research Center, 2016).

The 2011 Canadian Census counted 64 575 same-sex coupled families, up 42.4 percent from 2006 (Milan & Bohnert, 2012). Of these couples, 21 015 were same-sex married couples and 43 560 were same-sex common-law couples (Milan & Bohnert, 2012). The number of same-sex married couples nearly tripled between 2006 and 2011, reflecting the first five-year period for which same-sex marriage has been legal across the country (Milan & Bohnert, 2012). Same-sex common-law couples rose 15.0 percent, an increase slightly higher than the 13.8 percent increase for opposite-sex common-law couples (Milan & Bohnert, 2012).

For same-sex couples interested in having children, options are better than they were before. Couples may choose to use a surrogate, adoption, or a fostering agency, or they may choose to co-parent with another same-sex or straight couple or single person (Ryan & Berkowitz, 2009). Although same-sex couples can legally adopt children in Canada, some international agencies will not allow adoption by same-sex couples (Adoption Council of Canada, 2012). Same-sex couples wanting to adopt children, however, face a number of barriers that heterosexual couples do not face. In one study, a social worker employed in the adoption unit of the British Columbia Ministry of Children and Family Development commented that "the reality is that sometimes some workers in our system are reluctant to choose same-sex couples or single applicants; consequently, they tend to wait longer than do traditional couples in our system" (Sullivan & Harrington, 2009). Some of the perception that surrounds the prejudice against same-sex couples who want children focuses mainly on the children themselves and is manifested in three main arguments: the risk for the children to be homosexual themselves; the stigmatization of children with homosexual parents; and the belief that children need both a mother and father (Pennings, 2011).

Multiple studies challenge these misconceptions, and researchers have found that adopted teenagers' life satisfaction was positively related to attachment with adoptive parents but unrelated to their parents' sexual orientation. The results of this study found no significant differences between families based on sexual orientation, for variables including adolescent attachment to parents and peers, parent satisfaction with the relationship to their adopted children, and adolescent life satisfaction. Essentially, these findings challenge the misperception that children of homosexual parents have worse psychosocial outcomes than children of heterosexual parents (Erich et al., 2009). The Reading for this chapter, *Growing Up With Same-Sex Parents*, explores the experiences of three adult kids who have been raised by same-sex parents. What do you think? Where do the stereotypes and prejudicial beliefs surrounding same-sex parenting originate? How are they challenged or reinforced within society?

Though same-sex marriage and/or families may still be illegal and/or not accepted in many countries, these families continue to thrive—but they do so as minority subcultures. Access to healthcare has become a guarded issue because of heterosexism, fear of discrimination, poor past healthcare experiences, and insensitivity. Many LGBT people are hesitant to disclose their sexual identity when looking for healthcare services because of the perceived risk of discrimination, and this can prevent them and their families from accessing quality healthcare (Chapman

& Shields, 2012). Even in Canada, where attitudes toward homosexuality are more liberal, discrimination still affects those who live alternative sexual lifestyles. According to a major report compiled by Health Canada, LGBT people expressed many complaints about accessing healthcare, including the attitudes of healthcare workers, ignorance of issues, and a lack of supportive providers. Hence, disclosure of sexual orientation was a major problem in consulting healthcare professionals or gaining access to treatment (Ryan, Brotman, & Rowe, 2001).

DIVORCE AND REMARRIAGE

Although married families continue to be the predominant family form in Canada, some changes are happening within those families. In 1961, married couples accounted for 91.6 percent of census families; by 2011, this proportion had declined to 67.0 percent (Statistics Canada, 2012d). The trend in divorce rates is partially to blame for the decline, but since its peak in 1987, divorce rates have come down significantly (Ambert, 2009). But at the same time, marriage rates have come down and cohabitation rates have increased, so it is difficult to conclude that couple dissolution has actually decreased (Ambert, 2009). Divorce and remarriage rates vary by age and gender, but approximately 70 percent of men and 58 percent of women who divorce will remarry (Ambert, 2009). As a result of divorce and remarriage, complex **blended families** are created.

BLENDED FAMILIES

In Canada, the 2011 Census counted stepfamilies for the first time. Statistics Canada separates this category into simple and complex stepfamilies. A **simple stepfamily** is a family with the children of only one of the married or common-law partners. **Complex stepfamilies** can be any of the following:

- Families with child(ren) of both parents and child(ren) of one parent only—these families represent 32.2 percent of all stepfamilies in Canada.
- Families with child(ren) of each parent and no child(ren) of both parents—these families represent 7.7 percent of all stepfamilies in Canada.
- Families with child(ren) of both parents and child(ren) of each parent—these families represent 1.6 percent of all stepfamilies in Canada (Vanier Institute of the Family, 2012b).

Of the 3 684 675 couples with children in Canada, 12.6 percent were stepfamilies (Milan & Bohnert, 2012). In 2011, 7.4 percent of couples with children were simple stepfamilies, in which all children were the biological or adopted children of one married spouse or common-law partner. An additional 5.2 percent of couples with children were complex stepfamilies, most of which were comprised of at least one child of both parents as well as at least one child of one parent only (Milan & Bohnert, 2012).

In many ways, blended families are living in untested waters. Firmly entrenched societal norms that apply to these families simply do not exist, and in some areas, neither do laws. In some cases, policymakers and the courts are just now wrestling with decisions that surround the rights and responsibilities of stepparents (Preece, 2004). Living in a blended family often brings many changes and provides plenty of opportunities for new relationships to develop. Adjusting to a new family brings forth attachment, cultural, and conflict issues, which families must work together to overcome, as expectations and responsibilities are communicated.

SINGLEHOOD

In 1981, 60.9 percent of the Canadian population aged 15 and over was married, while 39.1 percent was unmarried (Milan, 2013). Fast forward 30 years later to 2011 and the percentage of the Canadian population aged 15 years and over that is unmarried (53.6 percent) has surpassed the percentage of those who are married (46.4 percent) (Milan, 2013). Although these statistics include those who have divorced, separated, and been widowed, along with those who have never married, there seems to be a demographic shift toward postponing marriage, choosing cohabitation as an alternative to marriage, or remaining single by choice. According to one study, about 1 in 10 single people in their 30s and early 40s doubt they will ever get married (Crompton, 2005). Of those, 54 percent believe that being in a relationship is still important, while

Blended families: family consisted of a couple and their children from previous relationships with or without children from current relationship.

Simple stepfamilies: Families with the children of one, and only one, of the married or common-law partners.

Complex stepfamilies: Families with children of both parents and children of one parent only, families with children of each parent and no children of both parents, or families with children of both parents and children of each parent.

46 percent stated that being a part of a couple is not very or not at all important (Crompton, 2005).

Canadians who remain single by choice represent a small but distinct group of individuals who opt for a life without a partner and/or children for a variety of reasons. But despite the demographic shift toward singlehood, the stigma associated with being single still exists. The misconception that in order to feel happy and fulfilled, one must seek a romantic relationship and have children is still prevalent in Canadian society. Some single people who truly enjoy their lives are hesitant to say so in fear of being perceived as if something is wrong with them (Pittaway, 2011). The assumption that if you are single, you must be lonely, is widespread, and it is not the only anti-single stereotype either. Bella DePaulo, author of *Single With Attitude: Not Your Typical Take on Health and Happiness, Love and Money, Marriage and Friendship*, argues that those who choose to live alone or not have children are often unfairly perceived as selfish or self-absorbed, lonely, and/or immature (Pittaway, 2011). If you are a single woman, DePaulo says, you might be stereotyped as a promiscuous party girl who cannot get a date; and if you are a single man, you could be mislabelled as a "slob living in your parents' basement, a closeted homosexual or a potential child predator" (Pittaway, 2011). Not surprisingly, the inherent assumption that one should always want or be seeking a romantic relationship leads those who choose singlehood to alter the truth when questioned about their relationship status. Many respond with something more socially acceptable, like "I'm not really looking right now" or "I just haven't met the right person" in order to avoid the stigma.

PARENTHOOD AND FAMILY

The traditional nuclear family that was the norm in Canadian society 50 years ago looks very different today. Families with children come in all shapes and sizes—in addition to the nuclear structure, we see family ties in the form of lone parent families, foster and adoptive families, families created through the use of assisted human reproduction, co-parenting families, and childfree families.

In 2001, the percentage of households comprised of couples with children exceeded the percentage of households comprised of couples without children (Statistics Canada, 2012b). In 2011, we see a reversal, with households comprised of couples without children surpassing the number of households comprised of couples with children (Statistics Canada, 2012b). According to the 2011 Canadian census, the largest group of couples with children were married couples (31.9 percent), a 5.5 percent decrease from 2001 when married couples with children made up 37.4 percent of census families; while married couples without children increased from 33.1 percent in 2001 to 35.1 percent in 2011 (Statistics Canada, 2012b). Common-law couples with children also increased over this decade from 6.2 percent to 7.3 percent, and so did common-law couples without children, from 7.6 percent in 2001 to 9.4 percent in 2011 (Statistics Canada, 2012b).

Whether married or living common law, the composition of and dynamics within the traditional Canadian family with children are also changing.

According to the 2011 census, 42.3 percent of the 4 318 400 young adults aged 20 to 29 lived in their parental home, either because they never left it or because they returned home after living elsewhere (Statistics Canada, 2012e). Unlike the once common **empty nest** phenomenon, the **cluttered or crowded nest** is now the new reality for many parents of adult children. Often referred to as the **boomerang generation**, many of these young adults are staying at home or returning home because of changes in relationship status, cultural preferences, difficulties finding or securing long-term employment, and the high costs associated with education (Bartlett & LeRose, 2012). No matter the reason and despite the increase in prevalence, there still seems to be a stigma attached to adult children living with their parents, but there is a bit of a double standard as well, in that it is more socially acceptable for women to return home than for men to do so.

Also on the rise are **multigenerational households** where grandparents, parents, and children are living together under one roof. Increased mobility, along with advances in home care, public transportation, and assistive technologies have afforded the Canadian senior population more choice in where and how they live, and many are choosing to live with their children and grandchildren (Battams, 2016). With over one million Canadians or 4.5 percent of the population living in multigenerational households, seniors are able to receive care from younger generations and also provide care to younger generations (Battams, 2016).

Empty nest: A household consisting of a parent or parents whose children have grown and moved out of the house.

Cluttered or crowded nest: A household consisting of a parent or parents whose children have either decided to stay living at home past the customary age for leaving home, or who have returned to live at home after leaving for a brief period of time.

Boomerang generation: Term used to describe those young adults who return to their parental home after leaving for education, work, or a relationship.

Multigenerational households: A household where grandparents, parents, and children live together under one roof.

This can have a significant impact on the health and well-being of the entire family—social isolation that is often common among seniors is challenged through familial interaction, and grandparents can often take care of grandchildren while their parents are at work, limiting the cost of daycare. The middle generation, sometimes referred to as the **sandwich generation**, can struggle to balance caring for both their parents and their children when they reside in separate households, so living together can eliminate or reduce those stressors (Battams, 2016).

Similar to multigenerational households, **skip-generation** households, where a child lives with their grandparent(s) without a parent present, have also increased. In 2011, 30 005 children or 0.5 percent of children aged 14 years and under lived in skip-generation families (Statistics Canada, 2012b). The majority or 57.8 percent of children living in skip-generation households live with a grandparent couple, while the remaining 42.2 percent live with one grandparent only (Statistics Canada, 2012b). In Canada, most skip-generation households were found in Nunavut (2.2 percent), in the Northwest Territories (1.8 percent), and in Saskatchewan (1.4 percent) (Statistics Canada, 2012b).

LONE-PARENT FAMILIES

Lone-parent families are not a new phenomenon in Canada, but the circumstances surrounding their existence have changed significantly over the years. In 1961, the majority or 65.1 percent of lone parents were widowed, 35.8 percent were divorced or separated, and 2.7 percent reported never having been married (Statistics Canada, 2012d). By 2011, the most common status for lone parents was divorced or separated at 50.8 percent, with 30.5 percent reporting never having been married, and 17.7 percent of all lone parents being widowed (Statistics Canada, 2012d). (See Figure 13.2.)

The makeup of lone-parent families has also changed. In earlier years, when a couple separated, custody of the children often went to the primary caregiver, which, at the time, was usually the mother. Recently, we have seen an increase in the number of lone-parent

Sandwich generation: The middle generation in a multigenerational household, which often struggles balancing caring for aging parents while caring for children.

Skip-generation: A household where a child or children live with their grandparents without a parent present.

Distribution (in percentage) of the Legal Marital Status of Lone Parents, Canada, 1961–2011

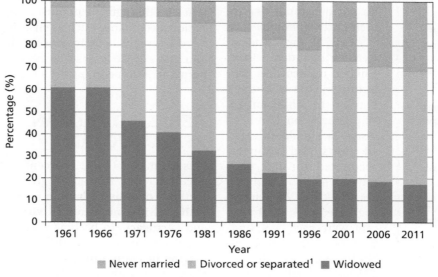

Legend: ■ Never married ■ Divorced or separated[1] ■ Widowed

Source: Statistics Canada. (2015). Distribution (in percentage) of the legal marital status of lone parents, Canada 1961–2011. Retrieved from Statistics Canada: https://www12.statcan.gc.ca/census-recensement/2011/as-sa/98-312-x/2011003/fig/fig3_1-2-eng.cfm. Date accessed: September 11, 2016. Reproduced and distributed on an "as is" basis with the permission of Statistics Canada.

families headed by males. Of the 16.3 percent of lone families in Canada in 2011, 12.8 percent were female lone-parent families, while male lone-parent families represented 3.5 percent of all census families (Milan & Bohnert, 2012). The increase in male lone-parent families was more than twice the increase in female lone-parent families between 2006 and 2011.

FOSTERING AND ADOPTION

A number of families in Canada choose to seek parenthood through fostering or adoption. **Foster care** is defined in the *Child and Family Services Act* as the placement of a child or young person in the home of someone who receives compensation for caring for the child but is not the child's parent. It requires a foster family to take a child into their home and provide them with necessities of life as well as emotional support through a confusing and difficult situation. The foster care system falls within the provincial/territorial jurisdiction and processes vary from province to province. Children end up in the foster care system when, for a variety of reasons, they do not or cannot live in their parental home.

In 2011, the Canadian Census included foster children in their data collection for the first time. Similar to children in skip-generation families, foster care children aged 14 years and younger accounted for 0.5 percent of children in private households—29.0 percent were aged 0–4, 29.9 percent were aged 5–9, and 41.1 percent were aged 10–14 (Statistics Canada, 2012b). A small majority or 52.5 percent of foster care children aged 14 and under in 2011 were boys, and 47.5 percent were girls (Statistics Canada, 2012b). Children may need foster care for a short period of time, for months, or even years. The preferred plan is to unite children or youth with their families, but in cases where that is not possible, more permanent alternatives like legal custody, adoption, and independent living are explored.

For individuals or families in Canada who wish to adopt, the process can be a long one. They can choose to adopt either publicly or privately, domestically or internationally. If the applicants choose a public domestic adoption, a government agency arranges and finances the adoption. The applicants would contact the agency in their province that is responsible for adoptions (Children's Aid or another licensed facility) and register to begin the adoption process. For applicants who wish to adopt privately, the rules differ from province to province, as does the meaning of "private adoption." Some applicants choose the option of "fostering to adopt," meaning they would foster a child for a period of time first, and if the child and foster parent(s) both agreed and circumstances allowed, the adoption would proceed.

International adoptions follow much the same process, in that they can be private or public, but they are subject to a much more rigorous course of action, since there are government regulations of two countries involved. Each country that allows international adoption has its own laws and regulations, which determine eligibility requirements and the adoption process (Focus on the Family, 2011). Canadian laws state that the adoption laws of the sending country have to be followed before the child is allowed to enter Canada (Paul-Carlson, 2012). The Hague Convention of 1993 was pivotal in international adoption—countries that sign this agreement must comply with strict international standards, such as ensuring that birth parents have truly given their consent, have not been paid, and that every effort has been made to find the child a permanent home in his or her country of origin.

ASSISTED HUMAN REPRODUCTION

It is estimated that 20 percent of Canadians experience infertility to some degree, and many are using methods of **assisted human reproduction (AHR)**, like **artificial insemination**, **in vitro fertilization (IVF)**, and **surrogacy**, to become parents (Haaf, 2012). The most common form of artificial insemination is **intrauterine insemination (IUI)**, which involves washing the sperm and then inserting it directly into the uterus at the time of ovulation. With in vitro fertilization, the woman's egg is fertilized with the man's sperm outside of the body and then placed back inside the woman's

Foster care: The placement of a child or young person in a home of someone who receives compensation for caring for the child, but is not the child's parent.

Assisted human reproduction (AHR): The use of medical techniques to assist in the conception and birth of a child, including artificial insemination, intrauterine insemination, in vitro fertilization, egg, sperm and embryo donation, and drug therapy.

Artificial insemination: The injection of sperm into the vagina or uterus other than by sexual intercourse.

In vitro fertilization (IVF): A medical procedure where sperm fertilizes an egg outside of the body and then is implanted into the uterus.

Surrogacy: When a woman becomes pregnant and gives birth to a baby in order to give it to another person or couple.

Intrauterine insemination (IUI): A medical procedure where sperm is washed and then inserted directly into the uterus at the time of ovulation.

Traditional surrogacy: Occurs when the surrogate's egg is fertilized by the father's sperm, making the child biologically related to the father and the surrogate.

AGENT OF CHANGE

Lucas Medina

Lucas Medina, left, executive director, and Chad Craig, operations director of Five/Fourteen, Canada's first foster care agency for LGBTQ youth.

As a former Crown ward, who endured an emotionally abusive foster home because of his sexual orientation, Lucas Medina has first-hand experience with a foster care system that is lacking the ability to meet the needs of its LGBTQ community (Caton, 2016). After living with the same foster family for eight years, Medina came home to a note on his bedroom door informing him that because he "had chosen a lifestyle that was directly in conflict with [his foster parents'] beliefs," he was disrespectful and could no longer refer to his parents as "Mom" and "Dad"—he was to call them by their first names and should be grateful to them for even allowing him to remain in their home as a tenant until he aged out of the system (CBC Radio, 2016). He spent as little time as he could at home and moved out at the age of 17 years old. In an interview with CBC, Medina talks about how isolating that was for him and how, as an LGBTQ youth in foster care, he never felt like he was a part of a community (CBC Radio, 2016). As an

adult, when Medina heard the stories of other LGBTQ foster kids in the *Youth Leaving Care Report*, released in May, 2012, he realized that little had changed and that the system was still lacking support for LGBTQ youth in care (CBC Radio, 2016). These stories ignited a feeling of responsibility in him to do something to improve the system. He went back to school for Public Administration and Governance at Ryerson University and focused his research on Child Welfare in Ontario (CBC Radio, 2016). Since then, through work with Dr. Shelley Craig, Associate Dean of the Faculty of Social Work at the University of Toronto, Medina and his partner, Chad Craig, opened Five/Fourteen, Canada's first and only foster care agency for LGBTQ youth (CBC Radio, 2016). Medina is the Executive Director and Craig is the Operations Director. Located in Windsor, Ontario, Five/Fourteen received licensing to offer 60 beds to LGBTQ youth from the Ontario Ministry of Children and Youth Services (CBC Radio, 2016). The name Five/Fourteen is significant because May 14th is "Children and Youth in Care Day" in Ontario, and May 14, 1969 was the day homosexual acts were decriminalized in Canada (Caton, 2016; CBC Radio, 2016). May 14th also holds special significance for Lucas Medina as it is his birthday: a day that was once a dreaded countdown for him while in care because it reminded him of how much closer he was to aging out of the system and losing those supports became a date of celebration, representing positive social change (CBC Radio, 2016). Although Five/Fourteen is based in Windsor, with offices in Toronto and Ottawa, the agency can bring in youth from any of the 47 Children's Aid Societies across Ontario (CBC Radio, 2016). For more details, visit the webpage at http://fivefourteen.ca/, or to listen to Lucas Medina's interview with CBC, visit http://www.cbc.ca/radio/candy/why-a-former-foster-kid-opened-a-foster-care-agency-for-gay-teens-1.3677299?autoplay=true.

Caton, M. "Canada's First Foster Care Agency for LGBTQ Youth Now Licensed to Open," *The Windsor Star* 16 Aug 2016; CBC interview: http://www.cbc.ca/radio/candy/why-a-former-foster-kid-opened-a-foster-care-agency-for-gay-teens-1.3677299?autoplay=true.

womb, where it grows and naturally develops into a baby. Successful pregnancies with IUI or IVF can take more than one or even several attempts or "cycles," and can be very costly. Although the cost varies from clinic to clinic, IVF can cost as much as $30 000 for three cycles of treatment (Fidelman, 2010). Across Canada, the portion of costs covered by the provincial healthcare plan differs by jurisdiction.

Another alternative available to those experiencing problems with fertility is surrogacy. With **traditional surrogacy**, the surrogate uses her own egg and is inseminated with the father's sperm, making the child biologically related to both the surrogate and the father. With **gestational surrogacy**, the sperm and the egg of the couple experiencing issues with fertility

Gestational surrogacy:
Occurs when the sperm and the egg from the couple seeking surrogacy is fertilized through in vitro fertilization and implanted into the surrogate, making the child biologically related to both parents but not the surrogate.

is fertilized through in vitro fertilization and implanted in the surrogate, making the child biologically related to both parents, but not biologically related to the surrogate.

Assisted Human Reproduction is challenging traditional concepts of parenthood and family by providing alternative paths to parenthood for those who need it. Some assisted reproductive technologies have raised complex issues with moral, legal, and social implications. As of October 2012, Health Canada is the federal authority responsible for developing policy and regulations around the **Assisted Human Reproduction Act**. The purpose of the Act, passed in 2004, was to regulate assisted human reproduction and related research—it outlines which activities related to AHR are prohibited in Canada, while identifying controlled activities that require a license and adherence to certain regulations.

CO-PARENTING

Another type of parenting that is challenging traditional concepts of family is **co-parenting**. Co-parenting involves two people having a child without being in a romantic relationship—the child is conceived through insemination or in vitro fertilization and then raised by two individuals who live in separate households through shared custody (Hunter, 2013). Parents who are co-parenting share equally in the financial, social, and emotional responsibilities in raising a child, but do so without ever having been in a relationship. Co-parenting is not a new concept for parents who have separated or divorced, but now, potential parenting couples are bypassing the marriage portion altogether (Hunter, 2013). Websites like Modamily.com, Coparents.com, MyAlternativeFamily. com, and CoParentMatch.com are providing platforms for individuals to learn more about creating a family in this non-traditional way (Hunter, 2013).

Author and Toronto-based child and family therapist, Jennifer Colari, recognizes that co-parenting is outside the traditional norm, but says that it is "not particularly different from a divorced couple who raise a child together after romance has been removed from their relationship" (Hunter, 2013). She argues that children are extremely adaptable and it really does not matter what shape or form a family takes, "as long as [children] are loved and as long as that love is predictable and consistent, then they'll be okay" (Hunter, 2013).

Assisted Human Reproduction Act: A law enacted by the Parliament of Canada to regulate assisted human reproduction and related research.

Co-parenting: Involves two people having a child without being in a romantic relationship. The child is usually conceived through assisted human reproduction and raised by both individuals who share custody but live in separate households.

CHILDFREE COUPLES

For the first time in 2006, there were more households comprised of couples without children (29.0 percent) than households comprised of couples with children (28.5 percent) (Statistics Canada, 2012d). That gap has continued to grow and, according to the 2011 Canadian census, 29.5 percent of households are comprised of couples without children, and 26.5 percent of households are comprised of couples with children (Statistics Canada, 2012d). Many of these couples have made a choice not to have children and perceive themselves as "childfree" rather than "childless." They resist the notion that having children is the automatic next step in a committed relationship and believe that they can lead happy and fulfilled lives without having kids.

Magenta Baribeau, director of the documentary film "Maman? Non, Merci! (No Kids For Me, Thanks!), has known for a long time that she does not want children, but says people keep telling her that she will change her mind one day (CBC News, 2015). In her award-winning documentary about people who are happily childless, Baribeau explores societal pressures to have children and the stereotypes and prejudicial beliefs about those who choose not to. She argues that it is condescending to assume that someone who does not want children will eventually change their mind and says she is tired of being called selfish because of her decision to be childfree (CBC News, 2015). Baribeau says she spent six years making the documentary, and over those six years has seen very little change with respect to stigma around those who make the choice not to have children. It was this that inspired her to organize a Child-Free Day in June 2015 in Montreal, to help promote respect for people's decisions to remain childless (CBC News, 2015). NoKidding! (NoKidding.net) is also promoting respect for adult couples and singles who do not have children. Founded by Jerry Steinberg in Vancouver, British Columbia, the international social club is a non-profit organization run by volunteers, with over 40 active chapters in Canada, New Zealand, and the United States (Vogels, 2010). NoKidding! organizes social activities and hosts an annual conference to help unite childfree and childless adults with other like-minded individuals, providing a place for people to share interests and experiences not involving children.

FAMILY VIOLENCE IN CANADA

The family is supposed to provide safety and security, but far too often it can be the source of violence. Family violence refers to abuse that takes place between family members, whether they are family through blood, marriage, common-law partnership, foster care, or adoption, and can come in the form of spousal abuse, child abuse, and elder abuse (Statistics Canada, 2016). As we discussed elder abuse in detail in Chapter 12, we will focus on spousal abuse and child abuse in the following sections.

Spousal Abuse

Spousal abuse, also referred to as domestic violence or intimate-partner violence, refers to any intentional act or series of acts by one or both partners in an intimate relationship that causes injury to either person (Statistics Canada, 2016). In 2014, 4 percent of Canadians reported having been physically or sexually abused by their spouse, common-law partner, or former spouse during the previous five years (Statistics Canada, 2016). Of those 4 percent, equal proportions of men and women reported being victims of spousal violence (Statistics Canada, 2016). According to the 2014 General Social Survey (GSS) on victimization, the most common form of spousal violence was being pushed, grabbed, shoved, or slapped, with 40 percent of women reporting victimization, compared with 31 percent of males (Statistics Canada, 2016). The GSS indicated that approximately 34 percent of women and 16 percent of men reported being sexually assaulted, beaten, choked, or threatened with a gun or a knife, and 35 percent of men, compared

to less than 10 percent of women, reported having been kicked, bit, hit, or hit with something (Statistics Canada, 2016). Further findings of the GSS show that 15 percent of men and 13 percent of women reported being emotionally or financially abused by a spouse, former spouse, or common-law partner at some point during their lifetime (Statistics Canada, 2016). The GSS revealed that women reported the most severe types of spousal violence more often than men, and they were more likely than men to have reported physical injuries (Statistics Canada, 2016). Findings from the GSS indicate that for the majority of spousal violence victims (70 percent), police were never made aware of the incidences of abuse (Statistics Canada, 2016).

If we look at police-reported incidences of family violence, regardless of age, females were twice as likely to be victims of family violence and more likely to be victimized by a spouse (31 percent), whereas males were more likely to be victimized by a parent (24 percent) or an extended family member (18 percent) (Statistics Canada, 2016).

Child Abuse

When it comes to family violence involving children and youth, 31 percent of child and youth victims of family violence were victimized by a parent, sibling, extended family member, or spouse, with the majority of victims

PICTURE THIS...

tomazl/Getty Images

While many people are aware of the devastating consequences of physical and sexual abuse of children and youth, verbal abuse is often overlooked. For children and youth, what are the emotional and psychological consequences of verbal abuse?

Since its creation in 1988, the Family Violence Initiative (FVI) in Canada seeks to address family violence and its impact on Canadian society (Statistics Canada, 2016). The FVI involves the collaboration of 15 different federal departments to develop, implement, test, and assess models, strategies, and tools that improve the justice system's response to family violence. Additionally, the FVI funds various projects and initiatives that seek to raise awareness and provide support to victims of family violence.

being victimized by parents (61 percent) (Statistics Canada, 2016). Female children under the age of 18 were nearly two times more likely than their male counterparts to be victims of police-reported family violence (Statistics Canada, 2016). At a rate of 134.4 incidences per 100 000 people under the age of 18 years, physical assault was the most common type of police-reported family violence against children and youth, followed by sexual offences at a rate of 73.9 per 100 000 children under the age of 18 years (Statistics Canada, 2016). While instances of physical assault against children and youth perpetrated by a family member were similar for males and females, cases of sexual assault against female children and youth (121.8 per 100 000) were at a rate four times higher than that of their male counterparts (28.5 per 100 000) (Statistics Canada, 2016). From 2004 to 2014, there were 316 family-related homicides in Canada (Statistics Canada, 2016). Among them, the most frequent causes of death were beating (25 percent), strangulations, suffocation or drowning (25 percent), and stabbing (17 percent), with 51 percent of homicide victims under the age of 4 years (Statistics Canada, 2016). The most common motive for family-related homicides against children and youth was frustration, anger, or despair (62 percent) (Statistics Canada, 2016).

INTERSECTIONALITY

Race, Ethnicity, and the Family

According to the 2011 National Household Survey, 360, 045 couples, or 4.6 percent of all married or common-law couples in Canada, are mixed union (Statistics Canada, 2014). Many culturally diverse families have experienced and continue to experience prejudice, racial discrimination, and oppression, both in their countries of origin and in Canada. In a precedent-setting case in Nova Scotia in 2011, two teens received jail time after erecting a six-metre-high cross with a noose on it, and lighting it on fire, in the front yard of an interracial couple (Canadian Press, 2011). Acts like this one serve to remind us that even though the number of mixed unions continues to grow in Canada, some still face prejudice and discrimination. The investigation of mixed unions is important: they reflect another facet of the changing backdrop of Canada's familial landscape. In a situation like the one in Nova Scotia, for example, people will talk about it for years. Will this child's couple be bullied? Will he or she always be known as the "kid who had the cross burned in the front yard"? Or will feelings change, and if so, when?

Gender and the Family

Historically, men have been the head of the household, but as society changes, so do the roles assigned to men and women. Today's father can be single or divorced, gay, straight, adoptive, foster, a stepfather, or a co-parent. He may be a non-residential father, or he may be a stay-at-home dad. In short, he may not resemble the traditional notion of the head of the household at all, but he can still remain emotionally and financially invested in his children, while maintaining a stable, healthy, and loving relationship with them (American Psychological Association, 2012).

Women's roles have also been changing within the family. Mothers are still mothers, but a shift in societal attitudes has found women marrying later, having fewer children, obtaining an education, and entering the workforce—all of which have profound effects on their roles as the traditional primary caregiver (Kohen, 1981).

Socio-Economic Status and the Family

Socio-economic status can impact a family's financial security and can influence access to opportunities. This chapter's In Their Shoes section explores the intersectionality of social class and family.

IN THEIR SHOES

If a picture can say a thousand words, imagine the stories your shoes could tell! Try this student story on for size—have you walked in this student's shoes?

I come from a lower-middle-class family. My father is a pastor and does not receive a salary worthy of his educational calibre. My mother attended college for one year, but quit to get married and follow her husband to graduate school. Since then, my mother has worked hard in various part-time positions at minimum wage to help make ends meet.

I know that my parents live paycheque to paycheque and often utilize their savings when they come up short at month's end. As a result, for my family, going out for dinner, including fast food, was a privilege. When my family did go out for dinner, we went to an

(Continued)

average-priced restaurant (Swiss Chalet) and we were expected to order lower-priced meals. We all understood that steak was off-limits and that water was to be ordered. Appetizers were not even on our radar and dessert was a rarity. My siblings and I also knew never to ask our parents for money for anything "extra" such as movie tickets or money for shopping. In fact, we never begged our parents for anything. We understood at a young age that our parents were making daily sacrifices for us and we were always grateful.

Although I was grateful to my parents, somewhere at mid-elementary school I began to take special note of other people's privileges and my wants. Other kids at school were more privileged than me. They had better and more cutting-edge toys and went on more luxurious vacations. There were the families who vacationed in Florida every year, and even worse were the families who went on cruises in Bermuda. My family's idea of a vacation was camping and we went every summer. I remember longing to visit Florida and being jealous of the kids with better clothes, better snacks in their lunches, and better Christmas presents. On some conscious level, I was aware that these kids were higher up on the social strata than me. However, it was not until I began to date my current husband that I realized the depth of differences between the lower middle class and upper middle class.

My husband comes from an upper-middle-class family. My father-in-law is a lawyer and owns his practice, and my mother-in-law is a school teacher. Together they must make a pretty hefty sum of money. My in-laws do not worry about money. They travel to different parts of the world for spring break and eat at fancy restaurants. In fact, they eat out frequently. Not only that, they order *whatever* they want, including steak, appetizers, and alcoholic beverages. With my in-laws, ordering appetizers and alcoholic beverages is normal and encouraged. Eating out with my in-laws is still a culture shock for me.

My parents and in-laws make very different lifestyle choices, particularly in relation to eating, drinking, and spending habits. I believe these differences are directly correlated with socio-economic status. My husband's family enjoys a range of foods of which my family has never heard, including calamari, quinoa, bulgur, couscous, rosé sauce, capers, and shallots. They also enjoy trying new recipes and types of foods. My parents eat a standard menu of meat and potatoes

and do not make much use of recipes. My in-laws eat their spaghetti with a spoon and a fork and twirl it around their fork rather than cut it with a fork and knife. I was not even aware that this alternative way of eating spaghetti was a possibility. My in-laws also have a different attitude toward alcohol in comparison with my parents. My in-laws enjoy drinking and drink significantly more than my parents. I believe that this is partly because it is a value of the upper class to drink socially, also because they can afford to drink wine at their meals and casually throughout the day. My family rarely drinks. For my family, alcohol is too expensive and is not high on their priority list of needs.

Lastly, my parents have a very different view of money in comparison to my in-laws. My family frowns upon frivolous and excessive spending. My parents taught me to be thrifty by weighing my needs and wants. As a result, I feel guilty for spending money in general. (How do you really define a *need* anyway?) In my family, one has to work to play and work very hard to deserve it, but in my husband's family, spending money is a pastime. My husband, as a result of his upbringing, enjoys "treating" me and having "nice things." I enjoy having nice things but I struggle with a guilty twinge. Since meeting my husband's parents, I have been exposed to a new world and have become aware of the differences between the lower-middle and upper-middle classes.

The lower-middle and upper-middle classes each have a different subculture and I have found these differences difficult to adjust to as a newlywed. I have often felt inadequate and out of place among my new family. I have felt jealous of the many privileges that my husband received as a child, such as his family vacations. My husband's family is the "Other" to me and I have struggled with "othering" them. It is easier for me to judge his family just because they are different, but that is not how I want to be.

After many years of struggle accepting the differences between my family of origin and my new family, I have learned to appreciate the best in both approaches. My husband and I have assimilated what we like from his family and mine, and left behind what we do not. I have learned from both sets of parents. My lower-class upbringing taught me to be deeply grateful for all I have been given, also to be wise in my spending. My husband's upbringing has taught me to *enjoy* what I have been given. I am thankful for both life lessons in my journey.

Reprinted with permission of Sarah Anne Tuckey, B.A. (Hons) Clinical Psychology.

ENDING THOUGHTS

What kind of family is the best kind of family? Or, is there a best kind of family? It seems that the family form has diversified over the years, indicating a shift away from the traditional picture of the nuclear family, characteristically presented as a mom and a dad, two children, and a house with a white picket fence. And while this diversification of family type has led to some social policy reform (same-sex marriage, for example) and a reexamination and redistribution of gender roles within the family units themselves, there are those that see the "dilution" of the traditional family form as threatening to the good of society as a whole. The Institute of Marriage and the Family argues that, "a declining marriage culture is the wrong trend line for Canada ... all research points to the fact that married-parent families offer more stability for children and decreased poverty rates" (Mrozek, 2012). Commenting on the results of a newly released report, author Joel Kotkin sees a worrying global trend in the decline of the traditional family form:

Today, in the high-income world and even in some developing countries, we are witnessing a shift to a new social model. Increasingly, family no longer serves as the central organizing feature of society. An unprecedented number of individuals—approaching upwards of 30% in some Asian countries—are choosing to eschew child bearing altogether and, often, marriage as well. (Kotkin, 2012, p. 1)

In Kotkin's view, this trend can lead to a diminished workforce and to fewer workers who will be able to contribute financially to those members of the population who will need funds in their retirement years. Additionally, he sees the traditional role of the family—something that we have built our society around for centuries—changing, as more and more people choose to be single and childfree.

In the end, does it really matter how you define a family? Whether it includes your best friend or your neighbour, your family consists of those people who have made a commitment to care and provide for each other, for an extended period of time. It is not about what a family looks like, but more about what a family does that makes a family. The bottom line is that your own family is the best kind of family—however you choose to define it.

GROWING UP WITH SAME-SEX PARENTS

By Andrea Gordon

Those who make history don't always do so by choice.

Over the last quarter century, many same-sex couples in Canada chose to, as they fought for equal rights to marry and raise families. In turn, their sons and daughters also broke new ground as the first wave of children born to gay and lesbian couples.

Many of those kids are now in their 20s, and behind them, their ranks are growing, especially in urban centres like Toronto. They're still a small minority, but today's toddlers and preschoolers being raised by two moms or two dads are much more likely to see families like theirs in playgrounds and classrooms.

"Something warms my heart about them all running around," says Sadie Epstein-Fine, 21, who grew up in Toronto with two moms and a large supportive network of family and close friends. "There really is safety in numbers."

At the same time, she says, it's important for people to recognize that kids like her have their own distinct identities, "because so often we're attached to (the identities of) our parents."

According to the 2011 Census, there were 64 575 same-sex married and common-law couples in Canada that year, up 42 percent from 2006 and double the number in 2001 when Statistics Canada started tracking them. Those couples had 6410 children living at home. Same-sex families with kids led by women outnumbered male couples by more than four to one.

In Toronto, prenatal and parenting classes and other events offered through the LGBTQ Parenting Connection are thriving. Camp Ten Oaks runs a one-week overnight camp every summer for children from alternative families, and a similar program called Project Acorn for youth 16 to 24.

Research has consistently shown that children raised by lesbian and gay couples fare just as well as kids with heterosexual parents. Groups like the American Academy of Pediatrics and the American Psychological Association support same-sex families, along with adoption and reproductive rights.

For a long time, it's been primarily parents in the public eye, with or without the kids at their side. Now those grown children are able to speak for themselves.

Zach Wahls of Iowa was among the first to gain public attention. Two years ago at age 19, he made a presentation to Iowa state legislators in support of same-sex marriage that instantly went viral. It garnered him a spot on national talk shows and led to his 2012 book, *My Two Moms*.

His remarks concluded, "the sexual orientation of my parents has had zero impact on the content of my character."

Everyone's experience is unique.

Robbie Barnett-Kemper

"Traditional" might not be the first word that comes to mind when describing a family headed by two moms. But Robbie Barnett-Kemper says his childhood was shaped by plenty of traditions.

For starters, there was the family meal most nights, a ritual that by most accounts is falling by the wayside in many modern households.

"One of the greatest gifts my parents gave me was the conversation at the dinner table," says Robbie, 21, who lives in east-end Toronto.

Kids being kids, you can be sure there was also the occasional whine about having to be dragged off the computer, or what was on the menu. But Robbie and his older sister Hannah could count on lively discussion with their moms—Alison

Kemper, who now teaches entrepreneurship at Ryerson University, and Joyce Barnett, a psychotherapist. Dinner topics included politics, Biblical interpretation, what happened at school, and which day the cleaning lady was coming.

The family attended a downtown Anglican church. Barnett is also a priest and Kemper is a deacon. Robbie went to high school at Royal St. George's College, a private boys' school, headed to camp each summer, and visited his grandparents' Muskoka cottage. He also inherited his late grandfather's love of bow ties.

"We're the standard lesbian-led family," says Robbie, who's going into his fourth year of a five-year program combining education and computer science at Queen's University and working as a web designer in Toronto this summer.

His moms were always after him to clean up his room. But those dinner table conversations taught him to express his opinion from an early age—especially on matters of injustice or intolerance.

His mothers were among the first lesbian couples married after Ontario gave same-sex marriage the go-ahead in 2003. The following year, when it was being challenged in the Supreme Court of Canada, 12-year-old Robbie wrote an affidavit that was read aloud as the panel considered whether to uphold his parents' right to marry.

"Now other kids can't say that I don't have a real family," he wrote. Not long after, Robbie sat at the front of his classroom and fielded questions about it from the other kids.

A 2011 survey of high school students by Egale Canada found that most heard anti-gay comments on a daily basis, and that many kids with queer parents felt unsafe. Robbie doesn't recall being teased or ostracized. He always knew his situation was unusual, but there were other kids with non-traditional and blended families. He does remember that hearing casual derogatory cracks like "that's so gay" would provoke him to talk back and bring up the fact that he was being raised by two moms.

There isn't one question he hasn't been asked, he says, as if to issue a challenge. People may tiptoe around it, but they're usually curious about how exactly his family came to be. The short answer is: two women in love and one sperm donor.

Barnett, also known as "Mummy," gave birth to Hannah in 1986. Six years later, Kemper (Mama) delivered Robbie. He knows his biological father. He gave Robbie his first guitar. They have lunch every couple of years.

"There was nothing hard about having two moms," says Robbie. He did have to learn to shave from another kid at camp. But he had male mentors in his life, including his godfather (who taught him to burp).

Nonetheless, having mothers who are great communicators didn't make the sex talk any easier. "It was awful for me as it is for every other kid. Because let's face it, they're still your parents."

He's straight, but didn't feel pressures or expectations when it came to his sexuality and saw it as a fluid thing.

"Just being able to be myself was the best thing."

Robbie describes his parents as a source of inspiration. "Being in a house with two mothers who are involved in their occupations and so passionate, it really sets you up as a child."

They taught him to stand up for what he believes, and told him they loved him every day. How can any parent do better than that, he says. It's what he'll remember when he's a dad.

Sadie Epstein-Fine

Sadie Epstein-Fine loved the term "queer spawn" from the moment she first heard it years ago.

She knows others may balk at the phrase, but "I thought, yeah, that's exactly what I am," she says, her face lighting up at the memory.

The 21-year-old theatre student at York University believes in the power of shared experience. And like her mothers, LGBTQ Parenting Network co-ordinator Rachel Epstein and playwright Lois Fine, she is active in the community. Sadie is a counsellor for teens at Camp Ten Oaks each summer. She regularly speaks about being the child of same-sex parents at workshops for couples planning families or already raising children. She's volunteered for political canvassing and spoken in high schools about homophobia.

To Sadie, the term "queer spawn," coined a decade ago, playfully captures the unique situation for kids like her who grow up "living in two worlds"—the gay community that surrounds their parents and the other one inhabited by most of their peers. And it's an identity that needs to be acknowledged, she says.

Rachel Epstein gave birth to Sadie in 1992 at home, surrounded by a dozen women, including her partner Fine. She became pregnant through a fertility clinic following insemination from an anonymous sperm donor.

"I grew up with a sense that family isn't just nuclear," says Sadie, who was constantly in the company of her mothers' closest friends and their children. She has an older brother from Fine's previous relationship. He is married with a child of his own.

Epstein and Fine split up when Sadie was 10. Fine moved across the street in their Wychwood neighbourhood of Toronto, and Sadie went back and forth between houses. She now lives with Epstein and their cat Wolverine.

As part of the widely watched first wave of offspring of gay parents, Sadie says she felt a subtle pressure to "turn out okay." It didn't come from her mothers. But she feared that talking about the tough times that are inevitable in every family might somehow reflect on them.

She knows rebellion is part of adolescence, and so are struggle, distress, and being driven crazy by your parents, but as a teen she was reluctant to share it.

Instead she felt it important to always talk about her family "in a way that let's people know I'm all right."

It's something her mother Rachel Epstein acknowledges in a collection of essays on queer parenting that she edited in 2009 called *Who's Your Daddy?* In her introduction, she notes the pressure kids too often feel to present themselves as "poster families" or act as "the ambassadors" for same-sex parents.

At 16, Sadie came out herself. "I always knew it was an option. I always knew I wouldn't get kicked out or be a shame to my family. I knew you could be happy and be gay."

Still, it wasn't easy and she didn't tell anyone for a year.

"I don't think my parents turned me gay," she says. "I think my parents gave me options and led me to who I am."

During talks with same-sex parenting groups, couples often ask Sadie for her biggest piece of advice.

She tells them to remember it's normal for kids to make mistakes and get into trouble; that's how they learn. Too many are so intent on raising perfect children that they forget family life is a rollercoaster, whether the parents are gay or straight. Don't worry so much, she says.

Zak Higgins

It's been a while since Zak Higgins was in the spotlight.

As the son of a lesbian couple at the forefront of the fight for parental rights, he was no stranger to cameras and interviews in his early years. Now 21, Zak remembers being a preschooler and building a Lego tower over and over as a television crew filmed him.

These days, his mind is on the things that preoccupy most recent university graduates—a career, travel plans, and what direction his life is about to take. This spring, Zak finished his Bachelor of Arts at Wilfrid Laurier University. He's now working as a bartender at Harbourfront, learning the ropes in the restaurant and bar business.

Zak has told the story of his family "a million times." But sure, he'll tell it again, he says agreeably at a coffee shop near the Leslieville home he shares with his mothers, who have been together for 25 years.

On December 17, 1991, Chris Higgins gave birth to him in a downtown Toronto hospital with Chris Phibbs at her side. Zak had been conceived through insemination using a sperm donor.

The women shared everything, from the euphoria of new parenthood to diaper duty, sleepless nights, and walks pushing the stroller. But it was three years before Chris Phibbs was officially recognized as Zak's second parent, after the couple and three others won their court case granting partners of parents in same-sex couples the right to adopt the children they are jointly raising.

Every now and then in school, his class would be watching a film about equal rights or diversity and Zak would end up staring at a clip of himself with his moms, or his 4-year-old self at the head of the Pride Parade holding a giant water gun and spraying the crowd.

"I'd always think here we go again," he shrugs. "I'd keep popping back up every few years."

In early childhood, he had a circle of playmates also being raised by same-sex partners. But his moms prepared him for the fact that they were a minority.

"I think from a young age my parents were smart about 'here's how our family is, and here's how your friends' families might look.'"

He remembers being aware that he was the only kid at school with two mothers, but he doesn't recall it being a big deal. However, before he brought a new friend home, he'd usually preface it with "just to let you know, I have two moms."

Later, he was like any other prickly adolescent who finds their parents annoying.

"As a teen, you're always angst-ing about stuff and pissed off at your parents. It took me awhile to see how incredible (it is) what they've done. But now I see they are really, really inspirational people."

His moms—"the Chrises," he calls them—are frank and open. They told him he always had the option of contacting his biological father. He wasn't interested until he turned 18.

The two met for the first time in a therapist's office. It went well. Zak ended up going over for dinner that night. Now, he drops over to help him and his male partner around the garden.

"I wasn't looking for two more parents," says Zak. Just an adult guy to hang out with every now and then.

Zak, who's straight and single at the moment, says he's been more influenced by his mothers' loving long-term relationship than by their gender. He's proud of their courage and determination, but he doesn't feel compelled to take a public stand.

"I fully support (equality and rights) but don't know if it's my fight," he says. "It's their fight."

Not long ago, the Chrises popped in to visit him at work. His boss came out to meet them and offered feedback any mom would be tickled to hear.

Something about a good kid, great work ethic. In other words, his parents raised him right.

DISCUSSION QUESTIONS

1. Why is same-sex marriage referred to as such? Why is it not simply called marriage? As a general rule, when a man and woman marry, we do not call it hetero-marriage, yet when homosexuals marry, it is usually referred to as gay or same-sex marriage. How does the distinction affect those who it defines? Is British Columbia moving in the right direction by redefining 'spouse,' rather than creating separate same-sex marriage legislation?

2. Sadie Epstein-Fine states that it is "important for people to recognize that kids like her have their own distinct identities, because so often [they're] attached to [the identities of their] parents." Why do you think children of same-sex parents are sometimes attached to the identities of their parents? Are children of hetero-marriages also attached to the identities of their parents in some way?

3. Rachel Epstein (Sadie's mother) talks about the pressure kids from same-sex families feel to present themselves as "poster families" or act as "the ambassadors" for same-sex parents. Why do you think this pressure exists for children of same-sex families in society and how do you think it impacts their experiences growing up?

KNOW IT AND OWN IT: WHAT DO I BRING TO THIS?

The "K" in the KWIP process involves examining aspects of your own identity and social location as the first step in becoming diversity competent.

ACTIVITY: JOURNAL

Do you come from a traditional family, or one that has taken on a newer, more modern form? How do you think your family of orientation has influenced, or would influence, your family of procreation? Do you think you would carry on, or have you carried on, certain family values or traditions with your family of procreation? Examine your feelings surrounding issues like same-sex marriage, foster parenting or adoption, interracial or religious unions, and common-law partnerships. Knowing what your family situation is in relation to many other kinds of families that exist will go a long way in understanding the diversity that has changed the face of families today.

WALKING THE TALK: HOW CAN I LEARN FROM THIS?

The "W" in the KWIP process presents a scenario or case study that challenges you to "walk the talk" through problem-based learning

ACTIVITY: CASE STUDY

You have been a member of a blended family for five years—your parents divorced ten years ago when you were just a teenager, and since, both your mom and dad have remarried. Neither of them have had children of their own, but your stepfather has two children from his previous marriage that are close to your age. You have always had a good relationship with your stepsiblings and extended stepfamily, especially with your step-grandfather. He had come to live with your mom and stepfather two years ago when he needed some extra care, as he lost mobility due to age and could no longer maintain a house of his own. Recently, your step-grandfather had an accident and did not make it through surgery. His death was very unexpected and your family was devastated. Some time had passed after the funeral and you received a call from your mom—your step-grandfather had left all of his grandchildren $5000, including you. At first, you were a little surprised that he included you in his will in the same way he had included his biological grandchildren, but you had spent a great deal of time together, and you loved and respected him as if he were your biological grandfather. Unfortunately, the rest of your stepfamily did not agree and was arguing about whether you should have been included in the same way the other grandchildren were. They were appalled that he had amended his will to include you after only knowing you for a short period of time. No one said anything to you directly, but your stepfather's brother communicated to him that most of the family did not think it was fair and believed that you should not accept the inheritance that was intended for you. Your stepfather told you not to worry about what they thought and to just ignore the situation.

1. How do you respond when your stepfather tells you about how the rest of your stepfamily feels? Do you accept the inheritance or turn it down?

2. The next time you are with your extended stepfamily, how do you feel? Do you discuss your feelings with them or do you do as your stepfather suggests and just ignore the situation?

IT IS WHAT IT IS: IS THIS INSIDE OR OUTSIDE MY COMFORT ZONE?

The "I" in the KWIP process requires you to honestly confront and identify ways in which our complex identities result in experiences of privilege and oppression, and to reflect on how we can learn to honour that privilege.

ACTIVITY: GOT PRIVILEGE?

As with other characteristics of diversity, family structure can sometimes create privilege for some and barriers for others. While families take on many forms in Canada, some people still idealize the traditional nuclear family and the traditional family values that come with it. As a result, alternative family structures often encounter barriers in the form of stereotypes, prejudice, or discrimination.

ASKING: Do I have privilege?

1. I can see families similar to mine represented in the media and in my everyday experiences.

2. I have never been questioned, judged, or criticized on my decision to get married or have children.

3. The government and my workplace take my family structure into account when creating legislation and policies affecting families.

4. If I choose to, I can have children without the use of assisted human reproduction.

REFLECTING: Honouring Our Privilege

Consider family diversity in Canada and how your family fits into the mix. Describe two circumstances in which you feel disadvantaged because of your family structure. Then describe two circumstances in which your family structure gives you privilege over someone else's. How can you use the advantages you experience to combat the disadvantages experienced by others?

PUT IT IN PLAY: HOW CAN I USE THIS?

The "P" in the KWIP process involves examining how others are practising equity and how you might use this.

ACTIVITY: CALL TO ACTION

The One Parent Family Association of Canada (OPFA) is the largest Canadian, non-profit organization dedicated to providing support for one-parent families in Canada. The six chapters across Ontario, in Toronto, York Region, Ajax-Pickering, Limestone, Ottawa, and Whitby-Oshawa, are run by lone parents who volunteer to organize activities in three areas: Family and Youth Activities, Parent Social Activities, and Educational Meetings. Each chapter of the organization also works toward subsidizing a wide range of activities, bursaries, and other direct benefits for both children and single parents. They cater to children of a variety of ages, from babies to teens, as well as to adults who are one, separated, divorced, or widowed parents looking to socialize in a safe, no-pressure environment. To learn more about the One Parent Family Association of Canada, visit https://oneparent-families.net/.

Study Tools
CHAPTER 13

Located at www.nelson.com/student

- Review Key terms with interactive **flash cards**
- Check your Comprehension by completing **chapter review quizzes**
- Gauge your understanding with *Picture This* and accompanying short answer questions
- Develop your critical thinking/reading skills through compelling **Readings** and accompanying short answer questions
- Apply your understanding to your own experience with **Connect A Concept** activities
- Evaluate Diversity in the Media with engaging *Video Activities*
- Reflect on your Understanding with *KWIP* activities

REFERENCES

Adoption Council of Canada. (2012). FAQ's. Retrieved from *Adoption Council of Canada*: http://www.adoption.ca/faqs

Ambert, A. (2009). Divorce: Facts, Causes, and Consequences. Retrieved on August 12, 2016 from *Vanier Institute of the Family*: http://www.thefamilywatch .org/doc/doc-0073-es.pdf

American Psychological Association. (2012). The changing role of the modern day father. Retrieved from *American*

Psychological Association: http://www.apa.org/pi/families/resources/changing-father.aspx

Bartlett, S., & LeRose, M. (2012, November 17). Generation boomerang. Retrieved from *Doc Zone*: http://www.cbc.ca/doczone/episode/generation-boomerang.html

Battams, N. (2016). Sharing a roof: Multi-generational homes in Canada. Retrieved from *Vanier Institute*: http://vanierinstitute.ca/multigenerational-homes-canada/

Bird, A. (2010). Legal parenthood and the recognition of alternative family forms in Canada. *University of New Brunswick Law Journal*, 264–293.

Canadian Press. (2011, January 11). Cross-burning brother gets 2 months in jail. Retrieved from *CBC News*: http://www.cbc.ca/news/canada/nova-scotia/story/2011/01/11/ns-justin-rehberg-sentencing.html

Caton, M. (2016). Canada's first foster care agency for LGBTQ youth now licensed to open. Retrieved on September 15, 2016, from *Windsor Star*: http://windsorstar.com/news/local-news/canadas-first-foster-care-agency-for-lgbtq-youth-now-licensed-to-open

CBC News. (2015). Child-free day celebrates choice to be childless. Retrieved on September 12, 2016, from *CBC News*: http://www.cbc.ca/news/canada/montreal/child-free-day-celebrates-choice-to-be-childless-1.3095137

CBC Radio. (2016). Why a former foster kid opened a foster care agency for gay teens. Retrieved on September 15, 2016, from *CBC Radio*: http://www.cbc.ca/radio/candy/the-candy-palmater-show-for-july-13-2016-1.3676999/why-a-former-foster-kid-opened-a-foster-care-agency-for-gay-teens-1.3677003

Chapman, R., & Shields, L. (2012). An essay about health professionals' attitudes to lesbian, gay, bisexual and transgender parents seeking healthcare for their children. *Scandinavian Journal of Caring Services*, 333–339.

Crompton, S. (2005). Canadian social trends: Always the bridesmaid: People who don't expect to marry. Retrieved September 15, 2016, from *Statistics Canada*: http://www.statcan.gc.ca/pub/11-008-x/2005001/article/7961-eng.pdf

Erich, S., Kanenburg, H., Case, K., Allen, T, & Bogdanos, T. (2009). An empirical analysis of factors affecting adolescent attachment in adoptive families with homosexual and straight parents. *Child and Youth Services Review*, 398–404.

Fidelman, C. (2010, March 5). IVF funding on way: Bolduc. Retrieved from *Canadian Fertility and Andrology Society*: http://www.cfas.ca/index.php?option=com_content&view=article&id=969:ivf-funding-on-way&catid=1:latest-news&Itemid=50

Focus on the Family. (2011, March 19). International adoption for Canadians. Retrieved from *Waiting to Belong*: http://waitingtobelong.ca/articles/international-adoption-canadians

Gordon, A. (2013). Growing up with same-sex parents. Retrieved on August 12, 2016, from *The Toronto Star*: https://www.thestar.com/life/2013/08/16/growing_up_with_samesex_parents.html

Haaf, W. (2012, March 12). IVF & infertility clinic costs across Canada – How much are we paying to get pregnant? Retrieved from *iVillage.ca*: http://www.ivillage.ca/pregnancy/fertility/ivf-costs

Hunter, P. (2013). Co-parenting: Hoping to become a dad without a romantic relationship. Retrieved on September 17, 2016, from *The Star*: https://www.thestar.com/news/insight/2013/02/22/coparenting_hoping_to_become_a_dad_without_a_romantic_relationship.html

Kazia, A. (2013). "4 myths about common-law relationships." Retrieved August 15, 2016, from *CBC News*: http://www.cbc.ca/news/canada/4-myths-about-common-law-relationships-1.1315129

Kohen, J. (1981). Housewives, breadwinners, mothers, and family heads: The changing family roles of women. *Advances in Consumer Research*, 576–579.

Kotkin, J. (2012, October 10). The rise of post-familialism: Humanity's future? Retrieved from *newgeography.com*: http://www.newgeography.com/content/003133-the-rise-post-familialism-humanitys-future

Milan, A. (2013). Marital status overview, 2011. Retrieved September 15, 2016, from *Statistics Canada*: http://www.statcan.gc.ca/pub/91-209-x/2013001/article/11788-eng.pdf

Milan, A., & Bohnert, N. (2012, September 18). Portrait of families and living arrangements in Canada. Retrieved October 2, 2012, from *Statistics Canada*: http://www12.statcan.gc.ca/census-recensement/2011/as-sa/98-312-x/98-312-x2011001-eng.cfm#a2

Milan, A., Maheux, H., & Chui, T. (2010, April 20). A portrait of couples in mixed unions. Retrieved November 25, 2012, from *Statistics Canada*: http://www.statcan.gc.ca/pub/11-008-x/2010001/article/11143-eng.htm

Mrozek, A. (2012, September 12). Canada's declining marriage rate spells increasing poverty. Retrieved from *Institute of Marriage and the Family*: http://www.imfcanada.org/issues/canadas-declining-marriage-rate-spells-increasing-poverty

Omand, G. (2015). "B.C. gets go-ahead to pursue polygamy charge against religious leader." Retrieved August 15, 2016, from *The Globe and Mail*: http://www.theglobeandmail.com/news/british-columbia/bc-gets-go-ahead-to-pursue-polygamy-charge-against-bountiful-bc-leader/article25113882/ Paul-Carlson, P. (2012, May). Intercountry adoption in Canada: Does it protect the best interests of the child? Retrieved from *Adoption Council of Canada*: http://www.adoption.ca/publications

Pennings, G. (2011). Evaluating the welfare of the child. *Human Reproduction*, 1609–1615.

Pew Research Center. (2016). Gay marriage around the world. Retrieved on August 22, 2016, from *Pew Research Centre*: http://www.pewforum.org/2015/06/26/gay-marriage-around-the-world-2013/#us

Pew Research Center. (2010). The decline of marriage and rise of new families. Retrieved from *Pew Research*

Center: http://www.pewsocialtrends.org/2010/11/18/the-decline-of-marriage-and-rise-of-new-families/6/

Pittaway, K. (2011). The joy of singlehood. Retrieved on September 16, 2016, from *The Star*: https://www.thestar.com/life/2011/02/12/the_joy_of_singlehood.html

Preece, M. (2004, December). When lone parents marry, the challenge of stepfamily relationships. Retrieved November 20, 2012, from *Vanier Institute of the Family*: http://www.vifamily.ca/library/transition/334/334.html

Ryan, B., Brotman, S., & Rowe, B. (2001). "Certain circumstances": Issues in equity and responsiveness in access to health care in Canada. Retrieved from *Health Canada*: http://www.hc-sc.gc.ca/hcs-sss/alt_formats/hpb-dgps/pdf/pubs/2001-certain-equit-acces/2001-certain-equit-acces-eng.pdf

Ryan, M., & Berkowitz, D. (2009). Constructing gay and lesbian parent families: Beyond the closet. *Qualitative Sociology*, 153–172.

Scoffield, H. (2012, September 18). Define the Canadian family? It'll require a flow chart. Retrieved from *Maclean's Online*: http://www2.macleans.ca/2012/09/18/define-the-canadian-family-itll-require-a-flow-chart/

Statistics Canada. (2012a, October 28). Catalogue no. 93-716, Vol. II (1971 Census of Canada: Families). Retrieved October 28, 2012, from *Statistics Canada*: http://www5.statcan.gc.ca/bsolc/olc-cel/olc-cel?catno=95M0025XCB&lang=eng

Statistics Canada. (2012b, September 18). Portrait of families and living arrangements in Canada. Retrieved October 28, 2012, from *Statistics Canada*: http://www12.statcan.gc.ca/census-recensement/2011/as-sa/98-312-x/98-312-x2011001-eng.cfm

Statistics Canada. (2012c, September 12). 2011 Census of population: Families, households, marital status, structural type of dwelling, collectives. Retrieved October 31, 2012, from *The Daily*: http://www.statcan.gc.ca/daily-quotidien/120919/dq120919a-eng.htm

Statistics Canada. (2012d, September 18). Fifty years of families in Canada: 1961 to 2011. Retrieved November 5, 2012, from *Statistics Canada*: http://www12.statcan.gc.ca/census-recensement/2011/as-sa/98-312-x/98-312-x2011003_1-eng.cfm

Statistics Canada. (2012e, September 18). Living arrangements of young adults aged 20 to 29. Retrieved November 5, 2012, from *Statistics Canada*: http://www12.statcan.gc.ca/census-recensement/2011/as-sa/98-312-x/98-312-x2011003_3-eng.cfm

Statistics Canada. (2014). Mixed unions in Canada – National household survey, 2011. Retrieved on August 12, 2016, from *Statistics Canada*: http://www12.statcan.gc.ca/nhs-enm/2011/as-sa/99-010-x/99-010-x2011003_3-eng.pdf

Statistics Canada. (2016). Family violence in Canada: A statistical profile, 2014. Retrieved on September 16, 2016, from *Statistics Canada*: http://www.statcan.gc.ca/pub/85-002-x/2016001/article/14303-eng.pdf

Sullivan, R., & Harrington, M. (2009). The politics and ethics of same-sex adoption. *Journal of GLBT Family Studies*, 235–246.

Vanier Institute of the Family. (2012a). Definition of family. Retrieved from *Vanier Institute of the Family*: http://www.vanierinstitute.ca/definition_of_family#.UIqWOVEwC-g

Vanier Institute of the Family. (2012b, November 5). Blended families: New challenges and opportunities. Retrieved from *Vanier Institute of the Family*: http://www.vanierinstitute.ca/modules/news/newsitem.php?ItemId=468#.UJlF5lGtu-g

Vogels, J. (2010). No kidding: Support for the childless by choice. Retrieved on September 15, 2016, from *The Star*: https://www.thestar.com/news/insight/2010/12/17/no_kidding_support_for_the_childless_by_choice.html

Glossary

A

ableism Refers to the set of ideas and attitudes that define "normal" abilities of people and that allocates inferior status and value to individuals who have developmental, emotional, physical, or psychiatric disabilities. p. 29

Aboriginal peoples A legal/administrative category that comprises the First Nations, Inuit, and Métis; often used to describe the Indigenous peoples within Canada's boundaries. p. 138

absolute homelessness Situation of individuals who live either in emergency shelters or on the street. p. 55

absolute poverty A situation where an individual lacks even the basic resources that are necessary for survival; people who live in absolute poverty live without food, clothing, or a roof over their heads. p. 52

accessibility The degree to which a product, device, service, or environment is available to as many people as possible. p. 244

acclimatization The process of beginning to adapt to a new environment. p. 177

accommodation The specialized design of products, environments, and individualized strategies, which are uniquely adapted to an identified limitation for the purpose of ensuring access. p. 241

acculturation The process of adaptation to a new culture, whereby prolonged contact between two cultures begins to modify both cultures. p. 180

acculturative stress The stress experienced by immigrants when there are difficulties resulting from the acculturation process. The stress of this process increases when immigrants experience discrimination, language difficulties, and incongruences in non-material aspects of culture. p. 182

adultcentrism The assumption that only adults can make substantial contributions to society. p. 267

adultism Refers to bias toward adults that often results in discrimination against youth. p. 267

ageism Stereotyping, prejudice, and discrimination against individuals or groups because of their age; for example, the tendency to view seniors as unable to work, confused, and fragile. pp. 29, 265

agender A person with no (or very little) connection to traditional systems of gender; or someone who sees themselves as existing without gender; sometimes called gender neutral or genderless. p. 81

agents of socialization The people and institutions in society that create the social contexts where socialization takes place. p. 281

age stratification The hierarchal ranking of people into age groups within a society. p. 259

Alzheimer Society of Canada A national charity for people living with and affected by Alzheimer's disease and other dementias. p. 269

androgyny A gender expression that has characteristics of both masculinity and femininity. p. 81

androsexual/androphilic Attraction to males, men, and/or masculinity. p. 99

Anglo- or Franco-conformity Policy that informed early Canadian immigration; immigrants were expected to conform to dominant British-based or French-based culture as opposed to retaining their own. p. 196

anti-psychiatry A term coined by David Cooper in 1967 and reflective of a movement that was critical of and opposed to many forms of psychiatric treatment, institutionalization of persons with mental illness and intellectual disabilities, the unequal power relationship between psychiatrist and patient, and the highly subjective process for diagnostic labelling. p. 234

Arranged marriages A type of marital union where partners are selected by family members. p. 283

artificial insemination The injection of sperm into the vagina or uterus other than by sexual intercourse. p. 288

Asian Heritage Month Observed annually in Canada (and the United States where it is called Asian-Pacific American Heritage Month) in May. The purpose of the month is to learn about and acknowledge the historical and contemporary contributions made by Canadians of Asian heritage. p. 128

asexual A person who does not experience sexual attraction to any group of people. p. 98

assimilation 1. A long-standing government policy, Indigenous assimilation was based on the premise that they would give up their own culture, languages, and beliefs, and live and act just like the British settlers/Canadian citizens. 2. Total re-socialization of individuals from one culture to another, such that traces of the former culture eventually deteriorate. pp. 147, 179

assimilationist policy Policy that encourages immigrants to integrate into the mainstream culture. p. 195

assisted human reproduction (AHR) The use of medical techniques to assist in the conception and birth of a child, including artificial insemination, intrauterine insemination, in vitro fertilization, egg, sperm and embryo donation, and drug therapy. p. 288

Assisted Human Reproduction Act A law enacted by the Parliament of Canada to regulate assisted human reproduction and related research. p. 290

asylum seeker A person seeking protection as a refugee in another country but has not yet been found to meet the definition of a refugee. p. 166

B

Baby Boomer Refers to individuals born between 1946 and 1964. p. 261

Balanced Refugee Reform Act Immigration legislation that, together with the Protecting Canada's Immigration System Act, makes changes to the Immigration and Refugee Protection Act and the refugee claimant process in Canada. p. 167

barriers Policies or practices that prevent full and equal participation in society; barriers can be physical, social, attitudinal, organizational, technological, or informational. p. 244

biased-based bullying Bullying resulting from bias against someone because of a real or perceived aspect of their identity. p. 9

bicultural Retaining two distinct cultural identities; in Canadian history, it often refers to the recognition of England and France as primary cultures within Canada. p. 197

bigender A person who fluctuates between traditional gender-based behaviours and identities, identifying with both genders or sometimes a third gender. p. 81

bilingual Able to speak two languages; in Canadian history, the official recognition of English and French as equal official languages. p. 197

biphobia The denial of bisexuality as a genuine sexual orientation, often reinforced by mainstream media. p. 97

birth rate The number of live births per year per 1000 population. p. 185

bisexual A person romantically and/or sexually attracted to people of their own gender as well as another gender. p. 97

bisexual invisibility The erasure and silencing of bisexual experiences, identities, and communities by presuming that individuals who identify as bisexual are in a temporary phase on their way to mature heterosexual or gay/lesbian identities. p. 97

blackface Makeup that was used historically in minstrel shows to impersonate black people and act out racist stereotypes. Beginning in the 1800s, white American actors would rub black shoe polish or greasepaint on their faces and then perform in ways that would demean and dehumanize black people. It is associated with Jim Crow racism. p. 121

Black History Month Observed during the month of February in Canada, the United States, and the United Kingdom to commemorate the important people and events and history of the African diaspora. p. 128

blended families A family consisted of a couple and their children from previous relationships with or without children from current relationship. p. 285

Blended Visa Office-Referred Refugees Refugees for whom UNHCR matches a private sponsor to share income support with the Government of Canada. p. 172

boomerang generation Term used to describe those young adults who return to their parental home after leaving for education, work, or a relationship. p. 286

bullying A form of direct or indirect aggression that involves a real or perceived power imbalance. p. 9

business immigrants A category of people in the economic class who invest in or start a business in Canada with the expectation that it contributes to the development of a strong Canadian economy. p. 171

C

Canadian Charter of Rights and Freedoms Referred to as the Charter, it is a bill of rights and the first part of the 1982 Constitution Act; it is considered the highest law of Canada and, as such, supersedes any other federal or provincial law that conflicts with it. p. 33

Canadian Experience Class (CEC) Program for people who want to apply to become permanent residents of Canada based on skilled work experience acquired in Canada while having legal status to work or study, such as temporary foreign workers and international student graduates with one year of full-time work experience. p. 170

Canadian Human Rights Act A federal law that protects all people who are legally in Canada from discrimination by federally regulated employers and service providers (Canadian Human Rights Commission, 2013). p. 33

Canadian Human Rights Commission An organization that was created by the Canadian Human Rights Act and is separate and independent from the Government of Canada and the Canadian Human Rights Tribunal (Canadian Human Rights Commission, 2013). p. 33

Canadian Multiculturalism Act Law passed in 1988 that aims to preserve and enhance multiculturalism in Canada. p. 198

Caregiver Program New immigration program replacing the Live-In Caregiver Program with two new pathways to permanent residence for caregivers: 1) Caring for Children for caregivers who have provided child care in a home; 2) Caring for People with High Medical Needs for caregivers who have provided care for the elderly or for those with disabilities or chronic disease. p. 172

cede Surrender or forfeit possession of something, usually by treaty. p. 142

chronic poverty Occurs when a person is in a state of poverty over an extended period of time, and barriers to well-being become cyclical in nature. p. 52

chronological age The number of years a person has lived. p. 258

cisgender A description of a person whose gender identity, gender expression, and biological sex align. p. 80

classism The systematic oppression of dominant class groups on subordinate classes in order to gain advantage and strengthen their own positions. p. 31

cluttered or crowded nest A household consisting of a parent or parents whose children have either decided to stay living at home past the customary age for leaving home, or who have returned to live at home after leaving for a brief period of time. p. 286

cohort Any group of people who share a time-specific period, such as graduating in the same year. p. 258

Common-law unions Two individuals living together that are not considered married. p. 283

complex stepfamilies Families with children of both parents and children of one parent only, families with children of each parent and no children of both parents, or families with children of both parents and children of each parent. p. 285

comprehensive claim Modern-day treaties to resolve claims of Aboriginal land rights not yet dealt with by past treaties or through other legal means. p. 145

compulsory heterosexuality The assumption that men and women are innately attracted to one another both emotionally and sexually and that heterosexuality is natural and normal. p. 96

compulsory sterilization Government program that forces or coerces the sterilization of persons so that they cannot have children; often part of a eugenics program to prevent the reproduction of members of a population who are deemed undesirable. p. 234

Convention Against Torture United Nations international human rights agreement signed on December 10, 1984, as a commitment against the use of torture in their country for any reason; signatory nations also agree not to use any evidence obtained under torture and not to deport or return people to countries where they are at risk of being tortured. p. 167

Convention refugee Individual who has been granted asylum by the 1951 Geneva Convention Relating to the Status of Refugees; someone who has reason to fear persecution in his or her country of origin due to race, religion, nationality, membership in a social group, or political opinion. p. 172

co-parenting Involves two people having a child without being in a romantic relationship. The child is usually conceived through assisted human reproduction and raised by both individuals who share custody but live in separate households. p. 290

critical social theory (also known as critical theory) A macro theory interested in those who are oppressed, which critiques social structures that exploit and marginalize members of a society and whose goal is liberation from oppression. p. 3

cross-dressing Wearing clothing that conflicts with the traditional gender expression of your sex and gender identity. p. 83

cultural ageism Occurs when one generation devalues the actions of another generation because it is different from what they are used to. p. 266

Cultural genocide The deliberate destruction of the cultural heritage and traditions of a group of people or a nation. p. 151

cultural greywashing Refers to the marginalization and overgeneralizations made about older adults that result in extreme stereotyping. p. 268

Cultural imperialism A form of oppression where the dominant group has made their beliefs and values the norms of a society. p. 26

cultural relativism Assumption that all cultures are essentially equally valuable and should be judged according to their own standards; no cultural standpoint is privileged over another. p. 196

cultural scripts Indicate appropriate sexual roles, norms, and behaviours in a given society; largely conveyed through mass media and other social institutions, such as government, law, education, family, and religion. p. 93

cyberbullying Involves the use of information and communication technologies such as the Internet, social networking sites, websites, email, text messaging, and instant messaging to intimidate or harass others (Royal Canadian Mounted Police, 2016). p. 9

D

deinstitutionalization A social movement that continues to reverse the institutionalization of persons with mental illness and intellectual disabilities that began in the late 19th and early 20th centuries in Canada. It involves a process of removing residents from long-term institutions to integrated community-based settings. p. 234

difference In a social context, a term used to refer to difference in social characteristics. p. 2

direct discrimination The unfair treatment of individuals or groups based on one or more of their protected characteristics, compared to other individuals or groups who do not have these characteristics in similar circumstances (Helly, 2004). p. 35

disability A universal human experience that anyone can experience at any time; disabilities can limit a person's ability to engage in daily activities; disabilities can be visible or invisible, temporary or permanent. p. 230

disability binarism Mode of thought that classifies people as either able-bodied or disabled. p. 231

discrimination The unequal treatment of individuals or groups based on their characteristics or behaviours. It involves actions or practices of dominant group members that have a harmful impact on members of a subordinate group. p. 32

discriminatory practice(s) Actions that are discriminatory, and in a legal context are based upon one or more grounds of discrimination. The Canadian Human Rights Act includes seven discriminatory practices that are prohibited by law. p. 33

discursive racism A form of racism that is expressed through written and spoken communication (discourse). It manifests in racial slurs and hate speech, and in words with racial meaning embedded in them. p. 118

distributive and redistributive justice A social justice model that is concerned with the fair distribution or redistribution of material and non-material resources between different groups within a society. p. 13

diversity An anti-oppression framework built on principles that value social equity, social justice, and social inclusion. p. 3

dominant groups Groups of people in a society who have power and privilege. p. 4

domination The systematic and continuous exertion of power by dominant groups over non-dominant groups. p. 4

drag king A person who consciously performs traditional masculinity, presenting an exaggerated form of masculine expression; often times done by a woman. p. 83

drag queen A person who consciously performs traditional femininity, presenting an exaggerated form of feminine expression; often times done by a man. p. 83

duty to accommodate Human rights legislation requires employers and service providers to accommodate peoples' needs, when those needs relate to one or more grounds of discrimination. p. 34

E

eagle feather A great honour among Aboriginal peoples; represents a mark of distinction. p. 158

economic ageism Refers to the ways in which people are discriminated against based on age in employment. p. 266

economic class An immigration category that includes federal and Quebec-selected skilled workers, federal and Quebec-selected business immigrants, provincial and territorial nominees, the Canadian Experience Class (CEC), and caregivers, as well as spouses, partners, and dependants who accompany the principal applicants in any of these economic categories (Government of Canada, 2016). p. 169

economic migrant Person who moves to another country for employment or a better economic future. p. 175

elder abuse A single or repeated act against an older person in which harm or distress is caused. p. 269

elder neglect Occurs when a person fails to provide an older person with the necessities of life, such as food, clothing, shelter, medical attention, personal care, and required supervision. p. 270

emigration Leaving one country with the intention of settling in another. p. 180

emphasized femininity The acceptance of gender inequality and a need to support the interests and desires of men; often associated with empathy, compassion, passivity, and focused on beauty and physical appearance. p. 72

employment equity Requirement under the Employment Equity Act of Canada that employers use proactive employment practices to increase representation of four designated groups: women, people with disabilities, Indigenous peoples, and racialized communities in the workplace. Employment equity mandates the accommodation of difference with special measures when needed. p. 122

Employment Equity Act Law that requires employers in Canada to eliminate barriers to and increase the hiring of women, people with disabilities, Aboriginal people, and visible minorities. p. 198

empty nest A household consisting of a parent or parents whose children have grown and moved out of the house. p. 286

Endogamy A marital union encouraged between people within the same or similar social groups. p. 283

enfranchisement The process whereby an individual gets the right to vote or become a citizen. p. 148

equality Fairness and justice achieved through same treatment. p. 13

equity Principle based on fairness, justice, access, opportunity, and advancement for everyone, while recognizing historically underserved and unrepresented populations, identifying conditions needed to provide effective opportunities for all groups, and eliminating barriers to their full participation. p. 13

ethnic cleansing The process or policy of eliminating unwanted ethnic, racial, or religious groups by deportation, forcible displacement, mass murder, or threats of such acts, with the intention of creating a homogenous population. p. 121

ethnic enclave Focused areas of homogeneous ethnic groups, usually coupled with some business and institutional activity. p. 200

ethnicity A person's or his or her ancestors' country of origin; includes material and non-material aspects associated with a culture and social identity. p. 126

Ethnocentrism Refers to a tendency to regard one's own culture and group as the standard, and thus superior, whereas all other groups are seen as inferior. p. 27

Exogamy A marital union encouraged between people of different social groupings. p. 283

expansionism A country's practice or policy of getting bigger, usually in terms of territory or currency. p. 198

exploitation The unfair use of people's time or labour without compensating them fairly. p. 26

Express Entry An online immigration application process that came into effect as of January 1, 2015, for persons applying to come to Canada through the Federal Skilled

Worker (FSW) program, the Federal Skilled Trades Program (FSTP), the Canadian Experience Class (CEC), and some provincial/territorial nominee programs. p. 170

F

family class Immigration category used to describe immigrants who have been sponsored to come to Canada as a spouse, partner, dependent child, parent, or grandparent. p. 169

family of orientation The family someone was born or raised in; gives us our heritage, social status, and history. p. 281

family of procreation The family we create when we become adults. p. 281

Federal Skilled Trades Program (FSTP) Program for people who want to apply to become permanent residents of Canada based on full-time work experience and qualifications in an eligible skilled trade. p. 170

Federal Skilled Worker (FSW) Program Program for people who want to apply to become permanent residents of Canada based on work experience and skills that are assessed through a point system. p. 170

fiduciary A person who holds power or property in trust for another person—usually for the benefit of the other person. p. 148

financial elder abuse Occurs when someone tricks, threatens, or persuades an older adult out of their money, property, or possessions. p. 270

First Nations A legal term used to refer to those Aboriginal people who are of neither Inuit nor Métis descent. First Nations refers to the ethnicity, while a band may be a grouping of individuals within that ethnicity (e.g., Seneca or Oneida). p. 138

fluid Characteristic of identity that describes it as something that can change and be shaped. p. 16

foster care The placement of a child or young person in a home of someone who receives compensation for caring for the child, but is not the child's parent. p. 288

Fourth Industrial Revolution or Industry 4.0 A term that describes the creation of new technologies that have the ability to combine the physical, digital, and biological worlds. p. 265

G

gay A person romantically and/or sexually attracted to a person with the same gender identity and/or biological sex as themselves. p. 94

gender A social construct that refers to a set of social roles, attitudes, and behaviours that describe people of different sexes. p. 70

gender confirmation surgery Refers to a group of surgical options that alter a person's biological sex; also referred to as sex alignment surgery or sex reassignment surgery. p. 85

gender expression The external display of gender that is generally measured on a scale of masculinity and femininity. p. 71

genderfluid Describes an identity that is a fluctuating mix of the options available. p. 81

gender identity The internal perception of an individual's gender and how they label themselves. p. 71

genderless A person who does not identify with any gender. p. 81

gender neutralism A social movement that calls for policies, language, and other social institutions to avoid distinguishing roles according to sex and gender in order to challenge gender prejudice and gender discrimination. p. 85

gender non-conforming Refers to a gender expression that indicates a non-traditional gender presentation. p. 81

genderqueer A blanket term used to describe people whose gender falls outside of the gender binary; a person who identifies as both a man and a woman, or as neither a man nor a woman; often used in exchange with transgender. p. 81

gender roles A set of behaviours that are considered acceptable, appropriate, and desirable for people based on their sex or gender. p. 71

gender socialization The process by which males and females are informed about gendered norms and roles in a given society. p. 71

generation A cohort or group of people who share similar chronological ages and are shaped by the events, trends, and developments of a particular span of time. p. 258

Generation Alpha Refers to individuals born in 2010 and later. p. 263

Generation X Refers to individuals born between 1964 and 1979. p. 261

Generation Y Refers to individuals born between 1980 and 1995. p. 262

Generation Z Refers to individuals born between 1996 and 2009. p. 263

genocide The intentional extermination or killing of an identifiable group. p. 121

gentrified Improved so as to appear more middle class. p. 201

gestational surrogacy Occurs when the sperm and the egg from the couple seeking surrogacy is fertilized through in vitro fertilization and implanted into the surrogate, making the child biologically related to both parents but not the surrogate. p. 289

Gini index A commonly used measure of income inequality, which measures inequality on a scale of 0 to 1; a Gini index of 0 represents exact equality (i.e., every person in a certain society or nation has the same amount of income), while a Gini index of 1 represents total inequality (i.e., one person has all the income and the rest of the society has none). p. 57

glass ceiling An invisible barrier that prevents women and minorities from advancement in organizations. p. 78

Government Assisted Refugee (GAR) A government-sponsored refugee selected overseas for resettlement in Canada. p. 172

grounds of discrimination Reasons a person may experience discrimination. Human rights legislation specifies specific reasons why employers and service providers cannot discriminate against people, such as race, religion, age, sexual orientation, marital status, family status, disability (these are only a few of many grounds used in human rights legislation). p. 33

gynesexual/gynophilic Attraction to females, women, and/or femininity. p. 99

H

hate crimes Crimes that are committed against people or property that are motivated by hate or prejudice against a victim's racial, ethnic, religious, or sexual identity; they can involve intimidation, harassment, destruction of property, vandalism, physical force or threat of physical force, or inciting hatred in other people. p. 123

head tax Tax imposed by the Canadian government on anyone immigrating to Canada from China between 1885 and 1923. p. 66

healthy immigrant effect Immigrants' health is generally better than the health of those born in Canada, but this declines as their years spent living in Canada increase. p. 183

hegemonic masculinity The version of masculinity that is set apart from all others and considered dominant or ideal within society; often associated with toughness, bravado, aggression, and violence. p. 72

heterogeneity Means having dissimilar characteristics. The opposite is homogeneity, which means having the same characteristics. p. 9

heteronormativity Examines the ways in which heterosexuality is produced as a natural, unproblematic, taken-for-grated phenomenon that is maintained and reinforced through the everyday actions of individuals and through dominant social institutions. p. 97

heterosexism The belief in the natural superiority of heterosexuality as a way of life and therefore its logical right to dominance. Comprised of a system of ideas and institutionalized beliefs, it leads to the oppression of any non-heterosexual form of behaviour, identity, relationship, or community. pp. 31, 97

heterosexual (or straight) A person romantically and/or sexually attracted to someone with the opposite gender identity and/or biological sex than they have. p. 96

hidden or concealed homelessness State of those without a place of their own who live in a car, with family or friends, or in a long-term institution such as a prison. p. 55

historical disadvantage Disadvantage related to past historical discriminatory actions combined with current disadvantage to contribute to systemic discrimination. p. 125

home-grown terrorism Acts or plans of mass violence for political purposes that are initiated and/or carried out by residents or citizens as opposed to foreign nationals. p. 194

homelessness Social category that includes anyone who cannot obtain and sustain long-term, adequate, and risk-free shelter, for any reason. p. 55

Homogamy A marital union where partners are similar to each other when it comes to age, personality, and cultural interests. p. 283

homophobia The irrational fear, dislike, hatred, intolerance, and ignorance toward and the marginalization of non-heterosexuals. p. 97

host country A country where representatives or organizations of another country co-exist, either because they have been invited by the government, or because an international agreement exists. p. 179

human rights Define how we are to be treated as human beings and what we are all entitled to, including a life of equality, dignity, and respect, free from discrimination. Human rights in Canada are protected by international, federal, provincial, and territorial laws (Canadian Human Rights Commission, 2013). p. 32

human rights commission A national or international organized body that investigates, protects, and advocates for the rights of human beings. p. 216

Human Rights Tribunal of Ontario An administrative tribunal in Ontario, Canada, that hears and determines claims based on the violation of rights protected under the Ontario Human Rights Code. p. 265

I

identity Social construction of a person's sense of self that is based upon social categories that influence self-perception and the perception of others. p. 15

ideological racism A form of racism rooted in the ideas, beliefs, and worldviews that reflect, reinforce, and advance notions of racial superiority or inferiority. p. 118

Immigrant Investor Venture Capital Pilot Program Pilot program launched by the government in 2015 for people who want to apply to become permanent residents of Canada, who have a personal net worth of $10 million CDN (acquired through lawful, private sector business or investment activities) and are willing to invest a minimum of $2 million CDN for fifteen 15 years in the Immigrant Investor Venture Capital Fund. p. 172

immigrants People residing in Canada who were born outside of Canada; this category excludes Canadian citizens born outside of Canada and people residing in Canada on temporary status, such as those with a student visa or temporary foreign workers. p. 166

immigration Entering into and becoming established in a new place of residence; usually means entering a country that one was not born in. p. 166

Immigration and Refugee Board (IRB) An independent tribunal established by the Parliament of Canada that is responsible for hearing refugee claims and appeals in accordance with the law. p. 172

Immigration and Refugee Protection Act (IRPA) Legislation whose mission is "respecting immigration to Canada and the granting of refugee protection to persons who are displaced, persecuted, or in danger." p. 166

Immigration, Refugees and Citizenship Canada (IRCC) The branch of the federal government that is responsible for immigration, settlement, and citizenship; formerly called Citizenship and Immigration Canada (CIC). p. 169

impairment According to the biomedical perspective, a medical condition that leads to disability; according to functional perspective, it is any loss or abnormality of physiological, psychological, or anatomical structure or function, whether permanent or temporary. p. 236

income gap A gap in income between one group and another, such as discrepancies in income between older and younger employees performing that same job. p. 265

Indian Act Canadian federal legislation, first passed in 1876, and subsequently amended, which details certain federal government obligations and regulates the management of reserve lands, money, and other resources. It was initiated to compel Indian assimilation into Canadian society. p. 138

Indian Register Series of documents established in the 1850s that kept track of all of the existing records of people recognized by the federal government as members of an "Indian" band. p. 148

Indigenous The descendants of groups of people living in the territory at the time when other groups of different cultures or ethnic origin arrived there. p. 138

indirect discrimination Refers to a rule, policy, practice, or requirement that applies to everyone, but has the effect of creating disadvantage for people with a protected characteristic. p. 35

individual racism Individual discriminatory attitudes or behaviour motivated by negative evaluation of a person or group of people using a socially constructed concept of race. p. 119

institutionalized Placed in or confined to a residential institution. p. 234

institutionalized ageism Policies or practices within institutions that discriminate based on age. p. 266

institutional racism Behaviour, policies, or practices that disadvantage racialized persons; can be intentional or unintentional. p. 119

integration The long-term, multidimensional process through which newcomers become full and equal participants in all aspects of society. As part of this process, newcomers interact with the larger society and also maintain their own identity. p. 177

interactional racism A form of racism that is expressed through social interaction with other people; more specifically, how those with privilege interact with those who are oppressed. p. 119

intergenerational poverty Often occurs when children and youth grow up in households experiencing chronic poverty, where limited access to opportunities can start another cycle of poverty for the next generation. p. 52

intergenerational programs Programs that have intentional and planned interactions of different age groups, which can be effective in breaking down barriers between generations by encouraging communication, cooperation, and promoting health and well-being. p. 271

internally displaced persons Persons forced to flee their home for safety but do not cross a border or leave their country; they seek safety in another location within their own country. p. 173

International Mobility Program A program that allows employers to hire temporary workers without a Labour Market Impact Assessment. p. 184

Internet of Things (IoT) Technology that allows for everyday objects to have network connectivity, allowing them to send and receive data. p. 265

interpersonal scripts Created when individuals use the general sexual guidelines they have learned from cultural scripts and adapt them to specific social situations. p. 93

interreligious, or interfaith, marriages Occur when two individuals who believe in different faiths marry. p. 219

intersex A person whose physical anatomy does not fit within the traditional definitions of male or female. p. 84

intimate partner violence Any intentional act or series of acts by one or both partners in an intimate relationship that causes injury to either person. It can include physical assault, emotional abuse, sexual violence, and sexual harassment. Sometimes referred to as spousal violence or domestic abuse. p. 78

intrapsychic scripts The ability to mentally rehearse sexual outcomes before they occur; internal and individual scripts based on previously adopted cultural and interpersonal scripts. p. 93

intrauterine insemination (IUI) A medical procedure where sperm is washed and then inserted directly into the uterus at the time of ovulation. p. 288

Inuit The Aboriginal peoples who live in the far north or Arctic regions of Canada. p. 138

invisible disability When a person has a disability that is not apparent to others. p. 231

in vitro fertilization (IVF) A medical procedure where sperm fertilizes an egg outside of the body and then is implanted into the uterus. p. 288

J

Jim Crow racism Anti-black racism that existed in the United States during the period of 1877–1960s; Jim Crow laws enforced racial segregation and a racialized social order that resulted in the subjugation, oppression, and death (through lynching and other violence) of African Americans. p. 118

L

Labour Market Impact Assessment Needed for temporary foreign workers to be granted work permits; assesses impact of hiring on the Canadian labour market. p. 184

lesbian A woman in a same-sex relationship who is romantically and/or sexually attracted to women. p. 94

low-income cut-off (LICO) Measure established by Statistics Canada annually that refers generally to what people call a poverty line; it represents the income level at which a family may face hardship because it has to spend a greater proportion of its after-tax income on food, shelter, and clothing than the average family of similar size. p. 50

low-income measure (LIM) A measure of poverty that is commonly used for making international comparisons. p. 51

lump of labour fallacy The false belief that the amount of work available to labourers is a fixed amount and, therefore, there is no capacity to absorb more labourers into an economy. p. 176

M

marginalization The process of pushing groups with less social power to the margins of society. p. 26

marginal poverty Occurs when a person lacks stable employment over an extended period of time. p. 52

market basket measure (MBM) A measure of low income based on the cost of a specified basket of goods and services representing a modest, basic standard of living in comparison to the standards of its community. p. 51

matrix of domination Term associated with the work of Patricia Hill Collins that refers to forms of oppression and resistance based on socially constructed differences shaped by cultural and historic contexts where an individual or group can experience both oppression and privilege as a result of their combined identities. p. 8

medicine wheel Ceremonial tool that symbolizes the interconnected, circular journey of all living things. p. 139

melting pot model Immigration and settlement model that is often used in reference to the United States, wherein newcomers are expected to dissolve that which makes them different into a "pot" mixed with everyone else. p. 194

Métis People indigenous to North America whose background is Aboriginal and European ancestry. p. 138

metrosexual A man with a strong aesthetic sense, who spends more time on appearance and grooming than is considered gender normative. p. 83

migration To move from one country, place, or region to another. p. 166

monogamy A relationship founded on sexual exclusivity. pp. 101, 283

mosaic model Immigration and settlement model that is often used in reference to Canada, wherein newcomers are encouraged to maintain their unique and distinct cultures and live alongside other Canadians with other distinct cultures. p. 194

multiculturalism A concept often used in different ways. It can refer to the demographic reality that we are a multicultural society made up of ethnoculturally and racially diverse groups of people. Multiculturalism can also refer to an ideology that uses ethnocultural and racial diversity and equity as its framework and a mosaic as its metaphor. It may also refer to formal policy and initiatives. p. 194

multigenerational households A household where grandparents, parents, and children live together under one roof. p. 286

N

National Aboriginal History Month Observed annually in Canada during the month of June to recognize the past and present contributions of First Nations, Inuit, and Métis peoples to the birth and development of Canada. p. 128

National Day of Action Day devoted to raising the awareness of serious issues facing Aboriginal people in Canada; first organized on June 29, 2007. p. 156

National Senior Strategy for Canada An organization working to promote and explore policy options that can support the development of a National Seniors Strategy. p. 268

Neo-Nazism A post–World War II movement related to the white nationalist and white power skinhead movements, which seeks to revive elements of Nazi ideology such as racism, xenophobia, homophobia, holocaust denial, and anti-Semitism. p. 124

newcomer Blanket term used to refer inclusively to all immigrants and refugees who have recently arrived in the country to settle on a permanent basis. p. 179

non-dominant groups Groups of people in a society without (or with less) power and privilege. p. 4

normative ageism The assumptions made about a person's capabilities, interests, strengths, or weaknesses based on age. p. 266

O

omnibus bills Proposed laws or legislature that can cover a number of different subjects, but are packaged together in one bill. p. 155

oppression The intentional and unintentional domination of non-dominant groups by powerful dominant groups that occurs on individual, cultural, and structural levels in society. p. 4

P

panhandler A person who approaches strangers on the street or in a public place and asks for money. p. 55

pansexual A person romantically and/or sexually attracted to members of all sexes and gender identities and/or expressions. p. 97

paternalistic ageism The idea that children and young adults lack maturity and therefore require older adults to make decisions for them. p. 266

pathologization Characterization as medically or psychologically abnormal. p. 238

patriarchy Historically, any social system that was based on the authority of the heads of the household, which were traditionally male; recently, the term has come to mean male domination in general. p. 76

pedophobic or ephebophobic ageism Denotes a fear or hatred of children or adolescents, and can be seen in the common misconceptions that young people are troublemakers, hormonally out of control, and a general burden to society. p. 266

permanent resident According to the Canadian Immigration and Refugee Protection Act (2002), a person who has come to Canada and successfully applied and received immigration status to live here permanently. p. 169

personal identity The part of a person's identity determined by their individual attributes and characteristics. p. 15

person of colour Considered to be an outdated term, which was originally intended to be more positive and inclusive of people than the terms "non-white" or "visible minorities"; was used to refer to people who may share common experiences of racism. p. 115

physical elder abuse Occurs when someone hits or handles a person roughly, even if there is no injury. p. 270

polyamory The state of being in love, or romantically involved with, more than one person at the same time. p. 102

polyandry Having multiple husbands. p. 102

polygamy Having multiple wives. pp. 102, 283

potlatch Organized meeting for special ceremonies, such as name-giving, birth, rites of passage, treaties, and weddings; practised mainly by First Nations of the west coast. p. 148

pluralism Existence of recognized diverse groups within a single (peaceful) society. p. 198

post-traumatic stress disorder (PTSD) A psychological disorder caused by a traumatic event involving actual or threatened death or serious injury to oneself or others; symptoms may include anxiety, depression, survivor guilt, sleep disturbances, nightmares, impaired use or loss of memory, concentration difficulties, hyper arousal, hypersensitivity, suspiciousness, fear of authority, and paranoia. p. 183

powerlessness Occurs when the dominant group has left the subordinate group with virtually no access to the rights and privileges that the dominant group enjoys. p. 26

precarious employment Employment that includes, but is not limited to, part-time, temporary, or contract work with uncertain hours, low wages, and limited to no benefits. p. 54

Prejudice A negative attitude based on learned notions about members of selected groups, based on their physical, social, or cultural characteristics. p. 27

Principle of Access One of the Five Principles of the National Senior Strategy for Canada recommending that service providers ensure that older Canadians, their families, and their caregivers can easily access supports in a timely and efficient way. p. 269

Principle of Choice One of the Five Principles of the National Senior Strategy for Canada recommending that older Canadians, their families, and their caregivers have as many choices as is possible and are empowered with the necessary information to make informed choices. p. 269

Principle of Equity One of the Five Principles of the National Senior Strategy for Canada recommending that older Canadians from different ethnocultural groups, those from LGBTQ communities, and those with varying abilities are equally supported. p. 269

Principle of Quality One of the Five Principles of the National Senior Strategy for Canada recommending that the quality of the programs and services is not sacrificed in an effort to save money. p. 269

Principle of Value One of the Five Principles of the National Senior Strategy for Canada recommending that tax dollars are spent in the most effective and efficient ways to protect the sustainability of programs and services. p. 269

protected person Someone who, according to the Immigration and Refugee Protection Act, meets the definition of a Convention refugee and can also mean a person in Canada who, if they were sent home, would be tortured or at risk of cruel and unusual treatment or punishment. p. 173

Protecting Canada's Immigration System Act Immigration legislation designed to make the review and determination of refugee claims faster and to expedite removals of those who do not qualify. p. 167

Provincial and Territorial Nominee programs Programs for people who want to apply to become permanent residents of Canada, who are selected by participating provinces or territories to live there. Selection criteria are based on streams that target specific skills, education, and work experience needed to contribute to the economy of that province or territory. p. 170

psychological age How a person feels, acts, or behaves with respect to maturity and emotional growth. p. 258

psychological/emotional elder abuse Refers to actions that limit a person's sense of identity, dignity, or self-worth, including insults, threats, intimidation, humiliation, treating an individual like a child, or isolating them from family, friends, or regular activities. p. 270

pull factors Things that influence a person to immigrate. p. 180

push factors Things that influence a person to emigrate. p. 180

Q

Quebec nationalism The belief that Quebec should be recognized as a sovereign and separate nation. p. 197

Quiet Revolution Period during the 1960s when Quebec saw rapid secularization and an increased sense of nationalism. p. 197

R

race A concept no longer recognized as valid, except in terms of its social consequences; in the past, the concept of race referred to biological divisions between human beings, based primarily on their skin colour. p. 115

race card Term that refers to the use of race to gain an advantage. p. 115

racial discrimination Behaviour that has a discriminatory effect based on race; there does not have to be an intention to discriminate. p. 120

racialization The process of social construction of race whereby individuals or groups are subjected to differential and/or unequal treatment based on their designation as a member of a particular "race." p. 115

racialized person or racialized group An individual or group of persons, other than Indigenous peoples, who are subjected to differential and/or unequal treatment based on their designation as a member of a particular "race." p. 115

racial prejudice Prejudgment or negative attitude based on a set of characteristics associated with the colour of a person's skin. p. 120

racial profiling Any action undertaken for reasons of safety, security, or public protection that relies on stereotypes to treat a person differently; while it is most often relevant in policing practices, it is not limited to the context of criminal justice. p. 125

racial stereotyping Using the concept of race or ethnicity to attach a generalized concept that all members of a group have a particular characteristic or ability. p. 117

racism An ideology that either directly or indirectly asserts that one group is superior to others, with the power to put this ideology into practice in a way that give advantages, privilege, and power to certain groups of people, and conversely, can disadvantage or limit the opportunities of racialized individuals or racialized groups. pp. 31, 117

rape culture Systemic attitudes reflected in victim blaming that minimize, ignore, and normalize sexual violence against women. p. 105

refugee A person who is forced to flee from persecution and is outside of his or her country of origin. p. 166

relative homelessness State of those who have housing, but who live in substandard or undesirable shelter and/or who may be at risk of losing their homes. p. 56

relative poverty A situation where an individual or group lacks basic resources for survival when compared with other people in the society as a whole; relative standard of living when measured to others. p. 52

religious accommodation Arrangements made by an employer so that employees can do their jobs and practise their faith at the same time. p. 211

representational racism A form of racism that uses imagery to depict racial stereotypes, often in popular culture and media, in a manner that reinforces perceived inferiority of racialized persons or groups. p. 118

reverse discrimination Discrimination against whites, usually in the form of affirmative action, employment equity, and diversity policies; the concept of reverse discrimination, specifically reverse racism, is considered by many to be impossible because of existing power structures in society. p. 115

reverse mentoring Programs that provide younger workers an ability to mentor and teach older colleagues. p. 266

Royal Proclamation of 1763 One of the most important documents pertaining to Aboriginal land claims. p. 142

S

salient Characteristic of identity that describes an aspect that is most noticeable or most important. p. 16

sandwich generation The middle generation in a multigenerational household, which often struggles balancing caring for aging parents while caring for children. p. 287

scientific ageism Refers to the ways that scientific fields, like biology, psychology, and psychiatry, are used to reinforce assumptions based on age. p. 266

secular Not related to anything religious or spiritual. p. 223

secularism The belief that religion should play no role in public life. p. 223

segregation Imposed separation of different groups, usually unequally. p. 195

separatist movement A movement that seeks separation from a nation to form a smaller independent nation, as with Quebec nationalism. p. 197

settlement The process of settling in another place or country to live; resettlement generally refers to newcomers' acclimatization and the early stages of adaptation. Settlement and resettlement are used interchangeably. p. 177

sex The biological components—chromosomal, chemical, and anatomical—that are associated with males and females. p. 70

sexism The belief that one sex is superior to the other, often resulting in the discrimination or devaluation of that gender and the roles related to it. p. 32

sexual double standard The belief that there are different rules and standards of sexual behaviour for women and men. p. 107

sexual elder abuse Occurs when an older adult is forced to engage in sexual activity; this may include verbal suggestive behaviour, sexual touching, sex without consent, or not respecting a person's privacy. p. 270

sexual expression The ways in which we engage in sexual behaviours. p. 93

sexual identity How we view ourselves as sexual beings. p. 93

sexuality All the ways in which individuals express and experience themselves as sexual beings. p. 71

sexual orientation The romantic, emotional, and sexual attractions that we experience. p. 93

sexual scripts Socially created guidelines that define how one should behave as a sexual being—communicated through culture and learned through social interaction. p. 93

sexual violence Any sexual act or an attempt to obtain a sexual act through violence or coercion. p. 78

simple stepfamilies Families with the children of one, and only one, of the married or common-law partners. p. 285

Singh Decision Landmark decision in refugee determination made by the Supreme Court of Canada in 1985; the decision protects the right of every refugee claimant to an oral hearing. p. 176

sizeism Prejudice against individuals based on their body size, including height and weight. p. 32

skip-generation A household where a child or children live with their grandparents without a parent present. p. 287

social age Based on how a person meets the set of expectations that people in a given culture have about when life's major events "should" occur. p. 258

social class Any group of people who share the same situations in a common social structure. p. 49

social identity The part of a person's identity that is determined by attributes and characteristics of groups the person aligns themselves with. p. 15

social inequality The unequal distribution of tangible or intangible goods or services to individuals or groups in society. p. 48

social justice Concept that challenges the social structures, processes, and practices associated with inequalities that lead to oppression. p. 12

social stratification Division of people into categories or strata that are rewarded unequally in terms of power, property, and prestige. p. 49

societal racism A framework for those concepts, ideas, images, and institutions that we use to interpret and give meaning to racialized thought (Ontario Human Rights Commission, 2016). p. 119

specific claim Process to deal with past grievances, mainly of First Nations, related to unfairly distributed treaty lands or mismanagement of First Nations funds by the Crown. p. 145

starlight tours Alleged police practice of picking up Indigenous people and taking them to some location far away and leaving them there to get home on their own. p. 125

stereotype A label that may have some basis in fact, but that has been grossly over-generalized and applied to a particular segment of the population or situation. p. 26

Sundance Known by the Niitsitapi people as the Okan, it is a communal year-renewal event held annually in mid-summer to reinforce intra-community bonds and interpersonal relationships. p. 149

surrogacy When a woman becomes pregnant and gives birth to a baby in order to give it to another person or couple. p. 288

systemic ageism Behaviour, policies, or practices that discriminate based on age; can be intentional or unintentional. p. 268

systemic or institutionalized discrimination Refers to the policies, practices, and patterns of behaviour of social institutions that can appear to be non-discriminatory and/ or unintentional, but in fact have a discriminatory effect on persons based on one or more protected grounds. p. 35

T

Temporary Foreign Worker Program A program for persons wanting to work temporarily in Canada. Work permits and Labour Market Impact Assessments are required to be allowed to work in Canada under this program (Government of Canada, 2016). p. 184

The –Ness Model A gender model that differentiates between gender identity, gender expression, biological sex, sexual attraction, and romantic attraction, and presents two spectrums for each concept ranging from "0," "Null," or "Nobody" on the one side to woman-ness/man-ness, femininity/masculinity, and female-ness/male-ness on the other. p. 80

third gender (1) A person who does not identify with the traditional genders of "man" or "woman," but identifies with a third gender; (2) the gender category available in societies that recognize three or more genders. p. 70

tokenism The practice of including one or a small number of members of a minority group to create the appearance of representation, inclusion, and non-discrimination, without ever giving these members access to power. pp. 13, 128

totem pole A pole or post usually carved from red cedar and painted with symbolic figures, erected by Indigenous people of the northwest coast of North America. p. 139

traditionalist Refers to individuals born in 1945 or earlier. p. 259

traditional surrogacy Occurs when the surrogate's egg is fertilized by the father's sperm, making the child biologically related to the father and the surrogate. p. 288

transgender A blanket term used to describe anyone who does not identify as cisgender. p. 81

transitional poverty Occurs when a person is living in poverty for a limited period of time; usually results from an event or life circumstance, such as the unexpected loss of employment. p. 52

transsexual A person who psychologically identifies with a sex/gender different from the one they were assigned at birth. Transsexuals often wish to transform their physical bodies, with puberty suppression, hormone therapy, or surgery, to align with their inner sense of sex/ gender. p. 84

treaty A formal agreement between two parties that has been negotiated, concluded, and ratified. p. 141

two-row wampum belt A treaty of respect for the dignity and integrity of the parties involved. In particular, it stresses the importance of Indigenous independence and mutual non-interference. p. 141

two-spirited An umbrella term used by Native American people to recognize individuals who possess qualities or fulfill roles of both genders. p. 70

U

underserved Disadvantaged because of structural barriers and disparities. p. 13

undertaking A contract signed by a sponsor with the Minister of Citizenship and Immigration, or with the Ministère de l'Immigration, de la Diversité et de l'Inclusion [MIDI] if you live in Québec, promising to provide financial support for basic requirements, including healthcare, for those relatives sponsored to come to Canada. p. 169

undue hardship Circumstances involving cost, or health or safety issues that would make it impossible or very difficult for an employer or service provider to meet the duty to accommodate (Canadian Human Rights Commission, 2013). p. 34

United Nations Convention on the Rights of Persons with Disabilities (CRPD) A human rights instrument adopted by the United Nations in 2006 for the purpose of protecting the human rights and dignity of persons with disabilities; signatories are required to ensure that people with disabilities are equal under the law. p. 236

universal design "The design of products, environments, programs and services to be usable by all people, to the greatest extent possible, without the need for adaptation or specialized design (but does not exclude assistive devices for particular groups of persons with disabilities where this is needed)" (United Nations, 2006). p. 245

V

victim blaming Suggesting that women invite sexual violence or deserve to be sexually assaulted because they dress or behave in ways that are considered to be sexually suggestive or inviting. p. 105

Violence The intentional use of physical force or power to cause injury, harm or death. p. 26

visible disability When a person has a disability that is apparent by looking at them. p. 231

visible minority Outdated term used primarily in Canada by Statistics Canada to refer to a category of persons who are non-Caucasian in race or non-white in colour and who do not report being Indigenous. p. 115

voluntary repatriation The process of returning voluntarily to one's place of origin or citizenship. p. 175

W

wampum White or purple shells that come from whelk (white ones) or quahog clams (purple ones). Wampum has a multitude of meanings and uses in Aboriginal cultures, including jewellery, healing, and decoration. p. 141

war crimes Serious violations of international law applicable during armed conflict, such as ill treatment of prisoners of war, killing of prisoners of war, and so on; as of 2002, those arrested for war crimes have been tried in the International Criminal Court. p. 121

War Measures Act 1914 statute that gives emergency powers to the federal government, which allows it to govern by orders when there's a real or perceived threat of war or invasion. Changed to the Emergencies Act in 1988. p. 146

white nationalism A political ideology that advocates for a racialized identity for "white people." p. 124

white supremacy An ideology that supports the superiority of whites over all others and is embraced by members of white supremacist organizations. p. 124

World Health Organization (WHO) United Nations authority on health issues. p. 230

worldview The ways in which a group of people perceive and understand their place in the universe, often in relation to their interconnectedness with others. p. 141

X

xenophobia Hostility to anything considered foreign. p. 121

National Household Survey, 212, 213, 292

National Seniors Strategy for Canada, 268–269

"Native Spirituality and Christian Faith - Beyond Two Solitudes" (Sinclair), 223–224

Nature of Prejudice, The (Allport), 27

Negative stereotypes, 5

Neoliberalism, 235

Neo-Nazism, 124

-Ness Model, The, 79–81, 84, 85

Netherlands, 195

Newcombers, 179

NoKidding!, 290

Non-dominant groups, 4, 5t
heterogeneity within, 8–9
narratives of, 15

Normative ageism, 266

Northern Gateway Pipeline project, 155

Norway, 57

Nott, Josiah, 116

Nuclear family, 286

Numbered Treaties, 142–145

Nunavut Food Security Coalition, 60

O

Obesity, 32

Oka Crisis, 146–147, 153

Omnibus bill, 155

One Drop Rule, 116

One Flew Over the Cuckoo's Nest, 235

Online dating, 102–103

Ontario (Human Rights Comm.) and O'Malley v. Simpsons-Sears Ltd., 35, 42

Ontario Disability Support Program, 237–238

Ontario Human Development and Sexual Health Curriculum, 103, 104t

Ontario Human Rights Commission, 126, 127, 271, 272

Ontario Student Assistance Program (OSAP), 48

Opportunities Fund, 239

Oppressed groups, heterogeneity within, 8–9

Oppression, 4–5
bullying as, 9
challenging, 36, 38
definition of, 4
forms of, 24–39
internalized, 6
intersectionality and, 6–8, 7f
matrix of, 7–8, 8f
multiple levels of, 4–6

PCS model of, 5f
theory of, 3–4

Oral histories, 138–140

Oreopoulos, Philip, 36

Organisation for Economic Cooperation and Development (OECD), 76

Orthodox Christians, 213

Ostry, Bernard, 196

O'Sullivan, Lucia, 101

P

Pain, 241

Panhandlers, 55, 61

Pansexuality, 97–98

Pansexuals, 82, 97–98

Parenthood, 286–287

Parents
co-parents, 290
same-sex, 294–296
single, 59

Parlby, Irene, 25

Part-time employment, 59, 76

Parvez, Aqsa, 221

Paternalistic ageism, 266

Pathologization, of disability, 238

Patriarchy, 4, 76

PCS model of oppression, 4–5, 5f

Pearson, Lester B., 197

Pedophobic ageism, 266

Peers, gender socialization and, 72

Perel, Esther, 101

Period Eight, 167

Period Five, 167

Period Four, 167

Period Nine, 167–168

Period One, 166

Period Seven, 167

Period Six, 167

Period Three, 167

Period Two, 166–167

Permanent resident, 168, 169

Personal identity, 15–16

Personal level, oppression at, 5, 6

Person of colour, 115

Persons Case, 25

Physical bullying, 9

Physical elder abuse, 270

Pickton, Robert, 126

Plazacic, Vesna, 250

Pluralism, 197–198, 214–215

Polyamory, 102

Polyandry, 102

Polygamy, 102, 283

Popular culture, gender socialization and, 73–74, 76

Portnick, Jennifer, 32

Post-racial society, 127–128

Postsecondary education, universal design and, 245–247

Post-traumatic stress disorder (PTSD), 183, 250

Potlatches, 148–149

Poverty, 48
absolute, 52
age and, 271–272
in Canada, 52–55
child, 52–54, 127, 156, 157f
chronic, 52
cultural, 52
disability and, 239t
economic, 52
homelessness, 55–57
intergenerational, 52
marginal, 52
measurement of, 50–52
people with disabilities and, 238–239
race and, 127
relative, 52
student, 54
transitional, 52
women in, 58–59
working poor, 54–55

Power, unequal distribution of, 26

Powerlessness, 26

Precarious employment, 54–55, 59, 76

Prejudice, 27–28, 27f, 120

Pride Toronto, 96

Principle of Access, 269

Principle of Choice, 269

Principle of Equity, 269

Principle of Quality, 269

Principle of Value, 269

Privilege, 9, 12, 36
white, 121–123, 221

Privy Council, 25

Protected persons, 172, 173

Protecting Canada's Immigration System Act, 167, 169

Protecting Canada's Seniors Act, 270

Protestants, 212

Provincial and Territorial Nominee programs, 169

Psychological age, 258

Psychological/emotional elder abuse, 270

Psychotherapy, 56

Pull factors, 180

Push factors, 180

Q

Quebec nationalism, 197

Queer, 99

Queer theory, 4

Queue jumping, 172

Quiet Revolution, 197

R

V

Values, fundamental, 211
Vancouver Island Treaties, 142
Venable, Victoria, 231
Verbal bullying, 9
Veterans with disabilities, 234
Victim blaming, 4, 105
Violence, 26
    cyber, 25
    in ethnic communities, 201
    family, 290–292
    intimate partner, 78–79, 103, 105, 291
    male, 74, 76
    sexual, 78–79, 103, 105–106
    against women, 25, 78–79, 105–106
Visible disabilities, 231
Visible minority, 115
Voluntary repatriation, 175
Volunteer Canada, 31, 267

W

Wage gap, 59, 78
Walk a Mile in Her Shoes, 79
Wallace, Mike, 127
Wallace, Patricia, 103
Wampum, 141
War Amps Canada, 234
War crimes, 121
War Measures Act, 146
WE Charity, 58
Whatcott, William, 124
White flight, 122
White nationalism, 124
White Paper, 151, 153, 155
White privilege, 121–123, 221
White Ribbon Campaign, 79
White supremacy, 123–124
Whitewashing, 31–32
Wilson, Cairine, 25
Wilson, Mitchell, 244
Wise, Tim, 121, 122
Witterick, Kathy, 85
Women
    educational attainment of, 76
    employment of, 76–78
    Indigenous, 153–155
    roles of, 292
    social inequality and, 58–59
    violence against, 25, 78–79, 105–106
    wage gap for, 59, 78
Women Against Violence Against
    Women (WAVAW), 105
Women's March Global, 25
Women's suffrage movement, 25
Wong, Lloyd, 203
Woodside, Blake, 238
Work ethic, 261, 263
Workforce, changes in, 184–185
Working class, 50
Working poor, 54–55
World Economic Forum, 2
World Health Organization (WHO),
    105, 230
World Pride, 96
Worldviews, 141
World War I, 234
World War II, 120, 168–169, 234
Wright, iO Tillet, 96
Wu, Zheng, 201

X

Xenophobia, 121, 168, 200

Y

Yi, Sun Kyung, 204
Young, Iris, 25
Youth
    ageism and, 266–268
    homeless, 60–61

Z

Znaimer, Moses, 272